Advances in Molecular Interactions

Advances in Molecular Interactions

Edited by **Warren Corrigan**

R CALLISTO REFERENCE

New York

Published by Callisto Reference,
106 Park Avenue, Suite 200,
New York, NY 10016, USA
www.callistoreference.com

Advances in Molecular Interactions
Edited by Warren Corrigan

International Standard Book Number: 978-1-63239-049-3 (Hardback)

Printed in the United States of America.

Contents

Preface

Materials science is a vast field of study. It mostly deals with interatomic connections within molecules, without paying much attention to fragile intermolecular interactions. However, the varied structures are, in fact, the consequence of weak ordering because of noncovalent interactions. For self-gathering to be feasible in soft materials, it is obvious that the forces between molecules must be more fragile than covalent bonds among the atoms of a molecule. The fragile intermolecular interactions accountable for molecular ordering in flexible materials consist of hydrogen bonds, management bonds in ligands and compounds, among others. Current developments in nano science and technology offer certain well-built arguments to sustain the significance of the subjects discussed in this book. The essential and productive features connected to molecular interactions are analyzed. This book intends to provide some useful information for students and experts so as to understand the subject better.

This book unites the global concepts and researches in an organized manner for a comprehensive understanding of the subject. It is a ripe text for all researchers, students, scientists or anyone else who is interested in acquiring a better knowledge of this dynamic field.

I extend my sincere thanks to the contributors for such eloquent research chapters. Finally, I thank my family for being a source of support and help.

Editor

Part 1

Physico-Chemical Aspects
of Molecular Interactions

Modeling of Volumetric Properties of Organic Mixtures Based on Molecular Interactions

Mirjana Lj. Kijevčanin, Bojan D. Djordjević,
Ivona R. Radović, Emila M. Živković,
Aleksandar Ž. Tasić and Slobodan P. Šerbanović
Faculty of Technology and Metallurgy,
University of Belgrade,
Serbia

1. Introduction

Mixing effects for thousands of compounds and their mixtures used in the process industry are rather difficult to be known, hence knowledge of thermodynamic properties such as densities, as well as excess molar volumes, V^E, of organic mixtures at various temperatures is of great importance. Selection of systems for analysis should be based on molecular structure of the individual components, as well as their industrial and ecological significance. Mixing of the compounds with different and complex molecular structure leads to various intermolecular interactions, resulting in non-ideal behaviour of mixtures. Therefore, knowledge of volumetric properties of individual components and their mixtures helps in understanding the complex structure of liquids.

In order to have good insight in molecular interactions present in mixtures of alcohols and organic solvents, this chapter presents the extension of our earlier studies (Djordjević et al., 2007, 2009; Kijevčanin et al., 2007a, 2007b, 2007c, 2009, 2010; Radović et al., 2008, 2009, 2010a, 2010b, 2011; Smiljanić et al., 2008a, 2008b; Šerbanović et al., 2006) of the thermodynamic properties and mixing effects present in following type of mixtures: alcohols + hydrocarbons (benzene and alkane), +chlorobenzene, + amine, + chloroform, + 2-butanone. Also, with the intention to present more general conclusion, systems of organic solvents with alkanediols have been included in the analysis.

Concerning the aforementioned tasks, our previous results have been extended with the experimental measurement and modelling of the density of the binary mixtures of acetone with the 1-propanol, 1,2-propanediol and 1,3-propanediol at atmospheric pressure over the whole composition range and at temperatures from 288.15 up to 323.15 K.

The reason of this additional selection was diverse industrial application of the proposed organic compounds.

Acetone is used for various applications in the pharmaceutical, cosmetic and food industries. For example, acetone is one of the agents used in the process of skin rejuvenation by chemical peeling. It is also listed as a component in food additives and food packaging. In chemical industry acetone is used as a solvent for most plastics and synthetic fibers (polystyrene, polycarbonate and some types of polypropylene), as a volatile component of some paints and varnishes and in production processes of solvents such as methyl isobutyl alcohol and methyl isobutyl ketone.

In laboratory, acetone has a wide range of practical applications; as a polar solvent in organic chemical reactions, in acetone/dry ice baths used to conduct reactions at low temperatures but also as a solvent for rinsing laboratory glassware and instruments. Acetone is fluorescent under ultraviolet light, and its vapor may be used as a tracer in fluid flow experiments.

Alcohols, including 1-propanol, alone or in combination with other solvents, are widely used in the pharmaceutical and chemical industry (Kijevčanin et al., 2007a; Riddick et al., 1988; Swarbrik & Boyland, 1993; Šerbanović at al., 2006), for production of pesticides, fats, oils, rubber, paints, varnishes, waxes, plastics, explosives, drugs, detergents, perfumes and cosmetics. Some alcohols (especially ethanol and 1-buthanol) are used as biofuel derived from biomass sources (Radović et al., 2010a).

Alkanediols such as 1,2-propanediol or 1,3-propanediol are basic structural units for polyhydroxy compounds. These compounds have a variety of industrial applications such as solvents, coolants, antifreezes, plasticizers, chemical intermediates, food additives, heat transfer fluids. Both 1,2-propanediol and 1,3-propanediol are of very low toxicity, biodegradable and can be obtained as a by-product of biodiesel production.

At the moment, 1,2-propanediol or propylene glycol is more widely used than its isomer 1,3-propanediol. Among other applications due to the similarity of its properties to those of ethylene glycol, 1,2-propanediol is recommended as a substitute for ethylene glycol when safer properties are desired. In the food industry 1,2-propanediol is used as a solvent for food colors and flavorings and as a food additive, labeled E1520. Some of its applications in the cosmetic industry are as a moisturizer in toothpaste, shampoo, mouth wash and hair care products, as the main ingredient in deodorant sticks, as a carrier in fragrance oils and as an ingredient in massage oils. 1,2-Propanediol is also used as a solvent in many pharmaceuticals, including oral, injectable and topical formulations.

1,3-Propanediol finds application in the chemical industry mainly in the production of polymers, composites, adhesives, laminates, coatings and moldings. It is also a solvent and used as an antifreeze and wood paint.

Additional insight in volumetric behaviour is performed by excess volume modeling, including correlation by the Redlich-Kister (Redlich & Kister, 1948) equation, partial volumetric properties determinations, and in addition, modeling of V^E performed by cubic CEOS and CEOS/G^E.

Using the measured density data, and results obtained by modelling of volumetric properties, the excess molar volumes were evaluated in terms of molecular interactions between different molecules of the investigated mixtures. Finally, detailed explanation of the molecular interactions typical for each of the investigated mixture was given.

2. Research methods

2.1 Experimental study

Densities ρ of the binary mixtures, and corresponding pure substances were measured with an Anton Paar DMA 5000 digital vibrating U-tube densimeter with a stated accuracy $\pm 5\cdot10^{-3}$ kg·m^{-3}. The temperature in the cell was measured by means of two integrated Pt 100 platinum thermometers with the stability better than ±0.002 K; temperature was regulated to ± 0.001 K with a built-in solid-state thermostat.

In order to minimize evaporation of the volatile solvents and to avoid the errors in composition, all mixtures presented in this paper were prepared using the cell and the procedure described previously (Radojković et al., 1976; Tasić et al., 1995). The mixtures were prepared by mass using a Mettler AG 204 balance, with a precision of $1\cdot10^{-4}$ g. The experimental uncertainty in the density was about $\pm1\cdot10^{-2}$ kg·m^{-3}, and the uncertainty in excess molar volume has been estimated at $\pm3\cdot10^{-9}$m^3·mol^{-1}.

2.2 Modeling of volumetric properties

The non-ideal binary mixtures behavior, presented by excess molar volumes V^E, were analysed from the experimentally determined density data using following equation:

$$V^E = \left(x_1 M_1 + x_2 M_2\right)/\rho - \left(x_1 M_1/\rho_1 + x_2 M_2/\rho_2\right) \tag{1}$$

where x_i is the mole fraction of component i (i =1, 2) in the mixture; M_i is its molar mass; ρ and ρ_i are the measured densities of the mixture and the pure component i, respectively.

The values of V^E were correlated with the Redlich-Kister (RK) equation (Redlich & Kister, 1948):

$$V^E = x_i x_j \sum_{p=0}^{k} A_p \left(2 x_i - 1\right)^p \tag{2}$$

where A_p, are the adjustable parameters, and the number of adjustable parameters (k + 1) has been determined using the F-test.

The V^E data for all systems, grouped by specific system of alcohol + organic solvents, are presented in Figure 1.

2.2.1 Derived volumetric properties

Partial molar properties are very important for the extensive properties analysis, particularly their changes with composition at constant pressure and temperature conditions demonstrating the volume contraction or expansion. Effect at infinite dilution appears to be of particular interest since at the limit of infinite dilution, solute-solute interactions disappear. Hence, the partial molar volume values at infinite dilution region provide insight into solute-solvent interactions (Iloukhani & Ghorbani, 1998).

(a)

(b)

(c)

(d)

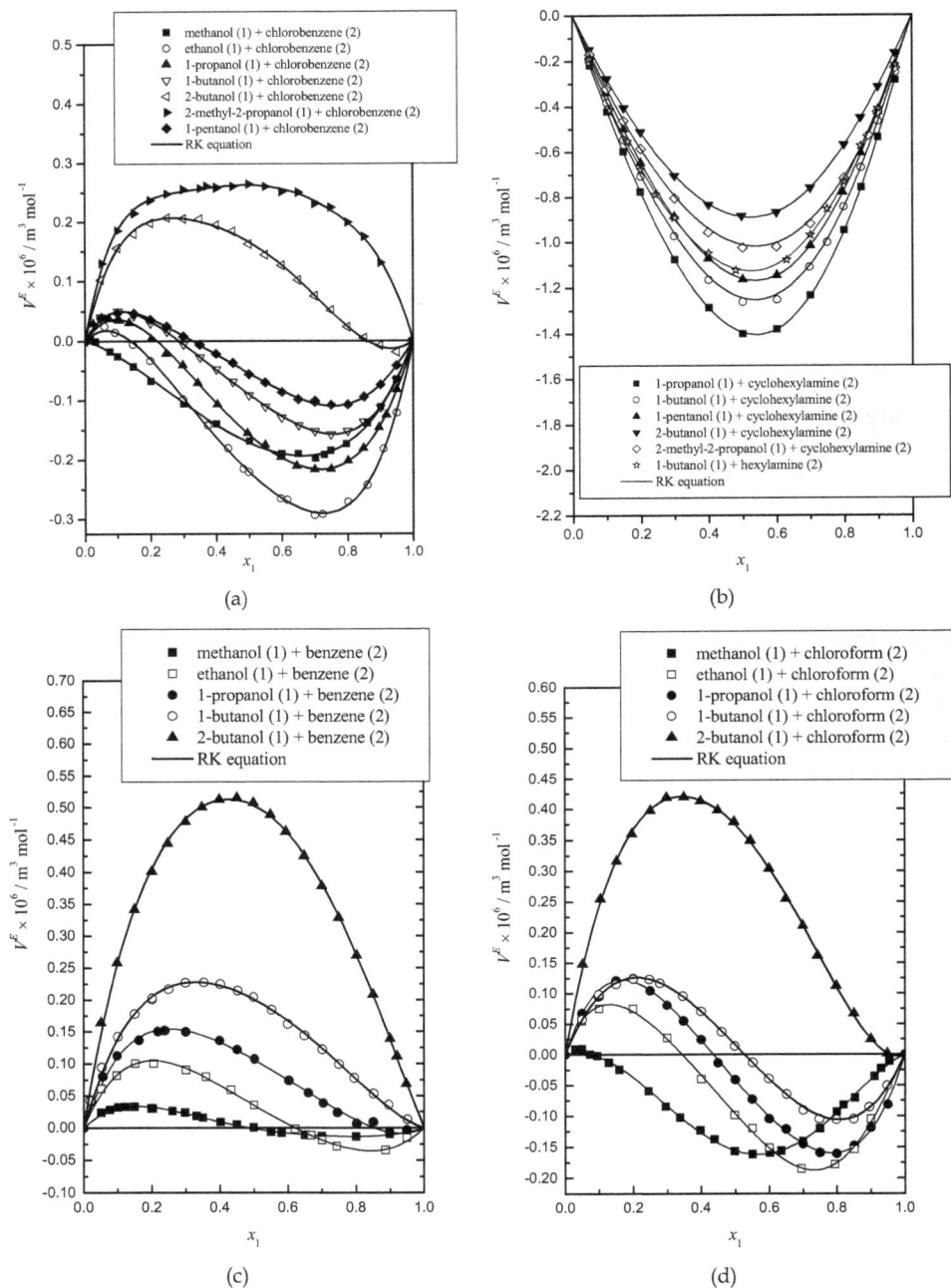

Fig. 1. Continued

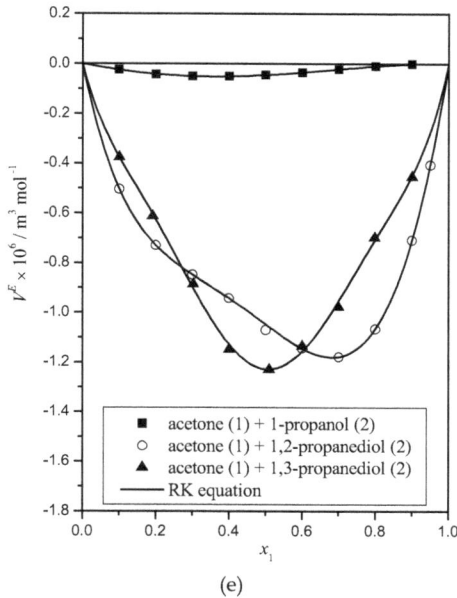

(e)

Fig. 1. V^E data: (a) alcohol (1) + chlorobenzene (2) systems at 303.15 K (Radović et al., 2008; Šerbanović et al., 2006); (b) alcohol (1) + amine (2) systems at 303.15 K (Radović et al., 2009, 2010a); (c) alcohol (1) + benzene (2) systems at 303.15 K (Kijevčanin et al., 2007b; Smiljanić et al., 2008a, 2008b; Šerbanović et al., 2006); (d) alcohol (1) + chloroform (2) systems at 303.15 K (Kijevčanin et al., 2007a, 2007b, 2007c); (e) acetone (1) + alcohol (2) systems at 298.15 K.

Partial molar volumes of component 1, $\bar{V}_1$ and component 2, $\bar{V}_2$ are calculated from the following relations:

$$\bar{V}_1 = V^E + V_1 + (1 - x_1)\left(\partial V^E / \partial x_1\right)_{p,T} \tag{3}$$

$$\bar{V}_2 = V^E + V_2 + x_1\left(\partial V^E / \partial x_1\right)_{p,T} \tag{4}$$

where V_1 and V_2 denote molar volumes of pure compounds 1 and 2. The derivative term of equations (3) and (4) is obtained by differentiation of equation (2) which leads to the following equations for $\bar{V}_1$ and $\bar{V}_2$:

$$\bar{V}_1 = V_1 + (1 - x_1)^2 \left[\sum_{p=0}^{k} A_p (2x_1 - 1)^p + x_1 \sum_{p=1}^{k} 2pA_p (2x_1 - 1)^{p-1}\right] \tag{5}$$

$$\bar{V}_2 = V_2 + (1 - x_2)^2 \left[\sum_{p=0}^{k} A_p (2x_1 - 1)^p + x_2 \sum_{p=1}^{k} (-2)pA_p (2x_1 - 1)^{p-1}\right] \tag{6}$$

Excess partial molar volumes $\bar{V}_1^E$ and $\bar{V}_2^E$ are then calculated by using the following relations:

$$\bar{V}_1^E = \bar{V}_1 - V_1 = V^E + (1 - x_1)\left(\partial V^E / \partial x_1\right)_{p,T} \tag{7}$$

$$\bar{V}_2^E = \bar{V}_2 - V_2 = V^E + x_1\left(\partial V^E / \partial x_1\right)_{p,T} \tag{8}$$

2.2.2 Excess molar volume correlation by CEOS and CEOS/ G^E models

For the excess molar volume calculation the two-parameter cubic equation of state (CEOS) was used. General form of the two parameter CEOS is given by the following equation:

$$P = \frac{RT}{V - b} - \frac{a(T)}{(V + ub)(V + wb)} \tag{9}$$

where P, T, V, and R denote pressure, temperature, molar volume and gas constant, respectively; the CEOS dependent constants u and w for the Peng-Robinson-Stryjek-Vera (PRSV) equation (Stryjek & Vera, 1986) applied here are: $u = 1 - \sqrt{2}$ and $w = 1 + \sqrt{2}$.

Parameters for the pure substance, the energy a_i and covolume b_i parameters are determined as:

$$a_i(T) = 0.457235 \frac{(RT_{ci})^2}{P_{ci}} \left\{1 + m_i\left(1 - T_{ri}^{1/2}\right)\right\}^2 \tag{10}$$

$$b_i = 0.077796 \frac{RT_{ci}}{P_{ci}} \tag{11}$$

$$m_i = k_{0i} + k_{1i}\left(1 + T_{ri}^{1/2}\right)(0.7 - T_{ri}) \tag{12}$$

$$k_{0i} = 0.378893 + 1.4897153\omega_i - 0.1713848\omega_i^2 + 0.0196554\omega_i^3 \tag{13}$$

where T_{ci} and P_{ci} are the critical temperature and critical pressure of component i, respectively, T_{ri} stands for the reduced temperature (T/T_{ci}), ω_i is the acentric factor, and k_{1i} represents the pure substance adjustable parameter (Stryjek & Vera, 1986).

For the determination of a and b parameters of a mixture, two different types of mixing rules were used: CEOS (vdW1) and CEOS/G^E (TCBT) models.

i.   The vdW1 mixing rule (Adachi & Sugie, 1986) is given by the following equations:

$$a = \sum_i \sum_j x_i x_j \left(a_i a_j\right)^{1/2}\left[1 - k_{ij} + l_{ij}\left(x_i - x_j\right)\right] \tag{14}$$

$$b = \sum_i \sum_j x_i x_j \left(b_i b_j \right)^{1/2} \left(1 - m_{ij} \right)$$ (15)

where k_{ij}, l_{ij} and m_{ij} are the binary interaction parameters.

ii.   Typical presentation of CEOS/G^E models could be given as:

$$\frac{a}{bRT} = f \left(a_i, b_i, x_i, G^E \text{ or } A^E \right)$$ (16)

The TCBT mixing rule (Twu et al., 1999) is developed from equation (16) for no reference pressure conditions and based on the van der Waals reference fluid (vdW). TCBT model is given as follows:

$$\frac{G^E}{RT} - \frac{G^E_{vdW}}{RT} + \left(Z - Z_{vdW} \right) = ln\left[\left(\frac{V^*_{vdW} - 1}{V^* - 1} \right) \left(\frac{b_{vdW}}{b} \right) \right] - $$
$$- \frac{1}{w - u} \left[\frac{a^*}{b^*} ln\left(\frac{V^* + w}{V^* + u} \right) - \frac{a^*_{vdW}}{b^*_{vdW}} ln\left(\frac{V^*_{vdW} + w}{V^*_{vdW} + u} \right) \right]$$ (17)

where G^E_{vdW} is calculated for the PRSV CEOS and $V^* = V/b = Z/b^*$ denotes the reduced liquid volume at P and T of the mixture. The compressibility factors Z and Z_{vdW} are calculated from Eq. (9) expressed in the Z form.

Parameters a_{vdW} and b_{vdW} are determined by using equations (14) and (15). NRTL equation (Renon & Prausnitz, 1968) proposed by Renon & Prausnitz was used as the G^E model:

$$\frac{G^E}{RT} = \sum_i x_i \frac{\sum_j x_j G_{ji} \tau_{ji}}{\sum_k x_k G_{ki}}$$ (18)

where:

$$G_{12} = exp\left(-\alpha_{12} \tau_{12} \right) \quad G_{21} = exp\left(-\alpha_{12} \tau_{21} \right)$$
$$\tau_{12} = \frac{\Delta g_{12}}{RT} = \left(g_{12} - g_{22} \right) / RT \quad \tau_{21} = \frac{\Delta g_{21}}{RT} = \left(g_{21} - g_{11} \right) / RT$$ (19)

The results for whole investigated temperature range were obtained by the CEOS and CEOS/G^E models with incorporated temperature dependent parameters as follows:

$$Y = Y_1 + Y_2 T$$ (20)

where $Y = k_{ij}$, l_{ij}, m_{ij}, Δg_{12} and Δg_{21}.

Optimization procedure for appropriate parameters or coefficients determination was performed using the methods described in our previous papers (Djordjević et al., 2007; Kijevčanin et al., 2007c; Šerbanović et al., 2006). Parameters of different CEOS and CEOS/G^E

mixing rules were treated as temperature independent or temperature dependent (only for whole temperature range correlation).

Modeling by CEOS and CEOS/G^E models was conducted by two and three parameter mixing rules. Models were established based on the following equations:

- equations (9) – (15), with optimized parameters k_{ij} and m_{ij}, while parameter l_{ij} was set to 0 - two parameters vdW1-2 model,
- equations (9) – (15), with optimized parameters k_{ij}, m_{ij}, and l_{ij} - three parameters vdW1-3 model),
- equations (9) – (19), with optimized parameters τ_{ij} and τ_{ji}, while parameters k_{ij}, m_{ij}, l_{ij} were set to 0; α_{ij} was set to 0.3 - two parameters TCBT-2 model,
- equations (9) – (19), with optimized parameters τ_{ij}, τ_{ji}, and k_{ij} while parameters m_{ij}, l_{ij} were set to 0; α_{ij} was set to 0.3 - three parameter TCBT-3 model.

Generally, considering the number of optimization parameters, the models with three parameters (vdW1-3 and TCBT-3) show better results than two parameter models (vdW1-2 and TCBT-2). But difference in quality of the obtained results strongly depends on type of the systems, present molecular interactions, deviation from ideal mixtures, shape of V^E -x curves (symmetric, asymmetric, S-shape) etc. Usually, very poor results of V^E correlation were obtained in the case of systems with very small V^E values. For example, in the case of methanol + benzene, or methanol + chlorobenzene mixtures (with S-shaped V^E curve) and acetone + 1-propanol system (with negative V^E values), only results obtained by TCBT-3 models could be assessed as acceptable.

Symmetrical systems, with positive V^E values, such as 2-butanol + benzene, presented at Figure 1c, usually show very good agreement between experimental and correlated data obtained by the temperature independent two parameters models (in this case errors were up to 1%). Therefore, two parameter models could be recommended for the excess volume modeling.

As a consequence of very specific interactions present in the systems of alcohols with benzene or chlorobenzene, as well as the systems alcohols + chloroform, almost all of them show S-shape V^E -x curves (except of the 2-methyl-2-propanol + chlorobenzene system). Consequently, only results obtained by three parameter models could be considered as more than acceptable. For methanol + benzene system, only TCBT-3 model is recommended. Also, as it can be seen in our previous paper (Radović et al., 2008), the results obtained for both models reproduce the tendencies of error of excess molar volume calculation increase as temperature rises (from 2 up to 5%). The vdW1-2 and TCBT-2 models, for the same systems gave very high deviations (vdW1-2 for low alcohols had errors close to 30%). Inadequate modeling of V^E by these models is a consequence of the fact that the two parameter model was not suitable for modeling of this type of V^E data.

Finally, it can be concluded that two parameter CEOS and CEOS/G^E models are not suitable to describe systems with S-shape of V^E - x curves.

Typical results of V^E calculation by the CEOS models for S-shaped V^E -x curve are presented graphically in Figure 2.

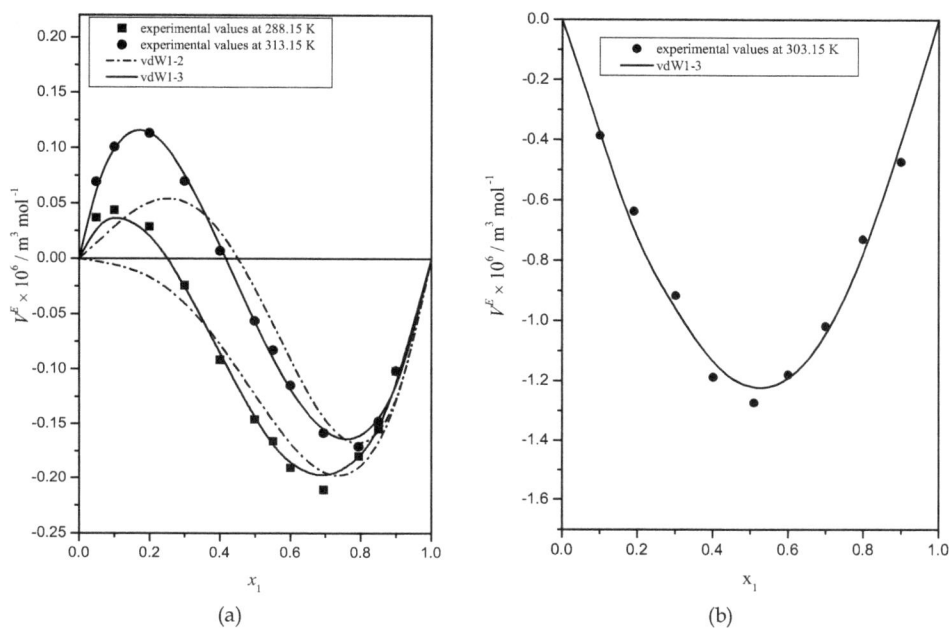

Fig. 2. Correlation of V^E data for the systems: (a) ethanol (1) + chloroform (2); (b) acetone (1) +1,3-propanediol (2).

Results of modeling of V^E in the case of negative symmetric, or slightly asymmetric V^E-x data, system 1-butanol+hexylamine (errors up to 1% for all models, except of 1.5% for vdW1-2) and alcohols + cyclohexylamine show that quality of correlation is very similar to those obtained for systems with symmetric positive values, and all investigated models could be recommended. However, in spite of the fact that 1,3-propanediol + acetone system shows quite symmetrical negative values of V^E, the TCBT-2 model, with error higher than 10%, could not be recommended for modelling of experimental values.

For the asymmetric 1-butanol + n-heptane system, with positive V^E-x values, the best results were obtained by the three parameter TCBT-3 model (errors up to 3%), especially comparing to two-parameter vdW1-2 and TCBT-2 models. The other three parameter model, vdW1-3 gave acceptable results for this system at lower temperature but errors increase with temperature rising (up to 2%).

In addition, the correlation of V^E data for all investigated systems was tested with linear temperature dependent CEOS and CEOS/G^E parameters, equation (20), but no significant improvement was obtained related to the quality of the fitting of each isotherm separately, therefore simpler models, with lower number of parameters, can be recommended.

For the correlation of V^E in whole temperature interval, the results achieved by the vdW1-3 and TCBT-3 (three parameters models) were superior comparing to the results obtained with two parameter models, but in some cases (for the S-shape curves) inclusion of temperature dependent parameters is mandatory. For example using temperature dependent parameters in the TCBT-3 model of correlation for 1-butanol + chloroform, error drops to 2.9%, while without temperature dependence it was 6.5%, or in the case of 1-butanol + heptane system, using temperature dependent parameters in vdW1-3 model, error drops from 6.6 to 2.15%.

3. Molecular interaction study

As can be seen from the number of studies, the experimental V^E values are treated in the following effects: a) physical, b) chemical and c) structural - geometrical contributions.

The magnitude and the positive sign of V^E can arise mainly from the factors: (i) Positive contribution indicates that there are no strong specific interactions between components of a mixture. It is a consequence of the rupture of the H bonds in the self-associated alcohol and the physical dipole-dipole interactions monomer and multimer of the substance. In addition it might be a result of breaking of chemical and/or nonchemical interactions among molecules in the pure components during the mixing process; (ii) Predominant intermolecular H bond stretching of the associated alcohol molecules in the presence of other substances; (iii) The steric hindrance as unsuitable interstitial accommodation due to different molar volumes and free volumes of unlike molecules.

The negative contributions are a consequence of the effects: (i) Strong intermolecular interactions attributed to the charge-transfer complex, dipole-dipole and dipole-induced dipole interactions and H-bonding between unlike molecules finally leading to the more efficient packing in the mixture than in the pure liquids; (ii) structural effects which arise from unsuitable interstitial accommodation giving more compact structure of mixtures. The nature and specific characteristics of different molecules in considered mixtures of ketone, alcohol and multifunctional alcohols determine the predominance of a particular type of interactions in the mixture.

For example, alcohol molecules are polar and self - associated through hydrogen bonding of their OH groups, however, multifunctional molecules are self - associated through inter and intra - hydrogen bonding.

Molecule of 1-propanol exists in 5 different conformations, or rotamers as a result of rotation about the central C-C and C-OH bonds. Information on the structures and relative energies of the separate rotational isomers and the potential barriers between them has been obtained by different experimental techniques including infrared, Raman, nuclear magnetic resonance and microwave spectroscopy, gas electron diffraction etc. In addition, this molecule has been studied by *ab initio* methods like molecular orbital theory (Radom et al., 1973). A microwave study (Abdurahmanov et al., 1970) of 1-propanol has shown the existence of two rotational isomers corresponding to gauche (G) and trans (T) arrangements about the C-C bond. *Ab initio* studies (Radom et al., 1973) have indicated that the two most stable rotamers have the Gt and Tt forms while the remaining three isomers are probably destabilized by steric effect.

It is a well known fact confirmed by experimental methods like infrared spectroscopy and *ab initio* studies that the alcohols and polyols have a tendency to form intermolecular hydrogen bonds. One of the main questions concerning the structural characteristics of polyols such as 1,2-propanediol or 1,3-propanediol is the question of the intramolecular hydrogen bonds existence. One of the most studied vicinal diols is ethylene glycol. There have been many experimental studies on ethylene glycol including infrared spectroscopy (Buckley and Giguere, 1967), gas phase electron diffraction (GED) (Kazerouni et al., 1997) and microwave spectroscopy (Christen et al., 2001) . At the same time numerous *ab initio* studies have been conducted on ethylene glycol based on density functional theory (DFT) with BP, B3P, BLYP and B3LYP functionals, Hartree-Fock (HF) or Moeller–Plesset (MP2 and MP4) levels of theory (Csonka & Csizmadia, 1995; Nagy et al., 1991, 1992) . Most of the authors agree that the two lowest energy conformers (g-Gt and g-Gg, where upper/lower case letters refer to rotations about C-C/C-O bonds and a positive sign designates counterclockwise motion of the forward group along the chain away from the gauche conformation) contain a weak intramolecular hydrogen bond configuration. However, there have also been different opinions. Theoretical studies by Klein (Klein, 2002, 2003, 2006) have concluded that an intramolecular hydrogen bond is not present in EG based on the hydrogen bonding criteria. Based on *ab initio* calculations using DFT methods Klein (Klein, 2002, 2003) suggested that 1,2-diols are not able to form an intramolecular hydrogen bond. The interaction observed in 1,2-diols is interpreted as a mixture of weak polarization and electrostatic effects. For all vicinal diols the most stable turns out to be the gauche configuration with the O–C–C–O dihedral close to ±60o. Diols in which the hydroxyl groups are spaced further apart, i.e., (n, n+2) or greater, are showing evidence of intramolecular hydrogen bonding with optimal ring structures achieved in (n, n+3) diols.

Only a few studies have investigated 1,3-propanediol molecule. *Ab initio* studies (Bultinck et al., 1995; Vazquez et al., 1988) based on the HF and MP2 calculactions have shown that the lowest energy conformers have gauche heavy-atom torsions, and that they are stabilized by intramolecular hydrogen bonding. The existence of intramolecular hydrogen bonds was also confirmed by gas-phase electron diffraction (GED) (Kinneging et al., 1982; Traetteberg & Hedberg, 1994). The results have shown that the molecule of 1,3 –propanediol exists exclusively in G+G- conformation with OH groups in 1,3 parallel positions nearly ideally placed to form an intramolecular hydrogen bond.

Alcohol + Chlorobenzene

The dependence of V^E on both composition and temperature, for the systems of alcohol and chlorobenzene, can be explained qualitatively as a balance between opposite effects (Letcher & Nevines, 1994; Saleh et al., 2005; Šerbanović et al., 2006): (i) positive contributions are predominantly attributed to hydrogen bond rupture and stretching of self-associated molecules of alcohol (in the alcohol lower region), (ii) negative contributions are considered to be a consequence of dipolar complexes involving alcohols and chlorobenzene (due to unlike intermolecular dipolar interactions), (iii) changes of free volume in real mixtures comprising alcohol monomers and multimers and chlorobenzene molecules and the geometrical fitting between components making this effect negative to V^E. The magnitude and sign of V^E are explained on the basis of some of above factors being predominant in certain mole fraction regions. The obtained results can be discussed in terms of the chain length of alcohols, degree of branching in the chain and relative position of the alkyl and OH group in the alcohol.

As can be seen from Figure 1a, the mixtures of chlorobenzene with 1-alcohols show sigmoidal curves v.s. mole fractions. The mixture of chlorobenzene with 1-alcohols as alcohol molecules were added to the large amount of chlorobenzene, would induce depolymerization of 1-alcohols resulting in positive values of V^E. The positive V^E values for these mixtures show that the effects due to the factor (i) are dominating over corresponding negative contributions.

Negative factors (ii) and (iii) are dominant over the larger parts of composition ranges of these mixtures. Namely, there is a formation of OH-Cl atom hydrogen bonded complexes between OH group of alcohols and a chlorine atom of chlorobenzene (Garabadu et al., 1997; Letcher & Nevines, 1994; Roy & Swain, 1988; Swain, 1984). The attraction between the unshielded proton of a hydrogen and an electronegative chlorine atom is relatively strong. In addition, the π-electronic cloud of the aromatic ring and OH group interactions take place influencing in relatively weak extent on contraction in volume (Lampreia et al., 2004). Structural effects also can lead to closer geometrical packing of unlike molecules although structure contribution is less characteristic with an increase in the chain length of 1-alcohols.

It is evident from plots of 1-alcohols with chlorobenzene that maxima of dissociation of the polymeric aggregates of alcohols differ slightly, since ordinary aliphatic alcohols are relatively poor proton donors. But, as mole fraction of higher 1-alcohol rises, the rupture of hydrogen bonds of alcohol increases from 1-propanol to 1-pentanol and positive volume changes appears at higher mole fraction of 1-pentanol.

The decrease in negative values of V^E with increase in chain length of 1-alcohols, could suggest that dipole-dipole interactions are weaker in higher 1-alcohols as a consequence of the decrease of their polarizability with rising chain length. Bearing in mind that complexation (Garabadu et al., 1997) is predominantly due to polarization interaction, trend of complex formation has the following order ethanol > 1-propanol > 1-butanol > 1-pentanol causing the rise of V^E from lower to higher 1-alcohols. Then relatively high electron donor capacity of chlorobenzene, as a consequence of introducing Cl atom into the benzene, interacts more strongly with ethanol than with other higher alcohols. Also, only smaller fraction of hydrogen bonds are ruptured in higher alcohols, while a steric hindrance of 1-pentanol is higher than that of ethanol, giving less negative values of V^E.

A possible explanation for an unexpected behaviour occuring with the system with methanol, bearing in mind the dominant assumption (iii), is that the molecules of chlorobenzene are accommodated interstitially in a network of bonded alcohol molecules leading to more dense packing of unlike molecules of methanol (van der Waals volume V_{mt}=21.71 cm³mol⁻¹) (Daubert & Danner, 1989) and chlorobenzene (V_{cb}=57.84cm³mol⁻¹) (Daubert & Danner, 1989). There are more methanol molecules in close contact, than larger alcohols leading to relatively weak interactions and decreasing the number of cross-associated OH–Cl bonds, that is the amount of possible formed complexes. As result, the contribution in volume decreases, having less negative values with respect to that of the ethanol + chlorobenzene mixture (Figure 1a).

Explanation for mixtures of chlorobenzene with branched alcohols (2-butanol, 2-methyl-2-propanol) compared to 1-butanol could be more difficult. The trend in positive values of V^E in the order: 1-butanol < 2-butanol < 2-methyl-2-propanol suggests that steric hindrance mainly due to branching, predominates over that of specific interaction of unlike molecules since complexes for three butanols (Roy & Swain, 1988) are of intensity in the order 1-

butanol > 2-butanol > 2-methyl-2-propanol. This degree of association could be in connection with a decrease in charge redistribution tendency within branched molecules. As a consequence, V^E values become increasing positive as the branching in the alcohol molecules increase in the order: 2-methyl-2-propanol > 2-butanol > 1-butanol. Therefore, the strength of interaction (H···Cl bonding) between chlorobenzene and these alcohols should follow the order 1-butanol > 2-butanol > 2-methyl-2-propanol.

When OH group is introduced at 2° carbon atom, the branching of the alkyl group in 2-butanol will create more steric hindrance for the proper orientation of chlorobenzene molecules to fit in associated network of alcohol. As a result V^E is more positive than for the system with 1-butanol.

The globular shape and relative position of OH group on the 2-methyl-2-propanol suggest that the effect of intermolecular interactive reduction and the structural effects of the close packing hindrance are predominant.

It is clear that the presence of one propyl group at 1° carbon atom in 1-butanol, one methyl and one ethyl group at 2° carbon atom in 2-butanol and three methyl groups at 3° carbon atom in 2-methyl-2-propanol create steric hindrance near OH group in the order 1-butanol < 2-butanol < 2-methyl-2-propanol. The increase of V^E for these systems is in accordance with this phenomenon. Finally, it could be indicated that plane structure of the chlorobenzene effects a smaller breaking of association in 1-alcohols than the steric arrangement of branched alcohols changing V^E towards higher values, as can be seen from Figure 1a.

Alcohol + Amines (cyclohexylamine or hexylamine)

The magnitude and the sign of V^E for the systems of alcohols and cyclohexylamine or hexylamine (Figure 1b) can arise from two opposing factors (Djordjević et al., 2009) (i) the positive contribution is a consequence of the disruption of the hydrogen bonds in the self-associated alcohol and the dipole–dipole interactions between alcohol monomer and multimer. The self-association of the amine molecules is rather small; (ii) negative contributions arise from strong intermolecular interactions attributed to charge-transfer, dipole–dipole interactions and hydrogen bonding between unlike molecules. Hence, the negative V^E values of the investigated systems assume that heteroassociates forming cross complexes in the alcohol + amine mixtures have stronger O–H ··· N bonds than O–H ·· O and N– –H ·· N bonds. This can be explained qualitatively by the fact that the free electron pair around the N atoms with less s and more p character has a higher polarizability and acts as a good proton acceptor for the donor –OH groups of the alcohols, which are more efficient than the –OH group itself.

The negative sign of V^E indicates a net packing effect contributed to the structural changes arising from interstitial accommodation. As can be seen from Figure 1b, the negative V^E values, are larger in the mixture with 1-propanol and decrease as the chain length of the 1-alkanol increases. This trend indicates that the strength of the intermolecular hydrogen bonding of cyclohexylamine with 1-propanol is much stronger than that with the other higher 1-alcohols in the following order: 1-propanol > 1-butanol > 1-pentanol. These strengths of the interaction OH ·· NH_2 bonds suggest that the proton donating ability of 1-alcohols is of the same order. Namely, longer 1-alcohols would increase the basicity of the oxygen and make the hydroxyl proton less available for H bonding. In addition, it means that the most efficient packing can be attributed to the lower alcohol, which decreases with increasing chain length of 1-alcohol, where the packing effects are the result of their lower self-association (higher breaking of their H bonds) and the fact that the crowded molecules of amine, as a consequence of steric hindrance, are better packed in the more open structure of the longer alcohols. Also, the effect of increasing chain length of the 1-alcohol for a given amine can be considered using the effective dipole moment. Bearing in mind the discussion given in previous works referring to 1-alcohols and various amines, it can be concluded that in the present systems the following behavior could be expected. In the systems of 1-alcohols with cyclohexylamine, the absolute value of the V^E decreases with decreasing effective dipole moment of the alcohol. The trend in the negative values of V^E for mixtures of cyclohexylamine with branched alcohols (2-butanol, 2-methyl-2-propanol) compared to 1-butanol is in the order: 1-butanol > 2-methyl-2-propanol > 2-butanol, suggesting that the interactions between tertiary alcohol and cyclohexylamine are stronger than between the secondary alcohol with cyclohexylamine, which in turn are stronger than the interactions between the primary alcohol and the amine (Dharmaraju et al., 1981). The results shown in Figure 1b indicate that V^E of the mixture of cyclohexylamine with 2-methyl-2-propanol are more negative than those with 2-butanol.

Qualitatively, this could be explained by the fact that the oxygen atom of 2-methyl-2-propanol should be regarded as a better acceptor towards the NH proton of the amine than the oxygen atoms of the 2-butanol. Also, the system with 2-methyl-2-propanol suggests that the steric hindrance of the *tert*-butyl group tends to hamper the complex less than with 2-butanol. This is a consequence of a predominating electrometric effect (+I effect) over steric effect in 2-methyl-2-propanol (Dharmaraju et al., 1981; Pikkarainen, 1982).

Alcohol + hydrocarbons (benzene, heptane, 2-butanone)

For the mixtures of alcohols and benzene the experimental data presented in Figure 1c can be explained qualitatively on the basis of the following resulting, opposite effects, predominated in a certain mole fraction region (Letcher & Nevines, 1994; Rodriguez et al., 1999; Tanaka & Toyama, 1997): (i) positive values (in the alcohol lower region) are attributed to rupture or stretch of the hydrogen bonding of self-associated molecules of alcohol, (ii) negative values are thought to be due to unlike specific interactions and (iii) the geometric fitting of benzene into remaining alcohol structure making this effect negative to V^E.

The magnitude and sign of V^E are a consequence of contributions taking place in the mixture. Thus, the positive values of V^E for the mixtures with benzene point out that there are no strong specific interactions between the components of mixtures, but this situation is a result of breaking down of the alcohol structure as alcohol molecules were added to the large amount of benzene. The negative volume changes in mixtures rich in alcohol could be result of predominant interaction between the -OH group of an alcohol and π electrons of aromatic ring of benzene.

Non-associating molecule n-heptane acts as an inert component, in the mixture with 1-butanol as associating components. For the binary system of 1-butanol + n-heptane (Kijevčanin et al., 2009) V^E values are positive in the entire composition range with asymmetric $V^E - x_1$ curves, shifted towards the lower 1-butanol mole fractions. Also, the V^E values increase with temperature rising from 288.15 to 323.15 K. As it was discussed previously (Treszczanowicz et al., 1981), the positive excess volumes in mixtures of n-alcohols and n-alkanes are the result of (i) the disruption of alcohol multimers due to breaking of hydrogen bonds (chemical contribution) and (ii) non-specific physical interactions between the real species in a mixture (physical contribution). Negative V^E values are mostly caused by interstitial accommodation and changes of free volumes (structural contribution). Since the chemical contribution of hydrogen bond breaking to V^E is negligible, except for small mole fractions, it is assumed that physical contribution comprises the major part of positive V^E values (Treszczanowicz et al., 1981) in this system. The sharp increases of the V^E in the dilute 1-butanol region suggest the dominance of the disruption of the H-bonds of alcohol multimers by unlike n-heptane molecules.

The effect of volume contractions obtained in the mixture of ethanol and 2-butanone (Grgurić et al., 2004) can be attributed to the association between the keto group of ketone and proton of the hydroxy group of an alkanol predominating the effect of dissociation of the alcohol molecules.

Alcohol + chloroform

As can be seen from Figure 1d, the systems alcohol (1) + chloroform (2) exhibit S-shape of the V^E-x_1 curves (maximum with positive and minimum with negative values of V^E). The shape of the V^E-x_1 curves might be influenced by three types of the interactions occurring in the mixture: (i) H-bondings between hydrogen atom of chloroform and OH group of an alcohol, (ii) intermolecular interactions between OH group and electronegative chlorine atom of chloroform and (iii) steric hindrance between alkyl group and chlorine atoms. The magnitude and sign of V^E are a consequence of some of above factors being predominant in certain mole fraction regions.

Acetone + 1-propanol, 1,2-propanediol and 1,3-propanediol systems

The negative values of excess partial molar volumes for both acetone + 1,2-propanediol and acetone + 1,3-propanediol systems suggest that the molar volumes of each component in the mixture are less than their respective molar volumes in the pure state that results in the contraction of volume over the entire composition range. Very small excess partial molar volumes obtained for the acetone + 1-propanol system support the trend observed in the V^E values shown in Figure 1e.

The thermodynamic behavior of solutions influenced by polyhydroxy compounds shows a noticeable non-ideality owing to the formation of hydrogen bonds. One of the most sensitive thermodynamic indicators of complex structure is probably the excess molar volume V^E. Measurements of this property could be very useful to determine a structural effect appearing in 1,2- and 1,3-diols.

Polyhydroxy compounds. Alkanediols are important structural units of polyhydroxy compounds. Great varieties of different approaches were employed to investigate these structures. Binary organic mixtures which include polyhydroxy compounds such as 1,2- and 1,3- propanediol are an important class of solutions. Unfortunately, the behavior of some of their thermodynamic properties is still not clear, particularly for highly associated systems. Interpretation of complex structure is extremely difficult because of various possible effects occurring in the process of mixing. Also, it should be emphasized that lack of experimental values of volumetric and other thermodynamic data may lead to an over-simplified explanation of molecular structure of this type of systems.

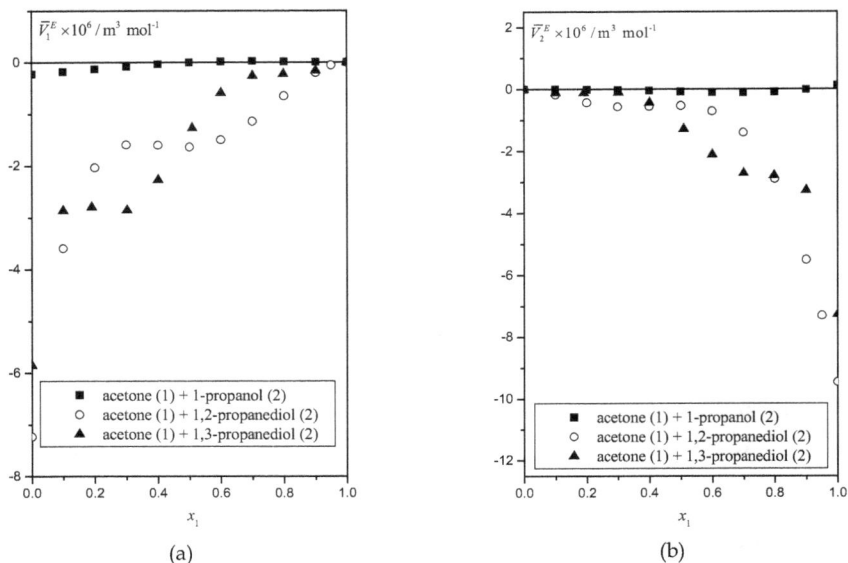

Fig. 3. Excess partial molar volumes : (a) $\overline{V}_1^E$ and (b) $\overline{V}_2^E$ at 303.15 K for the binary systems acetone (1) + alcohol (2)

Despite of most of the conformational studies of diols, accepting the importance of the intramolecular hydrogen bonding, it is very difficult to distinguish a particular conformer, namely those able to form an intramolecular H bond, bearing in mind that this bond plays main role in the conformational behavior of each molecule. Namely, spectral signatures of the conformers of a similar type are very comparable, and the number of possible conformers is large.

In a number of different theoretical and experimental studies, there is no definitive criterion which could be established for the characterizing the presence, nature and strength of intramolecular H bonds. As already mentioned, many factors have been employed as an evidence of existence of an intramolecular H bond such as frequency shifts observed in infrared (IR) and NMR studies, change in intensity ratio (low frequency/high frequency), abnormal OH bond lengths and interatomic distances or stabilization of gauche conformers. However, some of these parameters could be also quoted as the consequence of diverse nonbonding interactions.

More recently Fourier-transform near-infrared spectroscopic study (Haufa & Czarnecki, 2009) has confirmed conclusion of Klein (Klein, 2002, 2003) that 1,2-propanediol does not possess an intramolecular H bond.

Two hydroxyl groups substituted in 1,3 parallel ended positions of 1,3-propanediol are nearly ideally situated to form an intramolecular H bond. GED study of 1,3-propanediol (Kinneging et al., 1982) shows that the molecules exist in +gauche-gauche conformation (G+G-) with 1,3 parallel OH groups.

Acetone+1-propanol. Liquid state of acetone is determined by dipole-dipole interactions between neighboring molecules. The formation of liquid state through van der Waals interactions by the methyl groups also occurs but to a much smaller extent than by former interactions. 1-Propane has a three-dimensional structure forming different kind of clusters through H bonds. Under the influence of H bonding and (hydrophobic) association, 1-propanol predominantly appears in the gauche conformation. Data of V^E for the mixture acetone+1-propanol (Figure 1e) could be explained as a balance between positive and negative contributions. Positive contributions occur due to a H-bond rupture of 1-propanol associates and considerable difference in molecular volume between 1-propanol and acetone, that leads to important steric hindrance. Negative contributions are a result of intermolecular OH and C=O interactions and possible geometrical fitting between unlike molecules leading to H-bond effects and a weak dispersion type effect as weak forces of attraction, giving a slightly negative values of V^E over whole concentration range. This mixture shows almost ideal behavior.

Acetone + 1,2-propanediol or + 1,3-propanediol. Values of V^E for the mixtures of acetone with 1,2-propanediol and 1,3-propanediol are large and negative (Figure 1e). The curve for the mixture with 1,2-propanediol is skewed towards low 1,2-diol mole fraction with the minimum occurring near $x_1=0.7$, whereas a similar phenomenon was not observed for the curve with 1,3-diol where the minimum is near $x_1=0.5$. It is clear that V^E of acetone in the mixture with 1-propanol differs considerably from those of mixtures with diols.

Mixing of acetone with 1,2-propanediol would induce dissociation of H bonded structure present in the pure 1,2-diol and formation of H bonds (C=O···H-O) between proton-acceptor O atom (with two lone pair of electrons) of acetone C=O group and H atom of OH group of 1,2 -diol, leading to a contraction in volume as can be seen from Figure 1e.

Large displacement of minimum value of V^E is probably a consequence of the following factors: (i) presence of alkyl group of kind –CH_3 at a side chain of molecule, (ii) at low concentration of 1,2-diol, the OH groups are predominantly non-bonded and free for interaction with C=O group of acetone and (iii) when one of the OH groups is not located at the end of molecule as in the case of 1,2-propanediol the interstitial contribution, due to a possible cavity occupation by methyl group, results in larger volume loss with the minimum around $x_1=0.7$.

The negative values of V^E for the mixture of acetone with 1,3-propanediol (Figure 1e) indicate that interaction by hydrogen bonding between C=O group of acetone and OH groups of 1,3-propanediol that are present at extreme ends of molecule are more available to form hydrogen bonding. Namely, in the most stable structures of OH and C=O complex the intramolecular hydrogen bonding (G+G-) is broken into two almost linear intermolecular H bonds with the acetone molecule as proton acceptor. It is clear that the formation of such symmetrical structures under precipitation of both OH groups is excluded for the acetone + 1,2-propanediol complex from steric points of view because of the vicinal hydroxyl group.

The weak temperature effect on the V^E values of all investigated systems of acetone with 1-propanol, 1,2-propanediol and 1,3-propanediol, is most likely due to the fact that the increase in kinetic energy of molecules is not sufficient to change in an appreciable manner, the interaction of the H-bond type and other effects which occur in the mixtures.

4. Conclusion

Determination of excess volume, V^E, represents a very reliable method for better understanding and explanation of complex structures of numerous binary mixtures having decisive influence on many processes in chemical and related industries.

A survey of our articles, appearing in leading international journals, confirmed a great importance of thermodynamic method in explanation of molecular interactions present in mixtures of monohydroxy alcohols with various organic substances very frequently encountered in industrial practice.

Intention of the inclusion of complex alcohols (alkanediols) in the analysis, as well as the extension of applicability of the thermodynamic method to strongly nonideal binary systems, was to show that this method of the volumetric data determination, can contribute to the spectroscopic, computational and to the other theoretical methods in interpretation of the inter- and -intra molecular interactions, positions and shapes of conformers, as well as their mutual influence on system behavior.

5. Acknowledgment

The authors gratefully acknowledge the financial support received from the Research Fund of Ministry of Science and Technological Development (project No 172063), Serbia and the Faculty of Technology and Metallurgy, University of Belgrade.

6. References

Abdurahmanov, A.A.; Rahimova, R.A.; & Imanov L. M. (1970). Microwave spectrum of normal propyl alcohol. *Physics Letters* A, Vol.32, No.2, pp. 123-124, ISSN 0375-9601

Adachi, Y. & Sugie, H. (1986). A new mixing rule — modified conventional mixing rule. *Fluid Phase Equilibria*, Vol.28, No2., pp. 103-118, ISSN 0378-3812

Buckley, P. & Giguere, P.A. (1967). Infrared studies on rotational isomerism. I. Ethylene glycol. *Canadian Journal of Chemistry*, Vol.45, No.4, pp. 397-407, ISSN 0008-4042

Bultinck, P.; Goeminne, A. & Van de Vondel, D. (1995). Ab initio conformational analysis of ethylene glycol and 1,3-propanediol, *Journal of Molecular Structure (Theochem)*, Vol. 357, pp. 19-32, ISSN 0166-1280

Christen, D.; Coudert, L.H.; Larsson, J.A. & Cremer, D. (2001) . The Rotational–Torsional Spectrum of the $g'Gg$ Conformer of Ethylene Glycol: Elucidation of an Unusual Tunneling Path. *Journal of Molecular Spectroscopy* Vol. 205, No.2, pp. 185-196, ISSN 0022-2852

Csonka G. & Csizmadia I. (1995) . Density functional conformational analysis of 1,2-ethanediol. *Chemical Physics Letters*, Vol.243, pp. 419-428, ISSN 0009-2614

Daubert, T.E. & Danner, R.P. (1989). *Physical and Thermodynamic Properties of Pure Chemicals: Data Compilation*, Hemisphere Publishing Corporation, ISBN 0891169482, New York

Dharmaraju, G.; Narayanaswamy, G. & Raman, G. K. (1981). Excess volumes of (cyclohexylamine + an alkanol) at 303.15 K. *Journal of Chemical Thermodynamics,* Vol. 13, No.3, pp. 249-251, ISSN 0021-9614

Djordjević, B.D.; Šerbanović, S.P.; Radović, I.R.; Tasić, A.Ž. & Kijevčanin, M.Lj. (2007). Modelling of volumetric properties of binary and ternary mixtures by CEOS, CEOS/G^E and empirical models. *Journal of the Serbian Chemical Society,* Vol.72, No.12, pp. 1437–1463, ISSN 0352-5139

Djordjević, B.D.; Radović, I.R.; Kijevčanin, M.Lj.; Tasić, A.Ž. & Šerbanović, S.P. (2009). Molecular interaction studies of the volumetric behavior of binary liquid mixtures containing alcohols. *Journal of the Serbian Chemical Society,* Vol.74, No.5, pp. 477-491, ISSN 0352-5139

Iloukhani, H. & Ghorbani, R. (1998). Volumetric properties of N,N-dimethylformamide with 1,2-alkanediols at 20°C. *Journal of Solution Chemistry,* Vol.27, No.2, pp. 141-149, ISSN 0095-9782

Garabadu, K.; Roy, G. S.; Tripathy, S.; Swain, B. B. (1997). Study of some dipolar complexes involving long chain aliphatic alcohols. *Czechoslovak Journal of Physics,* Vol. 47, No.8, pp. 765-772, ISSN 0011-4626.

Grgurić, I.R.; Šerbanović, S.P.; Kijevčanin, M.Lj.; Tasić, A.Ž. & Djordjević, B.D. (2004). Volumetric properties of the ternary system ethanol + 2-butanone + benzene by the van der Waals and Twu–Coon–Bluck–Tilton mixing rules: experimental data, correlation and prediction. *Thermochimica Acta,* Vol.412, No.1-2, pp. 25-31, ISSN 0040-6031

Haufa, K.Z. & Czarnecki, M.A. (2009). Effect of temperature and water content on the structure of 1,2-propanediol and 1,3-propanediol: Near-infrared spectroscopic study. *Vibrational Spectroscopy,* Vol.51, No.1, pp. 80-85, ISSN 0924-2031.

Howard, D.L.; Jørgensen, P. & Kjaergaard, H. G. (2005). Weak Intramolecular Interactions in Ethylene Glycol Identified by Vapor Phase OH-Stretching Overtone Spectroscopy. *Journal of the American Chemical Society,* Vol.127, No.48, pp. 17096-17103, ISSN 0002-7863

Kazerouni, M. R.; Hedberg, L. & Hedberg, K. (1997). Conformational Analysis. 21. Ethane-1,2-diol. An Electron-Diffraction Investigation, Augmented by Rotational Constants and ab Initio Calculations, of the Molecular Structure, Conformational Composition, SQM Vibrational Force Field, and Anti-Gauche Energy Difference with Implications for Internal Hydrogen Bonding. *Journal of the American Chemical Society,* Vol.119, No.35, pp. 8324-8331, ISSN 0002-7863

Kijevčanin, M.Lj.; Đuriš, M.M.; Radović, I.R.; Djordjević, B.D. & Šerbanović, S.P. (2007a). Volumetric properties of the ternary system methanol + chloroform + benzene at temperature range (288.15-313.15) K, *Journal of Chemical & Engineering Data,* Vol.52, No.3, pp. 1136-1140, ISSN 0021-9568

Kijevčanin, M.Lj.; Purić, I.M.; Radović, I.R.; Djordjević, B.D. & Šerbanović, S.P. (2007b). Densities and Excess Molar Volumes of the Binary 1-Propanol + Chloroform and 1-Propanol + Benzene and Ternary 1-Propanol + Chloroform + Benzene Mixtures at (288.15, 293.15, 298.15, 303.15, 308.15, and 313.15) K. *Journal of Chemical & Engineering Data,* Vol.52, No.5, pp. 2067-2071, ISSN 0021-9568

Kijevčanin, M.Lj.; Šerbanović, S.P.; Radović, I.R.; Djordjević, B.D. & Tasić, A.Ž. (2007c). Volumetric properties of the ternary system ethanol + chloroform + benzene at temperature range (288.15–313.15) K: Experimental data, correlation and prediction by cubic EOS. *Fluid Phase Equilibria*, Vol.251, No.2, pp. 78–92, ISSN 0378-3812

Kijevčanin, M.Lj.; Radović, I.R.; Šerbanović, S.P.; Tasić, A.Ž. & Djordjević, B.D. (2009). Experimental determination and modelling of densities and excess molar volumes of ternary system (1-butanol + cyclohexylamine + n-heptane) and corresponding binaries from 288.15 to 323.15K. *Thermochimica Acta*, Vol.496, No.1-2, pp. 71-86, ISSN 0040-6031

Kijevčanin, M.Lj.; Radović, I.R.; Šerbanović, S.P.; Živković, E.M. & Djordjević, B.D. (2010). Densities and Excess Molar Volumes of 2-Butanol + Cyclohexanamine + Heptane and 2-Butanol + n-Heptane at Temperatures between (288.15 and 323.15) K. *Journal of Chemical & Engineering Data*, Vol.55, No.4, pp. 1739-1744, ISSN 0021-9568

Kinneging,A.J.; Mom,V.; Mijlhoff, F.C. & Renes, G.H. (1982). The molecular structure of 1,3-propanediol in the gas phase, an electron diffraction study, *Journal of Molecular Structure*, Vol.82, No.3-4, pp. 271-275, ISSN 0022-2860

Klein,R.A. (2002). Ab initio conformational studies on diols and binary diol-water systems using DFT methods. Intramolecular hydrogen bonding and 1:1 complex formation with water. *Journal of Computational Chemistry*, Vol. 23, No. 6, pp. (585-599), ISSN 0192-8651

Klein, R.A. (2003). Hydrogen Bonding in Diols and Binary Diol–Water Systems Investigated Using DFT Methods. II. Calculated Infrared OH-Stretch Frequencies, Force Constants, and NMR Chemical Shifts Correlate with Hydrogen Bond Geometry and Electron Density Topology. A Reevaluation of Geometrical Criteria for Hydrogen Bonding. *Journal of Computational Chemistry*, Vol. 24, No.9, pp.(1120-1131), ISSN 0192-8651

Klein, R.A. (2006). Hydrogen bonding in strained cyclic vicinal diols: The birth of the hydrogen bond. *Chemical Physics Letters*, Vol. 429, pp. 633-637, ISSN 0009-2614

Lampreia, I.M.S.; Dias, F.A. & Mendonca, A.F.S.S. (2004). Volumetric study of (diethylamine + water) mixtures between (278.15 and 308.15) K. *Journal of Chemical Thermodynamics*, Vol.36, No.11, pp. 993-999, ISSN 0021-9614

Letcher, T.M. & Nevines, J.A. (1994). Excess volumes of (a chlorinated benzene + an alkanol) at the temperature 298.15 K. *Journal of Chemical Thermodynamics*, Vol.26, No.7, pp. 697-702, ISSN 0021-9614

Nagy, P. I., Dunn III, W. J., Alagona, C., Chio, C. (1992) .Theoretical Calculations on 1,2-Ethanediol. 2. Equilibrium of the Gauche Conformers with and without an Intramolecular Hydrogen Bond in Aqueous Solution, *Journal of the American Chemical Society*, Vol. 114, No. 12, pp.(4752-4758), ISSN 0002-7863

Nagy, P.I.; Dunn III, W.J.; Alagona, C. & Chio, C., (1991). Theoretical Calculations on 1,2-Ethanediol. Gauche-Trans Equilibrium in Gas-Phase and Aqueous Solution. *Journal of the American Chemical Society*, Vol.113, No.18, pp. 6719-6729, ISSN 0002-7863

Pikkarainen, L. (1982). Excess volumes of (N-methylmethanesulfonamide + an aliphatic alcohol). *Journal of Chemical Thermodynamics*, Vol.14, No.6, pp. 503-507, ISSN 0021-9614

Radojkovic, N.; Tasic, A.; Djordjevic, B. & Grozdanic, D. (1976). Mixing cell for indirect measurements of excess volume. *Journal of Chemical Thermodynamics*, Vol. 8, No.12, pp. 1111-1114, ISSN 0021-9614

Radom, L.; Lathan W.A.; Hehre, W.J. & Pople, J.A.(1973). Molecular orbital theory of the electronic structure of organic compounds. XVII. Internal rotation in 1,2-disubstituted ethanes. *Journal of the American Chemical Society*, Vol.95, No.3, pp. 693-698, ISSN 0002-7863

Radović, I.R.; Kijevčanin, M.Lj.; Djordjević, E.M.; Djordjević, B.D. & Šerbanović, S.P. (2008) Influence of chain length and degree of branching of alcohol+chlorobenzene mixtures on determination and modelling of V^E by CEOS and CEOS/G^E mixing rules. *Fluid Phase Equilibria*, Vol.263, No.2, pp. 205-213, ISSN 0378-3812

Radović, I.R.; Kijevčanin, M.Lj.; Tasić, A.Ž.; Djordjević, B.D. & Šerbanović, S.P. (2009). Densities and Excess Molar Volumes of Alcohol + Cyclohexylamine Mixtures. *Journal of the Serbian Chemical Society*, Vol.74, No.11, pp. 1303–1318, ISSN 0352-5139

Radović, I.R.; Kijevčanin, M.Lj.; Šerbanović, S.P. & Djordjević, B.D. (2010a). 1-Butanol + hexylamine + n-heptane at temperature range (288.15-323.15) K: experimental density data, excess molar volumes determination. Modeling by cubic EOS mixing rules. *Fluid Phase Equilibria*, Vol.298, No.1, pp. 117-130, ISSN 0378-3812

Radović, I.R.; Kijevčanin, M.Lj.; Tasić, A.Ž.; Djordjević, B.D. & Šerbanović, S.P. (2010b). Derived thermodynamic properties of alcohol + cyclohexylamine mixtures. *Journal of the Serbian Chemical Society*, Vol.75, No.2, pp. 283-294 , ISSN 0352-5139

Radović, I.R.; Šerbanović, S.P.; Djordjević, B.D.; Kijevčanin, M.Lj. (2011). Experimental Determination of Densities and Refractive Indices of the Ternary Mixture 2-Methyl-2-propanol + Cyclohexylamine + n-Heptane at T) (303.15 to 323.15) K. *Journal of Chemical & Engineering Data*, Vol. 56, No.2, pp. 344-349, ISSN 0021-9568

Redlich, O. & Kister, A.T (1948). Thermodynamics of nonelectrolyte solutions, x-y-t relations in a binary system. *Journal of Industrial and Engineering Chemistry*, Vol.40, pp. 341-345, ISSN 0095-9014

Renon, H. & Prausnitz, J.M. (1968). Local compositions in thermodynamic excess functions for liquid mixtures. *AIChE Journal*, Vol.14, No.1, pp. 135-144, ISSN 0001-1541

Riddick, J.A.; Sakano, T.K. & Bunger, W.B. (1986). *Organic Solvents: Physical Properties and Methods of Purification*, 4th ed., Wiley, ISBN 0471084670, New York

Rodriguez, A.; Canosa, J. & Tojo, J. (1999). Physical Properties of the Ternary Mixture Dimethyl Carbonate + Methanol + Benzene and Its Corresponding Binaries at 298.15 K. *Journal of Chemical & Engineering Data*, Vol.44, No.6, pp. 1298-1303, ISSN 0021-9568

Roy, G. & Swain, B.B. (1988). Dielectric studies of hydrogen-bonded association complexes: butanols with chlorobenzene. *Acta Chimica Hungarica*, Vol.125, No.2, pp. 211-220, ISSN 0231-3146

Saleh, M.A.; Akhtar, Sh. &; Ahmed, M.Sh. (2005). Excess molar volumes of aqueous systems of some diamines. *Journal of Molecular Liquids*, Vol.116, No.3, pp. 147-156, ISSN 0167-7322

Šerbanović, S.P.; Kijevčanin, M.Lj.; Radović, I.R. & Djordjević, B.D. (2006) Effect of temperature on the excess molar volumes of some alcohol + aromatic mixtures and modelling by cubic EOS mixing rules. *Fluid Phase Equilibria*, Vol.239, No.1, pp. 69–82, ISSN 0378-3812

Smiljanić, J.D.; Kijevčanin, M.Lj.; Djordjević, B.D.; Grozdanić, D.K. & Šerbanović, S.P. (2008a). Densities and excess molar volumes of the ternary mixture 2-butanol + chloroform + benzene and binary mixtures 2-butanol + chloroform, or + benzene over the temperature range (288.15 to 313.15) K. *Journal of Chemical & Engineering Data*, Vol.53, No.8, pp. 1965-1969, ISSN 0021-9568

Smiljanić, J.D.; Kijevčanin, M.Lj.; Djordjević, B.D.; Grozdanić, D.K. & Šerbanović, S.P. (2008b). Temperature Dependence of Densities and Excess Molar Volumes of the Ternary Mixture (1-Butanol + Chloroform + Benzene) and its Binary Constituents (1-Butanol + Chloroform and 1-Butanol + Benzene). *International Journal of Thermophysics*, Vol.29, No.2 , pp. 586-609, ISSN 0195-928X

Stryjek, R. & Vera, J.H. (1986). PRSV: an improved Peng-Robinson equation of state for pure compounds and mixtures. *Canadian Journal of Chemical Engineering*, Vol. 64, No.2, pp. 323-333, ISSN 0008-4034

Swain, B.B. (1984). Dielectric properties of binary mixtures of polar liquids, I. Mutual correlation. *Acta Chimica Hungarica*, Vol.117, No.4, pp. 383-391, ISSN 0231-3146

Swarbrik, J.& Boyland, J.C. (1993) *Encyclopedia of Pharmaceutical Technology*, Marcel Dekker, Inc., ISBN 0824728181, New York

Tanaka, R. & Toyama, S. (1997). Excess Molar Volumes and Excess Molar Heat Capacities for Binary Mixtures of (Ethanol + Benzene, or Toluene, or o-Xylene, or Chlorobenzene) at a Temperature of 298.15 K. *Journal of Chemical & Engineering Data*, Vol. 42, No.5, pp. 871-874, ISSN 0021-9568

Tasić, A.Ž.; Grozdanić, D.K.; Djordjević, B.D.; Šerbanović, S.P. & Radojković, N. (1995). Refractive Indices and Densities of the System Acetone + Benzene + Cyclohexane at 298.15 K. Changes of Refractivity and of Volume on Mixing. *Journal of Chemical & Engineering Data*, Vol.40, No.3, pp. 586-588, ISSN 0021-9568.

Traetteberg, M. & Hedberg, K.(1994). Structure and Conformations of 1,4-Butanediol: Electron-Diffraction Evidence for Internal Hydrogen Bonding. *Journal of the American Chemical Society*, Vol.116, No.4, pp. 1382-1387, ISSN 0002-7863

Treszczanowicz, A.J.; Kiyohara, O. & Benson, G.C. (1981). Excess volumes for n-alkanols + n-alkanes. IV. Binary mixtures of decan-1-ol + n-pentane, + n-hexane, + n-octane, + n-decane, and + n-hexadecane. *Journal of Chemical Thermodynamics*, Vol.13, No.3, pp. 253–260, ISSN 0021-9614.

Twu, C. H.; Coon, J. E.; Bluck, D. & Tilton, B. (1999). CEOS/A^E mixing rules from infinite pressure to zero pressure and then to no reference pressure. *Fluid Phase Equilibria*, Vol. 158-160, pp. 271-281, ISSN 0378-3812

Vazquez, S.; Mosquera, R.A.; Rios, M.A. & Van Alsenoy, C. (1988). Ab initio-gradient optimized molecular geometry and conformational analysis of 1,3-propanediol at the 4-21G level. *Journal of Molecular Structure (Theochem)*,Vol.181, No.1-2, pp. 149-167, ISSN 0166-1280

2

Molecular Interactions of Some Free Base Porphyrins with σ- and π-Acceptor Molecules

Abedien Zabardasti
Lorestan University,
Iran

1. Introduction

The basic structure of porphyrin consists of four pyrrole units linked by four methine bridges Fig. 1. The porphyrin macrocycle is an aromatic system containing 22π electrons, but only 18 of them are involved in any one delocalization pathway. It obeys Hückel's rule of aromaticity (4n+2 pi electrons) and has been shown by X-ray crystallography to be planar.

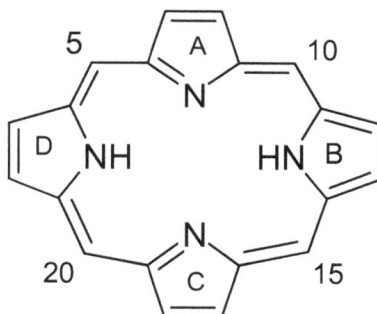

Fig. 1. Structure of free base porphyrin

1.1 ^{1}H NMR spectra of porphyrins

The aromatic character of porphyrins can also be seen by NMR spectroscopy. Studies performed in the last decades demonstrated that ^{1}H NMR spectra are very informative and adequately reflect the structural features of porphyrins.[1] The presence of the extended delocalized π-electron system of the porphyrin macrocycle gives rise to a strong ring current in the molecules placed in the magnetic field. The ring current causes anisotropic shielding of the protons located in the field of its action and (together with the diamagnetic component of paired σ-electrons) leads to a substantial shift of their signals in the ^{1}H NMR spectra. It can be stated that the ring current and the aromaticity of porphyrins change in a similar way in response to the analogous changes in the molecular structure of the porphyrin and the medium, which is most clearly seen on comparison of the spectra of porphyrins and their precursors. Due to the anisotropic effect from the porphyrin ring current, the NMR signals for the deshielded meso protons (protons on the bridging methine carbons) show up at low field, whereas the signals for the shielded protons on the inner nitrogen atoms show up at very high field. The theoretical analysis of the ^{1}H NMR spectra shows that the positions of the signals for the protons of porphyrins are determined primarily by the strength of ring π-electron currents (the macrocyclic current enclosing the molecule as a whole and local currents localized in the pyrrole nuclei (Fig. 2a,b). It should be noted that the macrocyclic current (Fig. 2a) possesses the major `bed' corresponding to the main 18-membered contour of conjugation and `arms' formed by the semi-isolated $C_{(\beta)}$-$C_{(\beta)}$ bonds in the pyrrolenine nuclei.

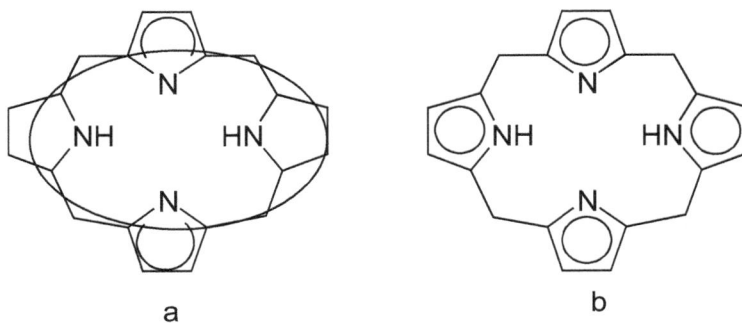

a   b

Fig. 2. Schematic representation of the circuits of the ring π-electron currents in the porphyrin molecule; (a) the macrocyclic current, (b) local currents.[1]

The protons at the β and meso positions are exocyclic with respect to the macrocyclic and local pyrrole currents and experience the deshielding influence of the latter. The signals for these protons are observed at low field (at δ from 9.7 to 11.2 ppm for the meso-protons and at δ from 8.5 to 9.9 ppm for the β-protons, Table 1). The protons of the NH groups are exocyclic with respect to the local ring currents and endocyclic with respect to the macrocyclic current. Undoubtedly, the shielding effect of the latter prevails and the signals for the protons of the NH groups are observed at very high field (at δ from -1.4 to -4.4 ppm). [1]

Compound	δ N-H	δ $H_{m,p}$	δ H_o	δ H_β	δ OCH_3, δ CH_3
H_2TPP	-2.71	7.81	8.26	8.87	-
H_2T(4-Cl)PP	-2.86	7.75	8.14	8.85	-
H_2T(4-CH_3)PP	-2.77	7.56	8.11	8.86	2.71
H_2T(4-CH_3O)PP	-2.82	7.23	8.07	8.79	3.95

Table 1. ^{1}H NMR chemical shift (δ/ppm) of the free base tetraarylporphyrins relative to $CHCl_3$.

The β-hydrogen atoms of the pyrrole and pyrrolenine fragments are in principle nonequivalent. However, rapid (within the NMR time scale) transformations occur at room temperature and the signals for the β-protons are averaged. Two different peaks corresponding to resonance absorption of the energy by the protons of the β-CH groups of the pyrrole and pyrrolenine fragments can be observed at low temperature, -80 °C. [1,2]

On going from H_2P to H_4P^{2+} it would expect that the signals for the protons of the NH groups in the ^1H NMR spectrum would be shifted downfield upon protonation of porphin. However, the signals are shifted upfield by ~0.7 ppm. Simultaneously, the signals for the β- and meso-protons are shifted downfield. This character of the spectral changes indicates that protonation of porphin leads to a substantial increase in the strength of the magnetic field induced by the macrocyclic ring current.

Protonation of porphin is accompanied by a change in the geometric structure of the molecule. The $C_{(\alpha)}$-NH-$C_{(\alpha)}$ fragments in the initial porphin are planar and the nitrogen atoms are sp^2-hybridized. In the protonated porphin, these fragments adopt a pyramidal structure (the nitrogen atoms have nearly sp^3 hybridization) and the degree of involvement of the nitrogen atoms in conjugation with the α-carbon atoms is substantially reduced. This fact is confirmed by an increase in the $C_{(\alpha)}$-N bond length observed upon protonation of porphin. The above-mentioned rearrangements lead to a change in the contour of macrocyclic conjugation. In the case of protonated porphin, this conjugation is realized primarily along the outer contour of the molecule. Therefore, protonation causes an increase in the diameter of the conjugation ring resulting in upfield shifts of the signals for the internal protons and downfield shifts of the signals for the external protons. In the case of β-alkyl substitution in porphin (on going to octaethylporphin and ethioporphin), the signals for the protons of the NH groups and the signals for the meso-protons are shifted downfield and upfield, respectively. Analogous, but more pronounced, changes are observed in the presence of a substituent in the meso position. It should be noted that the introduction of both electron-donating (meso-tetra-iso-butyl- and meso-tetra-n-pentylporphyrin) and electron-withdrawing (tetraphenylporphyrin and its derivatives) substituents gives the same results, viz., the signals for β-protons and for the protons of the NH groups are shifted upfield and downfield, respectively. It seems likely that the introduction of substituents both at the β- and meso positions leads to reduction in the strength of the aromatic macrocyclic current regardless of the electronic nature of the substituent.

The currents in the porphyrin molecule are substantially affected by complex formation.[1] The higher the degree of covalence of the N-H bond, the larger the decrease in the ring current due to the presence of the coordinated metal atom. Coordination to the medium-sized M^{2+} cations (M =Mg, Zn, Cd, Ni or Pd) is accompanied by upfield shifts of the signals for the meso-protons (the ring current is reduced). In the spectra of the complexes in which the M^{2+} metal ion deviates from the plane due to its large radius and weak coordination interactions (M =Sn or Pb) or in the spectra of the complexes with the M^{3+} or M^{4+} ions in which the metal atom deviates from the plane of the macrocycle under the action of the extra ligands, the signals for the meso-protons are shifted downfield relative to those in the spectra of the porphyrin ligand. These shifts may be due to an increase in the ring current. However, this effect is insignificant and depends on the nature of the ions serving as the extra ligands.[1]

1.2 Electronic absorption spectra of porphyrins

Electronic absorption spectra of porphyrins are very characteristic and contain one intense band in the near-ultraviolet region of the spectrum around 390-425 nm depending on whether the porphyrin is β- or meso-substituted with $\varepsilon > 2 \times 10^5$, the Soret band or B band, followed by four low-intensity absorption bands at higher wavelengths (480 to 650 nm) in the visible region (Q Bands), see Fig. 3 and Table 2. These absorptions giving rise to striking colours of porphyrins. Thus, free base porphyrins have four Q bands, denoted by increasing wavelength as IV, III, II, and I. The commonly accepted classification of these bands is as follows. The bands I and III in the visible region of the spectrum (Fig. 3) belong to quasi-forbidden electron transitions, whereas the bands II and IV are of electronic-vibrational origin, i.e., are vibrational satellites of the bands I and III, respectively. Although possessing a number of common features, the spectra of porphyrins show substantial variations, which reflect the changes in the molecular structure and the effect of the solvent.

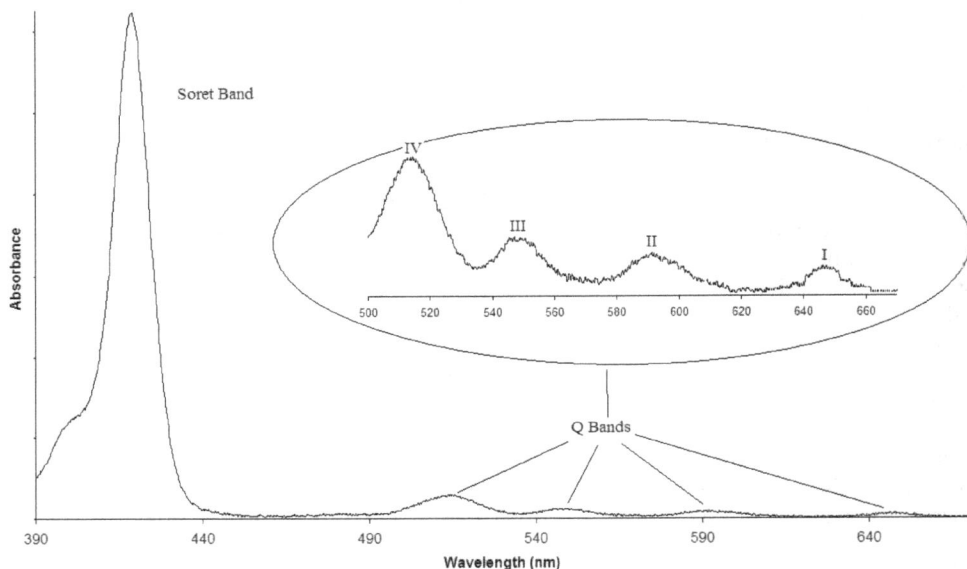

Fig. 3. Typical UV-vis absorption spectrum of a porphyrin.

Compound	Wavelength (nm)				
H$_2$TPP	417	514	549	589	646
H$_2$T(4-Cl)PP	418	514	550	590	646
H$_2$T(4-CH$_3$)PP	419	516	553	591	649
H$_2$T(4-CH$_3$O)PP	421	518	555	594	649

Table 2. UV-vis absorptions λ (CHCl$_3$/nm) of the free base tetraarylporphyrins.

According to the universally accepted concepts, light absorption is accompanied by excitation of the porphyrin molecule and the characteristic features of the absorption spectrum are determined by transitions of the π-electrons between two higher occupied and two lower unoccupied orbitals, the four-orbital Platt - Gouterman model.[1]

The intensity ratio of the absorption bands in the spectra of porphyrins depends on their structures in a peculiar fashion. When the relative intensities of these bands are such that IV > III > II > I, then the spectrum is said to be ethio-type after the ethioporphyrins in which the β-substituents are all alkyl groups, Fig. 4a. In practice, the ethio-type Q band spectrum is found in meso-tetraphenylporphyrin, ethio-porphyrin and all porphyrins in which six or more of the β-positions are substituted with groups without π-electrons, e.g. alkyl groups. For porphyrins characterized by the rhodo-type of the spectra (after rhodoporphyrin XV) the following sequence III > IV > II > I is realized, Fig. 4b. Among the latter compounds are rhodoporphyrin and other porphyrins containing an electron-withdrawing substituent (COOH, NO_2, Cl, etc.) in the pyrrole fragment. The rhodo-type spectrum has a "rhodoflying" or "reddening" effect on the spectrum by shifting it to longer wavelength. Phylloporphyrin and other porphyrins containing one or two meso-substituents characterized by the phyllo type of the spectrum (IV > II > III > I), Fig. 4c. Chlorin, tetrabenzoporphyrin and phthalocyanine give characteristic spectra of their own. In their spectra, the long-wavelength band I has the maximum intensity. Variations of the peripheral substituents on the porphyrin ring often cause minor changes to the intensity and wavelength of these absorptions. Protonation of two of the inner nitrogen atoms or insertion of a metal into the porphyrin cavity also changes the visible absorption spectrum. These absorptions can often be very helpful in determining certain features on a porphyrin.

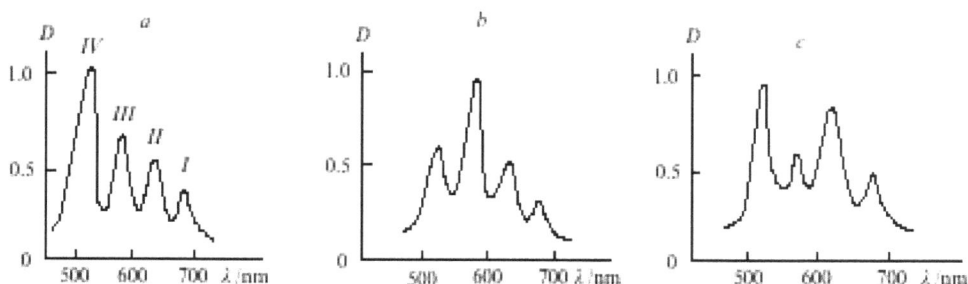

Fig. 4. Basic types of electronic absorption spectra of ethio- (a), rhodo- (b) and phylloporphyrins (c).

The meso-tetrasubstituted porphyrins are characterized by a distorted ethio type of spectra with rather large (25 - 40 nm) red shifts of the bands. The introduction of various substituents at the para positions of the phenyl rings of meso-tetraphenyl-porphyrin causes red shifts of the absorption bands in the visible region of the spectra. A comparison of the spectra of tetraphenylporphyrin and its para-substituted derivatives demonstrated that the intensities of the electron transition bands (I and III) rise as the electron-donating properties of the substituents increase. Simultaneously, the intensities of the vibrational bands (II and IV) are diminished.[3] The ethio type of the spectrum of tetraphenylporphyrin is distorted because the band I becomes more intense than the band II.

The presence of electron-donating substituents at the meta positions also leads to red shifts of the bands, the intensities of the vibrational bands being increased, while the intensities of the electron transition bands remaining virtually unchanged. The spectra of ortho-substituted tetraphenylporphyrins are characterized by lower intensities of the electron transition bands compared to those observed in the electronic absorption spectrum of tetraphenylporphyrin. In some cases, this gives rise to the spectra of the phyllo type, for example, in the case of tetra(2-halogenophenyl)porphyrins and, particularly, in the case of tetrakis(2,6-dihalogenophenyl)porphyrins.

The absorption spectra are also dependent on the solvent. Thus, the bands in the electronic absorption spectra of substituted meso-tetraphenylporphyrins are shifted redally and the intensities of the electron transition bands grow on going from nonpolar hexane to polar pyridine.

On going from porphyrin to metalloporphyrin, the symmetry of the planar macrocyclic fragment (of the π-electron cloud of the macrocycle) increases due to which the spectrum is simplified. The Soret band changes only slightly upon complex formation. The visible region of the spectra of metal complexes has two absorption bands, viz., the band I corresponding to the electron transition and the band II corresponding to the electronic-vibrational transition.

1.3 Sitting-Atop (SAT) complexes

A particularly attractive idea in the kinetic and mechanism of metalation of free base porphyrins is that of the Sitting-Atop (SAT) complexes.[4,5] In 1960, Fleischer and Wang [4] first proposed the so-called Sitting-Atop complexes of the protoporphyrin dimethyl ester with metal ions in chloroform on the basis of visible spectra, infrared spectra and composition of the SAT complex. According to the SAT idea the metalation of free base porphyrin begins with a preequilibrium step involving partial bonding of the metal ion to two of pyrrolenine nitrogens to form an intermediate, so-called Sitting-Atop complex. In this intermediate two protons on the pyrrole nitrogens still remain, $M(H_2P)^{2+}$ (Eq. 1). Then the overall metalation reaction will consist of at least two steps, i.e., the coordination step of two pyrrolenine nitrogens to form an intermediate, so-called Sitting-Atop complex, and the deprotonation step of the SAT complex to form the metalloporphyrin, as shown by following Equations, respectively:

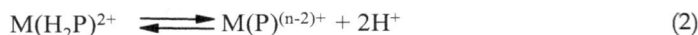

$$M^{n+} + H_2P \rightleftharpoons M(H_2P)^{2+} \tag{1}$$

$$M(H_2P)^{2+} \rightleftharpoons M(P)^{(n-2)+} + 2H^+ \tag{2}$$

The existence of the SAT complex could explain the experimental kinetics of the metalation for several porphyrins in N,N-dimethylformamide (DMF),[6-8] acetic acid,[9] dimethyl sulfoxide (DMSO),[10] and H_2O.[11,12] However, the formation constant of the SAT complex was reported as being on the order of 10^2 -10^4 mol $^{-1}dm^3$ for Cu(II) and Zn(II) ions in DMF.[13,14]

S. Funahashi et al.[15,16] have been succeeded to direct detection of SAT complexes of some tetraarylporphyrin with the Cu(II) ion in the acetonitrile as a solvent. Structural characterization of the Cu(II)-SAT complex in AN by the ^{1}H NMR and EXAFS methods clarified that two pyrrolenine nitrogens coordinate to the equatorial sites of the Cu(II) ion and H_2TPP act as a bidentate ligand.

They proposed the following overall mechanism for the metalation reaction of H_2tpp with an octahedrally solvated metal(II) ion (MS_6^{2+}) in the conventional basic solvent (S); which resolved into Eqs 3-8: the deformation of the porphyrin ring (eq 3), the outer-sphere association (eq 4), the rate-determining exchange of a coordinated solvent molecule with a first pyrrolenine nitrogen (eq 5), the chelate ring closure to form the SAT complex (eq 6), the first deprotonation on the pyrrole nitrogen in the SAT complex by the basic solvent molecule (S) (eq 7), and the second deprotonation to form the metalloporphyrin (eq 8),

$$H_2tpp \underset{}{\overset{K_D}{\rightleftharpoons}} H_2tpp \tag{3}$$

$$H_2tpp + MS_6^{2+} \underset{}{\overset{K_{OS}}{\rightleftharpoons}} H_2tpp * MS_6^{2+} \tag{4}$$

$$H_2tpp * MS_6^{2+} \underset{k_{-1}}{\overset{k_1}{\rightleftharpoons}} MS_5(H_2tpp)^{2+} + S \tag{5}$$

$$MS_5(H_2tpp)^{2+} \underset{}{\overset{K_{RC}}{\rightleftharpoons}} MS_p(H_2tpp)^{2+} + (5-p)S \tag{6}$$

$$MS_p(H_2tpp)^{2+} + S \underset{}{\overset{K_{-H1}}{\rightleftharpoons}} MS_q(H_2tpp)^+ + HS^+ + (p-q)S \tag{7}$$

$$MS_q(H_2tpp)^+ + S \underset{}{\overset{K_{-H2}}{\rightleftharpoons}} MS_r(tpp) + HS^+ + (q-r)S \tag{8}$$

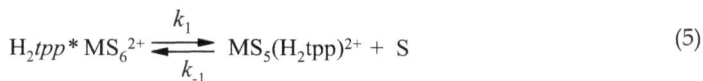

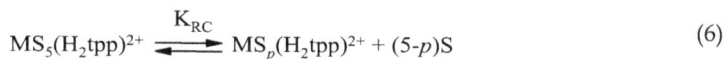

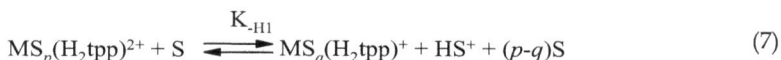

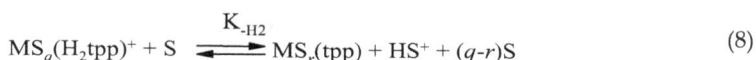

Where, $p \le 4$, $q \le p - 1$, and $r \le 2$. In the case of M = Cu, S = AN, and S = py, it was determined by fluorescent XAFS measurements that $p = 4$ and $r = 0$.

There are numerous reports on thermodynamic parameters of SAT complex of free base porphyrins with bivalent metal ions. M. Tanaka et al.[7] have studied the incorporation of Cu(II), Zn(II) and Cd(II) in the free base H_2TPP in DMF. They suggested a preequilibrium between metal ions and H_2TPP prior to metal insertion and proton release. Their values of K were 1.6×10^4 M^{-1} for Cu(II), 7.2×10^2 M^{-1} for Zn(II), and no kinetic evidence for complex formation was found for Cd(II). Pasternack's group[11] kinetically found a K 5 M^{-1} for Cu(II)-H_2TPP reaction in DMSO. P. Hambright and L.R. Rabinson [8] have studied the kinetic of Zn(II) incorporation into free base porphyrins in DMF as solvent. They obtained the following values of K for Zn(II) SAT complexes with tetraphenylporphyrin, tetrakis (2,6-diflourophenyl)porphyrin, tetra(4-aminophenyl)porphyrin, tetrakis(4-N,N-dimethylaminophenyl) porphyrin, and tetra(4-hydroxyphenyl)porphyrin, 12.3×10^3, 2.8×10^3, 7.7×10^3, 6.0×10^3, and 9.5×10^3 M^{-1}, respectively. This short review on Sitting-Atop complexes illustrated that these intermediates have been interest from 1960.

Concerning non-basicity and noncoordinating properties of some solvents such as chloroform and dichloromethane make them suitable solvents for obtaining SAT complex as a solid product. On the other hand, molecular Lewis acids such as organotin(IV) halides because of some special properties, i.e., a high solubility in noncoordinating solvents, a suitable stability in solution, a good tendency to adduct formation, and enough stability of their adducts for characterization are remarkable candidates for determination of SAT complexes.

1.4 Molecular adducts

1.4.1 Molecular adducts of organotin(IV) halides with free base *meso*-tetraarylporphyrins [17-23]

General procedure

On excess addition of organotin(IV) halides to a purple solution of free base tetraarylporphyrin in dry chloroform, its color changed to green. Evaporation of the solvent at room temperature results to shiny green powdery product. Elemental analysis, electronic absorptions spectra and ^{1}H NMR spectroscopy were used to characterization of product.

1.4.1.1 Molecular adducts of dimethyltin(IV) dichloride with free base meso-tetraarylporphyrins, [(Me$_2$SnCl$_2$)$_2$H$_2$T(4-X)PP] [22]

Table 3 shows the elemental analyses for adducts obtained from interaction of some organotin(IV) halides with free base tetraphenylporphyrin derivatives. These data are in agreement with a composition 2:1 of adducts in the solid state. On the basis of spectroscopic measurements a Sitting-Atop structure was suggested for these adducts. Of course, stoichiometry of adduct might differ in the solution from the solid product.

1.4.1.2 UV-Visible analysis

Interaction of tetraphenylporphyrins with sigma- and pi-acceptor species lead to green products that their UV-Vis spectra are comparable with diacid form of corresponding porphyrin. In the electronic absorption spectra of the diacid form of free base H$_2$TPP the intensity of band I is greater than band II. All the bands in the UV-visible spectra of H$_4$TPP^{2+} are shifted 20-40 nm to the red.[24-26] This is interpreted to be evidence of increased resonance interaction of the phenyl rings with the porphyrin nucleus in going from the free base to it's diacid form, which would be allowed by the tilting of the pyrroles observed in the solid state. The phenyl ring angle with plane of porphyrin in H$_2$TPP is within 60-85 °, which this angle reduces to 21 ° in H$_4$TPP^{2+}. The red shift in H$_4$TPP^{2+} (UV-Vis H$_4$TPP^{2+} (in CH$_2$Cl$_2$), 439, 602, 655 nm) relative to H$_2$TPP is consistent with the structural observation that the phenyl rings in H$_4$TPP^{2+} were more coplanar with the porphyrin nucleus than free base H$_2$TPP. With this coplanarity some resonance interaction of these rings with porphyrin nucleus is thus allowed, and may be the reason why the diacid form of H$_2$TPP is green in solution. Similar situations were observed for interaction of free base H$_2$T(4-X)PPs with organotin(IV) halide Lewis acids. During these interactions original peaks of free base H$_2$T(4-X)PPs (Soret band and Q band) were slowly changed to two new peaks which their position show about 20-40 nm red shift relative to the band V (Soret band) and the band I (of Q band) of free base H$_2$T(4-X)PPs, Table 4. It seems that deformation of porphyrin structure during its interaction with various acceptor species is similar to deviations of porphyrins skeleton in porphyrin diacids H$_4$T(4-X)PP^{2+} (Table 4). In the thermodynamic studies section we will review the spectral variation of free base meso-tetraarylporphyrins upon formation of their adducts.

Investigating the effect of adduct formation on the electronic absorption spectra of the free base porphyrins, the red shifts of absorption bands were observed on going from free bases to adducts. This is due to an increased in the resonance interaction of the peripheral phenyl rings with the porphyrin nucleus. The amounts of shift vary from 28 to 36 nm, depending on *the kind of acceptor, for the Soret band. In (MeSnBr$_3$)$_2$H$_2$TPP, Soret band (448 nm) and Q band I (666 nm) with 30 and 20 nm shift, respectively, we find the greatest red shift in the H$_2$TPP adducts. Also (MeSnBr$_3$)$_2$H$_2$T(4-Cl)PP with 34 and 25, in (MeSnBr$_3$)$_2$H$_2$T(4-CH$_3$)PP with 31 and 29 nm, in (MeSnBr$_3$)$_2$H$_2$T(4-CH$_3$O)PP with 35 and 46 nm red shift they have the greatest red shift in the corresponding adducts. It indicates that there is a direct relation between the acceptor property of the organotin(IV) halide Lewis acid and the red shift of the electronic absorption spectra of the coordinated porphyrin. Also, the red shifts of the absorption bands of the coordinated tetraarylporphyrins depend on the substituents at the para positions of their phenyl rings. By*

Adduct	Found			Calculated		
	C	H	N	C	H	N
$(Me_2SnCl_2)_2H_2TPP$	52.15	3.89	4.98	54.60	3.98	5.30
$(Me_2SnCl_2)_2H_2T(4\text{-}Cl)PP$	49.16	3.25	4.95	48.30	3.18	4.69
$(Me_2SnCl_2)_2H_2T(4\text{-}CH_3)PP$	59.10	4.72	5.30	56.87	4.50	5.04
$(Me_2SnCl_2)_2H_2T(4\text{-}CH_3O)PP$	51.13	4.46	4.50	53.40	4.25	4.77
$(Et_2SnCl_2)_2H_2TPP$	55.46	4.44	4.79	56.17	4.50	5.04
$(Et_2SnCl_2)_2H_2T(4\text{-}Cl)PP$	49.18	3.70	4.29	49.96	3.68	4.48
$(Et_2SnCl_2)_2H_2T(4\text{-}CH_3)PP$	56.86	4.73	4.55	57.59	4.97	4.80
$(Et_2SnCl_2)_2H_2T(4\text{-}CH_3O)PP$	55.51	4.98	4.56	54.60	4.71	4.55
$(Bu_2SnCl_2)_2H_2TPP$	55.79	4.96	4.63	58.97	5.41	4.59
$(Bu_2SnCl_2)_2H_2T(4\text{-}Cl)PP$	51.88	3.70	4.29	52.91	4.56	4.12
$(Bu_2SnCl_2)_2H_2T(4\text{-}CH_3)PP$	58.75	5.45	4.63	60.15	5.80	4.39
$(Bu_2SnCl_2)_2H_2T(4\text{-}CH_3O)PP$	55.65	5.49	3.83	57.28	5.52	4.18
$(MeSnCl_3)_2H_2TPP$	44.16	2.25	4.15	44.74	2.56	4.54
$(MeSnCl_3)_2H_2T(4\text{-}Cl)PP$	49.75	3.09	4.88	50.37	3.28	5.11
$(MeSnCl_3)_2H_2T(4\text{-}CH_3)PP$	51.62	4.12	4.63	52.10	3.82	4.86
$(MeSnCl_3)_2H_2T(4\text{-}CH_3O)PP$	48.53	4.04	4.20	49.35	3.62	4.61
$(Ph_2SnCl_2)_2H_2TPP.0.23CHCl_3$	61.57	3.91	3.98	61.62	3.81	4.21
$(Ph_2SnCl_2)_2H_2T(4\text{-}Cl)PP.0.45CHCl_3$	55.10	3.37	3.18	55.02	3.14	3.75
$(Ph_2SnCl_2)_2H_2T(4\text{-}CH_3)PP.0.6CHCl_3$	60.85	4.01	3.82	60.97	4.13	3.92
$(Ph_2SnCl_2)_2H_2T(4\text{-}CH_3O)PP.0.6CHCl_3$	58.30	3.76	3.35	58.35	3.95	3.75
$(Ph_2SnBr_2)_2H_2TPP$	55.60	3.80	4.20	55.19	3.38	3.78
$(Ph_2SnBr_2)_2H_2T(4\text{-}Cl)PP$	50.60	3.20	3.60	50.49	2.84	3.46
$(Ph_2SnBr_2)_2H_2T(4\text{-}CH_3)PP.0.8CHCl_3$	53.50	3.90	3.30	53.58	3.63	3.43
$(Ph_2SnBr_2)_2H_2T(4\text{-}CH_3O)PP.CHCl_3$	50.70	3.70	3.00	50.98	3.46	3.25

Table 3. Elemental analysis of $[(Me_2SnCl_2)_2H_2T(4\text{-}X)PP]$ adducts

increasing electron donation of the *para*-substituents on the phenyl rings greater red shift were observed for the absorption bands in the visible spectra of tetraarylporphyrin during adduct formation. Also the electronic absorption spectra of adducts are indicating that the intensity of the electron transition band I of the adducts rises as the electron donation properties of the substituents increase.

Considering λ_{max} of the electronic absorption bands (Soret and band I) of adducts it might be pointed out that the kind of acceptor have a minor effect on the position of the adduct bands. It seems positions of these bands depend on distortion and tilting of the porphyrin plane, which occurred during formation of 1:1 adduct for different acceptors. Therefore the kind of acceptor and also entering the second acceptor species isn't accompanied with a significant replacement in the position of the adduct bands. The same statement can be discussed for position of the isosbestic points of free base porphyrins adducts with various acceptors. The effect of distortion of porphyrin structure on electronic absorption bands and ^{1}H NMR chemical shift of various free base porphyrin protons will be better understood by comparison of electronic absorption spectra of 5,10,15,20-tetrabutylporphyrin as a planar free base porphyrin [UV-vis (CH$_2$Cl$_2$): Soret band (417 nm) and Q bands (520, 555, 600, 659 nm); ^{1}H NMR: N-H (-2.61 ppm) and H$_\beta$(9.45 ppm)] with 5,10,15,20-tetrakis(tert-butyl)porphyrin as a severely ruffled (distorted) free base porphyrin [Soret band (446 nm) and Q bands (552, 596, 628, 691 nm); ^{1}H NMR: N-H (1.52 ppm) and H$_\beta$(9.08 ppm)]. This example finely shows that the observed red shift mainly resulted upon distortion of porphyrin structure.

1.4.2 Molecular adducts of methyltin(IV) tribromide with free base *meso*-tetraarylporphyrins, [(MeSnBr$_3$)$_2$H$_2$T(4-X)PP]: [21]

On addition of methyltin(IV) tribromide to a solution of free base tetraarylporphyrins in dry chloroform its purple color changed to green, the electronic absorption spectra of adducts are given in the Table 4.

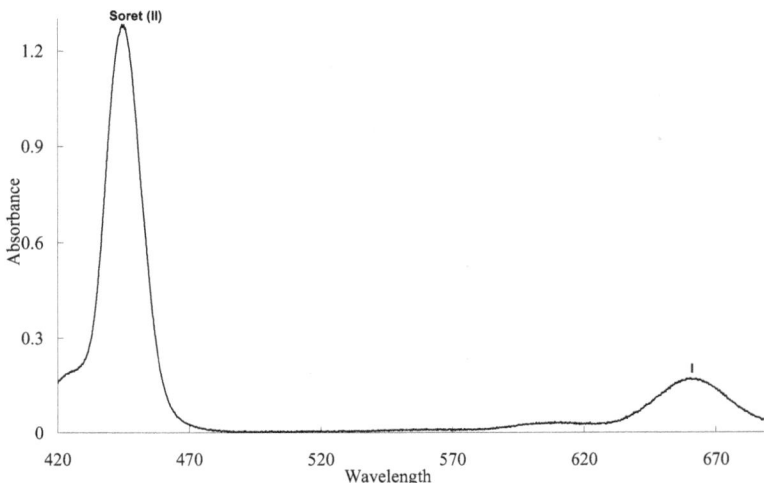

Fig. 5. Typical electronic absorption spectra of porphyrin diacid (H$_4$TPP^{2+}) and porphyrin molecular adducts [(R$_n$SnCl$_{4-n}$)$_2$H$_2$TPP].

Compound	λ(nm)		Compound	λ (nm)	
$(Me_2SnCl_2)_2H_2TPP$	445	660	$(Ph_2SnCl_2)_2H_2T(4-Cl)PP$	448	664
$(Me_2SnCl_2)_2H_2T(4-Cl)PP$	447	664	$(Ph_2SnBr_2)_2H_2T(4-Cl)PP$	452	668
$(Me_2SnCl_2)_2H_2T(4-CH_3)PP$	448	672	$(Ph_2SnCl_2)_2H_2TPP$	444	660
$(Me_2SnCl_2)_2H_2T(4-CH_3O)PP$	453	690	$(Ph_2SnBr_2)_2H_2TPP$	448	662
$(Et_2SnCl_2)_2H_2TPP$	444	660	$(Ph_2SnCl_2)_2H_2T(4-CH_3)PP$	448	670
$(Et_2SnCl_2)_2H_2T(4-Cl)PP$	448	664	$(Ph_2SnBr_2)_2H_2T(4-CH_3)PP$	450	676
$(Et_2SnCl_2)_2H_2T(4-CH_3)PP$	447	670	$(Ph_2SnCl_2)_2H_2T(4-CH_3O)PP$	453	688
$(Et_2SnCl_2)_2H_2T(4-CH_3O)PP$	453	688	$(Ph_2SnBr_2)_2H_2T(4-CH_3O)PP$	456	694
$(Bu_2SnCl_2)_2H_2TPP$	444	660	$(MeSnBr_3)_2H_2TPP$	448	666
$(Bu_2SnCl_2)_2H_2T(4-Cl)PP$	448	664	$(MeSnBr_3)_2H_2T(4-Cl)PP$	452	670
$(Bu_2SnCl_2)_2H_2T(4-CH_3)PP$	448	670	$(MeSnBr_3)_2H_2T(4-NO_2)PP$	458	662
$(Bu_2SnCl_2)_2H_2T(4-CH_3O)PP$	453	688	$(MeSnBr_3)_2H_2T(4-CH_3)PP$	450	676
$(MeSnCl_3)_2H_2T(4-Cl)PP$	452	670	$(MeSnBr_3)_2H_2T(4-CH_3O)PP$	456	696
$(MeSnCl_3)_2H_2TPP$	448	666	$H_4T(4-Cl)PP^{2+}$	444	661
$(MeSnCl_3)_2H_2T(4-CH_3)PP$	450	676	H_4TPP^{2+}	442	659
$(MeSnCl_3)_2H_2T(4-CH_3O)PP$	454	690	$H_4T(4-CH_3)PP^{2+}$	446	672
			$H_4T(4-CH_3O)PP^{2+}$	453	695

Table 4. UV-vis absorptions λ ($CHCl_3$/nm) of $[(Me_2SnCl_2)_2H_2T(4-X)PP]$ adducts.

1.4.2.1 ^{1}H NMR analysis

Up to getting a detailed description of ^{1}H NMR spectra of adducts, an overview on ^{1}H NMR of Cu(II)-SAT complex, S. Funahashi et al.,[16] will be informative. S. Funahashi et al. have studied the ^{1}H NMR variation of H_2TPP during a reaction between Cu(II) ion and H_2TPP in acetonitrile for detecting the SAT complex by spectroscopic method.

Table 5 summarized the chemical shift values of SAT complex together with the reported values for H_2TPP.[27, 28] Although the SAT complex contains the paramagnetic Cu(II) ion having $S = 1/2$, the β-pyrrole protons have been clearly observed. In the ^{1}H NMR spectra signal of the N-H protons in the SAT complex was clearly observed at -2.05 ppm relative to TMS strongly indicated that the N-H protons remain in the SAT complex $(Cu(H_2TPP)^{2+})$. In contrast to the case of H_2TPP, two kinds of β-protons for the SAT complex were observed with a ratio of 1:1, which had twice the area of the peak for N-H protons. The peaks, one a singlet and the other a doublet, were attributed to the protons in the pyrrolenine group coordinated and in the pyrrole group not coordinated to the Cu(II) ion, respectively. A similar splitting was reported for H_2TPP at the low temperature of -80 °C, because the N-H

Solute	N-H	Phenyl (H_o)	Phenyl ($H_{m,p}$)	Pyrrole (H_β)
H_2TPP	-2.87 (s)	7.80-7.87 (m)	8.22-8.25 (m)	8.85 (s)
$Cu(H_2TPP)^{2+}$	-2.05 (s)	7.31-7.50 (m)	8.10-8.12 (m)	8.66(d),8.77(s)

Table 5. Values of chemical shift (δ; relative to TMS/ppm) for H_2TPP and it's SAT complex in CD_3CN.

tautomerism at such a low temperature was frozen where the signal of the β-pyrrole protons of the pyrrole groups with N-H protons was observed at a lower field relative to that of the pyrrolenine groups without the N-H protons. The opposite trend in chemical shift was, however, observed for the SAT complex, i.e., the doublet peak assigned to the β-pyrrole protons appears at a higher field (see Fig. 6). This indicates that the two pyrrolenine nitrogens without an N-H proton bind to the paramagnetic Cu(II) ion, because the dipole-dipole and scalar coupling interactions with the paramagnetic ion lead to their downfield shift.[15,16] Furthermore, the peak for N-H protons was shifted downfield relative to that of H_2TPP. This is ascribed to both the distortion of the porphyrin ring and the dipole-dipole interaction with the paramagnetic ion. Because the N-H protons in H_2TPP are highly shifted to the upper field due to the ring current of the planar porphyrin ring, the distortion of the porphyrin ring in the SAT complex then leads to the downfield shift. The ^{1}H NMR spectrum determines a symmetrical structure with pyrrolenine nitrogens coordinating to the Cu(II) ion.

We used the low temperature ^{1}H NMR (-30 °C) to study the molecular interaction of free base porphyrins with organotin(IV) halides. ^{1}H NMR spectra of free base porphyrins undergo considerable variations during these interactions.

Data given in Table 6 are indicating that by interaction of free bases $H_2T(4-X)PP$ with R_nSnX_{4-n} the original bands of N-H, $H_β$, H_o, H_m, H_{CH3} and H_{OCH3}, were shifted and a new pattern were appeared for ^{1}H NMR spectrum of porphyrin moiety of adducts.

Compound	δ N-H	δ $H_{m,p}$	δ H_o	δ $H_β$	δ OCH_3,CH_3
$(Me_2SnCl_2)_2H_2TPP$	0.00	7.95	8.57	8.57	-
$(Me_2SnCl_2)_2H_2T(4-Cl)PP$	0.00	8.00	8.54	8.54	-
$(Me_2SnCl_2)_2H_2T(4-CH_3)PP$	0.00	7.94	8.50	8.50	2.89
$(Me_2SnCl_2)_2H_2T(4-CH_3O)PP$	0.00	7.50	8.48	8.48	4.11
$(Et_2SnCl_2)_2H_2TPP$	0.00	8.01-04	8.56-61	8.56-61	-
$(Et_2SnCl_2)_2H_2T(4-Cl)PP$	0.06	8.09-12	8.46-55	8.46-55	-
$(Et_2SnCl_2)_2H_2T(4-CH_3)PP$	0.15	7.81-85	8.49-59	8.49-59	2.80
$(Et_2SnCl_2)_2H_2T(4-CH_3O)PP$	0.20	7.49-56	8.46-53	8.46-53	4.15
$(MeSnCl_3)_2H_2TPP$	-0.2	7.99-8.02	8.59-8.62	8.59-8.62	-
$(MeSnCl_3)_2H_2T(4-Cl)PP$	0.13	8.01-8.05	8.52-8.59	8.52-8.59	-
$(MeSnCl_3)_2H_2T(4-CH_3)PP$	0.16	7.80-7.85	8.47-8.54	8.47-8.54	2.85
$(MeSnCl_3)_2H_2T(4-CH_3O)PP$	0.22	7.55-7.58	8.53-8.55	8.53-8.55	4.16
$(Ph_2SnCl_2)_2H_2TPP$	0.00	8.05-8.07	8.64	8.64	-
$(Ph_2SnBr_2)_2H_2TPP$	-0.45	7.92-8.00	8.54-8.57	8.54-8.57	-
$(Ph_2SnCl_2)_2H_2T(4-Cl)PP$	0.00	8.04-8.07	8.53-8.62	8.53-8.62	-
$(Ph_2SnBr_2)_2H_2T(4-Cl)PP$	0.00	8.18-8.21	8.37	8.52	-
$(Ph_2SnCl_2)_2H_2T(4-CH_3)PP$	0.00	7.83-7.86	8.49-8.59	8.49-8.59	2.80
$(Ph_2SnBr_2)_2H_2T(4-CH_3)PP$	-0.30	7.83-7.86	8.51-8.56	8.51-8.56	2.79
$(Ph_2SnCl_2)_2H_2T(4-CH_3O)PP$	0.00	7.41-7.56	8.37-8.42	8.37-8.42	4.04
$(Ph_2SnBr_2)_2H_2T(4-CH_3O)PP$	0.00	7.94	8.24-8.27	8.48-8.53	4.08

Table 6. ^{1}H NMR chemical shift (δ/ppm) of $[(Me_2SnCl_2)_2H_2T(4-X)PP]$ adducts relative to $CHCl_3$.

Comparing the chemical shifts of different protons of the free base $H_2T(4-X)PPs$ with the same protons of $H_2T(4-X)PPs$ in its complexes some useful information can be induced. The internal N-H signal moves downfield (≈2.7-2.8 ppm) and H_β protons signal move upfield (≈0.25-0.31 ppm). Both changes are discontinuous and are in the directions to be expected if the aromatic ring current decreases with the interaction of free base porphyrins with organotin(IV) halides. The two aromatic proton doublets related to the H_o and $H_{m,p}$ of the phenyl rings in H_2TPP, also H_o and H_m of phenyl of the $H_2T(4-X)PPs$, moves downfield (≈0.31-0.25 ppm) and $H_{CH3}(H_{OCH3})$ singlets of $H_2T(4-X)PPs$ (X= CH_3 and CH_3O) moves downfield (≈0.31-0.14 ppm). These changes can be described by decreasing the aromatic ring current because of the coordination of porphyrin to the organotin(IV) halide Lewis acids that led to deformation of porphyrin structure from planarity. Such changes were seen in the SAT complexes, let alone the paramagnetic property of Cu(II) ion which lead to some differences between ^{1}H NMR spectra of SAT complexes with our adducts. Also the ^{1}H NMR spectra of these adducts are comparable with related porphyrin dication.[25-27] Concerning the ^{1}H NMR spectra of 22,24-dihydro-5,10,15,20-tetraphenylporphyrin diperchlorate $[H_4TPP^{2+}][ClO_4^-]_2$; N-H (-2.47 ppm), $H_{m,p}$(8.04 ppm), H_o (8.61), H_β (8.77ppm) and 22,24-dihydro-5,10,15,20-tetraphenylporphyrin dihydrogen sulfate $[H_4TPP^{2+}][HSO_4^-]_2$; N-H (-1.21 ppm), $H_{m,p}$(7.82 ppm), H_o (8.45), H_β (8.51ppm) , (δ relative to TMS in $CDCl_3$) similar trends can be found between ^{1}H NMR chmical shift of porphyrin dication and our adducts.

According to ^{1}H NMR pattern these adducts have symmetrical structures, so that coordination to organotin(IV) halides couldn't differ between each class of free base porphyrin protons (N-H, H_β, H_o, and ...) upon adduct formation and these protons remained equivalent after complex formation. Referring to elemental analysis data these adducts have the mole ratio 2:1 of acceptor to donor, $[(R_2SnCl_2)_2(H_2T(4-X)PP)]$.

In Fig. 6 (a), because of attachment of tin atoms to two pyrrolenine nitrogens, splitting of ^{1}H NMR signal of pyrroles and pyrrolenine Hβs and producing a doublet band for beta hydrogens is expected, a singlet for pyrroles Hβs and a singlet for pyrrolenines Hβs. On the other hand, in Fig. 6 (b) and (c) Hβs have relatively identical environment, therefore we predict a singlet band for both pyrroles and pyrrolenine Hβs, but in Fig. 6 (c) because of high steric congestion of N-H pyrrole hydrogens with organotin(IV) halide upon adduct formation, doesn't seem it has a significant apart in adduct structure. Experimentally, in the low-temperature ^{1}H NMR spectra of adducts a singlet was appeared for Hβs, referring to S. Funahashi results on slowness of the N-H tautomerism in the SAT complex with respect to the time scale of the ^{1}H NMR at 21 °C,[16] it is confirming the Fig. 6 (b) structure for these adducts. On the basis of these results we suggest that free base porphyrin as a bidentate bridging ligand make a bridge between two molecules of the Lewis acid through it's nitrogenes lone pairs. It is probable that two neighbor nitrogen atoms (a pyrrole and a pyrrolenine nitrogen) of the porphyrin binded to one of the R_2SnCl_2 molecules which posited on the above of the porphyrin plane and the other two nitrogen atoms (a pyrrole and a pyrrolenine nitrogen) binded to the second R_2SnCl_2 molecule from below of this plane. Therefore we have a structure close to that suggested by K.M. Smith et al. for XHg-TPP-HgX, X= Cl- and CH_3COO^-.[29]

Interactions of organotin(IV) halides with free base $H_2T(4-X)PP$ are very sensitive to temperature and donor property of the solvent, *Fig. 7*. By increasing the temperature the green color of the adduct solution changes to brown and the color return to the primary color of free base porphyrin solution, eventually. Also these adduct formations were not observed in ligating solvents such as CH_3CN and DMSO under these conditions.

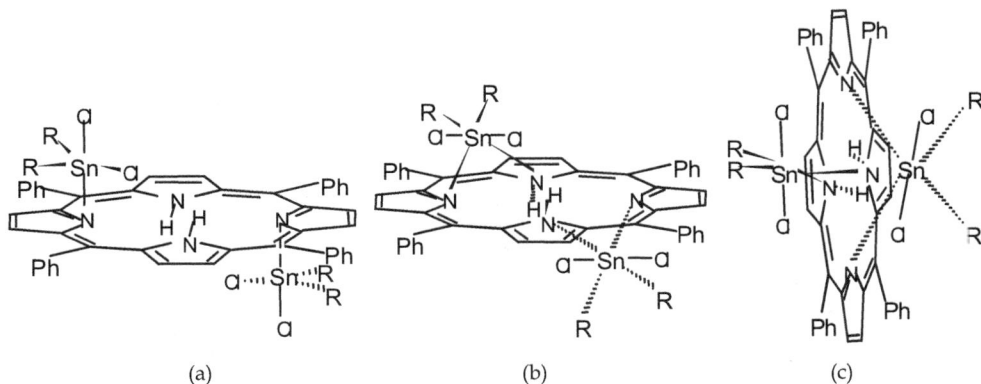

(a)   (b)   (c)

Fig. 6. Proposed structures of $[(R_2SnCl_2)_2H_2TPP]$ adducts. (a) H_2TPP as a monodentate bridging ligand made adducts with five-coordinated trigonal bipyramidal structure for tin atoms; (b) and (c) H_2TPP as a bidentate bridging ligand make adducts with six-coordinated octahedral structure around the tin atoms. Our ^1H NMR data are in consistent with structure (b).

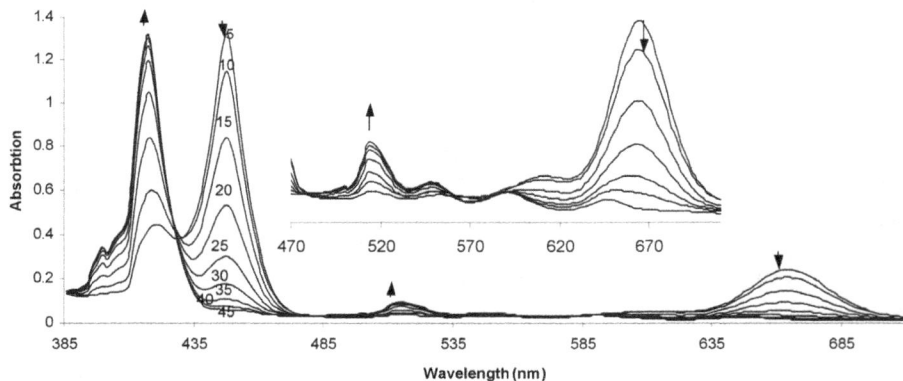

Fig. 7. The temperature dependence change of the absorption profile of the $[(Me_2SnCl_2)_2H_2T(4-Cl)PP]$ adduct in chloroform: By addition of Me_2SnCl_2 (0.0125 M, 0.5 ml) to the solution of $H_2T(4-Cl)PP$ (5×10^{-6} M, 2.5 ml) in chloroform in an UV-vis cell at 5 °C, the $(Me_2SnCl_2)_2H_2T(4-Cl)PP$ adduct was formed. Then the composition of the cell was remained constant and the temperature was raised to 45 °C, stepwisely.

1.4.2.2 Basicity of the free base porphyrins:

The basicity of the free base porphyrins toward proton is usually determined when possible by pK_3 measurements for the dissociation of the mono-cation $H_3T(4-X)PP^+$ into the free base $H_2T(4-X)PP$ measured in detergent solutions:

$$H_3T(4\text{-}X)PP^+ = H_2T(4\text{-}X)PP + H^+ \quad K_3 \qquad (9)$$

On the other hand, the basicity of the porphyrins have been related to the reduction potential (in volts) of the free base porphyrin to its radical anion form.[30] The value of $E_{1/2}(1)$ for some of the free base porphyrins in DMF is given in Table 7.

$$H_2T(4\text{-}X)PP + e \rightleftharpoons H_2T(4\text{-}X)PP^- \quad E_{1/2}(1) \qquad (10)$$

X	NO$_2$	Cl	H	CH$_3$	CH$_3$O
$E_{1/2}(1)$	-1.34	-1.47	-1.55	-1.57	-1.59

Table 7. Reduction potentials (in volts) for $H_2T(4\text{-}X)PPs$ (in DMF, 25 °C).[30]

A linear relationship between pK$_3$ and $E_{1/2}(1)$ has been reported for some of porphyrins:

$$pK_3 = -5.9\ E_{1/2}(1) - 5.2$$

according to this Equation the most basic porphyrins are the more difficult to reduce. Therefore, more negative values of $E_{1/2}(1)$ are indicating stronger basic properties of free base porphyrin.

1.4.2.3 The thermodynamic studies

The thermodynamic parameters are useful tools for studying the interactions between the donor and the acceptor molecules. The first step in these studies is the determination of formation constants, which make help to better understanding of relative stability of molecular adducts. In these studies we used the SQUAD program [31] for data refinement. This program is designed to calculate the best values for the equilibrium constants of the proposed equilibrium model by employing a non-linear least-squares approach and UV-vis data. The formation constants and thermodynamic parameters $\Delta H°$, $\Delta S°$ and $\Delta G°$ values for different systems are listed in Tables 8 and 9, respectively.

According to results presented for molecular interactions of organotin(IV) halides with free base porphyrins the following order was obtained for acceptor properties of organotin(IV) Lewis acids:

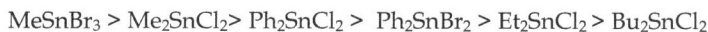

$$MeSnBr_3 > Me_2SnCl_2 > Ph_2SnCl_2 > Ph_2SnBr_2 > Et_2SnCl_2 > Bu_2SnCl_2$$

This trend shows that interaction of organotin(IV) halides with free base porphyrins become weaker by increasing the electron-releasing and the steric hindrance of substituents on the tin atom.

On the other hand, results show the effect of X substituents of phenyl rings of free base porphyrins on the stability of adducts, explicitly. Considering data given in Tables 8 and 9 for the interaction of organotin(IV) Lewis acids with free base porpyrins, $H_2T(4\text{-}X)PP$ (X= H, Cl, CH$_3$, CH$_3$O, NO$_2$), basicity of free base porphyrins for interaction with organotin(IV) halides decreases as follows:

$$H_2T(4\text{-}CH_3O)PP > H_2T(4\text{-}CH_3)PP > H_2TPP > H_2T(4\text{-}Cl)PP > H_2T(4\text{-}NO_2)PP$$

Of course, $H_2T(4\text{-}NO_2)PP$ under our working conditions (concentration and temperature) did not show a measurable interaction with diorganotin(IV) dihalides.

Adduct	lgK				
	5 °C	10 °C	15 °C	20 °C	25 °C
$[(Me_2SnCl_2)_2H_2TPP]$	7.61± 0.02	6.95± 0.01	6.46± 0.01	5.81± 0.01	5.05± 0.01
$[(Me_2SnCl_2)_2H_2T(4\text{-}Cl)PP]$	6.10± 0.01	5.66± 0.01	5.14± 0.01	4.64± 0.01	-
$[(Me_2SnCl_2)_2H_2T(4\text{-}CH_3)PP]$	10.06±0.01	9.30± 0.01	8.46± 0.01	7.81± 0.01	6.99± 0.01
$[(Me_2SnCl_2)_2H_2T$ (4-CH$_3$O)PP]	10.25± 0.01	9.43± 0.01	8.69± 0.01	7.96± 0.01	7.20± 0.01 -
$[(Et_2SnCl_2)_2H_2TPP]$	6.38 ± 0.02	5.87 ± 0.01	5.53 ± 0.02	5.08 ± 0.04	-
$[(Et_2SnCl_2)_2H_2T(4\text{-}Cl)PP]$	5.25 ± 0.02	4.76 ± 0.01	4.36 ± 0.02	4.04 ± 0.04	
$[(Et_2SnCl_2)_2H_2T(4\text{-} CH_3)PP]$	6.96 ± 0.01	6.51 ± 0.01	6.04 ± 0.04	5.45 ± 0.05	5.11 ± 0.02
$[(Et_2SnCl_2)_2H_2T$ (4-CH$_3$O)PP]	8.08 ± 0.03	7.34 ± 0.02	7.04 ± 0.04	6.38 ± 0.02	5.93 ± 0.01
$[(Bu_2SnCl_2)_2H_2TPP]$	5.69 ± 0.01	5.34 ± 0.02	4.88 ± 0.01	4.38 ± 0.02	4.04 ± 0.04
$[(Bu_2SnCl_2)_2H_2T(4\text{-}Cl)PP]$	4.70 ± 0.01	4.32 ± 0.02	3.82 ± 0.01	3.54 ± 0.01	3.11 ± 0.02
$[(Bu_2SnCl_2)_2H_2T(4\text{-}CH_3)PP]$	6.89 ± 0.01	6.40 ± 0.02	5.99 ± 0.01	5.32 ± 0.02	5.15 ± 0.02
$[(Bu_2SnCl_2)_2H_2T$ (4-CH$_3$O)PP]	7.85 ± 0.01	7.15 ± 0.02	6.76 ± 0.01	6.20 ± 0.02	5.80 ± 0.01
$(Ph_2SnCl_2)_2H_2TPP$	7.12±0.01	6.46±0.01	5.57±0.09	5.25±0.04	4.24±0.01
$(Ph_2SnCl_2)_2H_2T(4\text{-}Cl)PP$	6.32±0.02	5.90±0.03	5.40±0.09	4.85±0.04	4.10±0.01
$(Ph_2SnCl_2)_2H_2T(4\text{-}CH_3)PP$	8.02±0.01	7.66±0.01	6.69±0.09	5.94±0.03	5.10±0.02
$(Ph_2SnCl_2)_2H_2T$ (4-CH$_3$O)PP	9.90±0.02	9.04±0.02	8.20±0.09	7.50±0.02	6.63±0.04
$(Ph_2SnBr_2)_2H_2TPP$	6.45±0.01	6.00±0.02	5.36±0.09	4.83±0.07	4.07±0.08
$(Ph_2SnBr_2)_2H_2T(4\text{-}Cl)PP$	5.91±0.01	5.33±0.01	5.07±0.09	4.71±0.09	4.05±0.08
$(Ph_2SnBr_2)_2H_2T(4\text{-}CH_3)PP$	7.50±0.02	6.80±0.02	5.71±0.09	5.35±0.06	4.81±0.05
$(Ph_2SnBr_2)_2H_2T$ (4-CH$_3$O)PP	9.20±0.02	8.40±0.02	7.52±0.09	6.68±0.06	6.13±0.05
$(MeSnBr_3)_2H_2TPP$	12.51 ± 0.06	11.33± 0.03	10.46± 0.03	9.58 ± 0.03	8.45 ± 0.02
$(MeSnBr_3)_2H_2T(4\text{-}Cl)PP$	11.41 ± 0.03	10.56± 0.04	9.75 ± 0.02	9.15 ± 0.02	7.95 ± 0.02
$(MeSnBr_3)_2H_2T(4\text{-}NO_2)PP$	6.68 ± 0.01	6.11 ± 0.03	5.58 ± 0.05	4.90 ± 0.08	4.41 ± 0.04
$(MeSnBr_3)_2H_2T(4\text{-}CH_3)PP$	13.61 ± 0.06	12.49± 0.06	11.53± 0.04	10.45± 0.04	9.31 ± 0.05
$(MeSnBr_3)_2H_2T(4\text{-}CH_3O)PP$	14.28 ± 0.03	13.12± 0.03	11.96± 0.02	10.88± 0.02	9.76 ± 0.09

Table 8. The formation constants lgK for $H_2T(4\text{-}X)PP$ adducts in $CHCl_3$ solvent.

Adducts have negative values of $\Delta H°$, $\Delta S°$, and $\Delta G°$ (Table 9) which correspond to exothermic adduct formations between organotin(IV) Lewis acids and free base porphyrins.

Adduct	$-\Delta H°$	$-\Delta S°$	$-\Delta G°$ [b]
$(Me_2SnCl_2)_2H_2TPP$	199 ±11	569 ± 33	38 ±11
$(Me_2SnCl_2)_2H_2T(4\text{-}Cl)PP$	167 ± 7	483 ± 25	30 ±7
$(Me_2SnCl_2)_2H_2T(4\text{-}CH_3)PP$	242 ± 7	678 ± 26	50 ± 7
$(Me_2SnCl_2)_2H_2T(4\text{-}CH_3O)PP$	240 ± 8	668 ± 31	51 ± 8
$(Et_2SnCl_2)_2H_2TPP$	138 ± 5	373 ± 19	32 ± 5
$(Et_2SnCl_2)_2H_2T(4\text{-}Cl)PP$	128 ± 8	360 ± 28	26 ± 8
$(Et_2SnCl_2)_2H_2T(4\text{-}CH_3)PP$	150 ± 6	405 ± 23	35 ± 6
$(Et_2SnCl_2)_2H_2T(4\text{-}CH_3O)PP$	168 ± 10	451 ± 35	40 ± 10
$(Bu_2SnCl_2)_2H_2TPP$	136 ± 5	378 ± 19	29 ± 5
$(Bu_2SnCl_2)_2H_2T(4\text{-}Cl)PP$	124 ± 5	356 ± 18	23 ± 5
$(Bu_2SnCl_2)_2H_2T(4\text{-}CH_3)PP$	142 ± 7	379 ± 26	35 ± 7
$(Bu_2SnCl_2)_2H_2T(4\text{-}CH_3O)PP$	160 ± 7	427 ± 25	39 ± 7
$(Ph_2SnCl_2)_2H_2TPP$	216 ±17	644 ±58	34 ±17
$(Ph_2SnCl_2)_2H_2T(4\text{-}Cl)PP$	174 ±11	508 ±40	31 ±11
$(Ph_2SnCl_2)_2H_2T(4\text{-}CH_3)PP$	234 ±18	688 ±58	39 ±18
$(Ph_2SnCl_2)_2H_2T(4\text{-}CH_3O)PP$	251 ±5	717 ±16	48 ±5
$(Ph_2SnBr_2)_2H_2TPP$	184 ±11	538 ±37	32 ±11
$(Ph_2SnBr_2)_2H_2T(4\text{-}Cl)PP$	138 ±12	382 ±44	29 ±12
$(Ph_2SnBr_2)_2H_2T(4\text{-}CH_3)PP$	213 ±20	625 ±70	36 ±20
$(Ph_2SnBr_2)_2H_2T(4\text{-}CH_3O)PP$	244 ±8	707 ±30	44 ±5
$(MeSnBr_3)_2H_2TPP$	301 ±11	845 ±37	61 ±11
$(MeSnBr_3)_2H_2T(4\text{-}Cl)PP$	264 ±17	730 ±60	57 ±17
$(MeSnBr_3)_2H_2T(4\text{-}NO_2)PP$	182 ±5	528 ±17	32 ±5
$(MeSnBr_3)_2H_2T(4\text{-}CH_3)PP$	337 ±8	953 ±28	67 ±8
$(MeSnBr_3)_2H_2T(4\text{-}CH_3O)PP$	358 ±2	1014 ±8	71 ±2

[a] $\Delta H°$ (kJ.mol^{-1}), $\Delta S°$ (J.K^{-1}. mol^{-1}), and $\Delta G°$ at 10 °C (kJ. mol^{-1}).

Table 9. The thermodynamic parameters $\Delta H°$, $\Delta S°$ and $\Delta G°$ for $H_2T(4\text{-}X)PP$ adducts in CHCl$_3$.[a]

2. Molecular interactions of organic π-acceptors with $H_2T(4\text{-}X)PP$ (X= H, Cl, CH$_3$, CH$_3$O)

Intermolecular charge-transfer (CT) complexes are formed when electron donors and electron acceptors interact, a general phenomenon in organic chemistry. Mulliken [32] considered such complexes to arise from a Lewis acid-Lewis base type of interaction, the bond between the components of the complex being postulated to arise from the partial transfer of an π electron from the base to orbitals of the acid. TCNE, DDQ, TBBQ and TCBQ can form CT complexes when mixed with molecules possessing π-electrons or groups having atoms with an unshared electron pair [33-35]. The thermodynamic of CT complexes of several free base tetraaryl- as well as tetraalkylporphyrins with organic π-acceptors TCNE, DDQ, TBBQ and TCBQ have been investigated. It would be instructive to compare their thermodynamic parameters since it improve our insight about characteristics of their interactions and efficiency of their applications.

Adducts which have been prepared from mixing of organic π-acceptors with a solution of free base porphyrin in dry chloroform, those ¹H NMR and UV-vis data were been given in Tables 10-12.

Compound	δN-H	δH$_{m,p}$	δH$_o$	δH$_\beta$	δCH$_3$
H$_2$TPP	-2.76	7.75	8.24	8.85	-
(TCNE)$_2$H$_2$TPP	-1.26	8.04	8.64	8.81	-
(DDQ)$_2$H$_2$TPP	-0.40	7.98	8.67	8.59	-
(TCBQ)$_2$H$_2$TPP	-2.00	7.79	8.28	8.81	-
(TBBQ)$_2$H$_2$TPP	-0.13	8.06	8.67	8.81	
H$_2$T(4-Cl)PP	-2.86	7.75	8.14	8.85	-
(TCNE)$_2$H$_2$T(4-Cl)PP	-1.16	8.05	8.57	8.81	-
(DDQ)$_2$H$_2$T(4-Cl)PP	-0.29	8.00	8.56	8.56	-
(TCBQ)$_2$H$_2$T(4-Cl)PP	-2.60	7.76	8.20	8.83	
(TBBQ)$_2$H$_2$T(4-Cl)PP	-0.11	7.80	8.58	8.83	
H$_2$T(4-CH$_3$)PP	-2.77	7.56	8.11	8.86	2.71
(TCNE)$_2$H$_2$T(4-CH$_3$)PP	-1.20	7.83	8.52	8.76	2.80
(DDQ)$_2$H$_2$T(4-CH$_3$)PP	-0.38	7.80	8.55	8.55	2.77
(TCBQ)$_2$H$_2$T(4-CH$_3$)PP	-2.10	7.57	8.12	8.84	2.75
(TBBQ)$_2$H$_2$T(4-CH$_3$)	-0.73	7.57	8.12	8.84	2.75
PP H$_2$T(4-CH$_3$O)PP	-2.82	7.27	8.11	8.86	3.95
(TCNE)$_2$H$_2$T(4-OCH$_3$)PP	-0.92	7.55	8.57	8.68	4.17
(DDQ)$_2$H$_2$T(4-OCH$_3$)PP	-0.11	7.50	8.59	8.48	4.11
(TCBQ)$_2$H$_2$T(4-OCH$_3$)PP	-1.16	7.27	8.13	8.76	3.96
(TBBQ)$_2$H$_2$T(4-OCH$_3$)PP	-0.12	7.60	8.40	8.76	4.10

Table 10. ¹H NMR of free base tetraarylporphyrins and their adduct with DDQ and TCNE.

The schematic chemical equilibria for interaction of molecular π-acceptors with free base porphyrins could be written according to Equation 11: [36-38]

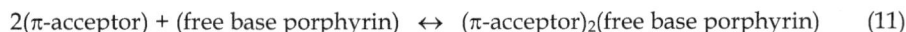

$$2(\pi\text{-acceptor}) + (\text{free base porphyrin}) \leftrightarrow (\pi\text{-acceptor})_2(\text{free base porphyrin}) \qquad (11)$$

2.1 ¹H NMR analysis

¹H NMR analysis: in the ¹H NMR spectra of adducts the signals correspond to N-H, H$_o$, H$_{m,p}$, and CH$_3$- or -OCH$_3$ protons of tetraarylporphyrin moved downfield, while H$_\beta$ has an upfield shift (Table 11). But tetraalkylporphyrins, the signals correspond to H$_\beta$ has an upfield shift the N-H signal in 5 and 11 moved downfield while for 6 and 12 upfield shift observed for this proton with adduct formation (Table 12).

Compound	δN-H	δCH$_3$	δCH$_2$	δCH$_2$	δCH$_2$	δH$_\beta$
H$_2$TnBP	-2.61	1.10	1.73-1.91	2.41-2.58	4.88	9.45
(TCNE)$_2$H$_2$TnBP	-3.45	1.22	1.93	2.54	4.83	9.03
(DDQ)$_2$H$_2$TnPP	-3.45	1.20	1.90	2.55	4.81	9.02
H$_2$TtBP	1.52	2.01	-	-	-	9.08
(TCNE)$_2$H$_2$TtBP	-0.4	2.13	-	-	-	8.14
(DDQ)$_2$H$_2$TtBP	-0.6	2.10	-	-	-	8.16

Table 11. ¹H NMR of free base tetraalkylporphyrins and their adduct with DDQ and TCNE.

2.2 Infra red spectra

The most significant difference which emerges from a comparison of the vibrational spectra of free base porphyrins and their molecular adducts with π-acceptor molecules is disappearance of the band $v(N-H)$ as a consequence of the complex formation. It seams that N-H was contributed in intermolecular hydrogen bonds that formed between its proton and N or O atoms of π-acceptor molecules. The IR spectra of free TCNE shows CN stretching frequencies at 2225 and 2270 cm^{-1}. The significant shift of these CN vibrations toward lower frequencies (2214 cm^{-1}) is indicating charge transferring from free base porphyrins to an π* orbital of CN group of TCNE. This charge transfer phenomenon leads to weakening of corresponding C≡N bond. Similarly, in the free DDQ the CO and the CN stretching frequencies were appeared at 1675 and 2245 cm^{-1}, respectively. Upon complex formation the CO vibrations shift to 1651 and 1696 cm^{-1} and CN vibration shift to 2229 cm^{-1}.

Compound	λ/nm		Compound	λ/nm				
(TCNE)$_2$H$_2$TPP	441	653	(TCNE)$_2$H$_2$T(4-OCH$_3$)PP	454				653
(DDQ)$_2$H$_2$TPP	445	663	(DDQ)$_2$H$_2$T(4-OCH$_3$)PP	453				688
(TCBQ)$_2$H$_2$TPP	445	662	(TCBQ)$_2$H$_2$T(4-OCH$_3$)PP	443				688
(TBBQ)$_2$H$_2$TPP	447	666	(TBBQ)$_2$H$_2$T(4-OCH$_3$)PP	445				694
(TCNE)$_2$H$_2$T(4-Cl)PP	443	652	H$_2$TnBP	417	520	555	600	659
(DDQ)$_2$H$_2$T(4-Cl)PP	448	652	(TCNE)$_2$H$_2$TnBP	425				636
(TCBQ)$_2$H$_2$T(4-Cl)PP	447	665	(DDQ)$_2$H$_2$TnPP	433				643
(TBBQ)$_2$H$_2$T(4-Cl)PP	450	669	H$_2$TtBP	448	552	596	628	691
(TCNE)$_2$H$_2$T(4-CH$_3$)PP	445	653	(TCNE)$_2$H$_2$TtBP	451				690
(DDQ)$_2$H$_2$T(4-CH$_3$)PP	447	677	(DDQ)$_2$H$_2$TtBP	455				692
(TCBQ)$_2$H$_2$T(4-CH$_3$)PP	448	672						
(TBBQ)$_2$H$_2$T(4-CH$_3$)PP	450	678						

Table 12. UV-vis peaks (λ/nm in CHCl$_3$) of FBPs and [(A)$_2$D] adducts.

2.3 UV-vis analysis

By interaction of free base porphyrin with π-acceptor molecules, their original peaks changed to a new pattern with two new peaks which their positions show about 20-40 nm red shift relative to the band V (Soret band) and the band I (of Q band) of free base H$_2$T(4-X)PPs, Table 10.

π-Acceptor molecules show different behaviors in their interactions with free base porphyrins. For DDQ the interactions were found exothermic and stability of adducts decreased at elevated temperatures, so that a solution of adduct at 5 °C goes to a solution containing dissociated free base and DDQ molecules when temperature were raised to 35 °C. It shows that by increasing of temperature the corresponding equilibrium in Equation 11 is shifted to the left.

On the other hand, for TCNE, TBBQ and TCBQ interactions are endothermic and at elevated temperatures the stability of adducts increased, so that a purple solution of reactant at 5 °C turns to a green solution of adduct at 35 °C. It means that, for these acceptors the equilibrium in Equation 11 is going to completion at higher temperatures.

2.4 Thermodynamic studies

The thermodynamic parameters were investigated for these interactions by UV-vis spectrometry method. The formation constants, K, were determined at several temperatures by analyzing the concentration and temperature dependence of UV-vis absorptions using the SQUAD program, Tables 13 and 14. Van't Hoff plots of these formation constants were used for obtaining the other thermodynamic parameters $\Delta H°$, $\Delta S°$, and $\Delta G°$ (Table 15).

The data of Table 13 shows that stability of adducts for all acceptors undergo a regular increase from $H_2T(4\text{-Cl})PP$ to H_2TPP, $H_2T(4\text{-CH}_3)PP$, and $H_2T(4\text{-OCH}_3)PP$; also adducts of H_2TtBP are more stable than H_2TnBP. These sequences are in agreement with electron releasing property of free base porphyrins.

Adduct	lg K				
	5 °C	10 °C	15 °C	20 °C	25 °C
$(TCNE)_2H_2TPP$	5.72±0.02	6.47±0.02	6.79±0.01	7.33±0.03	8.03±0.02
$(TCNE)_2H_2T(4\text{-Cl})PP$	4.40±0.01	5.01±0.02	5.42±0.01	6.20±0.03	6.45±0.02
$(TCNE)_2H_2T(4\text{-CH}_3)PP$	6.10±0.02	6.65±0.03	7.06±0.01	7.88±0.02	8.36±0.01
$(TCNE)_2H_2T(4\text{-OCH}_3)PP$	6.30±0.02	6.87±0.03	7.30±0.01	8.00±0.03	8.62±0.01
$(TCNE)_2H_2TnBP$	6.03±0.02	6.33±0.03	6.82±0.01	7.06±0.03	7.30±0.01
$(TCNE)_2H_2TtBP$	7.51±0.02	7.86±0.03	8.29±0.01	8.59±0.02	8.80±0.01
$(DDQ)_2H_2TPP$	9.66±0.02	9.41±0.01	9.24±0.01	9.06±0.03	8.83±0.01
$(DDQ)_2H_2T(4\text{-Cl})PP$	9.50±0.01	9.24±0.02	9.09±0.03	8.92±0.01	8.70±0.02
$(DDQ)_2H_2T(4\text{-CH}_3)PP$	10.11±0.01	9.85±0.01	9.64±0.03	9.45±0.01	9.24±0.02
$(DDQ)_2H_2T(4\text{-OCH}_3)PP$	10.34±0.02	10.12±0.02	9.93±0.04	9.69±0.01	9.42±0.03
$(DDQ)_2H_2TnPP$	9.45±0.02	9.30±0.02	9.13±0.04	8.95±0.01	8.81±0.03
$(DDQ)_2H_2TtBP$	10.17±0.02	10.01±0.02	9.81±0.04	9.62±0.01	9.48±0.03

Table 13. The formation constants lgK for $H_2T(4\text{-X})PP$ adducts in $CHCl_3$ solvent.

Adduct	lg K				
	15 °C	20 °C	25 °C	30 °C	35 °C
$(TCBQ)_2H_2T(4\text{-Cl})PP$	3.89±0.03	4.17±0.09	4.51±0.08	4.70±0.07	5.00±0.09
$(TCBQ)_2H_2TPP$	4.07±0.06	4.39±0.06	4.64±0.07	4.98±0.09	5.20±0.06
$(TCBQ)_2H_2T(4\text{-CH}_3)PP$	4.41±0.09	4.80±0.09	5.01±0.04	5.32±0.09	5.58±0.04
$(TCBQ)_2H_2T(4\text{-OCH}_3)PP$	4.97±0.04	5.35±0.04	5.70±0.08	6.05±0.03	6.34±0.03
$(TBBQ)_2H_2T(4\text{-Cl})PP$	5.66±0.06	6.01±0.09	6.42±0.08	6.67±0.08	7.00±0.09
$(TBBQ)_2H_2TPP$	6.70±0.06	7.20±0.06	7.69±0.09	8.20±0.08	8.50±0.08
$(TBBQ)_2H_2T(4\text{-CH}_3)PP$	6.86±0.07	7.45±0.07	7.85±0.08	8.67±0.09	9.00±0.08
$(TBBQ)_2H_2T(4\text{-OCH}_3)PP$	7.00±0.07	751±0.05	7.93±0.07	8.93±0.07	9.18±0.08

Table 14. The formation constants lgK for $H_2T(4\text{-X})PP$ adducts in $CHCl_3$ solvent.

Table 13 shows that formation constants for interaction of DDQ with free base porphyrins decreased at higher temperatures. In contrast for other acceptors (TCNE, TBBQ and TCBQ) formation constants increased at higher temperatures. This discrepancy might be due to presence of different mechanisms for caring out charge transfer between free bases and these π-acceptors. Since these π-acceptors are common oxidizing agents in organic chemistry, this dual influence of temperatures on their interactions with free base porphyrins is an interesting point that less has been attended previously. According to formation constants the following order was suggested for acceptor property of these molecules: DDQ 〉TCNE 〉 TBBQ 〉 TCBQ

Table 15 gives the thermodynamic parameters ($\Delta H°$, $\Delta S°$, and $\Delta G°$) for charge transfer adducts of DDQ, TCNE, TBBQ and TCBQ with free base porphyrins in chloroform. These results show that meso-group of free base porphyrins significantly affected strength of their interactions with acceptor species. While electron releasing meso-groups improves interactions of free base porphyrins with π-acceptors molecules, electron withdrawing meso-groups diminished these interactions.

Compound	$\Delta H°$	$\Delta S°$	$-\Delta G°$	Compound	$\Delta H°$	$\Delta S°$	$-\Delta G°$
(TCNE)₂H₂TPP	174±12	734±43	45±12	(TCBQ)₂H₂T(4-CH₃)PP	101±6	435±19	29±1
(DDQ)₂H₂TPP	-64±3	-45±10	50±3	(TBBQ)₂H₂T(4-CH₃)PP	168±14	714±45	45±1
(TCBQ)₂H₂TPP	98±7	416±23	27±1	(TCNE)₂H₂T (4-OCH₃) PP	183±10	784±33	51±10
(TBBQ)₂H₂TPP	160±6	683±23	44±12				
(TCNE)₂H₂T(4-Cl)PP	168±12	688±43	37±12	(DDQ)₂H₂T (4-OCH₃) PP	-72±4	-61±12	54±4
(DDQ)₂H₂T(4-Cl)PP	-61±3	-38±10	49±3				
(TCBQ)₂H₂T(4-Cl)PP	94±5	401±18	26±1	(TCBQ)₂H₂T (4-OCH₃) PP	109±3	509±9	33±1
(TBBQ)₂H₂T(4-Cl)PP	116±5	512±15	36±1				
(TCNE)₂H₂T (4-CH₃)PP	179±10	761±35	47±10	(TBBQ)₂H₂T (4-OCH₃) PP	197±37	816±124	46±1
(DDQ)₂H₂T (4-CH₃)PP	-68±2	-52±7	52±3	(TCNE)₂H₂TnBP	107±7	495±26	40±7
				(DDQ)₂H₂TnPP	-53±1	38±4	41±1
				(TCNE)₂H₂TtBP	109±6	529±22	49±6
				(DDQ)₂H₂TtBP	-63±3	-26±11	70±3

[a] $\Delta H°$ (kJ mol⁻¹) and $\Delta S°$ (J K⁻¹ mol⁻¹); $\Delta G°$ at 25 °C (kJ mol⁻¹).

Table 15. The thermodynamic parameters for [(Acceptor)₂(Free base porphyrin)] adducts in CHCl₃.[a]

For [(DDQ)₂FBP] adducts, $\Delta H°$ and $\Delta S°$ are negative. But other adducts ([(TCNE)₂FBP], [(TBBQ)₂FBP], [(TCBQ)₂FBP]) have positive values for $\Delta H°$ and $\Delta S°$ parameters. Since interactions of free base porphyrins with various acceptors are exothermic, endothermic interactions for these π-acceptors are interesting and unexpected. It seems presence of strong self π-stacking between these π-acceptor molecules make their dissolving endothermic, on the other hand dissociation of such π-stacks make their dissolving, entropically a favorable phenomena.

The standard Gibbs free energies, $\Delta G°$, of interactions are negative in all cases. It seems that for TCNE, TBBQ and TCBQ term $\Delta S° 〉 0$ is more effective than $\Delta H° 〉 0$ in the following Equation: $\Delta G° = \Delta H° - T\Delta S°$ so that negative values for $\Delta G°$ were occurred. Negative values of ΔG° show that interactions of these molecules with free bases are favorable with the same ordering as their electron releasing property.

The main goal in this work is "positive $\Delta H^{\circ \prime}$" for TCNE, TBBQ and TCBQ adducts. In view of the discrepancy, further studies are certainly necessary to clarify this essential issue. Our results are indicating that electron acceptor properties of these molecules depend on temperature and this might be important to consider it for facilitating their subsequent applications.

3. Hydrogen bond complexes of 2,4-dichloro-; 2,4,6-trichloro- and 4-nitrophenol with free bases *meso*-tetraarylporphyrins

Hydrogen bonds play a crucial role in many chemical, physical, and biochemical processes, and they are also very important in crystal engineering [39-42]. Hydrogen bonds usually designated as X-H...Y in which there is an X-H proton donating bond and an acceptor of protons (Y-center). Interaction of phenol derivatives (PD) such as 4-nitrophenol (4NP), 2,4-Dichlorophenol (24DCP), and 2,4,6-Trichlorophenol (246)TCP with biological systems are of interest. Porphyrins with two N-H proton donating sites and numerous nitrogen atoms as proton acceptor centers can form HB complexes with some hydroxylated compounds. Last studies were demonstrated that hydrogen bonded complexes of phenol derivatives with free base porphyrins has a 2:1 mole ratio of phenol to porphyrin.

On addition of phenol derivatives (PD; 4NP, 24DCP and 246TCP) to a solution of $H_2T(4-X)PP$ in chloroform, PD form HB complex with free base porphyrin according to following equation: [43, 44]

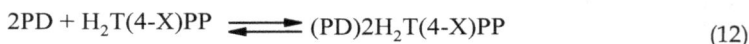

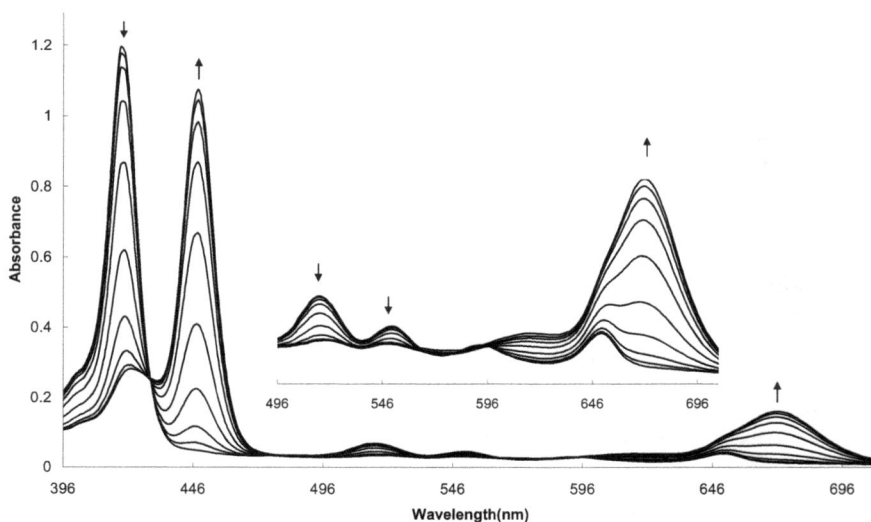

$$2PD + H_2T(4-X)PP \rightleftharpoons (PD)2H_2T(4-X)PP \qquad (12)$$

Fig. 8. Titration spectra for interaction of 24DCP with $H_2T(4-CH_3O)PP$ in chloroform. Bands appeared at 455 and 688 nm are related to adduct, isosbestic point at 428.

These interactions were studied by means of UV-vis spectrometry method and data refinement were carried out by SQUAD program. Interaction of PDs with free base porphyrins leads to a fundamental change in the porphyrins electronic absorption spectra, Fig.8 and Table 16. For example, when a solution of $H_2T(4-CH_3O)PP$ in chloroform was interacted with 24DCP, the original peaks of $H_2T(4-CH_3O)PP$ were vanished and adduct peaks' were appeared at 455 and 688 nm. A clear isosbestic point at 428 nm represents formation of hydrogen bond complex in a reversible reaction. For experimental illustration of an endothermic equilibrium adduct formation a solution of $[(PD)_2H_2T(4-CH_3)PP]$ adduct at 25 °C was cooled to 5 °C in a UV-vis cell. Observations show that upon cooling the adduct $[(PD)_2H_2T(4-CH_3)PP]$ with Soret band at 445 nm was dissociated to PD and free base porphyrin $H_2T(4-CH_3)PP$ with Soret band at 419 nm. By heating of solution to 25 °C the adduct was formed again and Soret band was returned to 445 nm (Fig. 9).

Compound	λ (nm)		Compound	λ (nm)	
$(24DCP)_2H_2TPP$	445	651	$(246TCP)_2H_2T(4-CH_3)PP$	448	669
$4NP-H_2TPP$	446	661	$4NP-H_2T(4-CH_3)PP$	450	671
$(24DCP)_2H_2T(4-Cl)PP$	448	658	$(24DCP)_2H_2T(4-CH_3O)PP$	445	688
$4NP-H_2T(4-Cl)PP$	449	660	$(246TCP)_2H_2T(4-CH_3O)PP$	454	687
$(24DCP)_2H_2T(4-CH_3)PP$	449	670	$4NP-H_2T(4-CH_3O)PP$	456	695

Table 16. UV-vis peaks (λ/nm in $CHCl_3$) of hydrogen bonded complexes of FBPs with PDs.

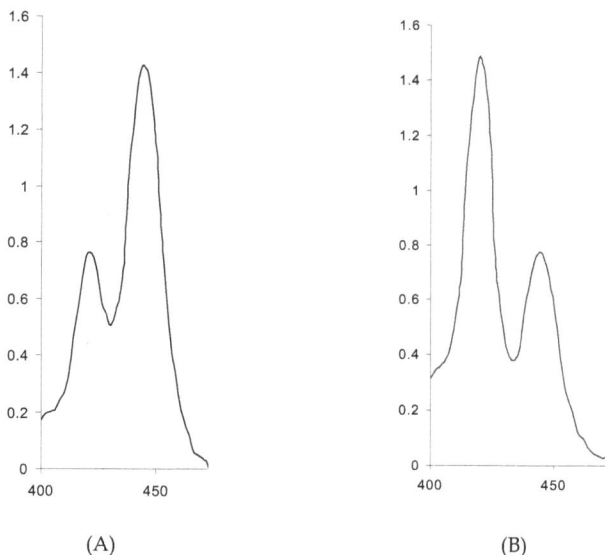

(A)   (B)

Fig. 9. In a UV-vis cell a solution of H_2TPP (5×10^{-6} M, 2 ml) and 24DCP (3 M, 90 mic.lit.) at 25 °C (profile A) was cooled to 5 °C (profile B). Observations show that upon cooling the adduct $[(24DCP)_2H_2TPP]$, Soret at 445 nm, has be dissociated to 24DCP and H_2TPP, Soret at 419 nm. By heating of solution to 25 °C the adduct $[(24DCP)_2H_2TPP]$ was formed again and Soret returned to 445 nm.

Stabilities of the adducts increase from $H_2T(4-Cl)PP$, H_2TPP, $H_2T(4-CH_3)PP$, to $H_2T(4-CH_3O)PP$ for each PD, Table 17. For example, at 20 °C we have the following order of the formation constants, K:

$(24DCP)_2H_2T(4-Cl)PP < (24DCP)_2H_2TPP < (24DCP)_2H_2T(4-CH_3)PP < (24DCP)_2H_2T(4-CH_3O)PP$

lg K: 1.14   <   1.51   <   1.75   <   1.93

246TCP did not show a measurable interaction with H_2TPP and $H_2T(4-Cl)PP$ under our experimental conditions also its interactions with $H_2T(4-CH_3)PP$ and $H_2T(4-CH_3O)PP$ are sizably weaker than corresponding interactions for 24DCP. It might be due to high steric hindrances which coincident to interaction of 246TCP with free base porphyrins. From obtained K for HB complexes the following order could be defined for HB formation ability of PDs: 4NP 〉 24DCP 〉 246TCP

Adduct	lgK				
	5 °C	10 °C	15 °C	20 °C	25 °C
$(24DCP)_2H_2T(4-Cl)PP$		0.99±0.02	1.08±0.02	1.14±0.02	1.20±0.02
$(24DCP)_2H_2TPP$		1.37±0.01	1.43±0.03	1.51±0.03	1.60±0.03
$(24DCP)_2H_2T(4-CH_3)PP$		1.58±0.03	1.69±0.04	1.75±0.04	1.86±0.05
$(24DCP)_2H_2T(4-CH_3O)PP$		1.72±0.01	1.81±0.03	1.93±0.05	2.03±0.02
$(246TCP)_2H_2T(4-CH_3)PP$		0.67±0.06	0.70±0.05	0.76±0.05	0.80±0.05
$(246TCP)_2H_2T(4-CH_3O)PP$		0.87±0.02	0.92±0.01	0.99±0.02	1.02±0.02
$(4NP)_2H_2T(4-Cl)PP$	0.97±0.02	1.09±0.02	1.24±0.02	1.35±0.02	1.44±0.02
$(4NP)_2H_2TPP$	1.02±0.07	1.16±0.05	1.29±0.05	1.47±0.05	1.63±0.08
$(4NP)_2H_2T(4-CH_3)PP$	1.90±0.02	2.12±0.01	2.35±0.02	2.48±0.02	2.70±0.03
$(4NP)_2H_2T(4-CH_3O)PP$	1.91±0.01	2.22±0.01	2.40±0.01	2.71±0.02	2.89±0.02

Table 17. The lgK for HB complexes $(PD)_2H_2T(4-X)PP$ in $CHCl_3$ as solvent.

Thermodynamic parameters ΔH°, ΔS° and ΔG° of $[(PD)_2H_2T(4-CH_3)PP]$ complexes are given in Table 18. The free energy changes (ΔG°) of HB complexes become more negative through the series $H_2T(4-Cl)PP$, H_2TPP, $H_2T(4-CH_3)PP$, to $H_2T(4-CH_3O)PP$ which indicates stronger HB interaction along this sequence.

Complex	ΔH°	ΔS°	$-\Delta G^\circ$
$(24DCP)_2H_2T(4-Cl)PP$	19.9±1.8	89.8±6.4	6.9±0.1
$(24DCP)_2H_2TPP$	25.7±2.3	116.7±7.9	9.1±0.1
$(24DCP)_2H_2T(4-CH_3)PP$	30.1±2.9	136.4±9.9	10.5±0.1
$(24DCP)_2H_2T(4-CH_3O)PP$	35.2±1.7	156.8±5.8	11.5±0.1
$(246TCP)_2H_2T(4-CH_3)PP$	15.1±1.4	65.9±4.7	4.5±0.1
$(246TCP)_2H_2T(4-CH_3O)PP$	17.5±1.3	78.4±4.6	5.9±0.1
$(4NP)_2H_2T(4-Cl)PP$	38±2	156±5	8±2
$(4NP)_2H_2TPP$	48±2	193±7	9±2
$(4NP)_2H_2T(4-CH_3)PP$	62±3	260±9	15±3
$(4NP)_2H_2T(4-CH_3O)PP$	78±4	316±13	17±4

[a] ΔH° (kJmol^{-1}), ΔS° (JK^{-1}mol^{-1}), ΔG° (kJmol^{-1}) at 25°C

Table 18. The thermodynamic parameters for HB complexes $(PD)_2H_2T(4-X)PP$ in $CHCl_3$[a].

Adducts have positive values of ΔH° and ΔS° but greater contribution of ΔS° relative to ΔH° leads to a negative value of ΔG°.

The positive value of ΔS° may return to association between PDs and free base porphyrin, which accompanied with releasing a greater number of chloroform molecules, that solvated the initial substances. The aggregation between PD and free bases in the selected range of temperatures is endothermic. The following sequence was obtained for interactions of $H_2T(4-X)PPs$ with phenol derivatives:

$$H_2T(4-CH_3O)PP > H_2T(4-CH_3)PP > H_2TPP > H_2T(4-Cl)PP$$

4. Effect of nonplanarity of free base porphyrins on their interactions with dimethyl- and diethyltin(IV) dichlorides

The nonplanarity of the free base porphyrin can be regulated by electronic and steric effects of the *meso-* and β-pyrrole substitution and it has profound consequences on spectral, electrochemical, and other properties of porphyrins. [45-48] Chemical properties that are known to be modified by nonplanar distortion include oxidation potentials, basicity of the inner nitrogen atoms, and axial ligand binding affinity, all of which can influence the biological function of porphyrin cofactors in proteins. A good result of nonplanar distortion of porphyrins appears in porphyrin metalation. For example, the predeformed octabromo-tetramesityl-porphyrin incorporates Zn^{2+} about 4000 times faster than the planar tetramesityl-porphyrin in DMF with the same rate law. Since the rates of metal ion incorporation are attributed to the conformational change of the rigid planar porphyrin core, thus the metalation of a nonplanar porphyrin is distinct in reaction rate from that of planar porphyrins.

This fact has be pointed out that nonplanarity enhances the basicity of porphyrins. This part do to comparison of the acceptor property of nonplanar *meso*-tetrakis-tert-butylporphyrin (H_2TtBP) with planar *meso*-tetra-n-propylporphyrin (H_2TnPP), in their interactions with dimethyl- and diethyltin(IV) dichloride.

On mixing of the R_2SnCl_2 (R= Me, Et) with a solution of free bases H_2TnPP and H_2TtBP Lewis acid-base interactions were occurred between them.[49] Porphyrins as Lewis bases coordinated to organotin compounds as Lewis acids. Extent of these interactions depends on acid-base properties of interacting components.

4.1 ^{1}H NMR analysis

The ^{1}H NMR spectra of porphyrins moiety in these adducts show clear differences relative to the free bases reactants. Upon complexation of free base porphyrins with organotin(IV) halides the H_β of pyrrole rings (H_2TnPP 9.5 and H_2TtBP 9.08 ppm) were shifted up field (9.03 and 8.1 ppm, respectively) while the N-H pyrrolic protons (H_2TnPP -2.6 and H_2TtBP 1.52 ppm) show down field shift and collapsed with alkyls protons at chemical shifts 1-2.5 ppm. These variations in ^{1}H NMR spectra result from enhancement of nonplanar distortion in porphyrin structure during adduct formation. Comparison of chemical shifts of ^{1}H NMR of free bases H_2TnBP and H_2TtBP especially for NH protons clearly shows the effect of nonplanarity on the ^{1}H NMR of free base porphyrins. In the 5,10,15,20-tetrabutylporphyrin (H_2TnBP) as a planar porphyrin the N-H and H_β signals were appeared at -2.61 and 9.45

ppm, respectively, while in the 5,10,15,20-tetrakis(tert-butyl)porphyrin (H₂TtBP) as a severely distorted porphyrin the N-H and H_β were appeared at 1.52 and 9.08 ppm.

4.2 UV-vis analysis

4.2.1 Interactions of R₂SnCl₂ with H₂TnPP

By interaction of R₂SnCl₂ with H₂TnPP Soret band of free base porphyrin (original peak at 417 nm) was shifted to 433 nm, see Table 19 and Fig.9. In addition other peaks of H₂TnPP were weakened while a new peak was appeared at 643 nm. Presence of a clear isosbestic point at 425 nm can aroused from an equilibrium in solution.

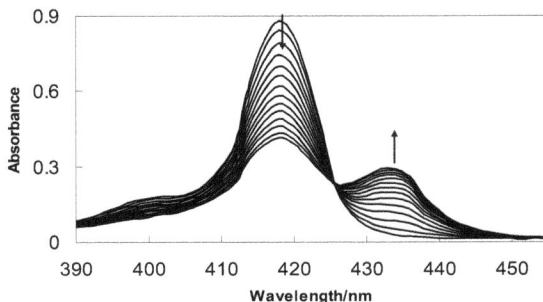

Fig. 9. Titration spectra for titration of H₂TnPP with Et₂SnCl₂ in chloroform. Band appeared at 433 nm is related to adduct, isosbestic points at 425 nm.

Compound	λ (nm)					Compound	λ (nm)				
H₂TnPP	417	520	555	600	659	H₂TtBP	446	552	596	628	691
(Me₂SnCl₂)₂H₂TnPP	433	-	-	-	643	(Me₂SnCl₂)₂H₂TtBP	453	-	-	-	695
(Et₂SnCl₂)₂H₂TnPP	433	-	-	-	643	(Et₂SnCl₂)₂H₂TtBP	453	-	-	-	695

Table 19. UV-vis peaks of free base porphyrins and their adducts with R₂SnCl₂ in CHCl₃.

4.2.2 Interaction of R₂SnCl₂ with H₂TtBP

By interaction of R₂SnCl₂ with H₂TtBP, the Soret band of free base porphyrin (original peak at 446 nm) was shifted to 453 nm, see Table 19 and Fig. 10. In addition, other peaks of H₂TtBP were weakened while a new peak was appeared at 695 nm. Presence of a isosbestic point at 454 nm can be argued by existence of equilibrium in solution.

Amount of replacement of Soret band in H₂TnPP (16 nm) is greater than H₂TtBP (7 nm). This aroused from greater nonplanar distortion in H₂TnPP skeleton comparing to H₂TtBP during adduct formation. Since the observed red shift mainly depends on the distortion of porphyrins structures during adduct formations.

For practical illustration of this subject, if in a batched UV-vis cell at 5 °C, a solution of H₂TnPP titrated with Me₂SnCl₂ the Soret band of free base (417 nm) will diminish and Soret band related to adduct will appear at 433 nm. By keeping the composition of solution unchanged and raising the temperature to 35 °C the Soret band will go back to 417 nm. This phenomenon which could be repeated successively shows that adduct dissociate to Me₂SnCl₂ and H₂TnPP at elevated temperatures and forms on cooling in equilibrium.

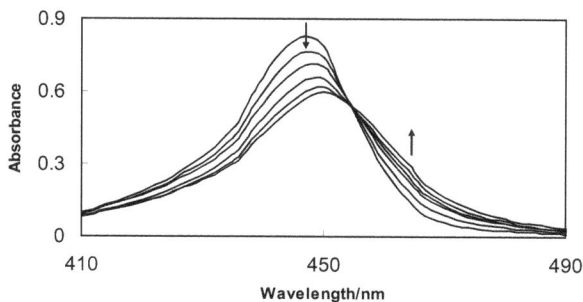

Fig. 10. Titration spectra, for titration of H$_2$TtBP with Me$_2$SnCl$_2$ in chloroform. Band appeared at 453 nm is related to adduct, isosbestic point at 454 nm.

About 20-30 nm red shift have been observed for Soret band of porphyrin moiety in the complexes of tetraarylporphyrins with organotin(IV) halides. A similar situation but with smaller red shifts (16 and 7 nm) were observed for interaction of H$_2$TnPP and H$_2$TtBP with R$_2$SnCl$_2$. Of course, with a considerable resonance that there is between peripheral aryl rings and porphyrin core in tetraarylporphyrins and lack of such resonance in tetraalkyl porphyrins this difference will be reasonable. Generally, porphyrin distortion is accompanied by bathochromic shift of its electronic absorption bands. [24,26] For example, in UV-vis spectra (in CH$_2$Cl$_2$) of planar porphyrin H$_2$TnPP Soret band and Q bands were observed at (417 nm) and (520, 555, 600, 659 nm), respectively. But in distorted H$_2$TtBP, Soret band and Q bands appeared at (446 nm) and (552, 596, 628, 691 nm). The effect of distortion of porphyrin plane on electronic absorption and ^{1}H NMR spectra of the free base porphyrin clearly observed from these data. Thus in these adducts greater red shifts in the UV-vis spectra of H$_2$TnPP may be due to more distortions in its structure during adduct formations.

4.3 Thermodynamic studies

The data given in Table 20 show that formation constants for adducts of H$_2$TtBP with dimethyl- and diethyltin(IV) dichloride are greater than related values for H$_2$TnPP. It shows that (R$_2$SnCl$_2$)$_2$H$_2$TtBP adducts are more stable than (R$_2$SnCl$_2$)2H$_2$TnPP. For example at 5 °C we have the following order of the formation constants:

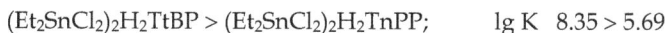

$$(Me_2SnCl_2)_2H_2TtBP > (Me_2SnCl_2)_2H_2TnPP; \quad lg \ K \ 9.70 > 6.58$$

$$(Et_2SnCl_2)_2H_2TtBP > (Et_2SnCl_2)_2H_2TnPP; \quad lg \ K \ 8.35 > 5.69$$

Also comparison of stability constants show that Me$_2$SnCl$_2$ adducts are more stable than their Et$_2$SnCl$_2$ counterparts.

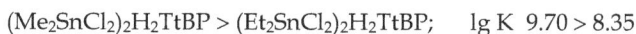

$$(Me_2SnCl_2)_2H_2TnPP > (Et_2SnCl_2)_2H_2TnPP; \quad lg \ K \ 6.58 > 5.69$$

$$(Me_2SnCl_2)_2H_2TtBP > (Et_2SnCl_2)_2H_2TtBP; \quad lg \ K \ 9.70 > 8.35$$

Table 21 shows the thermodynamic parameters obtained for the interactions of R$_2$SnCl$_2$ with free base porphyrins in chloroform. Adducts have negative values of ΔH°, ΔS°, and ΔG°. The negative values of ΔS° refer to association between donor and acceptor molecules.

Adduct	lgK				
	5 °C	10 °C	15 °C	20 °C	25 °C
$(Me_2SnCl_2)_2H_2TnPP$	6.58±0.02	6.17±0.01	5.72±0.01	5.30±0.04	4.90±0.01
$(Et_2SnCl_2)_2H_2TnPP$	5.69±0.02	5.33±0.03	4.84±0.09	4.52±0.04	4.18±0.01
$(Me_2SnCl_2)_2H_2TtBP$	9.70±0.01	9.15±0.01	8.66±0.09	8.01±0.03	7.51±0.02
$(Et_2SnCl_2)_2H_2TtBP$	8.35±0.02	7.87±0.02	7.52±0.09	7.11±0.02	6.73±0.04

Table 20. The formation constants (lg K) of adducts in $CHCl_3$.

Adduct	$-\Delta H^{\circ}$	$-\Delta S^{\circ}$	$-\Delta G^{\circ}$
$(Me_2SnCl_2)_2H_2TnPP$	134 ±1	356 ±5	28 ±1
$(Et_2SnCl_2)_2H_2TnPP$	121 ±4	328 ±40	24 ±4
$(Me_2SnCl_2)_2H_2TtBP$	170 ±6	426 ±20	43 ±20
$(Et_2SnCl_2)_2H_2TtBP$	123 ±2	284 ±5	39 ±2

a ΔH° (kJmol^{-1}), ΔS° (JK^{-1}mol^{-1}), and ΔG° at 10 °C (kJmol^{-1}).

Table 21. The thermodynamic parameters of adducts in $CHCl_3$.[a]

Both free energy and enthalpy of adducts of H_2TtBP are more negative than corresponding values for H_2TnPP. This indicates that interactions of H_2TtBP with organotin compounds are stronger than H_2TnPP. This trend is in according to the electron donation of the free base porphyrins. Since, in the pre-distorted H_2TtBP lone pair electrons of nitrogen atoms are more ready for coordination to R_2SnCl_2 Lewis acid. Therefore, we found greater formation constants for complexes of R_2SnCl_2 with H_2TtBP than H_2TnPP.

5. References

[1] N. Zh. Mamardashvili, O A Golubchikov, *Russian Chemical Reviews*, 70 (2001) 577 and references there in.
[2] H Ogoshi, E-i Watanabe, N Koketsu, Z-i Yoshida, *Bull. Chem. Soc. Jpn.*, 49 (1976) 2529.
[3] N. Datta-Gupta, T. J. Bardos, *J. Heterocycl. Chem.*, 3 (1966) 495.
[4] E.B. Fleischer and J.H. Wang, *J. Am. Chem. Soc.*, 82 (1960) 3498.
[5] F.R. Longo, E.M. Brown, W.G. Rau, and A.D. Alder in *"The Porphyrins"* Vol. V, D. Dolphin, Ed., Academic Press, New York, 1979, pp 459.
[6] Funahashi, S., Yamaguchi, Y., Tanaka, M., *Bull. Chem. Soc. Jpn.*, 57 (1984) 204.
[7] Funahashi, S., Yamaguchi, Y., Tanaka, M., *Inorg. Chem.*, 23 (1984) 2249.
[8] Robinson, L. R.; Hambright, P., *Inorg. Chim. Acta*, 185 (1991) 17.
[9] Bain-Ackerman, M. J.; Lavallee, D. K., *Inorg. Chem.*, 18 (1979) 3358.
[10] Funahashi, S.; Saito, K.; Tanaka, M., *Bull. Chem. Soc. Jpn.*, 54 (1981) 2695.
[11] Pasternack, R. F., Vogel, G. C., Skowronek, C. A., Harris, R. K., Miller, J. G., *Inorg. Chem.*, 20 (1981) 3763.
[12] Hambright, P. and Chock, P. B., *J. Am. Chem. Soc.*, 96 (1974) 3123.
[13] Turay, J., Hambright, P., *Inorg. Chem.*, 19 (1980) 562.
[14] Fleischer, E. B., Dixon, F., *Bioinorg. Chem.*, 7 (1977) 129.
[15] S. Funahashi, M. Inamo, N. Kamiya, Y. Inada, M. Nomura, *Inorg. Chem.*, 40 (2001) 5636.

[16] S. Funahashi, Y. Inada, Y. Sugimoto, Y. Nakano, Y. Itoh, *Inorg. Chem.*, 37 (1998) 5519.

[17] M. Asadi, A. Zabardasti, J. Ghasemi, Polyhedron 21 (2002) 683.

[18] M. Asadi, A. Zabardasti, J. Ghasemi, *Bull. Chem. Soc. Jpn.,75 (2002)1137.*

[19] M. Asadi, A. Zabardasti, *J. Chem. Research (S)* 2002, 611.

[20] A. Zabardasti1, M. Salehnassaje, M. Asadi, V.A. Karimivand, *Polish J. Chem.*, 80 (2006) 1473.

[21] M Asadi, A. Zabardasti, V. A. Karimivand, J. Ghasemi, Polyhedron 21 (2002) 1255.

[22] A. Zabardasti, M. Asadi, A. Kakanejadifard, *J. Heterocyclic Chem.*, 43 (2006) 1157.

[23] A. Zabardasti, M. Asadi, V.A. Karimivand and J. Ghasemi, *Asian J. Chem.*, 17 (2005) 711.

[24] E.B. Fleischer and A.L. Stone, *Chem. Soc. Chem. Commu.* 332 (1967).

[25] B. Cheng, O. Q. Munro, H. M. Marques,□and W. R. Scheidt, *J. Am. Chem. Soc.*, 119 (1997) 10732.

[26] M.O. Senge, I. Bischoff, N.Y. Nelson, K.M. Smith, *J. Porphyrins Phthalocyanines*, 3 (1999) 99.

[27] Storm, C. B.; Teklu, Y., *J. Am. Chem. Soc.*, 94 (1972) 1745.

[28] Storm, C. B.; Teklu, Y.; Sokoloski, A. *Ann. N. Y. Acad. Sci.*, 206 (1973) 631.

[29] K.M.Smith and M.F.Hudson, *Tetrahedron*, 32 (1976) 597.

[30] P. Worthington, P. Hambright, R.F.X. Williams, J. Reid, C. Burnham, A. Shamim, J. Turay, D.M. Bell, R. Kirkland, R.G. Little, N. Datta-Gupta, and E. Eisner, *J. Inorg. Biochem.*, 12 (1980) 281.

[31] D.J. Leggett, *"Computational Methods for the Determination of Formation Constant"*, Plenum Press, New York, 1985.

[32] R.S. Mulliken, W.B. Person, *Molecular Complexes: A Lecture and Reprint*, John Wiley and Sons: New York; 1969.

[33] N. Haga,; H. Nakajima, H. Takayanagi, K. Tokumaru, *J. Org. Chem.*, 63 (1998) 5372.

[34] H. Dehghani, M. Babaahmadi, *Polyhedron*, 27 (2008) 2739.

[35] H. Dehghani, M. Babaahmadi, *Polyhedron*, 27 (2008) 2416.

[36] A. Zabardasti, M. Mirzaeian, *Chem. Lett.*, 35 (2006) 1348.

[37] A. Zabardasti, M. Mirzaeian, *Asian J. Chem.*, 19 (2007) 4753.

[38] A. Zabardasti, L. Shaebani, A. Kakanejadifard, F. Faragi Morchegani, Bull. Chem. Soc. Ethiop., 25 (2011) 127.

[39] L. Guilleux, P. Krausz, L. Nadjo, C. Giannotti, R. Uzan, J. Chem. Soc. Perkin Trans. II (1984) 475.

[40] K. Kano, S. Hashimoto, Bull. Chem. Soc. Jpn. 63 (1990) 633.

[41] K. Kano, T. Hayakawa, S. Hashimoto, Bull. Chem. Soc. Jpn. 64 (1991) 778.

[42] D. Mohajer, H. Dehghani, J. Chem. Soc. Perkin Trans. II (2000) 199.

[43] A. Zabardasti, S. Farhadi, Z. Rezvani-Abkenari, Asian J. Chem., 19 (2007) 5488.

[44] A. Zabardasti, Z. Rezvani-Abkenari, *J. Iran. Chem. Soc.*, 5 (2008) 57.

[45] E.I. Sagun, E.I. Zenkevich, V.N. Knyukshto, A.M. Shulga, D.A. Starukhin, C. von Borczyskowski, *Chem. Phys.*, 275 (2002) 211.

[46] C.K. Mathews, K.E. Van Holde, K.G. Ahern, *Biochemistry*, Addison, Wesley Longman, San Francisco, 2000.

[47] B.O. Fernandez, I.M. Lorkovic and P.C. Ford, *Inorg. Chem.*, 43 (2004) 5393.

[48] P.D. Beer, D.P. Cormode and J.J. Davis, *Chem. Commun.*, (2004) 414.
[49] A. Zabardasti, S. Farhadi, M.H. Rahmati, J. Chem. Research, (2010) 538.

3

Molecular Interactions in Chromatographic Retention: A Tool for QSRR/QSPR/QSAR Studies

Vilma Edite Fonseca Heinzen[1*], Berenice da Silva Junkes[2],
Carlos Alberto Kuhnen[3] and Rosendo Augusto Yunes[1]
[1]*Department of Chemistry, Federal University of Santa Catarina,
University Campus, Trindade, Florianópolis, Santa Catarina,*
[2]*Federal Institute of Education, Science and Technology of Santa Catarina,
Florianópolis, SC,*
[3]*Department of Physics, Federal University of Santa Catarina, University Campus,
Trindade, Florianópolis, Santa Catarina,*
Brazil

1. Introduction

Molecular interactions play a fundamental role in the behavior of the chemical and physical properties of any physico-chemical system. In gas chromatography (GC) the chromatographic retention is a very complex process. It involves the interaction of molecules through multiple intermolecular forces, such as dispersion (or London forces), orientation (dipole–dipole or Keesom forces), induction (dipole–induced dipole or Debye forces), and electron donor–acceptor forces including hydrogen–bonding, leading to the partition of the solute between the gas and liquid phases (Kaliszan, 1987; Peng, 2000; Yao et al., 2002). Other factors, such as steric hindrance of substituent groups within the solute molecule, can also affect the chromatographic behavior (Fritz et al., 1979; Peng et al., 1988). It is clear that correlations between gas chromatographic retention indices (RIs) and molecular parameters provide significant information on the molecular structure, retention time and the possible mechanism of absorption and elution (Körtvélyesi et al., 2001). Several topological, geometric, electronic, and quantum chemical descriptors have been used in research on quantitative structure–property and structure–activity relationships (QSPR/QSAR) (Karelson et al., 1996; Katritzky & Gordeeva, 1993; Kier & Hall, 1990). The topological descriptors have shown their efficacy in the prediction of diverse physicochemical and biological properties of various types of compounds (Amboni et al., 2000; Arruda et al., 1993; Estrada, 2001a, 2001b; Heinzen & Yunes, 1993, 1996; Heinzen et al., 1999a; Kier & Hall, 1986; Randic, 1993, 2001; Ren, 2002a, 2002b, 2002c; Sabljic & Trinajstic, 1981). In general, these indices are numbers containing relevant information regarding the structure of molecules. Most of the measured physicochemical properties are steric properties, and therefore they may be reasonably well described by topological indices. However, in some cases, these indices also contain structural information related to the electronic and/or polar features of molecules (Galvez et al., 1994; Hall et al., 1991). The molecular size, shape, polarity, and ability to participate in hydrogen bonding are among the different factors that can contribute to the physicochemical properties or biological activities of a molecule. It is well known that these factors are related to intermolecular interactions such as van der Waals forces.

The use of graph–theoretical topological indices in QSPR/QSAR studies has sparked great interest in recent years. The topological indices have become a powerful tool for predicting numerous physicochemical properties and/or biological activities of compounds, as well as for molecular design. One of the most important properties that have been extensively studied is the chromatographic retention (Estrada & Gutierrez, 1999; Ivanciuc, O. et al., 2000; Ivanciuc T. & Ivanciuc, O., 2002; Katritzky et al., 1994; Katritzky et al., 2000; Pompe & Novic, 1999; Ren, 1999, 2002a). Quantitative structure–chromatographic retention relationship (QSRR) studies have been widely investigated by gas chromatography (GC) and high-performance liquid chromatography (HPLC) (Markuszenwski &. Kaliszan, 2002). Topological indices (TI) are obtained *via* mathematical operations from the corresponding molecular graphs of compounds (Ivanciuc, O. et al., 2002; Kier & Hall, 1976; Liu, S.-S. et al. 2002; Marino et al., 2002; Rios–Santamarina et al., 2002; Toropov & Toropova, 2002) in contrast to the physicochemical characterization used by traditional QSAR (García-Domenech et al., 2002). One of the main advantages of TI is that they can be easily and rapidly computed for any constitutional formula yielding good correlation abilities. However, important disadvantages should be noted including the difficulties encountered in encoding stereo–chemical information, for example, to distinguish between *cis*- and *trans*–isomers, and their lack of physical meaning. Many topological indices have been proposed since the pioneering studies by Wiener (Wiener, 1947) and by Kier on the use of QSAR (Kier & Hall, 1976). The TI developed for QSAR/QSRR studies can be illustrated by Estrada's approach to edge weights using quantum chemical parameters (Estrada, 2002) and by Ren's atom–type AI topological indices derived from the topological distance sums and vertex degree (Ren, 2002d).

Based on a chromatographic behavior hypothesis, our group developed a topological index called the semi–empirical topological index (I_{ET}). This index was initially developed to predict the chromatographic retention of linear and branched alkanes and linear alkenes, with the objective of differentiating their *cis*- and *trans*- isomers and obtaining QSRR models (Heinzen et al., 1999b). The excellent results achieved stimulated our group to extend the new topological descriptor to other classes of compounds (Amboni et al., 2002a, 2002b; Arruda et al., 2008; Junkes et al., 2002a, 2002b, 2003a, 2003b, 2004; Junkes et al., 2005; Porto et al., 2008). The equation obtained to calculate the I_{ET} was generated from the molecular graph and the values of the carbon atoms, and the functional groups were attributed observing the experimental chromatographic behavior and supported by theoretical considerations. This was carried out due to the difficulty in obtaining a complete theoretical description of the interaction between the stationary phase and the solute. Based only on theoretical equations or hypotheses it is not possible, for example, to estimate how the molecular conformation of the solute affects the intermolecular forces. In view of this, it seems reasonable to assume that from the experimental behavior we can obtain insights regarding these factors in order to apply them to other processes involved in QSPR studies. Thus, it can be noted that the semi–empirical topological index has a clear physical meaning.

The semi–empirical topological index (I_{ET}) allowed the creation of a new descriptor, the electrotopological index, I_{SET}, which was recently developed by our group and applied to QSPR studies to predict the chromatographic retention index for a large number of organic compounds, including aliphatic hydrocarbons, alkanes and alkenes, aldehydes, ketones, esters and alcohols (Souza et al., 2008, 2009a, 2009b, 2010). The new descriptor for the above series of molecules can be quickly calculated from the semi-empirical, quantum-chemical, AM1 method and correlated with the approximate numerical values attributed by the semi-empirical topological index to the primary, secondary, tertiary and quaternary carbon atoms. Thus, unifying the quantum-chemical with the topological method provided a three-dimensional picture of the atoms in the molecule. It is important to note that the AM1 method portrays more reliable semi-empirical charges, dipoles and bond lengths than those obtained from time-consuming, low-quality, *ab initio* methods, that is, when employing a minimal basis set in *ab initio* calculations. Despite the fact that the calculated partial atomic charges may be less reliable than other molecular properties, and that different semi-empirical methods give values for the net charges with poor numerical agreement, it is important to recognize that their calculation is easy and that the values at least indicate the trends of the charge density distributions in the molecules. Since many chemical reactions or physico-chemical properties are strongly dependent on local electron densities, net atomic charges and other charge-based descriptors are currently used as chemical reactivity indices.

For alkanes and alkenes, this correlation allowed the creation of a new semi-empirical electrotopological index (I_{SET}) for QSRR models based on the fact that the interactions between the solute and the stationary phase are due to electrostatic and dispersive forces. This new index, I_{SET}, is able to distinguish between the *cis-* and *trans*-isomers directly from the values for the net atomic charges of the carbon atoms that are obtained from quantum-chemical calculations (Souza et al., 2008). For polar molecules like aldehydes, ketones, esters and alcohols, the presence of heteroatoms like oxygen changes considerably the charge distribution of the corresponding hydrocarbons, leading to a small increase in the interactions between the solute and the stationary phase (Souza et al., 2009a, 2009b, 2010). An appropriate way to calculate the I_{SET} was developed, taking into account the dipole moment exhibited by these molecules and the atomic charges of the heteroatoms and the carbon atoms attached to them. By considering the stationary phase as non-polar material the interaction between these molecules and the stationary phase becomes electrostatic with the contribution of dispersive forces. These interactions were slowly increased relative to the corresponding hydrocarbons. Hence, the interactions between the molecules and the stationary phase were slowly increased as a result of the charge redistribution that occurred in presence of the heteroatom. This charge redistribution accounted for the dipole moment of the molecules. Clearly the main outcomes in terms of the charge distribution due the presence of the (oxygen) heteroatoms occur in the neighborhood, and the excess charge of these atoms leads to electrostatic interactions that are stronger relative to the weak dispersive dipolar interactions.

2. Semi-empirical topological index (I_{ET})

Three important factors led us to develop the semi-empiric topological index: (i) no topological index alone was able to differentiate between the *cis-* and *trans-* isomeric structures of alkenes; (ii) if all the carbon atoms have a value of 100 as indicated by Kovàts, from the experimental results it is not possible to determine a constant value for each of the different carbon atoms (secondary, tertiary and quaternary) of alkanes; (iii) when the Kovàts indices of retention for very branched hydrocarbons (alkanes) are correlated with the number of carbon atoms an unacceptable linearization is observed. It is known that the chromatographic process of separation results from the forces that operate between solute molecules and the molecules of the stationary phase. The retention of alkanes and alkenes is due to the number of carbon atoms and the interaction of each specific carbon atom with the stationary phase. The interaction of the stationary phase with the carbon atoms is determined by its electrical properties and by the steric hindrance to this interaction by other carbon atoms attached to it. The values attributed to the carbon atoms were based on the results of the experimental chromatographic behavior of the molecules that measure the real electrical and steric characteristics of the carbons. For this reason the index is denominated semi-empirical. The representation of the molecules was based on the molecular graph theory, where the carbon atoms are considered as the vertexes of the graph and the hydrogens are suppressed (Hansen & Jurs, 1988). Thus, it is called a topological index.

2.1 Calculation of I_{ET} for alkanes and alkenes

Values were attributed to the carbon atoms (vertex of the molecular graphs) according to the following considerations. (i) According to the Kovàts convention, the correlation between the retention index and number of carbon atoms is linear for the alkanes (Kovàts, 1968). However, branched alkanes do not present this linear relationship with the Kovàts index, since the retention of the tertiary and quaternary carbon atoms is decreased by the steric effects of their neighboring groups. It is evident that secondary, tertiary and quaternary carbon atoms have values of less than 100 u.i., as previously attributed by Kovàts. (ii) Observing the experimental chromatographic behavior, approximate numerical values were attributed: 100 u.i. for the carbon atom in the methyl group in agreement with Kovàts, 90 u.i. for the secondary carbon atoms, 80 u.i. for the tertiary and 70 u.i. for the quaternary. All values were divided by 100 to make them consistent with the common topological values. (iii) The contribution of these carbon atoms to the chromatographic retention is also dependent on the neighboring substituent groups due to steric effects. In order to estimate the steric effects, it was observed that the values for the experimental RI decreased as the branch increased, showing a log trend. Therefore, it was necessary to add the value of the logarithm of each adjacent carbon atom. Thus, the new semi-empirical topological index (I_{ET}) is expressed as:

$$I_{ET} = \sum_i (C_i + \delta_i) \tag{1}$$

$$\delta_i = \sum_{j \sim i} log\, C_j$$

where C_i is the value attributed to each carbon atom in the molecule and δ_i is the sum of the logarithm of the value for each adjacent carbon atom (C_1, C_2, C_3 and C_4) and $\sim$ means 'adjacent to'. (iv) For alkenes, the main interaction force between the solute and stationary phase is the dispersive force, which is reduced by neighboring steric effects, however, the electrostatic force is also involved. The influence of conformational effects on the intermolecular forces makes it very difficult to predict these effects based only on theoretical considerations. For this reason, the values attributed to the carbon atom of the double bond for alkenes were calculated by numerical approximation based on the experimental retention indices, as described in our previous publication (Heinzen et al., 1999b; Junkes et al., 2002a).

2.2 Calculation of I_{ET} for compounds with oxygen-containing functional groups

The values attributed to the carbon atoms and functional groups (vertex of the molecular graphs) were based on the following considerations: (i) For this group of compounds, the main intermolecular forces that contribute to their chromatographic behavior on low polarity stationary phases are dispersive and inductive forces. The values attributed to functional groups are also based on the experimental retention index. (ii) The –COO- (ester), C=O (ketone or aldehyde) and C-OH (alcohol) groups were considered as a single vertex of the molecular graph of the compounds studied. This was carried out due to the difficulty and the inconsistency associated with calculating the individual values of the carbon atoms and the oxygen atoms of these groups. Thus, better numerical approximations were obtained, capable of reflecting the experimental chromatographic behavior of these compounds, when these groups were treated as a single vertex. (iii) The same considerations that were taken into account during the development of the semi-empirical topological method for the prediction of retention indices of alkanes and alkenes (Heinzen et al., 1999b; Junkes et al., 2002a) were employed to develop the I_{ET} for oxo-compounds. (iv) The contribution of the carbon atoms and functional groups to the chromatographic retention was represented by a single symbol, C_i, as indicated in Equation 1. The semi-empirical topological index can be expressed by a general Equation, for the entire set of compounds included in this work, where: C_i = value attributed to the –COO- (ester), C=O (ketone or aldehyde), C–OH (alcohol) groups and/or to each carbon atom, i, in the molecule. δ_i = the sum of the logarithm of the values of each adjacent carbon atom (C_1, C_2, C_3, and C_4) and/or the logarithm of the value of the –COO- (ester), C=O (ketone or aldehyde), C-OH (alcohol) groups, and ~ means 'adjacent to'. In a first step, an approximate I_{ET} (I_{Eta}) was calculated for each compound. This was achieved using the equation previously obtained for linear alkanes containing from 3 to 10 carbon atoms and Kováts experimental retention indices of compounds (Heinzen et al., 1999b). (v) Subsequently, the values of C_i for primary and secondary carbon atoms, previously attributed to alkanes (Heinzen et al., 1999b), and the approximate I_{ET}, calculated above, were used in Equation 1 in order to calculate the values of –COO-, C=O and C-OH groups of linear compounds. Thus, values were attributed to each class of functional group according to the position of the group in the carbon chain. (vi) One of the fundamental factors taken into consideration for the development of this topological index was the importance of the steric and other mutual intramolecular interactions between the functional group and nearby atoms. Therefore, for branched molecules, different values were attributed to carbon atoms in the α, β, and γ position with respect to the functional groups compared to those previously attributed to alkanes (Heinzen et al., 1999b) as described in the literature (Amboni et al., 2002a, 2002b; Junkes et al., 2003b, 2004).

The values of C_i for the carbon atoms and the values attributed to the functional groups of esters, aldehydes, ketones and alcohols are listed in Table 1 of Junkes et al. (Junkes et al., 2004,).

2.3 Calculation of I_{ET} for alkylbenzene compounds

The same considerations employed in the generation of the semi-empirical topological index, I_{ET}, for linear and branched alkanes and alkenes (Heinzen et al., 1999b; Junkes et al., 2002a) were applied to this group of compounds (alkylbenzenes). Firstly, the molecules were represented by hydrogen-suppressed molecular graphs based on chemical graph theory (Hansen & Jurs, 1988) where the carbon atoms were considered as vertexes of the molecular graph of these compounds. The contribution of each carbon atom to the chromatographic retention is represented by a single symbol, Ci, as can be observed from Eq. (1) where Ci is the value attributed to (=C<) fragments and/or each carbon atom i in the molecule; and δ_i is the sum of the logarithm of the values for each adjacent carbon atom (C_1, C_2, C_3 and C_4). The values of Ci for the carbon atoms of linear, branched, *ortho*, *meta* and *para* substituted, tri-substituted and tetra-substituted alkyl benzenes can be seen in Porto et al. (Porto et al., 2008).

2.4 Calculation of I_{ET} for halogenated aliphatic compounds

The present approach is based on the representation of molecules by hydrogen-suppressed molecular graphs which, in turn, are based on chemical graph theory, where the carbon atoms (Ci) are the graph vertexes. As with the carbon atoms the C–X and X–C–X fragments (where X = chlorine, bromine, or iodine atom) are considered a vertex of the molecular graph of these compounds, as previously considered for the functional groups (Heinzen et al., 1999b). The I_{ET} is expressed as equation (1) where Ci is the value attributed to each carbon atom i and/or to C–X or X–C–X fragments in the molecule; and δ_i is the sum of the logarithm of the values for each adjacent carbon atom (C1, C2, C3, and C4) and/or the logarithm of the values of the adjacent C–X and X–C–X fragment. The values to be attributed to the carbon atoms, and to the functional group (Ci) for halogenated hydrocarbons, are calculated by numerical approximation based on the experimental retention index (RIExp) values and supported by theoretical considerations. The values of Ci for the carbon atoms of linear and branched halogenated aliphatic compounds can be obtained in Arruda et al. (Arruda et al., 2008).

2.5 Development of QSRR models using the I_{ET}

As the starting point, the I_{ET} was developed for alkanes on a low polarity stationary phase. These are the simplest compounds and their properties are almost completely dependent on topological features. Subsequently, this novel topological descriptor was extended to different classes of organic compounds with more complex structural features. A summary of the best simple linear regression models (RI = b + a I_{ET}) and the statistical data for each data set of compounds, obtained in previous QSRR studies, is given in Table 1.

3. The semi-empirical electrotopological index, I_{SET}

The semi–empirical topological index (I_{ET}) discussed in the previous section allows the creation of a new descriptor, the electrotopological index, I_{SET}, which was developed and applied to QSPR studies to predict the retention index, boiling points and octanol/water partition coefficient (Log P), for a large amount of organic compounds, including aliphatic hydrocarbons alkanes and alkenes, aldehydes, ketones, esters and alcohols (Souza et al.,

No	Data Set	a	b	N	r	SD	Ref.
01	Alkanes	116.8	-19.05	157	0.9901	26.20	Heinzen et al., 1999b
02	Cis-/trans- linear alkenes	122.8446	-41.7054	79	0.99996	2.35	Heinzen et al., 1999b
03	Branched alkenes	120.4671	-29.0457	59	0.9985	5.76	Junkes et al., 2002a
04	Methyl- branched alkanes	123.1610	-39.5251	178	0.99998	4.31	Junkes et al., 2002b
05	Esters	123.79	-48.14	81	0.9995	5.79	Amboni et al., 2002a
06	Aldehydes and ketones	123.4951	-45.6553	54	0.9999	5.01	Junkes et al., 2004
07	Alcohols	124.1239	-51.3739	44	0.9991	5.70	Junkes et al., 2003b
08	Alkylbenzene	123.0824	-39.7381	122	0.9998	5.5	Porto et al., 2008
09	Halogenated compounds	124.7788	-56.8944	141	0.9997	8.0	Arruda et al., 2008

Table 1. Summary of the best simple linear regressions ($RI_{Calc} = a + b\ I_{ET}$) found for different data set on low polarity stationary phases.

2008, 2009a, 2009b, 2010). This new descriptor for this series of molecules can be quickly calculated from atomic charges obtained through the semi-empirical quantum-chemical, AM1 method (Bredow & Jug, 2005; Smith, 1996), since it was found that atomic charges correlated with the approximate numerical values attributed by the semi-empirical topological index to the primary, secondary, tertiary and quaternary carbons atoms.

3.1 Calculation of I_{SET} for alkanes and alkenes

For alkanes and alkenes, the above-mentioned correlation allowed the creation of a new semi-empirical electrotopological index (I_{SET}) for QSRR models based on the fact that the interactions between the solute and the stationary phase are due to electrostatic and dispersive forces (Souza et al., 2008). This new index, I_{SET}, is able to distinguish between the cis- and trans-isomers directly from the values of the net atomic charges of the carbon atoms that are obtained from quantum-chemical calculations. More precisely, this new semi-empirical electrotopological index, I_{SET}, was developed based on the refinement of the previous semi-empirical topological index, I_{ET}. The values for the Ci fragments that were firstly attributed from the experimental chromatographic retention and theoretical deductions have an excellent relationship with the net atomic charge of the carbon atoms. Thus, the values attributed to the vertices in the hydrogen-suppressed graph of carbon atoms (Ci) are calculated from the correlation between the net atomic charge in each carbon atom, which is obtained from quantum chemical semi-empirical calculations, and the Ci fragments for primary, secondary, tertiary and quaternary carbon atoms (1.0, 0.9, 0.8 and 0.7, respectively) obtained from the experimental values. This shows that it is possible to calculate a new index, I_{SET} (the semi-empirical electrotopological index) through the net atomic charge values obtained from a Mulliken population analysis using the semi-empirical AM1 method and their correlation with the values attributed to the different types of carbon atoms. This demonstrates that the I_{SET} encodes information on the charge distribution of the solute which drives the dispersive and electrostatic interactions between the solute (alkanes and alkenes) and the stationary phase (Souza et al., 2008).

Since the interactions between the solute and the stationary phase are dispersive for alkanes and electrostatic for alkenes, the chromatographic retention is strongly dependent on the electronic charge distribution of each carbon atom of these molecules. A simple linear regression equation was obtained between the values of the carbon atoms, SETi values, based on experimental gas chromatography retention (for primary (1.0), secondary (0.9), tertiary (0.8) and quaternary (0.7) carbon atoms) and the net atomic charges (δi) of these atoms, as given in Equation (2).

$$\text{SET}_i = -1.77125\delta_i + 0.62417 \tag{2}$$

This indicates that the physical reality encoded by the semi-empirical topological index (I_{ET}) developed in our laboratory is completely related to net atomic charges which, as is well known, are important forces in intermolecular interactions. It is clear that the interactions between the non-polar stationary phases and the different compounds were determined predominantly through the electronic charge distribution of the molecular structures of the compounds analyzed by gas chromatography. From Equation (1) it is clear that knowledge of the net atomic charges is sufficient to calculate the SETi value for all kinds of carbon atoms and not only the values given by the carbon models (that is 1.0, 0.9, 0.8 and 0.7) or in specific tables. Hence, the above method of calculating the SETi values of the carbon atoms allows a new index to be created, denominated the semi-empirical electropological index, I_{SET}. Considering the steric effects of the neighboring carbon atoms, as was observed in the calculation of I_{ET}, this new index can be calculated according to Equation (3).

$$I_{SET} = \sum_{i,j}(\text{SET}_i + \log\text{SET}_j) \tag{3}$$

In the above expression the i sum is over all the atoms of the molecule (excluding the H atoms) and the j is an inner sum of atoms attached to the i atom. The cis-2-pentene and trans-2-pentene molecules represented in the graph below are taken as an example of the I_{SET} calculation.

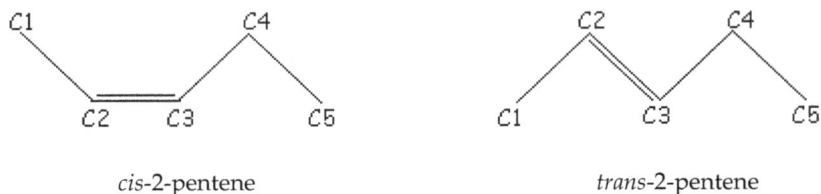

cis-2-pentene   trans-2-pentene

The net atomic charges and SET_i values for the above molecules are given in Table 2 below.

Carbon atom	cis-penten-2-eno		trans-penten-2-eno	
	δ_j	SET_j	δ_j	SET_j
C1	-0.187	0.9555	-0.185	0.9520
C2	-0.173	0.9307	-0.170	0.9254
C3	-0.165	0.9165	-0.167	0.9200
C4	-0.133	0.8598	-0.132	0.8580
C5	-0.209	0.9945	-0.207	0.9910

Table 2. The net atomic charge (δ_i) and the SET_i values for each carbon atom of cis-2-pentene and trans-2-pentene molecules.

The I_{SET} calculation now follows:

$I_{SETC1} = 0.9555 + \log 0.9307$ $I_{SETC1} = 0.9520 + \log 0.9254$

$I_{SETC2} = 0.9307 + \log 0.9555 + \log 0.9165$ $I_{SETC2} = 0.9254 + \log 0.9520 + \log 0.9200$

$I_{SETC3} = 0.9165 + \log 0.9307 + \log 0.8598$ $I_{SETC3} = 0.9200 + \log 0.9254 + \log 0.8580$

$I_{SETC4} = 0.8598 + \log 0.9165 + \log 0.9945$ $I_{SETC4} = 0.8580 + \log 0.9200 + \log 0.9910$

$I_{SETC5} = 0.9945 + \log 0.8598$ $I_{SETC5} = 0.9910 + \log 0.8580$

$I_{SET} = 4.3653$ $I_{SET} = 4.3481$

As expected on physical-chemical grounds, the AM1 calculation reveals that the optimized structures of the cis- and trans- isomers have slightly different charge distributions. As can be seen from the above results, the Mulliken population analysis gives the net atomic charges of the carbon atoms for each isomer, which implies that the difference in the values for the SETi fragments is sufficient to give different I_{SET} values.

3.2 Calculation of I_{SET} for compounds with oxygen-containing functional groups

3.2.1 Ketones and aldehydes

For polar molecules like aldehydes, ketones, esters and alcohols, the presence of heteroatoms like oxygen changes considerably the charge distribution of the corresponding hydrocarbons giving a small increase in the interactions between the solute and the stationary phase. An appropriate way to calculate the I_{SET} was developed that takes into account the dipole moment exhibited by these molecules and the atomic charges of the heteroatoms and the carbon atoms attached to them (Souza et al., 2009a). By considering the stationary phase as non-polar material, the interactions are slowly increased relative to the corresponding hydrocarbons due to the charge redistribution that occurs in presence of the heteroatom. This charge redistribution accounts for the dipole moment of the molecules. Thus, the dipolar charge distribution in such molecules leads to a small increase in the interactions of the solute with the stationary phase relative to hydrocarbons where the dipole moment is zero, or almost zero. Clearly the major effects on the charge distribution due the presence of the (oxygen) heteroatoms occur in the neighborhood and the excess charge of these atoms leads to electrostatic interactions that are stronger relative to the weak dispersive dipolar interactions (Christian, 1990).

In relation to the chromatographic retention it can be observed, for instance, that the molecules 2-hexanone, 3-hexanone and hexanal have experimental retention indices of 767, 764 and 776, respectively, and for the corresponding hydrocarbon molecule in the absence of the heteroatom, that is, the heptane, the retention index is 700. Due to the presence of the heteroatom (oxygen) there is an increase in the retention index of around 10%. Hence, the interactions between the molecules and the stationary phase are slowly increased and clearly this is due to the charge redistribution that occurs in the presence of the heteroatom. This charge redistribution accounts for the dipole moment of molecules like aldehydes and ketones. The dispersive force between these kinds of molecules and the stationary phase includes the charge-dipole interactions and dipole–induced dipole interactions which are weak relative to the electrostatic interactions. Thus, the dipolar charge distribution in such molecules leads to a small increase in the interactions of the solute with the stationary phase relative to hydrocarbons where the dipole moment is zero. Initially, it appears that the above-mentioned factors mean that the retention index can be calculated as in equation 3, and the same applies to the heteroatoms, but including subtle alterations that incorporate the effects of the dispersive dipolar interactions.

All of these factors can be included in the calculation of the retention index through a small increase in the SET_i values for the heteroatoms and the carbon atoms attached to them. This was carried out by multiplying the SET_i values of these atoms by a function A_μ which is dependent on the dipole moment of the molecule and the net charge at the oxygen and carbon atoms (to include both the electrostatic and dispersive interactions). Since we must have $A_\mu = 1$, when the dipole moment is zero or almost zero (as in the case of alkanes and alkenes) in a first attempt to achieve this function a linear dependence on the molecular dipole moment μ is considered, that is, $A_\mu = 1 + (\mu/\mu_F)$, where μ_F is a local function (in the units of the dipole moment) in the sense that it is dependent on the net charge of oxygen and carbon atoms. On the one hand this definition of A_μ works only if $\mu/\mu_F > 1$, since A_μ must reflect the small increase in the interactions due to dipolar dispersive forces. On the other hand good choices for the definition of μ_F for ketones and aldehydes (as we shall see below) means that the ratio μ/μ_F can be much greater than unity showing clearly that it is not possible to apply the above definition to A_μ. Considering that $\mu/\mu_F > 1$ then A_μ can not be a polynomial function of μ/μ_F. Thus, A_μ must have a weaker dependence on the dipole moment than the linear one and this weak dependence can be achieved through a logarithmic function since it is clear that the function $f(x) = x$ increases much faster than the function $f(x) = \log(1+x)$. Taking these factors into account it is possible to achieve a definition of A_μ that differs slightly from unity and is logarithmically dependent on the dipole moment of the molecule, as seen in equation 4

$$A_\mu = 1 + \log(1 + \frac{\mu}{\mu_F}),\qquad\qquad(4)$$

where μ is the calculated molecular dipole moment and μ_F is a local function which is dependent on the charges of the atoms belonging to the C=O bond. Clearly, μ_F must be directly related to the net charge of the oxygen atoms since it must reflect some contribution to the electrostatic interaction between these molecules and the stationary phase. In this regard, μ_F may also be related to the atomic charge of the carbon atom of the functional group C=O or related to the difference between the atomic charges of these atoms. Hence, μ_F

can be defined in different ways and some definitions of μ_F can be used in preliminary calculations. As expected, after some preliminary calculations, the best choice was for ketones $\mu_F = d\,|\,Q_C - Q_O\,|$ where d is the calculated C=O bond length and $|\,Q_C - Q_O\,|$ is the absolute value of the difference between the atomic charges at the carbon and oxygen atoms. This definition of μ_F is an attempt to take into account the contribution of the atomic charge of the oxygen atoms and the respective bonded carbon atom to the electrostatic interactions. For aldehydes the terminal carbon atom of the C=O bond is attached to a hydrogen and thus it is necessary to consider the net positive charge in this polar region of the molecule as the sum of the atomic charges of the carbon and hydrogen atoms. This means that for aldehydes the best choice for μ_F was $\mu_F = d\,|\,Q_C + Q_H - Q_O\,|$. Therefore, equation 4 indicates that there is an increase in the interaction between the molecules and the stationary phase due to the presence of the dipole moment and that this contribution may be screened by the charge located on the heteroatoms (oxygen atoms) if $\mu/\mu_F < 1$, or may be increased if $\mu/\mu_F > 1$. In the case of ketones and aldehydes the local function μ_F is less than the dipole moment showing that A_μ receives an appreciable contribution from the atomic charges of these atoms. This reveals the contribution of oxygen to the electrostatic interaction between the solute and the stationary phase. Therefore, to include the dispersive dipolar interactions in the calculation of the retention index we multiply the SET_i values for the heteroatoms (oxygen) and the carbon atoms attached to them by the dipolar function A_μ given in equation 4. That is, in this model the I_{SET} is calculated as in equation 5

$$I_{SET} = \sum_{i,j} (A_\mu SET_i + \log A_\mu SET_j),\qquad(5)$$

where the SET_i values are obtained using equation 2. As in equation 3, in the above expression the i sum is over the all the atoms of the molecule (excluding the H atoms) and the j is an inner sum of the atoms attached to the i atom. In the above expression, for the I_{SET} the dipolar function A_μ is taken as unity for the remaining carbon atoms of the molecules. Equation 4 reduces to equation 2 when the dipole moment of the molecule is zero or almost zero, as is the case for alkanes and alkenes since $A_\mu = 1$ for $\mu = 0$.

The 3-hexanone and hexanal molecules represented in the graph below are taken as an example of the I_{SET} calculation.

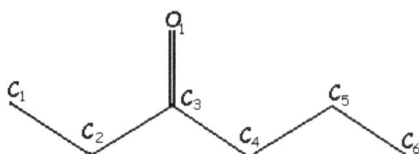

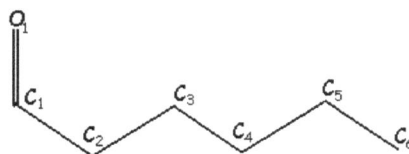

3-hexanone   hexanal $Q_H = +0.086$

$\mu_F = d\,|\,Q_C - Q_O\,|$ $\mu_F = d\,|\,Q_C + Q_H - Q_O\,|$

$\mu_F = 1.2342\,[0.224- [-0.288]] = 0.6319$ $\mu_F = 1.2314[+0.183+0.086-[-0.289]] = 0.6871$

$$A_\mu = 1 + \log(1 + \mu/\mu_F)$$

$A_\mu = 1 + \log(1+(2.6790/0.63191)) = 1.7193$ $A_\mu = 1 + \log(1+(2.7640/0.68712)) = 1.7009$

The net atomic charges and SET_i values are given in Table 3 below.

Atoms	3-hexanone			hexanal		
	δi	SET_i	$AuSETi$	δi	SET_i	$AuSETi$
O_1	-0.288	1.1346	1.9507	-0.289	1.1363	1.9328
C_1	-0.206	0.9892	-	+0.183	0.2995	0.5094
C_2	-0.212	0.9998	-	-0.233	1.0371	-
C_3	-0.224	0.2268	0.3899	-0.155	0.8988	-
C_4	-0.212	0.9998	-	-0.158	0.9041	-
C_5	-0.155	0.8988	-	-0.159	0.9059	-
C_6	-0.212	0.9998	-	-0.211	0.9980	-

Table 3. The net atomic charge (δi) and the SET_i values for each carbon and oxygen atom of 3-hexanone and hexanal molecules.

The I_{SET} calculation now follows:

$$I_{SET} = \sum_{i,j}(A_\mu SET_i + \log A_\mu SET_j)$$

3-hexanone

SET_{O1} = 1.9507 + log 0.3899 = 1.5416

SET_{C1} = 0.9892 + log 0.9998 = 0.9891

SET_{C2} = 0.9998 + log 0.9892 + log 0.3899 = 0.5860

SET_{C3} = 0.3899 + log 0.9998 + log 1.9507 + log 0.9998 = 0.6799

SET_{C4} = 0.9998 + log 0.3899+ log 0.8988= 0.5444

SET_{C5} = 0.8988 + log 0.9998 + log 0.9998 = 0.8986

SET_{C6} = 0.9998+ log 0.8988 = 0.9535

I_{SET} = 6.1931

hexanal

SET_{O1} = 1.9328 + log 0.5094 = 1.6398

SET_{C1} = 0.5094 + log 1.9328 + log 1.0371 = 0.8114

SET_{C2} = 1.0371+ log 0.5094 + log 0.8988 = 0.6978

SET_{C3} = 0.8988 + log 1.0371 + log 0.9041 = 0.8708

SET_{C4} = 0.9041 + log 0.8988 + log 0.9059 = 0.8148

SET_{C5} = 0.9059 + log 0.9041 + log 0.9980 = 0.8612

SET_{C6} = 0.9980 + log 0.9059 = 0.9550

I_{SET} = 6.6508

3.2.2 Esters

For esters the major effects related to the charge distribution are due to the presence of the two oxygen atoms and they occur on these atoms and in their neighborhood (their adjacent carbon atoms). The excess charge of these atoms leads to electrostatic interactions that are stronger than the weak dispersive dipolar interactions. For esters, all these factors were included in the calculation of the retention index through a small increase in the SET_i values for heteroatoms and the carbon atoms attached to them (Souza et al., 2009b). As in the case of ketones and aldehydes, it was verified that the introduction of the dipole moment of the molecule is not sufficient to explain the chromatographic behavior of these molecules. Thus, it was necessary to introduce an equivalent local dipole moment of the (-COOC-) group that contributes to the increase in the retention value. This was carried out by multiplying the SET_i values of the atoms belonging to the O=C-O-C group by the function A_μ which is dependent on the dipole moment of the molecule and the net charge of the oxygen and carbon atoms (to include both the electrostatic and dispersive interactions). The same approach used for ketones and aldehydes was applied to esters, that is, considering that A_μ has a weaker dependence on the dipole moment than the linear one, as given in equation 4. For esters μ_F is an equivalent local dipole moment (in the units of dipole moment) which is dependent on the charges of the atoms belonging to the O=C-O-C group. Clearly μ_F must be directly related to the net charge of the oxygen atoms since it must reflect some contribution to the electrostatic interaction between these atoms and the stationary phase. In this regard, μ_F may also be related to the atomic charge of the carbon atom of the functional group C=O or related to the difference between the atomic charges of these atoms. Hence, μ_F can be defined in different ways and some definitions of μ_F can be used in preliminary calculations.

Esters have two oxygen atoms and thus it is possible to define two local functions, one being dependent on the charges and bond length of the $C=O_1$ bond and another on the charges and bond length of the $C-O_2$ bond. Therefore, it was necessary to perform some calculations with different definitions for the equivalent local dipole moment. After the preliminary calculations it was found that for esters the charge difference, $Q_O - Q_C$, does not give reasonable results because the charges of the oxygen atoms mask the charge of the carbonyl carbon. As expected, our best choice was for the esters $\mu_{F1} = d_1 |Q_{O1}|$ and $\mu_{F2} = d_2 |Q_{O2}|$, where d_1 and d_2 are the calculated $C_1=O_1$ and C_1-O_2 bond lengths and $|Q_{O1}|$ and $|Q_{O2}|$ are the absolute values of the atomic charges of the oxygen atoms (O_1 and O_2). The equivalent local dipole moment is then calculated as the magnitude of the vectorial sum of two dipole moments, that is, $\mu_{F1} = (\mu^2_{F1} + \mu^2_{F2} + 2\,\mu_{F1}\,\mu_{F2}\cos\theta)^{1/2}$, where θ is the angle between the $C=O_1$ and $C-O_2$ bonds. For formates, a specific charge distribution occurs in the polar region of the molecules and the best mathematical model for the local moment was that which takes into account the contribution to the electrostatic interactions that originate from the atomic charges of the oxygen atoms, the carbon atoms and the H atom belonging to the $C_1O_1O_2C_{Al}$ group of the formate molecules (C_{Al} represents the carbon on the alcoholic side). Thus, the equivalent dipoles were built from the net charges of the HC_1O_1, HC_1O_2 and O_2C_{Al} groups of atoms. The equivalent dipoles associated with these net charges are:

$\mu_{F1} = d_1 |Q_H + Q_{C1} - Q_{O1}|$, $\mu_{F2} = d_2 |Q_H + Q_{C2} - Q_{O2}|$ and $\mu_{F3} = d_3 |Q_{CA1} - Q_{O2}|$ where d_1 and d_2 are the calculated $C_1=O_1$ and C_1-O_2 bond lengths and d_3 is the calculated $C_{Al}-O_2$ bond length. In a first approach, the local moments μ_{F2} and μ_{F3} are considered to be collinear and another equivalent dipole is obtained from the difference between μ_{F2} and μ_{F3}, that is, $\mu_{F4} = \mu_{F2} - \mu_{F3}$ and the final equivalent local moment is calculated as above, that is, $\mu_F = (\mu^2_{F1} + \mu^2_{F4} + 2 \, \mu_{F1} \, \mu_{F4} \cos\theta)^{1/2}$ where θ is the angle between the $C=O_1$ and $C-O_2$ bonds. Hence, for formates the charge of the hydrogen atom attached to the carbon atom of the COO functional group is also considered, as in the case of aldehydes, because the charge of the H atom contributes explicitly to the positive charge of the local polar region of the molecule. The above-mentioned best definitions for μ_F imply that the present approach to calculating the retention index considers important polar features of the organic functions, such as ketones, aldehydes and esters, through the information carried by the local moment μ_F. In other words, according to Equation (4) there is an increase in the interaction between the molecules and the stationary phase due to the presence of a dipole moment and this contribution may be screened by the charge located on the heteroatoms and the carbon atom of the $C=O$ group if , $\mu_F > \mu$ or may be increased if $\mu_F < \mu$. In the case of esters, the local function μ_F is less than the dipole moment showing that A_μ has an appreciable contribution from the atomic charges of those atoms. This verifies, for esters, the contribution of the oxygen atom to the electrostatic interaction between the solute and the stationary phase.

Therefore, in the case of esters the I_{SET} value is here calculated as in Equation (5), where the SET_i values are obtained using Equation (2) through AM1 calculations of the net atomic charges. As mentioned above, Equation (5) is calculated by multiplying the SET_i values of the atoms belonging to the $C_1O_1O_2C_{Al}$ group by the dipolar function A_μ which is taken as unity for the remaining carbon atoms of the molecules. Hence, Equation (5) is a general definition for the electrotopological index that can be applied to different organic functions, which are specified through appropriate definitions of the equivalent local moment μ_F. The preliminary applications of I_{SET} as given by Equation (5) showed that this expression overestimates the calculated retention index for branched esters and underestimates the results for methyl esters. This finding reveals the need to consider other definitions for the local moment μ_F for branched esters and methyl esters. However, another easy choice is to take into account the steric effects for the branched esters and methyl esters. The simplest way to do this is to consider the steric hindrance of the C_{Al} carbon atom of the $C_1O_1O_2C_{Al}$ group and the carbon atom attached to the acid side of the COO functional group (here named the C_{Ac} carbon). As seen in Equation (2), the $\log SET_j$ factor gives, precisely, the steric effect of atom j. Thus, to include a steric correction (sc) in Equation (5) for branched esters the term **sc = n logSET(C_{AC}) + n logSET(C_{Al})** was added, where **n** is the number of branches of the ester. On the other hand, for methyl esters the C_{Al} carbon is bound to three H atoms and it is necessary to remove the overestimated steric effects of the **logA$_\mu$SET$_j$** terms in Equation (5). For methyl esters this is easily achieved by including a second steric correction (ssc) by adding the term **ssc = -log SET(C_{Al})** to equation (5). Very good results were obtained using this approach, which reveals that in this model the complex steric effects in branched esters can be included simply by considering the steric hindrance using the net charge (through the SET_i values) of the two carbon atoms bound to the alcoholic and acid sides of the COO functional group. The calculation of I_{SET} for a large amount of molecules is easily carried out by means of a FORTRAN code developed in our lab that calculates I_{SET} by reading the output data (calculated net charges, dipole moment and atomic positions) from AM1 semi-empirical calculations.

3.2.3 Alcohols

As observed for the preceding compounds, for alcohols the major effects on the charge distribution are due the presence of the oxygen atom and they occur at the site of and close to their neighbors (adjacent carbon atoms). The excess charge at these atoms leads to electrostatic interactions that are stronger than the weak dispersive dipolar interactions. Thus, it is clear that it is necessary to introduce an equivalent local dipole moment for each of the organic functions that participate in increasing the retention value. For alcohols, as in the case of ketones, aldehydes and esters, this was achieved by multiplying the SET_i values of the atoms belonging to the C-OH group by a function A_μ as defined by equation 4, with μ_F being the equivalent local dipole moment which is dependent on the charges of the atoms belonging to the C-OH group (Souza et al., 2010). Clearly, μ_F is directly related to the net atomic charge of the oxygen atoms since it must reflect some contribution to the electrostatic interaction between these atoms and the stationary phase. Thus, μ_F may also be related to the atomic charge of the carbon atom of the functional group C-OH or to the difference between the atomic charges of these atoms. Hence, as with the other organic functions, μ_F can be defined in different ways and some of these definitions can be used in the preliminary calculations. For primary alcohols the terminal carbon atom of the C-O bond is attached to two hydrogen atoms and thus it is necessary to consider the net positive charge in this polar region of the molecule as the sum of the atomic charges of the carbon and hydrogen atoms. Thus, for primary alcohols we found that the best definition of the local moment is related to the charges of all atoms at the polar head of the molecules, that is, $\mu_F =$ $d \,|\, Q_C + (Q_{H1} + Q_{H2})/2 - Q_O - Q_{HO} |$ where d is the calculated C-O bond length and $|\, Q_C + (Q_{H1} + Q_{H2})/2 - Q_O - Q_{HO} |$ is the absolute value of the difference between the net atomic charge at the carbon (Q_C) plus the average charge of the hydrogen atoms attached to it $(Q_{H1} + Q_{H2})/2$ and the charges of the oxygen atom (Q_O) and the hydrogen attached to it (Q_{Ho}). For secondary, tertiary and quaternary alcohols the best choice for the local moment is related to the net atomic charge of the C and O atoms only, that is, $\mu_F = d \,|\, Q_C - Q_O |$, where d is the length of the C-O bond and $|\, Q_C - Q_O |$ is the absolute value of the difference between the charge of the carbon atom and oxygen atom attached to it. These definitions of μ_F attempt to take into account the contribution to the electrostatic interactions originating from the polar region of the molecules. Therefore, this shows again that Equation 4 represents a dipolar contribution to the interactions between the molecules and the stationary phase (which originates from the presence of a molecular dipole moment) and this contribution is decreased by the charge of the heteroatoms (oxygen atoms) when $\mu_F > \mu$, or increased when $\mu_F < \mu$. This reveals the contribution of oxygen to the electrostatic interaction between the solute and stationary phases.

For alcohols, the I_{SET} values are calculated as in Equation (5), where the SET_i values are obtained using Equation (2) through AM1 calculations of the net atomic charges.

3.3 Development of QSPR models using the I_{SET}

The molecular descriptor I_{SET} was developed first for alkanes and alkenes on a low polarity stationary phase and then extended to oxo-compounds through the inclusion of the molecular dipole moment and a local dipole moment in its definition. The models for the best simple linear regression between the retention index and the molecular descriptor (RI = b + a ISET) and the statistical data for each class of compounds, obtained in previous QSRR studies, are given in Table 4. For esters and alcohols good correlations between the retention index and the I_{SET} were obtained also for stationary phases with different polarities (not included in Table 4). The good statistical results achieved (Table 4) employing I_{SET} are better or equivalent to those obtained using multiple linear regression employing many molecular descriptors.

	Compounds	Phase	N	a	b	r^2	r^2_{CV}	SD	Ref.
01	Alkanes and Alkenes	SQ	179	120.8	-36.8	0.9980	0.9980	10.7	Souza et al., 2008
02	Aldehydes and ketones	HP-1	42	123.9	-13.2	0.9993	0.9993	11.7	Souza et al., 2009a
03	Esters	SE-30	100	115.0	-74.7	0.9981	0.9980	7.6	Souza et al., 2009b
04	Alcohols	SE-30	31	126.0	-186.6	0.9990	0.9977	9.3	Souza et al., 2010

Table 4. Summary of the best simple linear regressions ($RI_{Calc} = b + a\ I_{SET}$) found for different classes of compounds on low-polarity stationary phases.

As can be seen from Table 4, the QSRR models for 179 representative linear and branched alkanes and alkenes, obtained with the I_{SET} using the net atomic charge to calculate more precisely the Ci fragment values of I_{ET}, were of good quality for the statistical parameters obtained. This new descriptor I_{SET} contains information on the 3D features of molecules, and discriminates between geometrical isomers, such as cis- and trans-alkenes, and between conformers, and the elution sequence is correct for the majority of the compounds.

The results obtained for aldehydes and ketones are similar to those reported by Ren (Ren, 2003) in multiple linear regression models for 33 aldehydes and ketones using Xu and AI topological indices and by Héberger and co-workers using quantum-chemical descriptors (SW and μ) (Héberger et al., 2001) and physico-chemical properties (T_{Bp}, M_W, log P) (Héberger et al., 2000) for 31 and 35 compounds, respectively. For esters the results obtained by single linear regression using the I_{SET} are better than those reported by Lu et al. (Lu et al., 2006). For SE-30 and OV-7 stationary phases the results are also better than those found by Liu et al. (Liu et al., 2007) and for more polar stationary phases the statistical parameters differ only slightly. Both of these studies use multiple linear regression (MLR) between RI and the topological indices for 90 saturated esters on stationary phases with different polarities.

Several authors have developed QSRR models, based on MLRs, to predict the RI values for saturated alcohols. For example, Guo et al. (Guo et al., 2000), using the MLR analysis and artificial neural networks technique, obtained the statistical parameters r^2=0.9982, SD=8.21, N=19 for an SE-30 stationary phase. In a previous study, the best statistical parameters of the MLR models obtained by Farkas and Héberger (Farkas & Héberger, 2005), employing four molecular descriptors, were r=0.9804, SD=14.22, r^2_{CV}=0.9801 and N=44 for an OV-1 stationary phase. Therefore, our prediction results, on low polarity stationary phases, using the I_{SET} as a single descriptor, showed statistical quality comparable to similar studies reported by the above authors. Furthermore, the statistical parameters of the present approach have a good agreement with those obtained for alkanes and alkenes, aldehydes and ketones, for saturated esters and for alcohols using the semi-empirical topological index, I_{ET}, previously developed. These results show clearly that I_{SET} is a molecular descriptor that embodies in an appropriated manner the net atomic charges and charge distribution of molecules since the retention index embodies the intermolecular interactions between the stationary phase and the molecules.

The fact that properties that are determined by intermolecular forces can be adequately modeled by the I_{SET} descriptor can be easily seen in its relationship with the boiling point (BP). For alcohols a good correlation was obtained through a simple linear model (BP = b + a I_{SET}), as can be seen in Table 5. The QSPR model obtained for the experimental BP of 134 compounds showed high values for the coefficient of determination and cross validation coefficient showing the good predictive capacity of the model.

	Alcohols	N	a	b	r^2	r^2_{CV}	SD	Ref.
01	Boiling Point	134	22.0	-32.9	0.9818	0.9811	5.4	Souza et al., 2010

Table 5. The coefficients and statistical parameters for linear regression between experimental boiling point and I_{SET}.

This model can explain 98.20% of the variances in the experimental values and most of these compounds (N=101) are not included in the initial model used to build the I_{SET}, showing the external stability of the model. These results are similar to those obtained using I_{ET} for 146 aliphatic alcohols and can be compared with those obtained by Ren (Ren, 2002b), but using MLR models, for 138 compounds with five descriptors.

The octanol-water partition coefficient (log P) of compounds, which is a measure of hydrophobicity, is widely used in numerous Quantitative Structure-Activity Relationship (QSAR) models for predicting the pharmaceutical properties of molecules. The partition coefficient is a property that is determined by intermolecular forces and thus it is expected that it can be described by a molecular descriptor such as I_{SET}. The results obtained in the statistical analysis of the single linear regression between experimental log P values and I_{SET} are shown in Table 6 for each class of compounds.

Compound	N	a	b	r^2	r^2_{CV}	SD
Hydrocarbon	23	3.90×10^{-3}	0.9997	0.9971	0.9964	0.104
Alcohols	60	3,24	0,6394	0.9876	0.9870	0.183
Aldehydes	9	1.60×10^{-3}	1.0014	0.9972	0.9961	0.058
Ketones	19	-2.7182	0.6693	0.9864	0.9831	0.158
Esters	14	-3.1575	0.6587	0.9903	0.9838	0.118

Table 6. The coefficients and statistical parameters for linear regression between experimental log P values and I_{SET}.

The results in Table 6 indicate that the theoretical partition coefficients calculated using the I_{SET} method give good agreement with the experimental partition coefficients. The QSPR models obtained with I_{SET} showed high values for the correlation coefficient (r > 0.99), and the leave-one-out cross-validation demonstrates that the final models are statistically significant and reliable (r^2_{cv} > 0.98). As can be observed, this model explains more than 99%

of the variance in the experimental values for this set of compounds. Among the various classes of compounds the best results obtained with the I_{SET} method are for hydrocarbons (Table 6), which is related to the fact that the present model was developed initially for this class of organic compounds. As can be seen in Table 6, the lowest standard deviation was obtained for the correlation of aldehydes and for alcohols the correlation was stronger. The range of standard deviations obtained verifies the applicability of the present approach to different classes of organic compounds. For alcohols, the earlier approach of Duchowicz et al. (Duchowicz et al., 2004), based on the concept of flexible topological descriptors and on the optimization of correlation weights of local graphic invariants, is applied to model the octanol/water partition coefficient of a representative set of 62 alcohols, resulting in a satisfactory prediction with a standard deviation of 0.22. Recently, Liu et al. (Liu et al. 2009) carried out a QSPR study to predict the log P for 58 aliphatic alcohols using novel molecular indices based on graph theory, by dividing the molecular structure into substructures obtaining models with good stability and robustness, and values predicted using the multiple linear regression method are close to the experimental values (r = 0.9959 and SD = 0.15). The above results show the reliability of the present model calculation based on the semi-empirical calculation of atomic charges and local dipole moments using only one descriptor, I_{SET}. This new approach to polar molecules, with the introduction of the remodeled I_{SET} index including the contribution of the dipole moment of the molecule and an effective local dipole moment associated with the net charges of the atoms of the carbonyl group, opens new possibilities for studies on the chromatographic and other properties of different organic functions.

4. Conclusions

It is known that the chromatographic process of separation results from the forces that operate between solute molecules and the molecules of the stationary phase. These forces are called van-der-Waals forces since van der Waals recognized them as the reason for the non-ideal behavior of the real gases. Intermolecular forces are usually classified according to two distinct categories: i) the first category corresponds to the directional, induction and dispersion forces which are non-specific; and ii) the second group corresponds to hydrogen bonding forces and the forces of charge transfer or electron-pair donor-acceptor forces which are specific.

In the development of the semi-empirical topological index (I_{ET}) it was considered that the retention of alkanes is due to the number and interaction of each specific carbon atom with the stationary phase, considered as non-polar, which is determined by its electrical characteristic and by the steric hindrance by other carbon atoms attached to it. In this case only dispersion forces due to the continuous electronic movement, at any instant, result in a small dipole moment which can fluctuate and polarize the electron system of the neighboring atoms or molecules. For the alkenes, some carbon atoms with greater electronegativity give the molecules a dipole moment and for this reason besides the dispersion forces, electrostatics forces play an important role. However, in this method the behavior of this kind of carbon atom is determined from the experimental data and indicated in specific tables. As the values were obtained from the experimental data they encode the real physical interaction force. In the case of oxo-compounds, the presence of atoms with different carbon atom electronegativity introduces a dipole moment in the functional group and a change in the dipole moment of the whole molecule. These factors were considered in order to obtain the different values for the functional groups and they were able to encode the physical force involved in the chromatographic separation.

The new semi-empirical electrotopologiocal index (I_{SET}) demonstrated that the values for the carbon atoms that are not tetrahedral and functional groups (considering the new local dipole created by the heteroatom) can be calculated from the net atomic charges that are obtained from quantum-chemical calculations. In the case of esters, the major effects are due to the presence of the two oxygen atoms and their adjacent carbon atoms. As in the case of aldehydes and ketones it was verified that the introduction of the dipole moment of the molecules is not sufficient to explain the chromatographic behavior. Thus, it was necessary to introduce an equivalent local dipole moment of the ester group that contributes to the increase in the retention value. In the case of esters two local functions must be considered according to the charges and the bond lengths of the C=O and C-O bonds. Thus, the semi-empirical electrotopological index was developed based on the refinement of the previously developed semi-empirical topological index, unifying the quantum-quantum chemical with the topological method to provide a three-dimensional picture of the atoms in the molecule.

The I_{ET} and I_{SET} were generated to predict the chromatographic retention indices and other physical-chemical properties and to obtain the quantitative structure-retention relationship (QSRR/QSPR). The efficiency and the applicability of these descriptors were demonstrated through the good statistical quality and high internal stability obtained for the different classes of compounds studied.

5. Acknowledgement

The authors acknowledge the support from CNPq and CRQ-XIII (Brazil) for this research.

6. References

Amboni, D. M.; Junkes, B. S.; Yunes, R. A &, Heinzen, V. E. F. (2000). *Journal of Agricultural Food Chemistry*, Vol.48, pp. 3517–3521.

Amboni, R. D. M. C; Junkes, B. S.; Yunes, R. A. & Heinzen, V. E. F. (2002a). Semi-empirical topological method for prediction of the chromatographic retention of esters. *Journal of Molecular Structure (Theochem)*, Vol.579, pp. 53–62.

Amboni, R. D. M. C; Junkes, B. S.; Yunes, R. A. & Heinzen, V. E. F. (2002b). Quantitative structure-property relationship study of chromatographic retention indices and normal boiling points for oxo compounds using the semi-empirical topological method. *Journal of Molecular Structure (Theochem)*, Vol.586, pp. 71–80.

Arruda, A. C. S.; Souza, E. S.; Junke, B. S.; Yunes, R. A. & Heinzen, V. E. F. (2008). Semi-empirical topological index to predict properties of halogenated aliphatic compounds. *Journal of Chemometrics*, Vol.22, pp. 186-194.

Arruda, A. C.; Heinzen V. E. F. & Yunes, R. A. (1993). *Jornal of Chromatography A*, Vol.630, pp. 251–256.

Bredow, T. & Jug, K. (2005). Theory and range of modern semiempirical molecular orbital methods. *Theorical Chemistry Accounts*, Vol.113, pp. 1–14.

Christian, R. (1990). Solvents and Solvent effects in Organic Chemistry, 2nd ed. VCH, Germany.

Duchowicz, P. R.; Castro, E. A.; Toropov, A. A.; Nesterova, A. I. & Nabiev, O. M. (2004). QSPR Modeling of the octanol/water partition coefficient of alcohols by means of optimization of correlation weights of local graph invariants. *Journal of the Argentine Chemical Society*, Vol.92, pp. 29-42.

Estrada, E. & Gutierrez, Y. (1999). Modeling chromatographic parameters by novel graph theoretical sub-structural approach, *Journal of Chromatography A*, Vol.858, pp. 187-199.

Estrada, E. (2001). Generalization of topological indices. *Chemical Physics Letters*, Vol.336, pp. 248-252.

Estrada, E. (2001). Recent Advances on the Role of Topological Indices in Drug Discovery Research. *Current Medicinal Chemistry*, Vol.8, pp. 1573-1588.

Estrada, E. (2002). The Structural Interpretation of the Randić Index. *Internet Electronic Journal of Molecular Design*, Vol.1, pp. 360-366, http://www.biochempress.com.

Farkas, O. & Héberger, K. (2005). Comparison of ridge regression, partial least-squares, pairwise correlation, forward-and best subset selection methods for prediction of retention indices for aliphatic alcohols. *Journal of Chemical Information and Modeling*, Vol.45, pp. 339-346.

Fritz, D. F.; Sahil, A. & Kováts, E. (1979). Study of the adsorption effects on the surface of poly-(ethylene glycol)-coated column packings. *Journal of Chromatography*, Vol.186, pp. 63-80.

Galvez, J.; Garcia, R.; Salabert, M. T. & Soler, R. (1994). *Journal of Chemistry Information Computational and Science*, Vol.34, pp. 520-525.

García–Domenech, R. ; Catalá–Gregori, A.; Calabuig, C.; Antón–Fos, G.; del Castillo, L. & Gálvez, J. (2002). Predicting Antifungal Activity: A Computational Screening Using Topological Descriptors. *Internet Electronic Journal of Molecular Design*, Vol.1, pp. 339-350, http://www.biochempress.com.

Guo, W.; Lu, Y. & Zheng, X. M. (2000). The predicting study for chromatographic retention index of saturated alcohols by MLR and ANN. *Talanta*, Vol.51, pp. 479-488

Hall, L. H.; Mohney, B. & Kier, L. B. (1991). *Journal of Chemistry Information Computational and Science*, Vol.31, pp. 76-82.

Hansen, P. J. & Jurs, P. C. (1988). Chemical applications of graph theory. Part I. Fundamental and topological indices. *Journal of Chemical Education*, Vol.65, pp. 574-580.

Heinzen, V. E. F. & Yunes, R. A. (1993). *Jornal of Chromatography A*, Vol.654, pp. 83-89.

Heinzen, V. E. F. & Yunes, R. A. (1996). *Jornal of Chromatography A*, Vol.719, pp. 462-467.

Heinzen, V. E. F.; Cechinel Filho V. & Yunes, R. A. (1999a). *IL Farmaco*, Vol.54, pp. 125-129.

Heinzen, V. E. F.; Soares, M. F. & Yunes, R.A. (1999b). Semi-empirical topological method for prediction of the chromatographic retention of cis- and trans- alkene isomers and alkanes. *Journal of Chromatography* A, Vol.849, pp. 495-506.

Héberger, K.; Görgényi, M. & Sjörström, M. (2000). Partial least squares modeling of retention data of oxo compounds in gas chromatography. *Chromatographia*, Vol.51, pp. 595-600.

Ivanciuc, O.; Ivanciuc, T., Carbol–Bass, D. & Balaban, A. T. (2000). Comparison of Weighting schemes for molecular graph descriptor: Application in quantitative structure-retention relationship models for alkylphenols in gas–liquid chromatography, *Journal of Chemistry Information Computational and Science*, Vol.40, pp. 732–743.

Ivanciuc T. & Ivanciuc, O. (2002). Quantitative Structure–Retention Relationship Study of Gas Chromatographic Retention Indices for Halogenated Compounds, *Internet Electronic Journal of Molecular Design*, Vol.1, pp. 94–107, http://www.biochempress.com.

Ivanciuc, O.; Ivanciuc, T.; Cabrol–Bass, D. & Balaban, A. T. (2002). Optimum Structural Descriptors Derived from the Ivanciuc–Balaban Operator. *Internet Electronic Journal of Molecular Design, Vol.1*, pp. 319–331, http://www.biochempress.com.

Junkes, B. S.; Amboni, R. D. M. C; Yunes, R. A. & Heinzen, V. E .F. (2002a). Use of a semiempirical topological method to predict the chromatographic retention of branched alkenes. *Chromatographia,* Vol.55, pp. 75–81.

Junkes, B. S.; Amboni, R. D. M. C; Yunes, R. A. & Heinzen, V. E .F. (2002b). Quantitative structure-retention relationships (QSRR), using the optimum semiempirical topological index, for methylbranched alkanes produced by insects. *Chromatographia,* Vol.55, pp. 707–715.

Junkes, B. S.; Amboni, R. D. M. C; Yunes, R. A. & Heinzen, V. E .F. (2003a). Semi-empirical topological index: a novel molecular descriptor for quantitative structure-retention relationship studies. *Internet Electronic Journal of Molecular Design*, Vol.2, pp. 33–49.

Junkes, B. S.; Amboni, R. D. M. C; Yunes, R. A. & Heinzen, V. E .F. (2003b). Prediction of chromatographic retention of saturated alcohols on stationary phases of different polarity applying the novel semi-empirical topological index. *Analytica Chimica Acta,* Vol.477, pp. 29–39.

Junkes, B. S.; Amboni, R. D. M. C; Yunes, R. A. & Heinzen, V. E .F. (2004). Application of the semiempirical topological index in the QSRR of aliphatic ketones and aldehydes on stationary phases of different polarity. *Journal of Brazilian Chemical Society*, Vol.15, pp. 183–189.

Junkes, B. S.; Arruda, A. C. S.; Yunes, R. A.; Porto, L. C. & Heinzen, V. E. F. (2005). Semiempirical topological index: application to QSPR/QSAR modeling. *Journal of Molecular Modeling*, Vol.11, pp. 128–134.

Kaliszan, R. (1987). *Quantitative structure–chromatographic retention relationships*, Wiley-Interscience, New York, USA.

Karelson, M.; Lobanov, V. S. & Katritzky, A. R. (1996). Quantum-Chemical Descriptors in QSAR/QSPR Studies. *Chemical Reviews*, Vol.96, pp. 1027–1044.

Katritzky, A. R. & Gordeeva, E. V. (1993). Traditional topological indices VS electronic, geometrical and combined molecular descriptors in QSAR/QSPR research. *Journal of Chemistry Information Computational and Science*, Vol.33, pp. 835–857.

Katritzky, A. R.; Ignatchenko, E. S.; Barcock, R. A.; Lobanov, V. S. & Karelson, M. (1994). Prediction of gas chromatographic retention times and response factors using a general quantitative structure–property relationship treatment. *Analytical Chemistry*, Vol.66, pp. 1799–1807.

Katritzky, A. R.; Chen, K.; Maran, U. & Carlson, D. A. (2000). QSRR correlation and predictions of GC retention indexes for methyl–branched hydrocarbons produced by insects, *Analytical Chemistry*, Vol.72, pp. 101–109.

Kier L. B. & Hall, L H. (1976). Molecular connectivity chemistry and drug research, Academic Press, New York, USA.

Kier, L. B. & Hall, L. H. (1986). Molecular connectivity in structure activity studies. Research Studies Press, Letchworth, UK.

Kier, L. B. & Hall, L. H. (1990). An Electrotopological State Index for Atoms in Molecules. *Pharmaceutical Research*, Vol.7, pp. 801–807.

Körtvélyesi, T., Görgényi, M. & Hérberger, K. (2001). Correlation between retention indices and quantum–chemical descriptor of ketones and aldehydes on stationary phases of different polarity. *Analytica Chimica Acta*, Vol.428, pp. 73–82.

Kováts, E. (1968). Zu fragen der polarität. Die method der linearkombination der wechselwirkungskräfte (LKWW). *Chimie*, Vol.22, pp. 459-462.

Liu, S.-S.; Liu, H.-L. ; Shi, Y.-Y. & Wang, L.-S. (2002). QSAR of Cyclooxygenase-2 (COX–2) Inhibition by 2,3-Diarylcyclopentenones Based on MEDV–13. *Internet Electronic Journal of Molecular Design*, Vol.1, pp. 310–318, http://www.biochempress.com.

Liu, F.; Liang, Y.; Cao, C. & Zhou, N. (2007). QSPR study of GC retention indices for saturated esters on seven stationary phases based on novel topological indices, *Talanta*, Vol.72, pp. 1307–1315.

Liu, F.; Cao, C. & Cheng, B. (2011). A quantitative structure-property relationship (QSPR) study of aliphatic alcohols by the method of dividing the molecular structure into substructure. *International Journal of Molecular Sciences*, Vol.12, pp. 2448-2462.

Lu, C.; Guo, W. & Yin, C. (2006). Quantitative structure-retention relationship study of gas chromatographic retention indices of saturated esters on different stationary phases using novel topological indices, *Analytica Chimica Acta*, Vol.561, pp. 96–102.

Marino, D. J. G.; Peruzzo, P. J.; Castro, E. A. & Toropov, A. A. (2002). QSAR Carcinogenic Study of Methylated Polycyclic Aromatic Hydrocarbons Based on Topological Descriptors Derived from Distance Matrices and Correlation Weights of Local Graph Invariants. *Internet Electronic Journal of Molecular Design*, Vol.1, pp. 115–133, http://www.biochempress.com.

Markuszenwski, M. &. Kaliszan, R. (2002). Quantitative structure–retention relationships in affinity high–performance liquid chromatography, *Journal of Chromatography B*, Vol.768, pp. 55–66.

Peng, C. T.; Ding, S. F.; Hua, R. L. & Yang, Z. C. (1988). Prediction of retention indexes I. Structure–retention index relationship on apolar columns. *Journal of Chromatography*, Vol.436, pp. 137–172.

Peng, C. T. (2000). Prediction of retention indices V. Influence of electronic effects and column polarity on retention index. *Journal of Chromatography A*, Vol.903, pp. 117–143.

Pompe, M. & Novic, M. (1999). Prediction of gas–chromatographic retention indices using topological descriptors, *Journal of Chemistry Information Computational and Science*, Vol.39, pp. 59–67.

Porto, L. C.; Souza, E. S.; Junkes, B. S.; Yunes, R. A. & Heinzen ,V. E. F. (2008). Semi-Empirical Topological Index: development of QSPR/QSRR and optimization for alkylbenzenes. *Talanta*, (Oxford). Vol.76, pp. 407-412.

Randic, M. (1993). Comparative structure-property studies: Regressions using a single descriptor. *Croatica Chemica Acta* (CCACAA), Vol.66, pp. 289–312.

Randic, M. (2001). The connectivity index 25 years after. *Journal of Molecular Graphic and Modeling*, Vol.20, pp. 19–35.

Ren, B. (1999). A new topological index for QSRR of alkanes. *Journal of Chemistry Information Computational and Science*, Vol.39, pp. 139–143.

Ren, B. (2002a). Application of novel atom–type AI topological indices to QSRR studies of alkanes, *Computer & Chemistry*, Vol.26, pp. 357–369.

Ren, B. (2002b). Novel Atomic-Level-Based AI Topological Descriptors: Application to QSPR/QSAR Modeling. *Journal of Chemical Information and Computer Sciences*, Vol.42, pp. 858–868.

Ren, B. (2002c). Novel atom-type AI indices for QSPR studies of alcohols. *Computer & Chemistry*, Vol.26, pp. 223–235.

Ren, B. (2002d). Application of novel atom-type AI topological indices in the structure-property correlations. *Journal of Molecular Structure*, (Theochem), Vol.586, pp. 137–148.

Ren, B. (2003). Atom-type-based AI topological descriptors for quantitative structure-retention index correlations of aldehydes and ketones. *Chemometrics Intelligent Laboratory Systems*, Vol.66. pp. 29–39.

Rios–Santamarina, I.; García–Domenech, R.; Cortijo, J.; Santamaria, P.; Morcillo, E. J. & Gálvez, J. (2002). Natural Compounds with Bronchodilator Activity Selected by Molecular Topology. *Internet Electronic Journal of Molecular Design*, Vol.1, pp. 70–79, http://www.biochempress.com.

Sabljic, A. & Trinajstic, N. (1981). Quantitative structure activity relationships: The role of topological indices. *Acta Pharmaceutica Jugoslavica*, Vol.31, pp. 189–214.

Smith, W. B. Introduction to Theoretical Organic Chemistry and Molecular Modeling. VCW: New York, 1996; 142–143.

Souza, E. S.; Junkes, B. S.; Kuhnen, C. A.; Yunes, R. A. & Heinzen, V. E. F. (2008). On a new semiempirical electrotopological index for QSRR models. *Journal of Chemometrics*, Vol.22, pp. 378-384.

Souza, E. S.; Kuhnen, C. A.; Junkes, B. S.; Yunes, R. A. & Heinzen, V. E. F. (2009a). Modelling the semiempirical electrotopological index in QSPR studies for aldehydes and ketones. *Journal of Chemometrics, Vol.*23, pp. 229-235.

Souza, E. S.; Kuhnen, C. A.; Junkes, B. S.; Yunes, R. A. & Heinzen, V. E. F. (2009b). Quantitative structure retention relationship modelling of esters on stationary phases of different polarity. *Journal of Molecular Graphics and Modelling*, Vol.28, pp. 20-27.

Souza, E. S.; Kuhnen, C. A.; Junkes, B. S.; Yunes, R. A. & Heinzen, V. E. F. (2010). Development of semi-empirical electrotopological índex using the atomic charge in QSPR/QSRR models for alcohols. *Journal of Chemometrics*, Vol.24, pp. 149-157.

Toropov, A. A. & Toropova, A. P. (2002). QSAR Modeling of Mutagenicity Based on Graphs of Atomic Orbitals. *Internet Electronic Journal of Molecular Design*, Vol.1, pp. 108–114, http://www.biochempress.com.

Wiener, H. (1947). Structural determination of paraffin boiling points. *Journal of the American Chemical Society*, Vol.69, pp. 17–20.

Yao, X.; Zhang, X.; Zhang, R.; Liu, M.; Hu, Z. & Fan, B. (2002). Prediction of gas chromatographic retention indices by the use of radial basis function neural networks. *Talanta*, Vol.57, pp. 297–306.

Part 2

Interactions at Interfaces, Molecular Design and Nano-Engineering of Multifunctional Materials

4

Molecular Interactions at Interfaces

Raj Kumar Gupta and V. Manjuladevi
Department of Physics, Birla Institute of Technology & Science, Pilani
India

1. Introduction

The study on molecular organization and structure formation at the nanometer length scale is important due to its vast application in the field of nanoscience and nanotechnology. Molecular interactions play a pivotal role in the process of molecular assembly. The properties of materials can be maneuvered precisely by manipulating the structures at the nanometer length scale. The field of thin films science and technology has been growing remarkably due to its enormous industrial applications. The properties of thin films depend on the nature of the adsorbate and the structures on the surface. The structures of the thin films on a surface leads to the growth of bulk material, and hence the material properties can be controlled by manipulating the structures of the thin films. The form of such structures depends on the molecule-substrate and intermolecular interactions. The development of thin films science and technology has influenced the field of nanoscience and nanotechnology significantly. In this chapter, we discuss the role of molecular interactions in ultrathin films at air-water (A-W) and air-solid (A-S) interfaces. We form monomolecular thick films on the surface of water and study the film stability, surface phases, and other thermodynamical parameters. We found that the stability of the films at the A-W interface primarily depends on the molecular-surface and intermolecular interaction. Amphiphilic molecules, when spread on the water surface, form a monomolecular thick film at A-W interface. Such monomolecular thick film is known as Langmuir monolayer. An amphiphilic molecule has two parts : hydrophilic (water loving) and hydrophobic (water hating) part. When such molecules with a proper balance between hydrophilic and hydrophobic parts are dispersed on water surface, the hydrophilic part gets anchored to the water surface whereas the hydrophobic part stays away from the water surface. Under such condition, the anchored molecules are constrained to move on the two dimensional smooth water surface. The surface density can be varied and a corresponding change in surface tension is recorded. A Langmuir monolayer has proved to be an ideal two dimensional system not only for studying the thermodynamics but also for depositing the films on different types of substrates by vertical deposition mechanism in a highly controlled manner. Such films at A-S interface are known as Langmuir-Blodgett (LB) films.

In this chapter, we discuss the effect of amphiphilicity (a balance between the hydrophobic and hydrophilic part of the amphiphilic molecule) on the Langmuir monolayer. We discuss the stability of Langmuir monolayer of a hydrophobic molecules (e.g. thiocholesterol) in the matrix of a very stable Langmuir film of cholesterol. The stability of such hydrophobic molecules is due to the attractive interaction between the molecule which arises due to overall gain in entropy of the system. There are few reports on the Langmuir monolayer of purely hydrophobic nanoparticles at A-W interface. Since the particles are hydrophobic, the stability of such systems are questionable. We changed the chemistry of gold nanoparticles such that it attains an amphiphilic nature. The Langmuir monolayer of such amphiphilic gold nanoparticle (AGN) exhibited variety of surface phases like low ordered liquid phase, high ordered liquid phase, bilayer of high ordered liquid phase and a collapsed state. Based on the thermodynamical studies of Langmuir monolayer of AGN, a phase diagram has been constructed. The Langmuir films at A-W interface can be transferred to solid substrates by LB technique. Here also, the nature of aggregation and deposition depends on molecule-substrates and intermolecular interaction. Interestingly, it was found that repeated process of adsorption and partial desorption of cholesterol molecules from hydrophobic substrates lead to the formation of uniformly distributed torus shaped domains.

2. Experimental techniques

2.1 Monolayer and multilayer at air-water interface

2.1.1 Surface manometry

Surface manometry is a standard technique to study the thermodynamics and the surface phases in a Langmuir monolayer. The presence of a monolayer at the A-W interface reduces the surface tension of water. Such reduction in the surface tension is defined as surface pressure (π) (Gaines, 1966). In surface manometry, the surface density of the molecules adsorbed at the interface is varied and the surface pressure (π) is recorded at a constant temperature. This yields a surface pressure - surface density isotherm. The area per molecule (A_m) is defined as the inverse of the surface density. The instrument for the measurement of $\pi - A_m$ isotherm is shown in Figure 1. It consists of a teflon trough and barriers. The subphase was ultrapure ion-free water having a resistivity greater than 18 megaΩ-cm

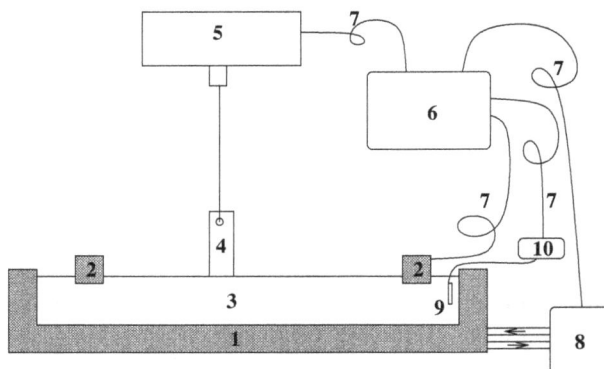

Fig. 1. A schematic diagram showing the experimental setup for the measurement of surface pressure (π) - area per molecule (A_m) isotherms. The basic parts of the setup are as follows: (1) teflon trough, (2) teflon barriers, (3) subphase (ion-free water), (4) Wilhemy plate (filter paper), (5) surface pressure sensor, (6) computer, (7) connecting wires, (8) thermostat, (9) resistance temperature detector (RTD) and (10) digital multimeter for measuring resistance

obtained by passing distilled water through filtering and deionizing columns of a Milli-Q Millipore unit. The sample was dissolved in an appropriate solvent to obtain a solution of known concentration. The solution of the samples was spread on the water subphase between the barriers using a precisely calibrated microsyringe (obtained from Hamilton). The solvent was allowed to evaporate for 15 minutes before starting the compression. The surface density of the molecules in the monolayer is varied by changing the area available for the molecules by moving the barriers laterally. The barriers are driven by motors which are controlled by a computer. They are coupled to each other so that it ensures a symmetric compression of the monolayer. The surface pressure was measured using a Wilhemy plate method (Gaines, 1966; Adamson, 1990). We have used filter paper of appropriate size as the Wilhemy plate. The surface pressure and area per molecule at a constant temperature were recorded simultaneously using a computer. This gives the π-A_m isotherm of the spread molecule. The isotherms at different temperatures are obtained by controlling the temperature of the subphase. This was achieved by circulating water at the required temperature inside the chamber of the trough using a thermostat. The temperature of the subphase was measured by a resistance temperature detector (RTD) and a digital multimeter. A typical π - A_m isotherm is shown in Figure 2(a). A plateau in the isotherm represents the coexistence of two phases. A

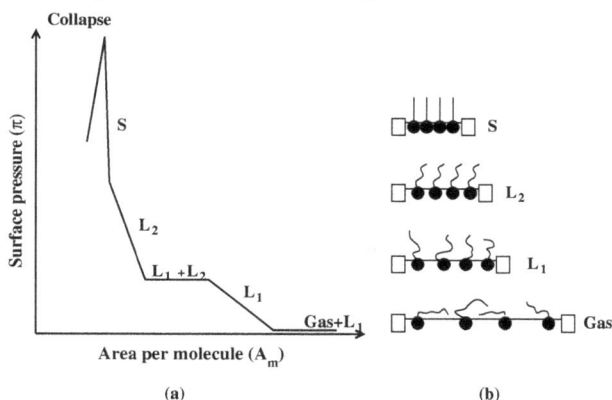

(a)   (b)

Fig. 2. (a) shows a typical isotherm indicating the different phases in a Langmuir monolayer. (b) shows the molecular arrangement in the different phases. The symbols L_1, L_2 and S represent liquid expanded, liquid condensed and solid phases, respectively

kink in the isotherm indicates a phase transition. At a very large area per molecule (A_m), the molecules are far apart and do not exert any force on each other. This is a 2D gas phase. On compression, the molecules condense to a low density liquid state (L_1). There are no positional and orientational orders in the molecules in this phase. On further compression, the L_1 phase transforms to a high density liquid state (L_2) accompanied by a two phase (L_1+L_2) coexisting region. This is known as condensed (L_2) phase. In the L_2 phase, molecules exhibit a long range orientational order and a quasi-long range positional order. On compression, the L_2 phase transforms to a 2D solid (S) phase. On further compression, the monolayer collapses. This is indicated by a sharp decrease in surface pressure.

The area per molecule at which the isotherm indicates very small and finite values of the surface pressure (e.g. 0.2 mN/m) is known as lift-off area per molecule (A_i). The average area occupied by the molecules in a phase is determined by extrapolating the corresponding region of the isotherm to the zero surface pressure on the A_m axis. The extrapolation of the steep region (e.g. S phase in Figure 2) of the isotherm to zero surface pressure is called limiting area per molecule (A_o). This is the minimum area to which the molecules can be compressed on the water surface without collapsing the monolayer. The orientational state (tilt or untilt) of the molecules in a phase can be estimated qualitatively by comparing the extrapolated area per molecule with that of molecular cross-sectional area in the bulk single crystal.

The isothermal in-plane elastic modulus E is an appropriate quantity for distinguishing very weak phase transitions. The isothermal in-plane elastic modulus Mohwald (1995) is defined as

$$E = -(A_m)\frac{d\pi}{dA_m} \tag{1}$$

The relaxation in the molecular area with time at a given surface pressure can indicate the nature of stability of the monolayer at the interface. A fast reduction in the area with time may indicate an unstable monolayer where the instability can be attributed to dissolution of the molecules to subphase or evaporation. A slow reduction can be attributed to the relaxation of the molecules in the monolayer. The nature and the mechanism of collapse can be studied by monitoring the change in molecular area with time at a constant surface pressure Smith & Berg (1980). In some special cases, kinetics of adsorption of the molecules from the subphase to the monolayer at the interface can be studied. In such cases, the area was found to increase or decrease with time due to complex formation of the molecules in the subphase with the molecules at the interface Ramakrishna et al. (2002). It provides a method to study kinetics of the surface chemistry.

The equilibrium spreading pressure (ESP) is a surface pressure of a monolayer coexisting with its bulk phase at the interface (Gaines, 1966). When a speck of crystallite is placed on the water surface, the molecules from the bulk crystallites elude out and form a monolayer at the interface spontaneously. After a certain period of time, the system reaches an equilibrium state where the rate of elution of molecules from the crystallites is equal to the rate of molecules binding to the crystallites. The variation in surface pressure with time shows an initial increase in surface pressure due to the formation of monolayer. On reaching the equilibrium, the surface pressure value saturates. The saturated value of surface pressure is known as ESP. The studies on ESP indicate the spreading capability of the molecules at an interface (Gaines, 1966). The finite value of ESP can suggest the monolayer to be stable against dissolution or evaporation of the molecules. The ESP values of some molecules like stearic acid, octadecanol and dipalmitoyl phosphatidylcholine are 5.2, 34.3 and 1 mN/m, respectively Smith & Berg (1980).

There are many two dimensional systems in nature that occur in mixed state. For instance, a bio-membrane constitutes of various kind of phospholipids, cholesterol and fatty acids. The Langmuir monolayer of phospholipids mimic the biological membranes and has been extensively studied (Mohwald, 1995). The stability and miscibility of the component molecules will be studied by estimating the excess Gibbs free energy of the system. For an ideal case of complete miscibility or immiscibility of two-component monolayer system, the area of the mixed monolayer is given by the rule of additivity,

$$A_{id} = A_1 X_1 + A_2 X_2 \tag{2}$$

where X_1 and X_2 are the mole fractions of the components 1 and 2, respectively. A_1 and A_2 are the A_m of the individual pure component monolayers. However, for a mixed system, the monolayer area can deviate from the ideal case. Such deviation in the monolayer area depends on the nature of the interaction between the component molecules and is known as the excess area per molecule, A_{ex}. The A_{ex} is defined as

$$A_{ex} = A_{12} - A_{id} \tag{3}$$

where A_{12} is the experimentally determined values of the A_m of the mixed monolayer. A_{id} is the ideal A_m value calculated from Equation 2. The positive or the negative value of the

A_{ex} indicates a repulsive or an attractive interaction, respectively between the component molecules in the mixed monolayer.

The stability and the degree of miscibility of a mixed monolayer were studied by calculating the excess Gibbs free energy (ΔG) Goodrich (1957); Seoane et al. (2001). The ΔG is given by

$$\Delta G = N_a \int_0^\pi A_{ex} d\pi \tag{4}$$

where N_a is the Avogadro number and π is the surface pressure. The negative or positive values of ΔG indicate stable or unstable mixed monolayer system respectively.

2.1.2 Epifluorescence microscopy

In 1981, Tscharner and McConnell Tscharner & McConnell (1981) have developed the epifluorescence microscopy technique to visualize a monomolecular layer at the A-W interface. The schematic diagram of the experimental setup for the epifluorescence microscopy of the Langmuir monolayer is shown in Figure 3. Here, the monolayer (S) was doped with very small quantity (≤ 1 mole%) of an amphiphilic fluorescent dye molecule (D). The dye doped monolayer was observed under an epifluorescence microscope (Leitz Metallux 3). The microscope was equipped with a high pressure mercury lamp (H) and a filter block (F). The filter block consists of an emission filter (Em), an excitation filter (Ex) and a dichroic mirror (DM). The excitation filter allows the light of appropriate wavelength to excite the dye molecules in the monolayer. The reflected light from the interface and emitted light from the dye doped monolayer was allowed to pass through the emission filter (Em) which allows only the emitted light. The emitted light was collected using an intensified charge coupled device camera (ICCD) and the images were digitized using a frame grabber (National Instruments, PCI-1411). The intensity of the emitted light depends on the miscibility of the dye molecules in a particular phase of the monolayer. The gas phase appears dark due to the quenching of the dye molecules. The liquid expanded phase appears bright in the epifluorescence images. On the other hand, the solid phase appears dark. This is due to the expulsion of dye molecules from the highly dense solid domains. In all the epifluorescence microscopy experiments, we have utilized 4-(hexadecylamino)-7-nitrobenz-2 oxa-1,3 diazole (obtained from Molecular Probes) as an amphiphilic fluorescent molecule.

2.1.3 Brewster angle microscopy

The epifluorescence microscopy on the Langmuir monolayer has certain disadvantages. For instance, the fluorescent dye acts as an impurity and may alter the phase diagram. There are difficulties in determining the surface phases at the higher surface pressures where the dye molecules are practically insoluble in the domains. Also the photo-bleaching can result in the decomposition of the dye molecules. To overcome such disadvantages, a microscope which works on the principle of Brewster angle, was developed. The microscope is known as the Brewster angle microscope Hönig & Möbius (1991); Hénon & Meunier (1991). The angle of incidence at which an unpolarized light acquires a linearly polarized state after reflection from the plane of an interface is known as Brewster angle of the reflecting material. The state of polarization of the reflected light at the Brewster angle (θ_B) is perpendicular (s-polarized) to the plane of incidence. At Brewster angle of incidence

$$tan(\theta_B) = n_2/n_1 \tag{5}$$

(a)

(b)

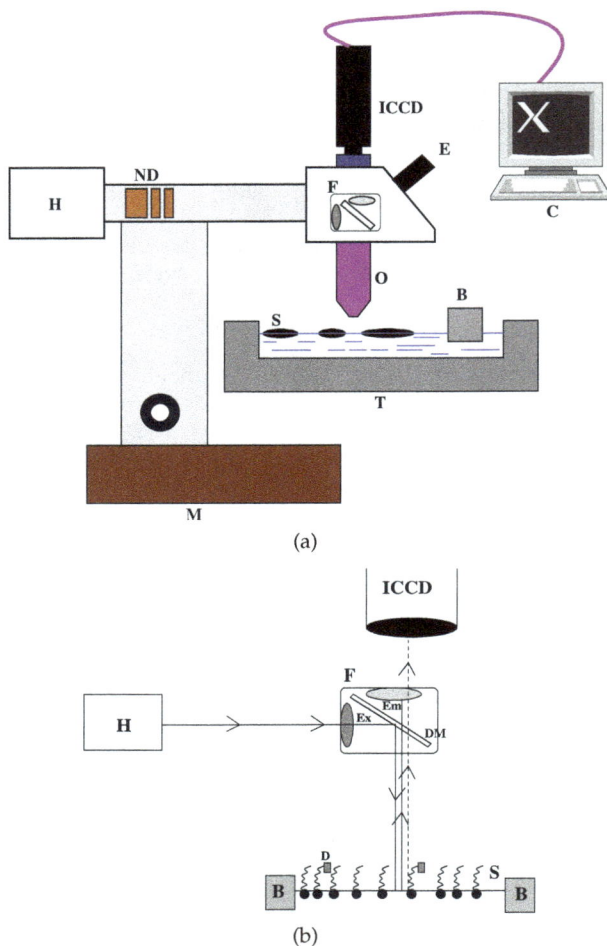

Fig. 3. Schematic diagram showing the experimental setup for the epifluorescence microscopy. (a) shows a complete microscope setup. (b) shows the working principle of the microscope. The different parts are as follows: high pressure mercury lamp (H), neutral density filter (ND), filter block (F) with emission (Em) and excitation (Ex) filters and a dichroic mirror (DM), objective (O), eye-piece (E), microscope stand (M), intensified charge coupled device (ICCD) camera, computer (C), trough (T), dye doped monolayer (S), dye molecule (D) and barrier (B)

where n_2 is the refractive index of the reflecting material and n_1 is the refractive index of the medium through which the light is incident. We have utilized a commercial setup, MiniBAM Plus from Nanofilm Technologie for BAM imaging. In the microscope, a polarized light source from a 30 mW laser of wavelength 660 nm falls on the water surface at the Brewster angle ($\sim 53°$). The reflected light is allowed to pass through a polarizer which allows only the p-component of the reflected light to enter a CCD camera, as shown in Figure 4. Since the

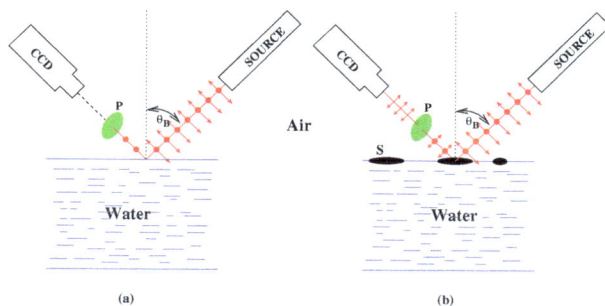

Fig. 4. Schematic diagrams showing the working principle of a Brewster angle microscope. θ_B is the Brewster angle of water with respect to air. The different parts are as follows: polarizer (P), charge coupled device (CCD) camera and monolayer at the air-water interface (S). (a) and (b) are the BAM setups without and with monolayer at the A-W interface, respectively

angle was set for the Brewster angle of water, the reflected intensity in the CCD camera was minimum for pure water. Therefore, any domain of the monolayer changes the refractive index at the interface and hence the Brewster angle for A-W interface gets altered. This in turn reflects some light which was collected by the CCD to form the images of the monolayer domains. The intensity of the reflected light depends on the thickness of the film and the surface density of the molecules. The optical anisotropy in the BAM images arises due to a difference in tilt-azimuth variation of the molecules in the monolayer and the anisotropy in the unit cell Rivière et al. (1994).

2.2 Monolayer and multilayer at air-solid interface

Langmuir-Blodgett (LB) technique can be employed to form monolayer or multilayer films at the solid-air interface. In this technique, the monolayer in a particular phase at the A-W interface can be transferred layer by layer onto a solid substrate by vertically moving the substrate in and out of the subphase. During deposition, the surface pressure is fixed at some target value known as target surface pressure (π_t). The experimental setup for LB deposition is shown in Figure 5. The setup is similar to that of the Langmuir trough except it possesses a well in the teflon trough and a dipper. For the LB film transfer, the trough has a motorized dipper which holds the substrate and it can be moved up and down very precisely. Such a motion makes the substrate to dip in and out of the subphase with a monolayer at the interface. During deposition, the surface pressure is fixed at π_t by a feedback mechanism. The barriers compress the monolayer until the π_t is attained. Any increase or decrease in surface pressure is fed back to the computer which in turn moves the barrier to maintain the required target surface pressure. The mechanism of building multilayer by LB technique is shown in Figure 6(a). The efficiency of transfer of the monolayer on the solid substrate is estimated by measuring the transfer ratio (τ). The transfer ratio (τ) is defined as Roberts (1990)

$$\tau = \frac{\text{area of monolayer transferred from the A} - \text{W interface}}{\text{area of the substrate to be deposited}} \quad (6)$$

The value of τ equal to one indicates defectless LB film. There are different types of LB deposition. If the monolayer transfers during both the upstroke and downstroke of the dipper, such deposition is known as Y-type of LB deposition. On the other hand, if the deposition

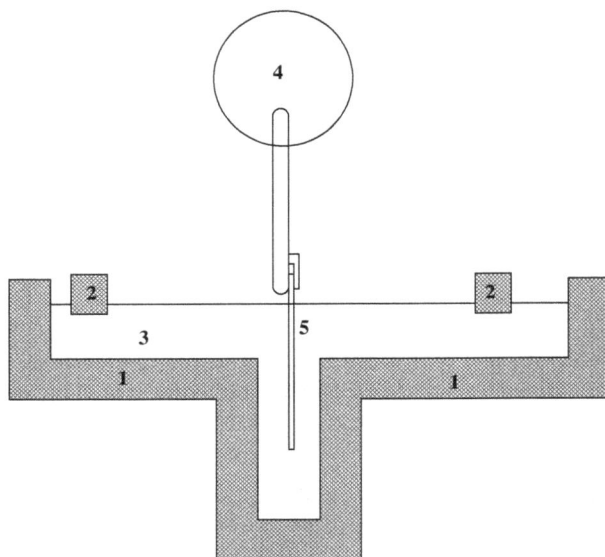

Fig. 5. A schematic diagram showing the experimental setup for forming Langmuir-Blodgett films. The parts are as follows: (1) teflon trough with a well in the center, (2) barriers, (3) subphase, (4) dipper and (5) substrate

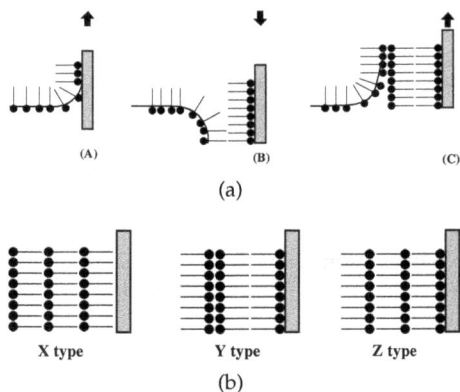

Fig. 6. (a) depicts the mechanism of formation of multilayer by the Langmuir-Blodgett technique. (A) first upstroke, (B) first downstroke and (C) second upstroke. (b) shows the structure of different types of multilayer obtained by LB technique, namely X-, Y- and Z-type

takes place only with either downstrokes or upstrokes, they are termed as X or Z-type of LB deposition, respectively (Figure 6(b)). Sometimes a combination of these depositions are also observed Roberts (1990). The negative values of τ indicate desorption. The LB depositions are dependent on the nature of interaction between the substrate and the molecules, dipper speed, target surface pressure, ion contents of the subphase and temperature. We have employed a

Langmuir-Blodgett trough procured commercially from NIMA (model 611M) for both surface manometry studies and LB film deposition.

The monolayer at the A-W interface can be transferred onto substrates by two other different techniques. These are called horizontal transfer method and Schaefer's method. In the horizontal transfer method, the substrate is immersed horizontally in the subphase and then the monolayer is formed at the A-W interface. Then the aqueous subphase is siphoned out very slowly from the other side of the barriers. The monolayer gets adsorbed onto the substrate as the water drains out. In the Schaefer's method, a hydrophobic substrate is allowed to touch the monolayer on the water surface. The hydrophobic part of the molecules get adsorbed to the substrate. To facilitate the drainage of water, the substrates in either case can be tilted by a small angle prior to the adsorption.

2.2.1 Scanning probe microscopy

The discovery of scanning probe microscopes by Binning and coworkers has revolutionized the field of nanoscience and nanotechnology. Its versatile range of applications have made it an indispensable tool in the fields of surface science and other branches of soft condensed matter. It is useful in the study of the surface topography, electronic properties of the film, film growth, adhesion, friction, lubrication, dielectric and magnetic properties. It has also been used in a molecular or atomic manipulation. Among the various scanning probe microscopes, scanning tunneling Binning et al. (1982) and atomic force microscopes Binning et al. (1986) are widely used to study nanostructures in thin films.

The schematic diagram of the STM is shown in Figure 7. In scanning tunneling microscope, a very sharp metallic tip (T) is brought very close to a conducting substrate (S). The substrate may be coated with a film (F). When a low bias voltage ($\sim$1 V) is applied between the tip and the substrate, a tunneling current flows between them. STM works in two modes – constant current (CC) mode and constant height (CH) mode. While scanning the surface in CC mode, the tunneling current between the tip and the sample is kept constant. In this mode, the height of the tip is adjusted automatically using a feedback circuit to achieve the constant tunneling current. The variation in tip height (z) as a function of lateral coordinates (x,y) gives the topographic information of the sample. In CH mode, the height of the tip is kept constant and the tunneling current is recorded as a function of (x,y). The CH mode provides the electronic information of the sample. Atomic force microscope (AFM) gives the topographic images by sensing the atomic forces between a sharp tip and the sample. A schematic diagram of an AFM is shown in Figure 8. Here, the tip is mounted on a cantilever. The head of the tip is coated with a reflecting material like gold and it is illuminated by a laser light. The reflected light is collected on a quadrant photodiode. Any deflection in the tip due to its interaction with the sample is monitored by measuring a distribution of light intensity in the photodiode. There are numerous modes of operation of AFM. In contact mode, the tip is brought into direct contact with sample and the surface is scanned. It is also known as constant force mode. Here, the force between the tip and the sample is kept constant and by monitoring the bending of the cantilever as a function of (x,y), a topographic image can be obtained. In tapping mode, the tip is allowed to oscillate nearly to its resonant frequency on the sample at a given amplitude and frequency. Due to interaction between the tip and the sample, the amplitude and phase of oscillation of the tip change. The change in amplitude can be used to obtain a topographic map of the sample. The phase change gives an insight about the chemical nature of the sample.

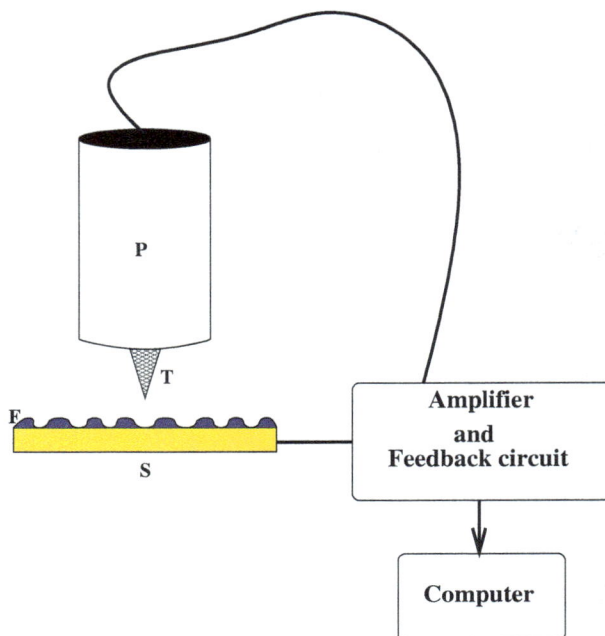

Fig. 7. Schematic diagram of a scanning tunneling microscope (STM). The parts are as follows: conducting substrate (S), film (F), metallic tip (T) and piezo tube (P)

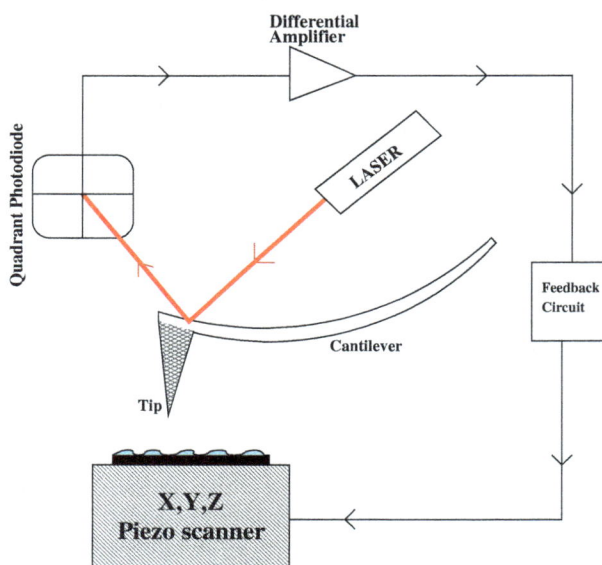

Fig. 8. Schematic diagram of an atomic force microscope (AFM)

3. Molecular interaction at air-water interface

An amphiphilic molecule has two parts : 1) hydrophilic headgroups like acid (-COOH), alcohol (-OH), amine(-NH$_2$) or ketone and 2) hydrophobic tail part like saturated or unsaturated aliphatic chain, and cyclic rings. The stability of the Langmuir monolayer at A-W interface essentially depends on the intermolecular and molecular - subphase interactions. In a stable Langmuir monolayer, the molecules get adsorbed to an A-W interface thereby reduces the surface energy. The energy at the air-water interface is reduced as the polar headgroups can form hydrogen bonds easily with the water molecules. The strength of molecular anchoring to the water surface depends on the nature of molecular interaction at the interface. The polar headgroups can interact with each other by coulombic, dipolar and hydrophobic interactions.

The stability of a monolayer at A-W interface depends on the nature of the aqueous subphase and the amphiphilic balance of the molecules. The Langmuir monolayer fatty acid are known to exhibit several phases. The isotherms of octadecanoic acid ($C_{17}H_{35} - COOH$), octadecanol ($C_{18}H_{37} - OH$), octadecanthiol ($C_{18}H_{37} - SH$), and octadecylamine ($C_{18}H_{37} - NH_2$) are shown in Figure 9. All the isotherms were obtained under the similar experimental conditions. In the present case, the molecules differs only in its hydrophilic headgroups. The polarity

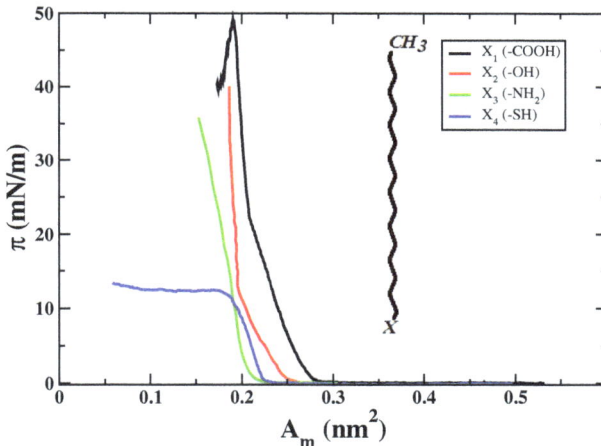

Fig. 9. Isotherms of Langmuir monolayer of octadecanol, octadecanoic acid, octadecylamine, and octadecanethiol

of the headgroups of these molecules can be arranged in decreasing order as : $-OH \rightarrow -COOH \rightarrow -NH_2 \rightarrow -SH$. The significant effect of polarity of the headgroups can be observed from the isotherms. The monolayer of the molecules with highly polar headgroup is more compressible before it collapses as compared to the molecules with lower polarity in the headgroups. The variation of isothermal inplane elastic modulus (E) as a function of A_m of above molecules are shown in Figure 10. The maximum values of E in the condensed state for X being -OH, -COOH, -NH$_2$ and -SH were found to be 475, 380, 175 and 80 mN/m respectively. This indicates that due to increase in polarity of headgroups of the amphiphilic molecules at the A-W interface, the condensed phase exhibits a higher degree of crystallinity.

The hydrophobic interaction between the molecules also plays a significant role in stabilizing the Langmuir monolayer. Such hydrophobic interaction arises due to an overall increase

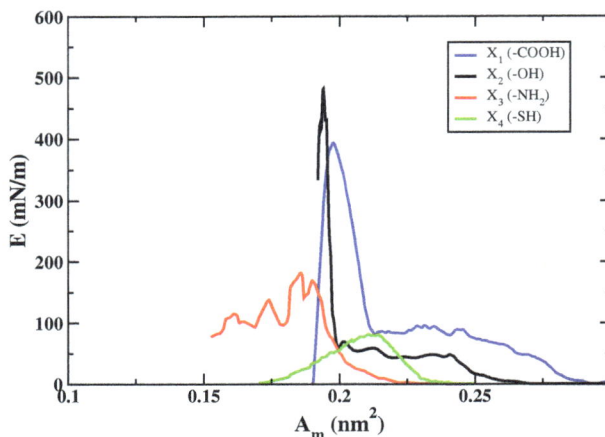

Fig. 10. Inplane elastic modulus E Vs A_m of octadecanol, octadecanoic acid, octadecylamine, and octadecanethiol.

in entropy of the system leading to an stabilization effect on the monolayer. It has been observed that monolayer of even pure hydrophobic materials can be stabilized provided the overall entropy increases. In some special cases, purely hydrophobic materials are also known to form a stable Langmuir monolayer Tabe et al. (2002). For instance, semifluorinated alkanes being purely hydrophobic molecules, show a stable Langmuir monolayer at the A-W interface Gaines (1991); Abed et al. (2002). The formation of such monolayers are attributed to the hydrophobic interaction between the molecules and the increase in overall entropy of the system. There is also a report on the formation of a non-traditional Langmuir monolayer of disubstituted urea lipid molecule where the hydrophilic part stays away from the water surface, whereas the hydrophobic part stays near to the water surface. The stabilization of such monolayer was attributed to the hydrogen bonding network of the urea moiety of the molecule Huo et al. (2000).

In an interesting work, we tried to stabilize hydrophobic analogous of cholesterol (Ch) molecules known as thiocholesterol (TCh) at the A-W interface. The stability of the TCh monolayer is improved by incorporating them in Ch monolayer matrix. Thiocholesterol is predominantly a hydrophobic molecule which can be obtained by the substitution of -OH group with -SH group in the cholesterol molecule. The -SH group is weakly acidic, and the amphipilicity of the TCh is not sufficient enough to form a stable Langmuir monolayer. There is a report on the formation of defect rich self assembled monolayer (SAM) of TCh on gold substrate (yang et al., 1996). The size of the defects were found to be of the order of 5-8 Å. Such defect rich SAM can be utilized for the fabrication of ultramicroelectodes selective permeation devices. It can also be used for electroanalytical and biosensors applications. However, the limitation of the formation of SAM of such organosulfur compounds is the substrate which has to be of coinage metals like gold, silver and copper. Another method of obtaining monolayer and multilayer on different types of substrate is LB technique. Hence, the formation of a stable Langmuir monolayer is important for the formation of controlled and organized LB films. It is well known that the Ch molecules form a stable monolayer at the A-W interface Slotte & Mattjus (1995); Lafont et al. (1998). Though -SH group is weakly acidic in nature Bilewicz & Majda (1991), the amphiphilic balance of the TCh for the formation of a stable insoluble monolayer at the A-W interface is not sufficient. The TCh molecule is mostly hydrophobic in nature and does not spread to form a monolayer at the A-W interface. Since the hydrophobic skeletons of the TCh and Ch molecules are identical, TCh molecules can be mixed to the Ch monolayer and can be transferred on to different substrates by LB technique. There are numerous reports in literature indicating the formation of mixed monolayer of a non-amphiphilic component doped into the monolayer of amphiphilic molecules (Silva et al., 1996; Wang et al., 2000; Viswanath & Suresh, 2004). In all these cases, the monolayer has been stabilized upto certain value of surface pressure due to hydrophobic interaction of the tail groups of the different components. Further compression leads to the squeezing out of the non-amphiphilic components. In this work we reported that the mixed monolayer to be stable upto 0.75 mole fraction of TCh in the Ch monolayer. Our analysis of the excess area per molecule for the mixed monolayer system suggests that the monolayer is stabilized by attractive interaction between the Ch and TCh molecules.

The surface pressure (π) - area per molecule (A_m) isotherms for the different mole fractions of TCh in the mixed monolayer of Ch and TCh (X_{TCh}) are shown in Figure 11. The isotherm of

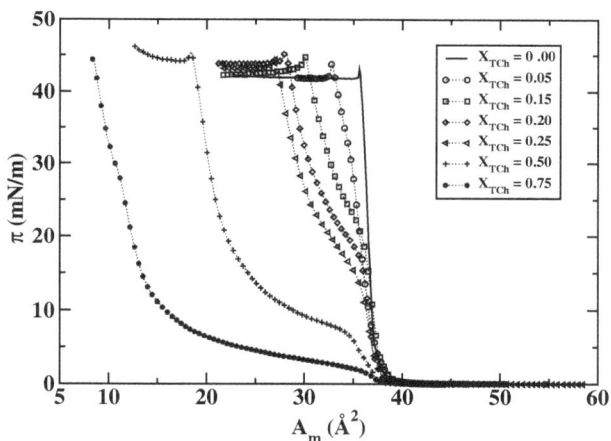

Fig. 11. The surface pressure (π) - area per molecule (A_m) isotherms of the mixed monolayer for different mole fractions of thiocholesterol in cholesterol (X_{TCh}). Reprinted with permission. (Gupta & Suresh, 2008)

pure cholesterol indicates the coexistence of a gas and an untilted condensed (L_2) phase at a large A_m Lafont et al. (1998); Viswanath & Suresh (2003). At an A_m of 39 Å^2, there is a steep rise in the surface pressure indicating a transition from the coexistence of gas and L_2 phases to the L_2 phase. The limiting area per molecule (A_o) is equal to 38 Å^2. This value approximately corresponds to the cross-sectional area of the Ch molecule for its normal orientation at the A-W interface. The monolayer collapses at an A_m of 37 Å^2 with a collapse pressure of 43 mN/m. The Ch monolayer shows a spike-like collapse and a plateau thereafter. The TCh molecules are predominantly hydrophobic in nature, and they do not spread at the A-W interface to form a stable Langmuir monolayer. We have attempted to form the TCh monolayer on different subphases obtained by changing the pH and adding salts in the ultrapure ion-free water. However, we were not able to form a stable TCh monolayer over such aqueous subphases. We have mixed the TCh and Ch molecules at different proportions, and formed the monolayer at the A-W interface. The isotherms of the mixed monolayer (Figure 11) show a sharp rise in surface pressure at around 39 Å². However, the presence of TCh in Ch monolayer changes the nature of the isotherm by introducing an additional change in slope of the isotherms. Such change in slope can be considered as an initial collapse of the two-component monolayer system. The isotherms also show a final collapse. The final collapse of the mixed monolayer reveals spike-like feature followed by a plateau. This is characteristic of the collapse for the cholesterol monolayer. Hence, the final collapse indicates a collapse of the Ch rich monolayer. The A_o values for the mixed monolayer were found to be nearly invariant with X_{TCh}. It lies in the range of 37.5 - 38.5 Å². The values of A_o for various X_{TCh} suggest a normal orientation of the molecules in the phase corresponding to the steep region of the isotherms (Figure 11). This phase may represent untilted condensed (L_2) phase.

For an ideal case of the two-component system of non-interacting molecules, the area per molecule of the mixed monolayer is a given in Equation 2. If one of the components is a non-amphiphilic molecule (for instance component 2), then the Equation 2 can be modified as

$$A_{id} = X_1 A_1 = (1 - X_2)A_1 \tag{7}$$

The variation of excess area per molecule (A_{ex}) as a function of mole fraction of TCh in Ch for different surface pressures are shown in Figure 12. The values of A_{ex} are negative for all the compositions. The negative values of A_{ex} suggest an attractive interaction between the Ch and TCh molecules in the mixed monolayer. Such attractive interaction leads to a stabilization effect on the mixed monolayer of Ch and TCh.

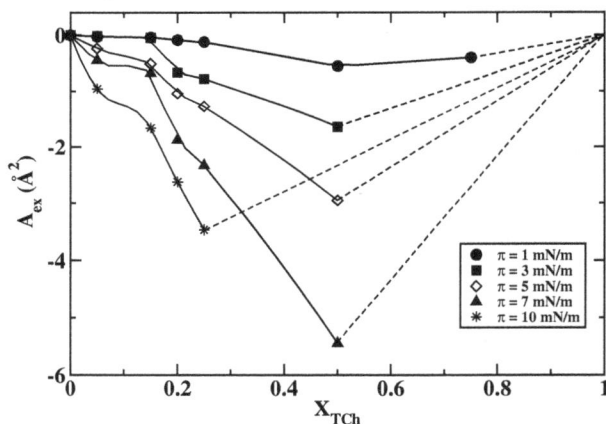

Fig. 12. Variation of excess area per molecule (A_{ex}) with respect to the mole fraction of TCh in Ch (X_{TCh}) at different surface pressures (π). The points are computed using Equation 3. The dashed lines are extrapolated to the zero value of the excess area for the pure TCh. Reprinted with permission. (Gupta & Suresh, 2008)

There are many reports on pure hydrophobic materials forming a stable monolayer at the A-W interface Tabe et al. (2002); Li et al. (1994). Tabe et al Tabe et al. (2002) have reported the formation of condensed monolayer of pure hydrophobic mesogenic molecules at the A-W interface. They attributed the stability of the monolayer to the gain in entropy on adsorption of the molecules at the interface. Li et al Li et al. (1994), have suggested that the stability of purely hydrophobic materials like a long chain alkane can be due to the van der Waals force. There is another report which indicated the stability of the monolayers of alkanes at the A-W interface.

This has been attributed to the gain in entropy due to strong fluctuations of the alkanes on a plane normal to interface as compared to their states in the bulk solid Tkachenko & Rabin (1996). Further, there are possibilities for the gain in entropy due to the rearrangement of the interfacial water molecules, and the fluctuation of the molecules as compared to their state in the bulk solids. Jensen *et al.* Jensen et al. (2003; 2004), have shown experimentally and also by simulation that the orientation of interfacial water molecules changes differently due to the presence of hydrophobic or hydrophilic molecule at the interface. The dipole moment associated with the water molecule faces on an average towards the hydrophobic alkane. However, for the hydrophilic monolayer covered interface, it faces away from the interface. In the case of mixed monolayer of Ch and TCh, we find an attractive interaction between the molecules which may arise due to the overall gain in entropy of the mixed monolayer system or due to van der Waals interaction between the molecules. In Ch - TCh mixed monolayer system, the nature of polarity of the head groups of the two mixed components are different. Hence, we can assume that each component molecules orient the interfacial water molecules differently. This will lead to a frustration in orientation of the water molecules at the interface which reduces the ordering of the interfacial water dipole moments. This increases the entropy of the system which in turn may help in stabilizing the mixed monolayer. Hence, we can assume the entropy gain in our system can be due to the reorientation of the water molecules at the interface and the fluctuations of the iso-octyl chain of the Ch and TCh molecules. On reducing the intermolecular distances by compressing the monolayer, the steric repulsion among the molecules becomes strong enough to overcome the van der Waals attraction and the entropy gain which may lead the non-amphiphilic component (*i.e.* TCh) to squeeze out of the mixed monolayer.

Because of the potential industrial applications, the field of nanoscience and nanotechnology is growing enormously. One among the major challenges involved in this area is assembly of nanomaterials such that it can exhibit extraordinary physical properties. One of the important parameters determining the properties of the nanoparticle crystals is the interparticle separation. Other parameters are charging energy of the particles, strength of interaction between the particles and the symmetry of the lattice formed by the particles Israelachvili (1992); Zhang & Sham (2003). The interparticle separation can precisely be maneuvered by forming the Langmuir monolayer of the nanoparticles at an interface and changing the surface density. Interestingly, Collier and coworkers have reported a metal-insulator transition in the Langmuir monolayer of hydrophobic silver nanoparticles at the A-W interface Collier et al. (1997); Markovich et al. (1998). They observed a quantum interference between the particles which was governed by interparticle distances. There are few studies on Langmuir monolayer of functionalized metal nanoparticles at the A-W interface Heath et al. (1997); Swami et al. (2003); Greene et al. (2003); Brown et al. (2001); Fukuto et al. (2004). However, the particles studied so far did not show a stable Langmuir monolayer. The important criterion for the formation of stable Langmuir monolayer at the A-W interface is that the particles should be amphiphilic in nature with a proper balance between its hydrophilic and hydrophobic parts Gaines (1966). Accordingly, we synthesized amphiphilic gold nanoparticles (AGN) functionalized with hydroxy terminated alkyl-thiol, and studied the Langmuir film of the particles by surface manometry and microscopy techniques. We find a stable Langmuir monolayer of the AGN at A-W interface. The monolayer exhibits gas, low ordered liquid (L_1), high ordered liquid (L_2), bilayer of L_2 (Bi) and the collapsed states.

The amphiphilic functionalized gold nanoparticles (AGN) were synthesized in the laboratory. The size of the particles was estimated from transmission electron microscope (TEM) images (Figure 13). The particles had a mean core diameter of 5.5 nm with a standard deviation of

± 1 nm. The diameter of the AGN for the fully stretched ligands was calculated as 8.4 nm. Hence, the mean cross-sectional area of each particle thus calculated was 55.4 nm^2.

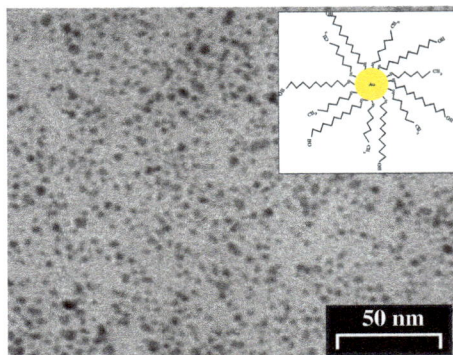

Fig. 13. Transmission electron microscope (Hitachi, H7000, 100 kV) image of the AGN. The solution of the sample of AGN was spread on carbon evaporated copper grid, and the image was scanned after 2 hours. The inset shows structure of an AGN

The surface pressure (π) - area per particle (A_P) isotherms of AGN at different temperatures (Figure 14) show zero surface pressure at very large A_P indicating a coexistence of gas and

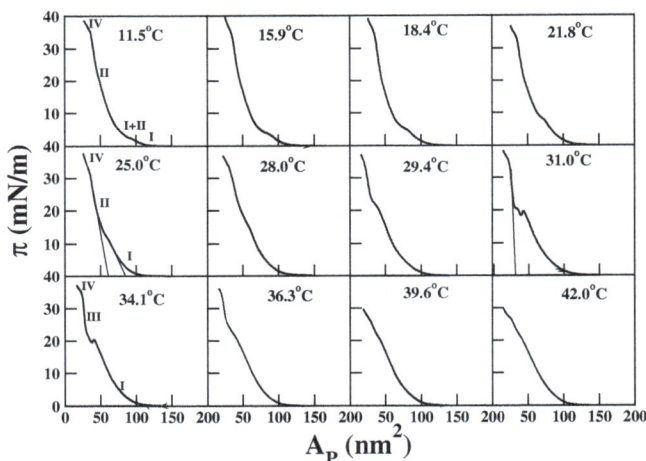

Fig. 14. Surface pressure (π) - area per particle (A_P) isotherms of AGN at different temperatures. The symbols **I, II, III and IV** denote L$_1$, L$_2$, Bi and collapsed regions in the isotherms, respectively. The straight lines in the isotherms for temperatures 25 and 31 °C are shown as the extrapolated lines for calculating average area occupied by each particle in the respective phases. The error incurred in such calculation is $\pm 1\%$. Reprinted with permission. (Gupta & Suresh, 2008)

a low density liquid (L$_1$) phase. On reducing A_P (monolayer compression), the gas phase disappears, and the isotherms show lift-off area per particle to be around 120 nm^2. On compressing the monolayer, the isotherms show a slow and gradual rise in surface pressure. This is the pure L_1 phase. The extrapolation of the region of the isotherm to the zero surface pressure on the A_P axis yields the average area occupied by the particles in that particular phase Gaines (1966). The average particle-area in the L_1 phase at 25 °C was 85 nm^2. On compressing the monolayer below 28 °C, the slow and gradual rise in surface pressure is accompanied by a plateau in the isotherms. On further compression, the surface pressure rises sharply till the monolayer collapses at A_P ∼ 36 nm^2. The region of the isotherm with the sharp rise in surface pressure can correspond to a high density liquid (L_2) phase. At 25 °C, the average particle-area in the L_2 phase was around 61 nm^2. This value nearly corresponds to the mean cross-sectional area of the particles determined through TEM images. The small difference in the value may be due to the poly-dispersity in the size-distribution of the particles. The appearance of plateau in the isotherms suggests a first order phase transition between L_1 and L_2 phases. The coexistence of two phases is denoted as **I+II** in the isotherms. The phase coexisting plateau region decreases with increasing T and finally vanishes above critical temperature (T_c). The T_c was calculated by plotting the enthalpy of the L_1-L_2 transition with respect to T, and extrapolating it to a zero value on the T−axis Callen (1985). We obtain a T_c value of 28.4 °C for this transition. These behaviors have been observed and understood in the case of the standard amphiphilic molecules like phospholipids Mohwald (1995); Birdi (1989); Baoukina et al. (2007). The isotherms indicate a smooth collapse of the monolayer. This may be due to the isotropic shape of the particles. Such smooth collapses were also observed in the case of spherical core functionalised fullerene derivatives Leo et al. (2000); gallani et al. (2002). The reversibility of the isotherms was studied by successive compression and expansion of the monolayer at 25 °C. We find the monolayer to be highly reversible. The stability of the phases were also tested qualitatively by holding the surface pressure at a given value and monitoring the drop in A_P. We find almost no variation in A_P over a long period of time. These observations suggest the monolayer of AGN at the A-W interface to be very stable.

The π-A_P isotherms obtained in the range 28.4≤ T ≤ 36.3 °C show interesting behaviors. The isotherms show the usual L_1 phase. On reducing the A_P, the L_1 phase appears collapsing by introducing an unstable region in the isotherm. However, on further reducing A_P, the isotherms indicate a steep rise in surface pressure and then smooth collapse like feature. Extrapolating the steep region to the zero surface pressure yields a value of ∼ 31 nm^2 at 31 °C. This value is nearly twice than that obtained for the L_2 phase (61 nm^2). Hence, this phase can presumed to be a bilayer (Bi) of L_2 phase. The unstable region of isotherms leading to the transition from L_1 to Bi phase can be considered as a coexistence region. The coexistence region initially increases upto 31 °C but decreases thereafter until it vanishes above 36.3 °C. The Bi phase also disappears above 36.3 °C. Thereafter, the monolayer showed gas, L_1 and the collapsed states.

In Brewster angle microscopy, the intensity levels in images depend on surface density and thickness of the films Rivière et al. (1994). The BAM image (Figure 15(a)) at a large A_P shows a coexistence of dark and gray region. The dark region vanishes on compression leading to a uniform gray texture (Figure 15(b)). The gray domains appeared fluidic and mobile under the microscope. The dark region represents gas phase whereas the gray domains correspond to the low density liquid (L_1) phase. The image (Figure 15(c)) corresponding to the plateau region of the isotherms shows bright domains in the gray background. On compression, the bright domains grow at the expense of the gray region leading to a uniform bright texture (Figure 15(d)). The uniform bright texture, obtained in the region **II** of the isotherm, corresponds to the high density liquid (L_2) phase. The coexistence of L_1 and L_2 domains in the

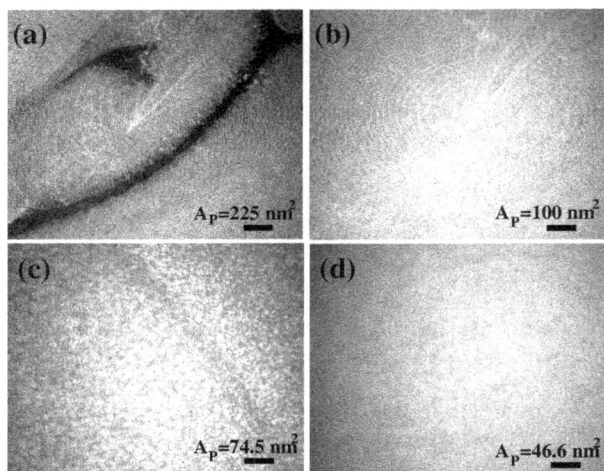

Fig. 15. The BAM images of the monolayer of AGN at 18.4 °C taken at different A_P. (a) a coexistence of dark and gray regions, (b) a uniform gray region, (c) a coexistence of bright domains on the gray background and (d) a uniform bright region. Scale bar = 500 μm. Reprinted with permission. (Gupta & Suresh, 2008)

plateau region of the isotherm establishes a first order phase transition between the phases. The BAM images in the collapsed state showed a wrinkled structure. The wrinkled structure might appear due to folding of the AGN monolayer Lu et al. (2002); Gopal & Lee (2001).

The BAM images of the monolayer of AGN at 31 °C showed the gas (dark region) and L_1 phases (gray region) at large A_P. The gas phase disappears on compressing the monolayer leading to a uniform L_1 phase. The L_1 phase continues to exist till monolayer destabilizes at $A_P \sim 44$ nm^2. Such destabilization leads to the nucleation and growth (Figure 16) of the very bright domains on the gray background. The bright domains grow at the expense of the gray region, finally leading to a uniform bright region. The bright region was obtained in the region of the isotherm corresponding to steep rise in the surface pressure after the destabilization. The bright region corresponds to the bilayer of the L_2 (Bi) phase. The bright region (Bi phase) also collapses on compression leading to a wrinkled structure.

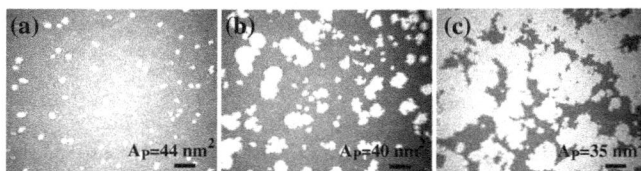

Fig. 16. The BAM images of the monolayer of AGN at 31 °C taken at different A_P. (a)-(c) nucleation and growth of bright domains on the gray background. Scale bar = 500 μm. Reprinted with permission. (Gupta & Suresh, 2008)

Based on our studies, we construct (Figure 17) a phase diagram of the Langmuir monolayer of AGN at the A-W interface.

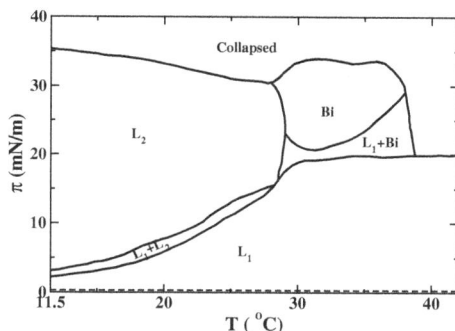

Fig. 17. The phase diagram showing the observed 2D phases of the Langmuir monolayer of AGN at A−W interface. The L_1 and gas phase coexistence was seen even at very large A_P and zero π. This is shown by the dotted line. The L_1, L_2, Bi are the low ordered, high ordered and the bilayer of L_2 phases, respectively. The coexistence of the phases are shown by "+" symbol. Reprinted with permission. (Gupta & Suresh, 2008)

We observed a stable Langmuir monolayer of AGN at the A-W interface. The monolayer exhibits variety of phases like gas, low order liquid (L_1), high ordered liquid (L_2), bilayer of L_2 (Bi) and the collapsed states. The electronic and optical properties of the phases on the different solid substrates will be probed in order to acquaint a better understanding about the structural dependence of the films on such properties.

In the above works, we have observed that an appropriate chemistry of the molecules yields variety of surface phases. Different functional groups in the molecules interact differently. By changing the amphiphilicity of molecules leads to the formation of a stable Langmuir monolayer.

4. Molecular interaction at air-solid interfaces

The variety of surface phases as obtained in the Langmuir films at an A-W interface can be transferred to air-solid (A-S) interface by Langmuir-Blodgett (LB) technique. Such LB films can potentially utilized for the fabrication of devices. The nature of aggregation of the molecules on the surface is important for the device applications. The surface can induce an ordering in the bulk, and hence the material properties can be varied by varying the structure of aggregates on the surface (Ruffieux et al., 2002; Friedlein et al., 2003). The structure of the aggregates on a surface primarily depends on intermolecular and molecule-substrate interactions. In a study we probed the effect of molecule-substrate interaction on the aggregation and nucleation of molecules on substrates which were deposited by LB techniques. We changed the nature of the substrates by chemically treating them to yield a hydrophilic or hydrophobic surfaces. We deposited the LB films of cholesterol on such substrates and studied their nucleation and aggregation using atomic force microscope. The hydrophilic and hydrophobic treatments to the substrates are discussed in details in the article (Gupta & Suresh, 2004). The LB films were transferred to these substrates at a target surface pressure of 30 mN/m. The Langmuir monolayer of Ch exhibits condensed phase at this pressure.

We find that the first layer gets deposited efficiently on both kind of substrates. However, we were not able to transfer more than one layer on hydrophilic substrates. This was due to the fact that for every upstroke and downstroke of the dipper, the molecules get adsorbed and desorbed by equal amounts. Ch molecule possesses a large hydrophobic sterol moiety. Hence, the interaction between the hydrophilic substrates and the Ch molecules are weak to support the multilayer formation during the LB deposition. On the other hand, the Ch molecules exhibited much better adhesion on hydrophobic glass substrates. During LB transfer on the hydrophobic substrates, we observed the X-type of deposition, where adsorption takes place only during the downstroke. The transfer ratio (τ) data is shown in Figure 18. In each cycle, the two bars represent the τ of one downstroke and one upstroke of deposition. The positive and negative transfer ratio data indicate the adsorption and desorption of the molecules, respectively. We find that the very first monolayer gets deposited efficiently where the τ was always 1 ± 0.02. Contrarily, for upstrokes there was some desorption. During each cycle of deposition, it is clear from the transfer ratio data that downstrokes adsorb the molecules efficiently while the upstrokes desorb fraction of the earlier deposited layers. The desorption increases with increasing number of cycles of deposition. This shows that though

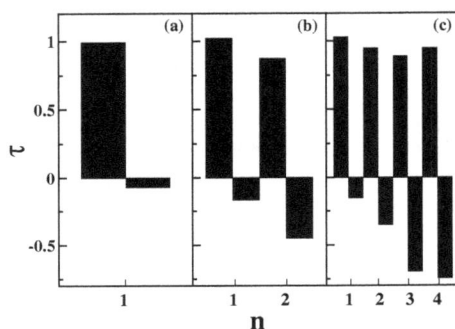

Fig. 18. The transfer ratio (τ) data as a function of number of cycles (n) of LB deposition of Ch on hydrophobic substrates. (a), (b) and (c) represent one, two and four cycles of LB deposition. Reprinted with kind permission of The European Physical Journal. (Gupta & Suresh, 2004)

the interaction between the hydrophobic substrate and the Ch molecule was strong, the layer to layer interaction was weaker to support multilayer formation. We define the effective adsorption on the substrate as $\eta_n = S_n/\rho_t$ where S_n is the net surface concentration on the substrate after n^{th} cycle and ρ_t is the surface concentration (inverse of A_m) of the monolayer at A-W interface at π_t. The variation of the effective adsorption with the number of cycles of deposition is shown in Figure 19. It is evident from the transfer ratio data (Figure 18) that the desorption becomes significant after the first cycle of deposition. Also, beyond four cycles of deposition, the desorption was almost equal to that of adsorption, resulting in the saturation of effective adsorption, η_n. Hence, it can be inferred that the interaction between the Ch-Ch layers is weaker as compared to that of the hydrophobic substrate-first layer of Ch. The adsorption of one layer of Ch on another layer during LB transfer can lead to defects due to weak layer to layer interaction. These defects can act as nucleation sites for further desorption of the molecules during upstrokes. With increasing cycles of deposition, the filling of the defects and reorganization of the molecules may yield patterns observed in AFM imaging. The AFM images for LB films of Ch on hydrophobic substrates deposited at different cycles show evolution of interesting patterns. The image for first cycle of deposition yields a very uniform texture indicating a uniform film on the surface. The AFM images for second

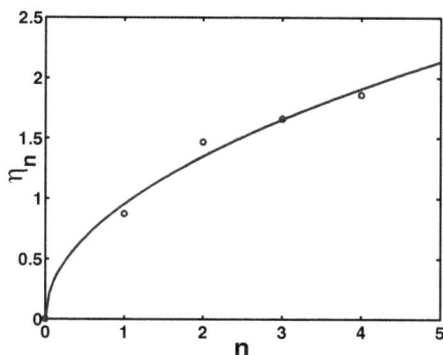

Fig. 19. The effective adsorption η_n is plotted with respect to number of cycles of deposition (n). The experimental data are represented by the open circles (o). The continuous line is a guide for the eye. Reprinted with kind permission of The European Physical Journal. (Gupta & Suresh, 2004)

cycle of deposition show elongated domains whereas for the three cycles of deposition, the AFM image shows partial torus shaped domains. Interestingly, the four cycles of deposition reveals uniformly distributed torus-shaped domains (doughnuts) having the outer diameter of about 65 nm and annular width of about 22 nm. The analysis of the phase images indicates the flipping of the Ch molecules. Thus the nature of molecule-substrate interaction play a dominant role not only for multilayer deposition but also the nature of nucleation and aggregation of the molecules on the surface.

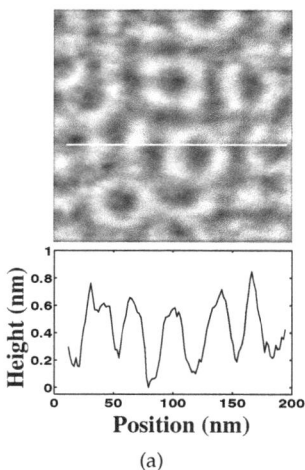

(a)

Fig. 20. AFM image of LB films of cholesterol (Ch) molecules on the hydrophobic substrate deposited during four cycles. The white line on the image is drawn to measure the height variation along it. The corresponding height profiles are shown below the respective images. The image size in each case is 200×200 nm^2. Reprinted with kind permission of The European Physical Journal. (Gupta & Suresh, 2004)

5. Conclusion

We have observed that the stability of Langmuir film can be achieved by changing the amphiphilic balance of the molecules appropriately. In a non-conventional approach, the Langmuir monolayer of hydrophobic molecules can be stabilized by incorporating them in a matrix of a stable monolayer. However, the choice of the stable Langmuir monolayer is such that the dopant mixes readily in the monolayer matrix. The shape and size of nanostructures at A-S interface can be controlled by exploring the possible molecule-substrate and intermolecular interactions.

6. Acknowledgement

We are thankful to University Grants Commission, India for its support under Special Assistance Programme.

7. References

Gaines, G. L. Jr. (1966). *Insoluble Monolayers at Liquid-Gas Interfaces*, Wiley-Interscience, New York.

Adamson A. W. (1990). *Physical Chemistry of Surfaces*, Wiley-Interscience, New York.

Mohwald, H., (1995). *Handbook of Biological Physics*, Lipowsky, R. & Sackmann, E., (Ed.), Elsevier Science, Amsterdam, Vol. 1, Chapter 4.

Smith, R. D. & Berg, J. C. (1980). The collapse of surfactant monolayers at the air-water interface. *J. Colloid Interface Sci.*, Vol. 74, 273-286.

Ramakrishnan, V.; Costa, M. D.; Ganesh, K. N. & Sastry, M. (2002). PNA-DNA hybridization at the air-water interface in the presence of octadecylamine Langmuir monolayers. *Langmuir*, Vol. 18, 6307-6311.

Goodrich, F. C. (1957). *Proceedings of the 2nd International Congress on Surface Activity*. Schulman J. H. (ed.), Butterworths, London, Vol. 1, 85.

Seoane, R.; Dynarowicz-tstka, P.; Miñones Jr., J. & Rey-Gòmez-Serranillos, I. (2001). Mixed Langmuir monolayers of cholesterol and essential fatty acids.*Colloid Polym. Sci.*, Vol. 279, 562-570.

Tscharner V. and McConnell, H. M. (1981). An alternative view of phospholipid phase behavior at the air-water interface. Microscope and film balance studies.*Biophys. J.*, Vol. 36, 409-419.

Hönig, D. & Möbius, D. (1991). Direct visualization of monolayers at the air-water interface by Brewster angle microscopy. *J. Phys. Chem.*, Vol. 95, 4590-4592.

Hénon, S. & Meunier, J. (1991). Microscope at the Brewster angle: Direct observation of first-order phase transitions in monolayers. *Rev. Sci. Instrum.*, Vol. 62, 936-939.

S. Rivière, S. Hénon, J. Meunier, D. K. Schwartz, M.-W. Tsao, and C. M. Knobler, C. M. (1994). Textures and phase transitions in Langmuir monolayers of fatty acids. A comparative Brewster angle microscope and polarized fluorescence microscope study. *J. Chem. Phys.*, Vol. 101, 10045-10051.

Roberts, G. (1990). *Langmuir-Blodgett Films*, Plenum, New York.

Binning, G.; Rohrer, H.; Geber, Ch. & Weibel, E. (1982). Surface studies by scanning tunneling microscopy. *Phys. Rev. Lett.*, Vol. 49, 57-61.

Binning, G.; Quate, C. F. & Gerber, Ch. (1986). Atomic force microscope. *Phys. Rev. Lett.*, Vol. 56, 930-933.

Tabe, Y; Yamamoto, T.; Nishiyama, I.; Aoki, K. M.; Yoneya, M. & Yokoyama, H.(2002). Can hydrophobic oils spread on water as condensed Langmuir monolayers? *J. Phys. Chem. B*, Vol. 106, 12089-12092.

Gaines, G. L. (1991). Surface activity of semifluorinated alkanes: F(CF2)m(CH2)nH. *Langmuir*, Vol. 7, 3054-3056.

Abed, A. El; Fauré, M-C.; Pouzet, E. & Abillon, O. (2002). Experimental evidence for an original two-dimensional phase structure: An antiparallel semifluorinated monolayer at the air-water interface. *Phys. Rev. E*, Vol. 65, 051603-4.

Huo, Q.; Russev, S.; Hasegawa, T.; Nishijo, J.; Umemura, J.; Puccetti, G.; Russell, K. C. & Leblanc, R. M. (2000) A Langmuir monolayer with a nontraditional molecular architecture. *J. Am. Chem. Soc.*, Vol. 122, 7890-7897.

Yang,Z. P.; Engquist, I.; Kauffmann,J. -M.; & Liedberg, B. (1996). Thiocholesterol on gold: A nanoporous molecular assembly . *Langmuir*, Vol. 12, 1704-1707.

Slotte, J. P. & Mattjus, P. (1995). Visualization of lateral phases in cholesterol and phosphatidylcholine monolayers at the air water interface - a comparative study with two different reporter molecules. *Biochim. Biophys. Acta*, Vol. 1254, 22-29.

Lafont, S.; Rapaport,H.; Sömjen, G. J.; Renault, A.; Howes, P.B.; Kjaer, K.; Als-Nielsen, J.; Leiserowitz, L. & Lahav, M. (1998). Monitoring the nucleation of crystalline films of cholesterol on water and in the presence of phospholipid. *J. Phys. Chem. B*, Vol. 102, 761-765.

Bilewicz, R. & Majda, M. (1991). Monomolecular Langmuir-Blodgett films at electrodes. Formation of passivating monolayers and incorporation of electroactive reagents, *Langmuir*, Vol. 7, 2794-2802.

Silva, A. M. G.; Guerreiro, J. C.; Rodrigues, N. G. & Rodrigues, T. O. (1996). Mixed monolayers of heptadecanoic acid with chlorohexadecane and bromohexadecane. Effects of temperature and of metal ions in the subphase,*Langmuir*, Vol. 12, 4442-4448.

Wang, H.; Ozaki, Y. & Iriyama, K. (2000). An infrared spectroscopy study on molecular orientation and structure in mixed Langmuir-Blodgett films of 2-octadecyl-7,7,8,8-tetracyanoquinodimethane and deuterated stearic acid: Phase separation and freezing-in effects of the fatty acid domains, *Langmuir*, Vol. 16, 5142-5147.

Viswanath, P. & Suresh, K. A. (2004). Photoinduced phase separation and miscibility in the condensed phase of a mixed Langmuir monolayer, *Langmuir*, Vol. 20, 8149-8154.

Viswanath, P. & Suresh, K. A. (2003). Polar head group interactions in mixed Langmuir monolayers, *Phys. Rev. E*, Vol. 67, 061604-8.

Gupta, R. K. & Suresh K. A. (2008). Stabilization of Langmuir monolayer of hydrophobic thiocholesterol molecules. *Colloids & Surfaces A:Physicochemical and Engineering Aspects*, Vol. 320, 233-239.

Li, M. Y.; Acero, A. A.; Huang, Z. & Rice, S. A. (1994). Formation of an ordered Langmuir monolayer by a non-polar chain molecule. *Nature*, Vol. 367, 151-153.

Tkachenko,A. V. & Rabin, Y. (1996). Fluctuation-stabilized surface freezing of chain molecules. *Phys. Rev. Lett.*, Vol. 76, 2527-2530.

Jensen, T. R.; Jensen, M. Ø.; Reitzel, N.; Balashev, K.; Peters, G. H.; Kjaer, K. & Bjørnholm, T. (2003). Water in contact with extended hydrophobic surfaces: Direct evidence of weak tweeting. *Phys. Rev. Lett.*, Vol. 90, 086101-4.

Jensen, M. Ø.; Mauritsen, O. G. & Peters, G. H. (2004). The hydrophobic effect: Molecular dynamics simulations of water confined between extended hydrophobic and hydrophilic surfaces. *J. Chem. Phys.*, Vol. 120, 9729-9744.

Israelachvili, J. N., *Intermolecular and Surface Forces : With Applications to Colloidal and Biological Systems* (Academic Press, London, 1992).

Zhang, P., & Sham, T. K. (2003). X-Ray studies of the structure and electronic behavior of alkanethiolate-capped gold nanoparticles: The interplay of size and surface effects. *Phys. Rev. Lett.*, Vol. 90, 245502-4.

Collier, C. P.;Saykally, R. J.; Shiang, J. J.; Henrichs, S. E. & Heath, J. R. (1997). Reversible tuning of silver quantum dot monolayers through the metal-insulator transition. *Science*, Vol. **277**, 1978-1981.

Markovich, G.; Collier, C. P. & Heath, J. R. (1998). Reversible metal-insulator transition in ordered metal nanocrystal monolayers observed by impedance spectroscopy. *Phys. Rev. Lett*, Vol. 80, 3807-3810.

Heath, J. R.; Knobler, C. M. & Leff, D. V. (1997). Pressure/Temperature phase diagrams and superlattices of organically functionalized metal nanocrystal monolayers: The influence of particle size, size distribution, and surface passivant. *J. Phys. Chem. B*, Vol. 101, 189-197.

Swami, A.; Kumar, A.; Selvakannan, P. R.;Mandal, S.; & Sastry, M. (2003). Langmuir-Blodgett films of laurylamine-modified hydrophobic gold nanoparticles organized at the air-water interface. *J. Col. Int. Sci.*, Vol. 260, 367-373.

Greene, I. A.; Wu, F.; Zhang, J. Z. & Chen, S. (2003). Electronic conductivity of semiconductor nanoparticle monolayers at the air-water interface. *J. Phys. Chem. B*, Vol. 107, 5733-5739.

Brown, J. J.; Porter, J. A.; Daghlian, C. P. & Gibson, U. J. (2001). Ordered arrays of amphiphilic gold nanoparticles in langmuir monolayers. *Langmuir*, Vol. 17, 7966-7969.

Fukuto, M.; Heilmann, R. K.; Pershan, P. S.; Badia, A. & Lennox, R. B. (2004). Monolayer/bilayer transition in Langmuir films of derivatized gold nanoparticles at the gas/water interface: An x-ray scattering study. *J. Chem. Phys.*, Vol. 120, 3446-3459.

Callen, H. B. (1985). *Thermodynamics and an Introduction to Thermostatistics*, John Wiley & Sons, New York.

Birdi, K. S. (1989). *Lipid and Biopolymer Monolayers at Liquid Interfaces*, Plenum, New York.

Baoukina, S.; Monticelli, L.; Marrink, S. J. & Tieleman, D. P. (2007). Pressure-area isotherm of a lipid monolayer from molecular dynamics simulations. *Langmuir*, Vol. 23, 12617-12623.

Leo, L.; Mele, G.; Rosso, G.; Valli, L.; Vasapollo, G.; Guldi, D. M. & Mascolo, G. (2000). Interfacial properties of substituted fulleropyrrolidines on the water surface. *Langmuir*, Vol. 16, 4599-4606.

Gallani, J.; Felder, D.; Guillon, D.; Heinrich, B. & Nierengarten, J. (2002). Micelle formation in Langmuir films of C_{60} derivatives. *Langmuir*, Vol. 18, 2908-2913.

Gupta, R. K. & Suresh, K. A. (2008). Monolayer of amphiphilic functionalized gold nanoparticles at an air-water interface. *Phys. Rev. E*, Vol. 78, 032601-4.

Lu, W.; Knobler, C. M.; Bruinsma, R. F.; Twardos, M. & Dennin, M. (2002). Folding Langmuir monolayer. *Phys. Rev. Lett.*, Vol. 89, 146107-4.

Gopal, A. & Lee, K. Y. C. (2001). Morphology and collapse transitions in binary phospholipid monolayers. *J. Phys. Chem. B*, Vol. 105, 10348-10354.

Ruffieux, P.; Gröning, O.; Bielmann, M.; Simpson, C.; Müllen, K.; Schlapbach, L. & Gröning, P. (2002). Supramolecular columns of hexabenzocoronenes on copper and gold (111) surfaces. *Phys. Rev. B*, Vol. 66, 073409-4.

Friedlein, R.; Crispin, X.; Simpson, C. D.; Watson, M. D.; Jäckel, F.; Osikowicz, W.; Marciniak, S.; de Jong, M. P.; Samori, P.; Jönsson, S. K. M.; Fahlman, M.; Müllen, K.; Rabe, J. P. & Salaneck, W. R. (2003). Electronic structure of highly ordered films of self-assembled graphitic nanocolumns. *Phys. Rev. B*, Vol. 68, 195414-7.

Gupta, R. K. & Suresh, K. A. (2004). AFM studies on Langmuir-Blodgett films of cholesterol. *Euro. Phys. J. E*, Vol. 14, 35-42.

5

Molecular Interactions in Natural and Synthetic Self-Assembly Systems

Liane Gloria Raluca Stan, Rodica Mirela Nita and Aurelia Meghea

University Politehnica of Bucharest,
Romania

1. Introduction

Tailoring nature is a continuous rediscovered paradigm of scientific community despite of the so called 'independence of the artificial creations' of the new age of technology. The last century discoveries in terms of chemistry science achievements could be easily projected into a design of a perfect elliptic model in continuing creation related to report to nature: an imaginary circle which never close to its natural model, even those new circles became asymptotic smaller and smaller to the imaginary point of model.

Reverences are need to Nature for its ability of using very limited (few) materials to build all variety of forms of lives. Natural materials are self-generated, hierarchical organized, multifunctional, nonlinear, composite, adaptive, self-repairing and biodegradable.

The effort made by chemistry science community during the last century in development of new technologies and materials is enormous, if we are counting the total number of scientific publications only for the last 25 years, but the integration of the new developed theoretical concepts and theories into the body of classical chemistry science is not always keeping the rhythm, and became difficult to find a classical concept and a unitary theory able to explain the new achievements of research results.

Molecular principles of new discovered bio-inspired or bio-hybrid materials and technologies are the new goals of science and our research goes deeply into its creative base: molecules, atoms and sub-atoms world.

Biomimetics or bioinspiration are concepts which have a big dependence of the specialization and discipline: electronic community is using these principles in learning the way biological systems are processing information; biomedical engineers consider the principle as a means of engineering tissues which are natural materials; material scientists view the concept as the only tools available in learning to synthesize materials under ambient conditions and with less impact on environment.

2. Silafins and deoxyuridine, the smart molecules which act as transcriptors in bio-silicas

Marine organisms, including sponges and diatoms, can produce solid silica structures with organized nanoscale morphologies, at ambient and neutral pH using a templated enzymatic mechanism initiated and conducted by a peptide class of substances – silicateins.

Naturally occurring, biomolecular machinery provides excellent platforms to assemble inorganic nanoparticles and two major players are DNA and proteins. The molecular recognition or specific interaction also exists between proteins and DNA, such as transcriptional factors and operator segments of DNA (Mock et al., 2008). The information in living systems is received, stored and transmitted by means of the specific interactions of biomolecules. The accuracy and precision of such interactions can be proved by any creature on the earth.

The ability of using the direct available elementary silicon from see water for building cell walls composed of amorphous, hydrated silicon dioxide (silica) embedded with a specific small organic molecules (Field et al., 1998; Nelson et al., 1995) is the main function of these organisms. The silica-based patterns of nano-sized pores and other cell wall structures of diatoms are so detailed and precisely replicated in more than 105 to 106 species with distinguish taxonomies. Silica patterning in diatoms has been hypothesized to depend on both self-assembly transcription processes and controlled silica polymerization (Brzezinski & Nelson, 1996) by cytoskeletal interactions (Leynaert et al., 2001).

The ability of diatoms to manipulate silicon at the nanoscale exceeds that of human nanotechnology, making the genetic and biochemical underpinnings of biosilicification of great interest in material science.

Marine sponges and diatoms produce silica made skeletons consisting of individualized elements (spicules) of lengths ranging from micrometers to centimeters, which can subsequently fuse or interlock one another. They differ from a skeletal point of view in the number of symmetry axes of their megascleres (ex. monaxons and tetraxons in demosponges and monaxons and triaxons in hexactinellids). The high diversity of spicule shapes and sizes in both fossil and living sponges has been repeatedly reported (Cha et al., 1999; Croce et al., 2004; Field et al., 1998; Fuhrmann et al., 2004; Krasko et al., 2004).

The process of biosilicification in diatoms begins with the active transport of dissolved silicic acid from the aqueous environment into the cell. This is mediated by silicon transporters, or SITS, whose expression and activity also show tight regulation during the cell cycle (Kröger et al., 1996, 1997). Intracellular concentrations of dissolved silicon exceed the solubility of silicic acid (Kröger et al., 2002), suggesting *the presence of an organic silicon complex* that may maintain silicon in solution and shepherd it to the silicon deposition vesicle (SDV).

The secretion of spicules in Demospongiae occurs in intracellularly specialized cells, the sclerocytes, where silica is deposited around an *organic filament* (Kröger et al., 2002; Schröder et al., 2006). It should be noted at this point that silica is not only deposited in spicules, but also as ribbons in the nucleus (Schröder et al., 2008). The latter observation is interesting insofar as siliceous spikes show a variable size and shape suggesting that silica in those structures is subjected to a high turnover. If the formation of siliceous spicules is inhibited the sponge body collapses (Sumper & Kröger, 2004a). The synthesis of spicules is a rapid process; in *Ephydatia fluviatilis* a 100 mm long megasclere is formed within 40 h (Sumper, 2004b).

Inhibition studies revealed that skeletogenesis of siliceous spicules is enzyme-mediated (Shimizu et al., 1998; Sumper &Lehmann, 2006) process. The siliceous spicules contain a definite axial filament and their synthesis is genetically controlled (Pouget et al., 2007). Experimental evidence suggested that deposition of silica particles induces biological reactions, most of them similar with that of collagen fibrillogenesis (Poulsen et al., 2003). A major step towards elucidating the formation of siliceous spicules at the molecular level was the finding that the formerly termed `axial organic filament' of siliceous spicules is in reality an enzyme, silicatein, which mediates the apposition of amorphous silica and hence the formation of spicules (Patwardhan et al., 2005; Perry & Keeling-Tucker, 2000). These studies have been performed with the demosponge *T. aurantia*. In contrast to the formation of collagen fibrils disconnected from the synthesis of spicules, those collagen polypeptides which are formed in connection with the silicification are produced in sclerocytes and exopinacocytes (Sumper, 2004b). The sclerocytes secrete the axial filaments, while the exopinacocytes secrete the collagen-like spongin, which functions as an organic sheath around the spicules (Krasko et al., 2000; Menzel et al., 2003).

Long-chain polyamines and phosphoproteins known as silaffins are the only diatom molecules thus far shown to have a direct impact on silica precipitation *in vitro*, with the resulting pore sizes of the formed structures determined by relative proportions of polyamines and silaffins (Cha et al., 1999; Croce et al., 2004; Field et al., 1998; Fuhrmann et al., 2004; Leynart et al., 2001). Diatom transporters have been sequenced and characterized and have been shown to interact directly with silicon to actively transport silicic acid against a large concentration gradient, although the mechanism for intracellular storage of soluble silicic acid is not known (Kröger et al., 1996; 1997).

The elegant work performed by the group of Morse clarified that the axial filaments around which silica is deposited are composed of the enzyme silicatein (Morse, 1999). In that study the silicatein cDNA was isolated from the sponge *S. domuncula* and was found to be closely related to the *T. aurantia* silicatein. As reported earlier (Sumper & Kröger, 2004a) silicatein is closely related to the enzyme cathepsin L. The phylogenetic analysis of the two sponge silicateins shows that they branched off from a common ancestor with the cathepsin L molecules prior to the appearance of sponges. This finding could imply that animals which evolved earlier than sponges already contained silicatein. As silicatein has not been identified in other metazoan phyla it has to be concluded that this gene has been lost during the evolution from sponges to the invertebrate and vertebrate phyla.

Recently it has been shown that the recombinant silicatein catalyzes the reaction of tetraethoxysilane to silica and silicone (Boyd, 2007). The data presented indicate that together with the induction of the gene expression for the axial filament/silicatein which leads to the initial process of spicule formation, the collagen gene is expressed in parallel and in a second cell (Shimizu et al., 1998). Based on inhibition studies it could be demonstrated that the formation of siliceous spicules can occur in the absence of the synthesis of such collagen polypeptides which form the perispicular spongin sheath (Sumper & Lehmann, 2006). This finding suggests that the expression of collagen is not directly connected with the formation of the siliceous spicules.

Silicateins are unique enzymes of sponges (phylum Porifera) that template and catalyze the polymerization of nanoscale silicate to siliceous skeletal elements. These multifunctional spicules are often elaborately shaped, with complex symmetries and together with a protein silintaphin-1, seem to guide silica deposition and subsequent spicular morphogenesis.

There are some incipient studies, both in biochemistry and synthetic chemistry assisted by different *in vivo* or *in situ* characterization techniques dealing with the mechanism revealed.

The results indicate that silicatein-α-mediated biosilicification depends on the concomitant presence of silicatein-α and silintaphin-1. Accordingly, silintaphin-1 might not only enhance the enzymatic activity of silicatein-α, but also accelerate the nonenzymatic polycondensation of the silica product before releasing the fully synthesized biosiliceous polymer (Kröger et al., 1996; 1997; 2002).

Little is known about the course and the control of spicule development, in either marine and freshwater sponges, besides electron microscopic studies, light microscopic investigations using sandwich cultures have been performed to study the spicule. The formation of spicules starts in sclerocytes within a specific vesicle. After the production of an axial organic filament, silicon is deposited around it, and the whole process of forming a spicule (190 μm in length and 6 to 8 μm in diameter) is completed after 40 h, at 21° C. The main actor in templating this mechanism seems to be deoxyuridine, a nucleoside type compound.

Recent works of Prof. Christopher F. van der Walle (Fairhead et al., 2008) show that silicatein α exists into its predominantly β- sheet structure. This conformational structure is produced in a soluble form with mixed α-helix/β-sheet structure akin to its cathepsin L homologue. Conformational transition studies found that β- sheet intermediate structure for silicatein α in marine sponge spicules represents a stable structural intermediate for silica formation process in spicules (Fig. 1).

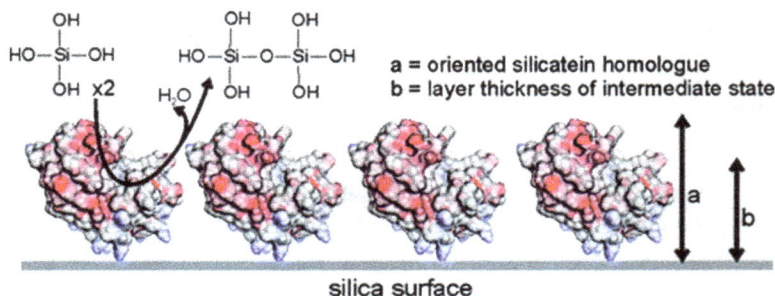

Fig. 1. Illustration of intermediate structure for silicatein α in marine sponge spicules

Very recently (Li et al., 2011) a new nucleoside derivative has been identified and characterized, named 3-acetyl-5-methyl-2'-deoxyuridine (**1**), along with two known compounds 3,5-dimethyl-2'-deoxyuridine (**2**) and 3-methyl-2'-deoxyuridine (**3**), isolated from the cultures of *Streptomyces microflavus*. This strain was an associated actinomycete isolated from the marine sponge *Hymeniacidon perlevis* collected from the coast of Dalian (China).

Sequence analysis of silicatein α showed that the protein was very similar to the human hydrolase, cathepsin L. It was later proven by site directed mutagenesis that silicatein α is a serine hydrolase (Cha et al., 1999; Shimizu et al., 1998). The active site of silicatein α consists of a hydrogen bond that is formed between a histidine and a serine residue. It is proposed that this hydrogen bond between the hydroxyl group of the serine and the imidazole group of the histidine increases the nucleophilicity of the serine residue and thus allows it to catalyze the hydrolysis of silica precursors (Figure 2). Hydrolysis of silica in proximity of the proteins leads to the proteins themselves acting as templates for the condensation of silica. The silicification process of this marine sponge serves as an excellent model for biomimetic inorganic materials due to its relatively simple idealized reaction mechanism and mild reaction conditions.

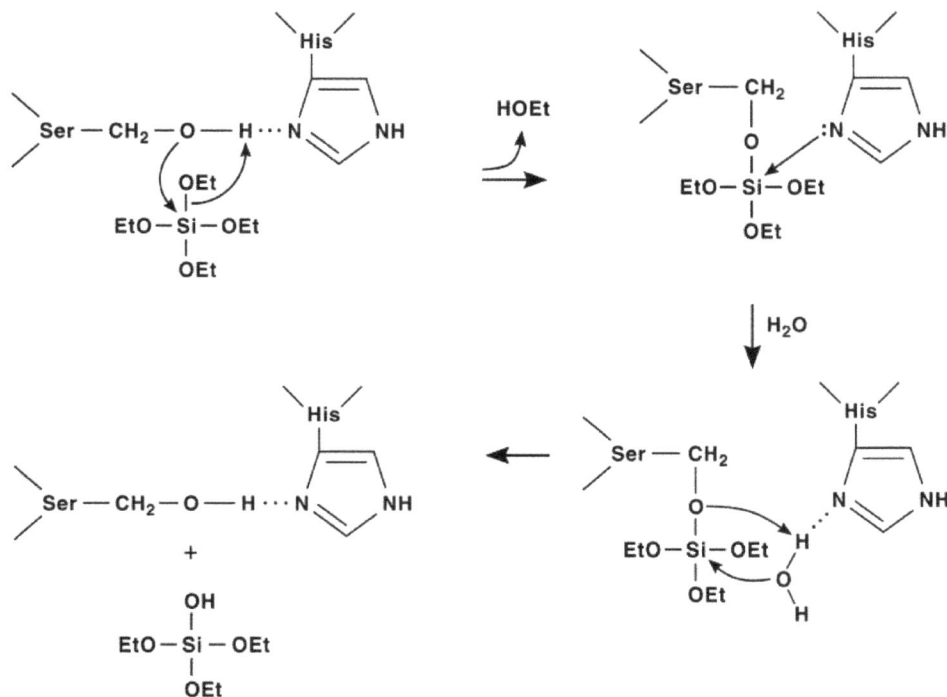

Fig. 2. Proposed hydrolysis reaction of TEOS in the active site of silicatein α (Zhou et al., 1999)

Moreover, the studies using biocatalytical active recombinant silicatein showed that silica formation in sponges is an enzymatic process (Krasko et al., 2000; Schröder et al., 2006). Furthermore, recent studies confirmed the mechanism of self-assembly of silicatein monomers to oligomers and long protein filaments (Croce et al., 2004; Croce et al., 2007; Murr & Morse, 2005;). The results from SAXS examination indicated that the axial filament is formed from a very high degree of organization (hexagonal) of the protein units (Croce et al., 2004; 2007). Müller and coworkers found that silicatein is present not only in the axial filament but also on the surface of the spicules (Müller et al., 2006), supporting the view that growth of spicules occurs through apposition of lamellar silica layer with a distance of 0.2-0.5 μm from each other (Müller et al., 2005).

Using milder extraction conditions (a slightly acidic, aqueous ammonium fluoride solution), native silaffins (termed natSil) have been isolated from C. fusiformis. (Kröger et al., 2002).

A complex approach in systematizing this theoretical information will have a significant added value in understanding and developing bio-inspired or bio-similar synthesis. Tailoring nature is the target of this century, so called smart molecules acting as real transcriptors in bio-silica synthesis.

The biomimetic silicatein systems that have been investigated so far have all been able to produce silica at higher rates than reactions without both the nucleophilic and the hydrogen bond donating groups; however, they have not been able to approach the hydrolytic activity of silicatein. It is clear that any biomimetic silicatein system needs to have a structure that can template the condensation, both a nucleophilic and cationic group need to be within hydrogen bonding distance of one another, and they need to have many more active sites per volume to have any chance at replicating the synthesis ability of the native protein. Such a system that could meet all of these requirements is based on peptide amphiphile micelles.

These biomimetic and bioinspired systems could be considered in the same manner as theoretical steps in explaining the mechanisms involved in real biological organisms and as simplified models in new approaches of modern synthetic chemistry. Using concepts and definitions from genetic molecular transcription mechanism we translated those principles to synthetic chemistry as follows:

1. the existence of a "template" unit for initiation of construction process which is not only a complementary building bloc but also having energetic and conformational complementarily as those involved into RNA complementary copy of DNA molecules;
2. decoding of structural information during hole synthesis process which allows the ordered silica network formation, by specific steric effects and similar enzymatic control;
3. the existence of a surface specific intermediary complex stabilized at the silica surface with minimum of kinetic energy involved;
4. the involvement of a hydrogen bonding and electrostatic effects.

During the silica synthesis process in diatoms have been already identified the deposition of silica around an organic filament into sclerocytes cells, those organic filaments partial identified as a complex of proteins. Using the identified proteins, especially cathepsin L. many successfully studies have been conducted in biomimetic silica synthesis as already presented.

Numerous approaches attempt to mimic and derivatize biomineralization processes, including poriferan biosilicification, in order to develop novel biomedical and biotechnological applications (Field et al., 1998). In conducting such studies it was imperative to identify silicatein-binding proteins required for pattern formation. Wiens et al. (2009) described the discovery of silintaphin-1, a unique protein that directs the assembly of silicatein filaments by forming a silicatein-binding scaffold/backbone. They shown that silintaphin-1 also interacts with silicatein that had been immobilized on functionalized g-Fe_2O_3 nanoparticles, thus directing the formation of a composite with distinct rod-like morphology. The resulting hybrid biomaterial combines the unique properties of magnetic iron oxide nanoparticles, a silica-polymerizing enzyme, and a cross-linking protein.

3. Comparison between structural and electronic effects induced by different classes of transcriptors in nano-silica

The silica-anabolic and silica-catabolic enzymes, silicatein and silicase are of extreme interest for a variety of applications in nanobiotechnology. Silica-based materials are widely used in industry and medicine. Therefore, the mechanism(s) underlying biosilica formation in the organisms are of high interest for the design, in particular on the nano-scale, of novel biosilicas to be used in nano(bio)technology.

The study of structural and electronic effects induced by different classes of transcriptors in nano-silica bio-inspired materials is the aim of this chapter, the theoretical aspects being sustained by a complex experimental survey.

Among biogenic minerals, silica appears rather singular. Whereas widespread carbonate and phosphate salts are crystalline iono-covalent solids whose precipitation is dictated by solubility equilibria, silica is an amorphous metal oxide formed by more complex inorganic polymerization processes. Biogenic silica has mainly been studied with regard to the diversity of the species that achieve this biomineralization process and at the level of diversity in the morphology of silica structures (Nelson et al., 1995). It is only recently that chemists turned their attention to the formation process.

Approaches that make use of current biological knowledge to investigate new chemical systems are certainly of great interest. Silica patterning in diatoms appears to rely on proteins that are able to catalyze silicate polymerization and to act as templating agents through self-assembly process, therefore synthetic models that exhibit both properties have been designed.

Cha *et al.* (2000) have synthesised a series of block copolypeptides of cysteinelysine that have the ability to mimic silicatein (a protein found in the silica spicules of the sponge *Tethya aurantia*). The synthesized block copolymers provide the first example of polymers which can hydrolyse and condense an inorganic phase (tetraethoxysilane) and also provide a template in the process. They were able to produce hard mesoporous silica spheres and assemblies of columnar amorphous silica. The same authors determined the specific conditions under which the columnar structures were stabilized and also elucidated the mechanisms involved in hydrolysis and condensation of the inorganic phase. The studies have attracted attention since the inorganic phase was produced from tetraethoxysilane under ambient conditions at a neutral pH.

The interaction of proteins with solid surfaces is not only a fundamental phenomenon but is also key to several important and novel applications. In nanotechnology, protein–surface interactions are pivotal for the assembly of interfacial protein constructs, such as sensors, activators and other functional components at the biological/electronic junction. Because of the great relevance of the protein–surface interaction phenomenon, much effort has been done into the development of protein adsorption experiments and models. The ultimate goals of such studies would be to measure, predict and understand the protein conformation, surface coverage, superstructure and kinetic details of the protein–surface interactions.

Sarikaya et al. (2003) have recently reviewed such molecular biomimetics, including a summary of 28 short peptide sequences that have been found to bind to solid surfaces ranging from platinum to zeolites and gallium arsenide.

Related ideas are explored in a recent review of the design of nanostructured biological materials through self-assembly (Zhang et al., 2002). In another study, biotinylated peptide linkers were attached to a surface *via* streptavidin to bind fibronectin in an oriented manner (Klueh et al., 2003). Such a material is then hoped to solicit a desired biological response (i.e. tissue integration). Perhaps the most exciting application is optical switching and modulation behavior based on proteins affixed to a substrate (Ormos et al., 2002).

There are different models proposed for explaining the protein transcription interactions under silica synthesis. Colloidal-scale models represent the protein as a particle and can accurately predict adsorption kinetics and isotherms. These colloidal-scale models include explicit Brownian dynamics type models (Oberholzer & Lenhoff, 1999; Oberholzer et al., 1997), random sequential adsorption models (Adamczyk et al., 2002; 1990; 1994; Adamczyk & Weronski, 1997), scaled particle theory (Brusatori & Van Tassel, 1999; Van Tassel et al., 1998), slab models (Stahlberg & Jonsson, 1996) and molecular theoretic approaches (Fang & Szleifer, 2001; Satulovsky et al., 2000; Szleifer, 1997). Most of these approaches treat the electrostatics and van der Waals interactions between the colloidal 'particle' and the surface, and thus can capture dependencies on surface charge, protein dipole moment, protein size or solution ionic strength.

Several researchers are exploring detailed atomic representations of proteins. The earliest studies to use protein crystal structures to simulate the adsorption process assumed a completely rigid protein and calculated screened coulomb and Lennard–Jones interactions over all protein rotations and distances (Lu & Park, 1990; Noinville et al., 1995). In rigid atomistic models with electrostatic treatments it was found that a net positively charged protein (lysozyme) could adsorb on a positively charged surface, due to the nonuniformity of the charge distribution on the protein. The orientations of an adsorbed antibody on a surface using a united residue model, whereby each amino acid is represented by a group with averaged electrostatic and van der Waals interactions. Molecular dynamics (MD) was used to simulate 5 ns of multipeptides interacting with gold. Also, MD-based simulations were used to find minimal energy orientations and unfolding trajectories of albumin subunits on graphite.

Latour & Rini, (2002) determined the free energy of individual peptides (in the context of a protein) interacting with a self-assembly monolayer (SAM) surface as a function of distance. These parameterizations could be used in protein–surface energy calculations.

The results clearly support the view that

- the presence of a minimal number of the hydrogen-binding sites is indispensable for transcription of the self-assembled structure into silica network,
- the helicity of silica can be accurately transcribed from that of the template, and
- this method will be applicable for the efficient transcription of self-assembled superstructures into inorganic materials.

Therefore, a variety of superstructural silica materials such as double-helical, twisted-ribbon, single fiber and lotus-like structures are created by a template method with the aid of the hydrogen-bonding interaction of the oligomeric silica species in the binary gel systems. The amino group of added *p*-aminophenyl aldopyranosides in gel fiber acted as the efficient driving force to produce novel structures of the silica nanotubes. The present system could be useful for transcription of various self-assembled superstructures into silica materials which are eventually applicable to catalysts, memory storage, ceramic filters, etc.

The building of complex structures is promoted by specific links due to the three-dimensional conformations of macromolecules, showing topological variability and diversity. Efficient recognition procedures occur in biology that imply stereospecific structures at the nanometer scale (antibodies, enzymes and so on). In fact, natural materials are highly integrated systems having found a compromise between different properties or functions (such as mechanics, density, permeability, colour and hydrophobia, and so on), often being controlled by a versatile system of sensor arrays. In many biosystems such a high level of integration associates three aspects: miniaturization, with the role to accommodate a maximum of elementary functions in a small volume, hybridization between inorganic and organic components optimizing complementary possibilities and functions and hierarchy.

Synthetic pathways currently investigated concern

- transcription, using pre-organized or self-assembled molecular or supramolecular moulds of an organic (possibly biological) or inorganic nature, used as templates to construct the material by nanocasting and nanolithographic processes;
- synergetic assembly, co-assembling molecular precursors and molecular moulds *in situ*;
- morphosynthesis, using chemical transformations in confined geometries (microemulsions, micelles and vesicles) to produce complex structures, and
- integrative synthesis which combines all the previous methods to produce materials having complex morphologies.

Self-assembling structures will be used to denote complex nano- or micro-structures such as protein aggregates, protein with lipid membranes, and certain intracellular organelles such as vesicles that form spontaneously from the constituent molecules in solution and that are thermodynamically stable. This does not apply to all structures, especially more complex aggregates, cells and organs. Such structures do not normally self-assemble spontaneously in solution, but are assembled step-by-step by energy requiring mechanisms of the cell or organ. Their maintenance also usually requires a continuous input of energy.

By using similar principles of self-assembly, it is now easy to reach to artificially construct 'biomimetic' structures in the laboratory.

Non-specific forces are those that arise between many different types of atoms, molecules, molecular groups or surfaces, and that can usually be described in terms of a generic interaction potential or force-function.

Specific interactions arise when a unique combination of physical forces or bonds between two macromolecules act together cooperatively in space to give rise to a (usually) strong but non-covalent bond. Because such specific interactions typically arise from a synergy of multiple geometric, steric, ionic and directional bonds they are also referred to as 'complementary', 'lock-and-key', 'ligand and receptor (LR)' and 'recognition' interactions.

4. Electronic enhanced effects in biomaterials

This section will identify all important mechanisms involved in bio-similar materials called "electron induced effects" as a part of complex understanding path in transferring key information from biology to smart materials.

Different classes of metallo-proteins in important life mechanism are presented, especially highlighting the spectacular and "strange" combinations which are not easy accepted by scientists, but having a huge potential in complex understanding of nature inspired solutions.

Vanadium is a biologically relevant metal, employed by a variety of organisms: it is in the active centre of two groups of enzymes, such as vanadate-dependent haloperoxidases and vanadium-nitrogenases. In addition, vanadium is accumulated by certain life forms such as sea squirts (*Ascidiaceae*) and *Amanita* mushrooms, e.g. the fly agaric. More generally, vanadium appears to be involved in the regulation of phosphate-metabolising enzymes also in plants and animals; the insulin-mimetic potential of many vanadium compounds (i.e. their anti-diabetic effect) is related to this action.

Such an example is **Vanabins** (also known as **vanadium-associated proteins** or **vanadium chromagen**), a specific group of vandium-binding metalloproteins found in some ascidians and tunicates (sea squirts). Vanabin proteins seem to be involved in collecting and accumulating this metal ion. At present there is no conclusive understanding of why these organisms collect vanadium, and it remains a biological mystery. It has been assumed that vanabins are used for oxigen transport like iron-based hemoglobin or cooper-based hemocyanin. From this point to a complex comparison of special relationship of this metal ion with oxygen at the level of bio- or other important chemical mechanisms, we are interested in presenting the electronic effects induced by a metallic center on the specific interest areas: biology, drugs, new material synthesis.

Vanabins are a unique protein family of vanadium-binding proteins with nine disulfide bonds. Possible binding sites for VO^{2+} in Vanabin2 from a vanadium-rich ascidian *Ascidia sydneiensis samea* have been detected by nuclear magnetic resonance study, but the metal selectivity and metal-binding ability of each site was not examined in detail.

Vanadium accumulated in the ascidians is reduced to the +3 oxidation state *via* the +4 oxidation state and stored in vacuoles of vanadocytes (Michibata et al., 2002). From the vanadocytes of a vanadium-rich ascidian, *Ascidia sydneiensis samea*, were isolated some vanadium binding proteins, designated as Vanabin. Recently, five types of Vanabins have been identified, Vanabin1, Vanabin2, Vanabin3, Vanabin4 and VanabinP that are likely to be involved in vanadium accumulation processes as so-called metallochaperone. Multi-dimensional NMR experiments have revealed the first 3D structure of Vanabin2 in an aqueous solution which shows novel bow-shaped conformation, with four α-helices connected by nine disulfide bonds (Hamada et al., 2005). There are no structural homologues reported so far. The 15N HSQC perturbation experiments of Vanabin2 indicated that vanadyl cations, which are exclusively localized on the same face of the molecule, are coordinated by amine nitrogens derived from amino acid residues such as lysines, arginines, and histidines, as suggested by the EPR results (Fukui et al., 2003).

Under physiological conditions, vanadium ions are limited to the +3, +4 and +5 oxidation states. When vanadate ions (V^V) in the seawater are accumulated in the ascidians they are firstly reduced to (V^{IV}) in vanadocytes and then stored in the vacuoles where (V^{IV}) is finally reduced to the +3 oxidation state.

Vanabin2 is selectively bound to V(IV), Fe(III), and Cu(II) ions under acidic conditions. In contrast, Co(II), Ni(II), and Zn(II) ions were bound at pH 6.5 but not at pH 4.5. Changes in pH had no detectable effect on the secondary structure of Vanabin2 under acidic conditions, as determined by circular dichroism spectroscopy, and little variation in the dissociation constant for V(IV) ions was observed in the pH range 4.5-7.5, suggesting that the binding state of the ligands is not affected by acidification. These results suggest that the reason for metal ion dissociation upon acidification is attributable not to a change in secondary structure but, rather, is caused by protonation of the amino acid ligands that complex with V(IV) ions.

Numerous *in vitro* and *in vivo* studies have shown that vanadium has insulin-like effects in liver, skeletal muscle, and adipose tissue (Fukui et al., 2003; Hamada et al., 2005; Michibata et al., 2002). Vanadium at relatively high concentrations *in vitro* and *in vivo* inhibits phosphotyrosine phosphatases (PTPs), thus enhancing insulin receptor phosphorylation and tyrosine kinase (IRTK) activity. However, some studies have demonstrated that vanadium can stimulate glucose uptake independently of any change in IRTK activity, suggesting that there are additional mechanisms for its insulin-mimetic effects. In liver, vanadium compounds have been reported to inhibit lipogenesis and gluconeogenesis and to stimulate glycolysis and glycogen synthesis. In skeletal muscle, vanadium augments glucose uptake, primarily by stimulating glycogen formation. Free oxygen radicals seem to be the key factors acting in a still unclear certain pathologic mechanisms as those presented with relation to vanadium. Vanadium compounds are relatively well known both as free oxygen radicals generators in some inorganic and bio-reaction mechanisms and as complex activators or inhibitors of many enzymes involved in carbohydrate or lipid metabolic pathways.

Special fluorescent properties of some encapsulated vegetal active substances, induced by vanadium sub-oxide species effect, used as co-activators have been considered as first step in evaluation of biocompatibility and biomimetic degree of some nano-materials using these effects induced by vanadium species in bio-systems. The key point for the design of new hybrids encapsulated materials is to extend the accessibility of the inner interfaces for including a new pre-designed vanadium complex able to remain bonded at the surface in the active bio-mimetic conditions and to transfer its electronic effect to the active reaction center. The nature of the interface or the nature of the links and interactions that the organic and inorganic components exchange has been used to design these hybrids.

5. Silica nanotubes – Fabrication and uses

Inorganic materials of nanometric size gained an increased interest for biomedical and biotechnological applications for drug/gene carriers, disease diagnosis and cancer therapy. Although organic structured materials as liposomes, dendrimers and biodegradable polymers are still playing a key role in nanomedicine, especially due to safety reasons, recent findings regarding the relation between the surface functionalization of inorganic nanomaterials (*e.g.* carbon nanotubes) and biocompatibility (Sayes et al., 2006) have inclined the balance in favor of the later. Other advantages of inorganic nanomaterials are represented by availability due to facile synthetic pathways and the possibility to control shape and size. Inorganic nanomaterials present interesting optical, electrical and physical properties that may help to solve the problem of physical barriers of the cell and therefore to extend the biomedical applications. Finally, their variety of shapes, mesoporous nanoparticles, hollow spheres, nanotubes, offers the possibility to accommodate a large amount of drug/genes inside the pores or cavities that facilitate the control release providing a "gate" whose opening may be triggered by physical or chemical stimuli. Encapsulation of biological material or drugs into inorganic nanoparticles or nanotubes provide an isolation from the environment and prevent hydrolytic degradation or aggregation (Son et al., 2007) and has the potential to enhance drug availability, reduce its toxicity and enable precision of drug targeting. Drug - delivery to desired physiologic targets is influenced by many anatomic features as the blood brain barrier, branching pathways of the pulmonary system, the tight epithelial junctions of the skin, etc. (Hughes, 2005). Optimum dimensions for the carriers (inorganic or organic) are established: for example the greatest efficacy for delivery into pulmonary system is achieved for particle diameters ~100nm (Courrier et al., 2002). Particles around 50-100 nm in size provide a greater uptake efficacy for gastrointestinal absorption (Desai, 1996a) and transcutaneous permeation (Hussain et al., 2001).

Silica based nanostructured materials were successfully involved as nanocontainers in drug-release applications due to their biologically stable shell surfaces which are slightly hydrophilic. The blood plasma proteins called opsonins absorb onto the particles with hydrophobic surfaces acting as binding enhancer for the process of phagocytosis and the entire system is cleared by the immune system.

Spherical silica nanoparticles with ordered pore structure, MCM type, were synthesized using microemulsion method or by template sol-gel technique using a large variety of structure directing agents, from the common cetyltrimethylamoniumbromide (CTAB) to more complex systems, nonionic surfactants and amphiphilic polymers (Bagshaw et al., 1995; Beck et al., 1992a; Botterhuis et al., 2006; Kresge et al., 1998; Vallet-Regi et al., 2001; Zhao et al., 1998). Although they represent the large majority of silica based mesoporous materials involved in biomedical and biotechnological applications they present limited options in terms of surface modification especially when multifunctionality is required.

Therefore, in the last years, silica nanotubes, with their hollow core structure, ultralarge specific surface areas, very narrow inner pores and catalytic surface properties were considered as attractive alternatives for some applications which require multifunctionality because they present two surfaces with hydroxyl groups (inner and outer) easy to tailor with different functional groups using commercial silane derivatives. By changing the nature of the interior of the nanotube or the outer surface, specific biomolecules for host-guest reactions may be attached onto the surface and the interiors can be loaded with hydrophobic drugs. Multifunctionality may be also achieved by loading and/or coating with superparamagnetic particles and fluorescent molecules necessary for different imaging techniques used in diagnosis.

5.1 Preparation of silica nanotubes by sol-gel polycondensation mediated by organogelators

Templating methods based on fibrous organogels (or hydrogels) represent versatile chemical strategies to prepare nanofibrous materials exhibiting a large variety of morphologies (such as tubes, fibers, rods, ribbons, belts and helices). From the first reports of Shinkai group (Van Bommel et al., 2003a; Jung & Shinkai, 2004b; Jung et al., 2001c; Jung et al., 2000d; Tamaru et al., 2002e) the strategy was successfully employed and the reported experimental data are summarized and discussed in excellent reviews (Jung et al., 2010; Yang et al., 2011; Llusar & Sanchez, 2008).

Low molecular mass organic molecules capable of forming thermoreversible physical (supramolecular) gels at very low concentrations (ca. 10^{-3} mol dm^{-3}, typically lower than 2% w/w) in a wide variety of organic solvents have attracted in the last twenty years considerable interest, an impressive and steadily growing number of gelator molecules are synthesized and characterized in the literature (Banerjee et al., 2009; Sangeetha & Maitra, 2005; Terech et al., 1997; Weiss & Terech, 2006). An organogelator is capable of self-organizing into finely dispersed anisotropic aggregates (nanofibers) by noncovalent interactions such as hydrogen bonding, van der Waals, π-π -stacking, electrostatic and charge-transfer interactions. Noncovalent crosslinks among the nanofibers and/or mechanical entanglements create a three-dimensional network which includes the solvent and so gelation occurs. Upon gelation, the organized self-assembly of these molecules results in the formation of highly anisotropic 3-D structures, mostly in the shape of fibers, but also as ribbons, platelets, tubular structures, or cylinders. The network is commonly destroyed by heating but is reformed on cooling thus rendering the system thermoreversible.

A classification of the existing organogelators could be made according to several different criteria (Llusar & Sanchez, 2008):

a. *chemical nature* – main structural scaffold – cholesterol-, amide-, urea-, amino-acid-, peptide-, cyclohexane-, sugar- based organogelators.
b. *level of complexity of their structure* - organogelators with one, two or more heteroatoms, two-component and hybrid.
c. *type of supramolecular forces involved in their self-assembly into fibrous networks* - H-bonding, van der Waals, hydrophobic/solvophobic, aromatic or π-π stacking, electrostatic/ ionic, charge transfer coordinating bonding, or usually a combination of some of them.
d. *fibrous morphologies of their Self-Assembled Fibrous Networks (SAFINs)*- fibrous, twisted or helical, ribbonlike, hollow fibers or tubular, lamelar, vesicular, etc.
e. *structural element or motif enabling an efficient transcription for template synthesis* - covalently or noncovalently attached positive charges, H-donating groups, coordinating or binding groups, etc.
f. *properties and applications* - polymerizable, liquid-crystalline or birrefringent, luminescent or fluorescent organogels, photoresponsive, metal-sensitive, etc.

A typical transcription procedure consists in catalyzed hydrolysis of a silica precursor - tetraethyl orthosilicate (TEOS) or other reactive silane derivative in an organogel formed by the structure-directing agent and the reaction solvent. The hydrolysis of TEOS followed by the polycondensation of silicate around the gel fibril leads to a silica gel which is dried. The subsequent removal of the organic materials by washing with a suitable organic solvent or calcination yields porous silica. An efficient transcription of the gel fiber structure onto silica requires an effective interaction between the organogel fibers and silica precursor. Thus for sol-gel polycondensation in acidic conditions, when anionic silicate species are present in the system, organogelator should be either with cationic structure or doped with an appropriate amount of cationic analogue (Jung et al., 2000a; Jung et al., 2000b). Under basic conditions the H-bonding groups present on the gel fibrils can interact with the silica precursor and promote good transcription. In some cases the solvent used in the sol-gel polycondensation plaies an important role in the transcription process because the interactions gel fiber-silica precursor may be affected by silica-solvent and gelator-solvent interactions. The use of a long chain phosphonium salt as template in a protic solvent (ethanol) yielded only plate-like silica due to solvent ability to form H-bonds with silicate species, decreasing the strength of the silicate-template electrostatic interactions. The use of an aprotic solvent (benzene) yielded fibrous silica (Huang & Weiss, 2006).

The template synthesis of inorganic nanostructured materials through organogelator approach involves three main synthesis pathways:

a. *in-situ coassembly (IC)* – formation (upon cooling) of an organogel in the appropriate organic solvent and in the presence of the silicate oligomers (and controlled amounts of water and catalysts when necessary), which upon condensation gives rise to an inorganic siloxane-based gel consisting of organogel fibrils coated with partially condensed inorganic species. The organogel is removed either by washing or thermal procedures.

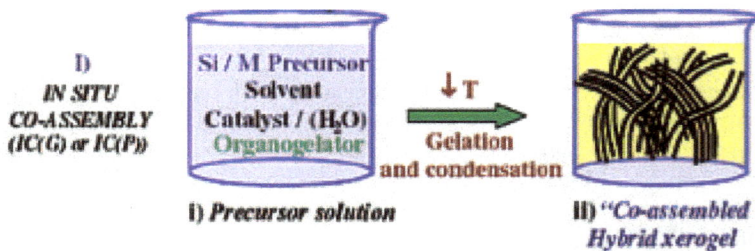

Fig. 3. Strategy for template synthesis of nanotubes mediated by organogelators *in-situ cossembly*. [Reprinted with permission from (Llusar & Sanchez, 2008), Copyright 2008, American Chemical Society]

b. *two-step simple post-transcription (PT)* – the preformed organic xerogel is subsequently impregnated or subjected to postdiffusion of solution of precursor together with water and catalyst in an appropriate solvent.

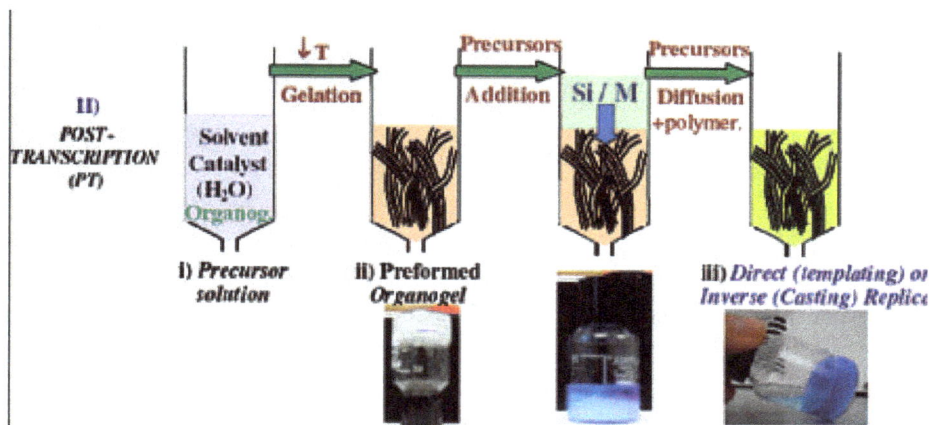

Fig. 4. Strategy for template synthesis of nanotubes mediated by organogelators post-transcription. [Reprinted with permission from (Llusar & Sanchez, 2008), Copyright 2008, American Chemical Society]

c. *self-assembly* (**SA**) process - hybrid (organic–inorganic) organogelators, in which the precursors of the inorganic species are already bonded to the organogelator molecules through covalent or coordinative bonds for example sylilated hybrid organogels.

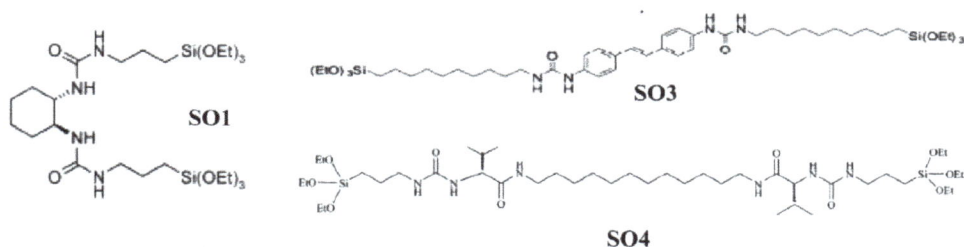

For any strategy involved, the morphology of the resulted nanostructured materials depends upon a variety of factors: the type and concentration of the organogelator, the nature of the solvent, pH value, the use of certain additives, aging time and thermal treatment. In this way variety of silica nanotubes may be produced, such as single walled, double walled, helical, lotus and mesoporous structures.

Single-walled silica nanotubes

The first report on the fabrication of silica tubular structures using an organogelator as template was made by Shinkai group (Ono et al., 1998) involving two cholesterol based organogelators without and with cationic charges in acidic conditions.

1: R=H , 2: R = NMe₃⁺Br⁻

Fig. 5. TEM micrograph of the tubular silica obtained from organogel **2**. [Reproduced from (Ono et al., 1998) by permission of The Royal Society of Chemistry]

Analizing the morphology of the obtained materials using SEM and TEM microscopy, fibrous material with open cavities at tube edges of inner diameter 10-200 nm yielded for the charged template, while the neutral organogelator yielded only granular silica. This is the exemple of how the structure of the template and hence the interaction with silica precursor may influence the competition between the condensation of silica species in the bulk liquid (solution mechanism) and that onto the surface of organogel fibrils (surface mechanism) leading to silica tubes.

Lotus shape silica nanotube

A series of α- and β- glucose derived organogelators with benzene rings substituted with amino or nitro groups were synthesized and used as neutral templates for sol-gel polycondensation of TEOS in the presence of benzylamine as catalyst (Jung et al., 2000). The presence of the amino grup provides the necessary H-binding site that favors the transcription in neutral conditions.

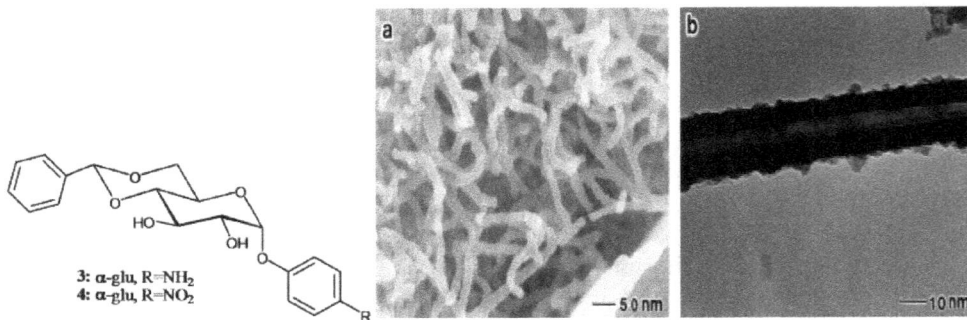

3: α-glu, R=NH₂
4: α-glu, R=NO₂

Fig. 6. SEM (a) and TEM (b) micrograph obtained from ethanol organogel **3** after calcinations. [Reproduced from (Jung et al., 2000) by permission of The Royal Society of Chemistry]

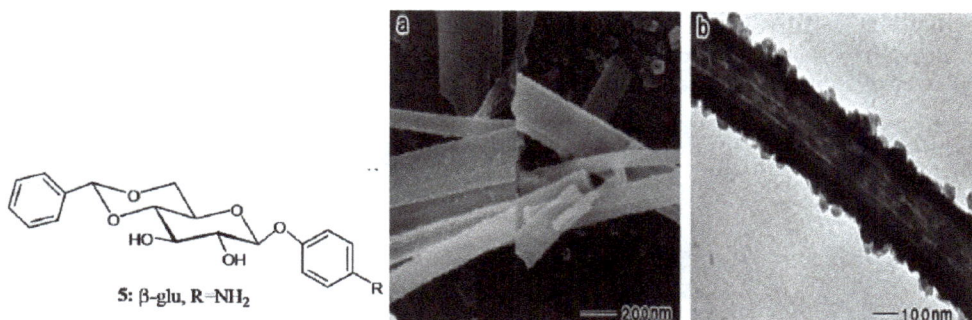

5: β-glu, R=NH₂

Fig. 7. SEM (a) and TEM (b) micrograph obtained from ethanol organogel **5** after calcinations. [Reproduced from (Jung et al., 2000) by permission of The Royal Society of Chemistry]

For all the amino substituted templates silica nanotubes were obtained: in the case of gelator **3** a tubular structure with a 20–30 nm outer diameter and 350–700 nm in length, and lotus shaped nanotubes of 50–100 nm inner diameters and 150–200 nm outer diameters composed of micro-tubes with diameters of 5–10 nm similar to the lotus root.

Double helical nanotubes

The importance of an appropriate amount of H-bonding sites on an organogelator fibril to achieve a good transcription is illustrated by the results reported by Jung et al. (2002) who associated a sugar-based gelator, **6** and an aminophenyl glucopyranoside **7**. The additional H-bonding sites provided by the amino-groups as well as π-π stacking of phenyl moieties suggested that the self-assembled superstructure in the gel obtained from de mixture of **6+7** was oriented into a more explicit chiral packing (Jung et al., 2002). That influenced the morphology of the xerogel obtained from 1:1 mixture in H₂O/MetOH, (10:1 v/v), as compare with that generated from pure gelator in the same mixture of solvents. Additional non-covalent bonds brought by the association of **6** and **7** transformed the 3-D network of bundles (20–500 nm) of partially twisted helical (left-handed) ribbons 20–100 nm width and ca. 315 nm pitch (Jung et al., 2001) observed for the xerogel of **6** into well defined several micrometer long double helical fibers with diameters of 3-25 nm and dimensions comparable to those of double-helical DNA and RNA structures (Jung et al., 2002).

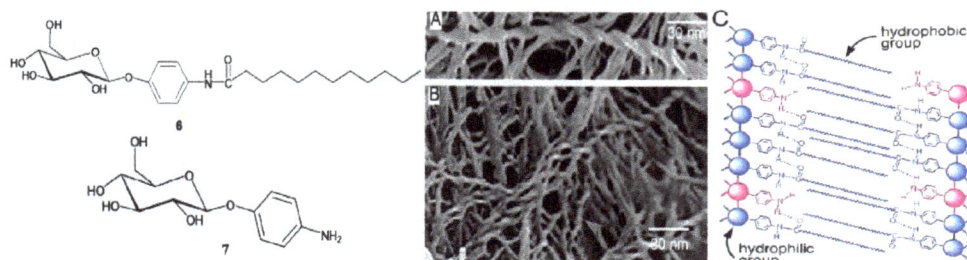

Fig. 8. (A and B) FE-SEM images of the xerogel prepared from the mixed gel of **6** and **7** (1:1 w/w) in H₂O/MetOH (10:1 v/v). (C) A possible self-assembling model in the bilayered chiral fiber from the mixed gel of **6** and **7**. [Reprinted with permission from (Jung et al., 2002), Copyright 2002, American Chemical Society]

Sol–gel polymerization of TEOS in a H_2O/MetOH (10 : 1 v/v) gel of **6** + **7** as a template produced a well defined double-helical nanotube with a diameter of 50-80 nm and a pitch of 50-60 nm as shown by SEM and TEM micrographs.

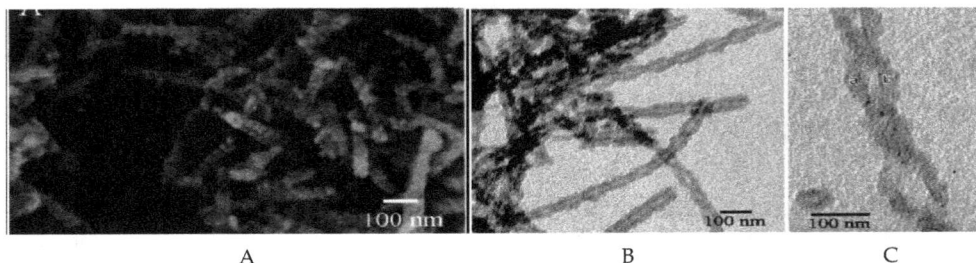

A   B   C

Fig. 9. (A) FE-SEM and (B and C) TEM images of the double-helical silica nanotube obtained from the mixed gel of **6** and **7** (1:1 w/w) after calcinations. [Reprinted with permission from (Jung et al., 2002), Copyright 2002, American Chemical Society]

The success of the transcription of the double-helical structure of the gel fibrils of **6** and **7** into silica nanotube structure may be explained by H-bonding between the amine moieties and negatively charged oligomeric silica particles. Transcription experiments with the mixture of compounds in different ratios revealed that for lower concentrations of the amino derivative **7** ([**7**/**6**+**7**]<0.2) only conventional granular silica was obtained, supporting the idea of a minimal number of H-binding sites indispensable for the template to achieve transcription.

Right-handed and left-handed single chiral silica nanotubes

Association of chiral neutral and cationic gelators derived from cyclohexanediamine **8-11** were used to fabricate right and left-handed chiralities in silica nanotubes (Jung et al., 2000) using electrostatic interactions. Cationic charge in organogelator is considered as indispensable in the polycondensation of TEOS but the ability to form gels was diminished by the positive charge. As in the case of H-binding sites there is a minimal ratio between charged and neutral gelators that ensures transcription of chirality. Thus, mixtures of urea based, neutral (*1R, 2R*)-**9** and amide-based cationic (*1R, 2R*)-**8** with molar ratios [**8**/**8**+**9**]=20-80% yielded right-handed helical silica structure with outer diameter of 90-120nm. The left-handed helical silica was fabricated with (*1S, 2S*)-enantiomers of gelators **10** and **11**.

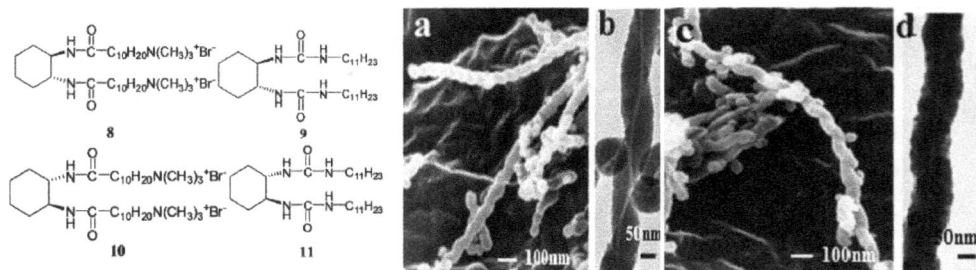

Fig. 10. (a and c) SEM and (b and d) TEM images of (a and b) the right-handed and (c and d) the left-handed helical silica nanotubes from organogels (a and b) **8+9** and (c and d) **10+11**. [Reprinted from (Jung et al., 2000) with permission from John Wiley and sons].

In a recent paper Hyun et al., (2009) suggested that each of the participants in the association has a distinct role in the transcription process: the neutral gelator **12** determines de shape of the organogel fibril and by consequence the shape of the silica nanotube, whilst the cationic amphiphile, **13** influences the polycondensation of TEOS after covering the surface of the organogel fibril formed by the neutral gelator (Hyun et al., 2009). For a better understanding of the mechanism for the transfer of helix and chirality to silica nanotubes Kim et al. (2011) performed crossover experiments: the neutral gelator *(1S, 2S)*-**12** was constant and the cationic amphiphiles varied from *(1S, 2S)*-**13**, *(1R, 2R)*-**13** and (±)-**13**.

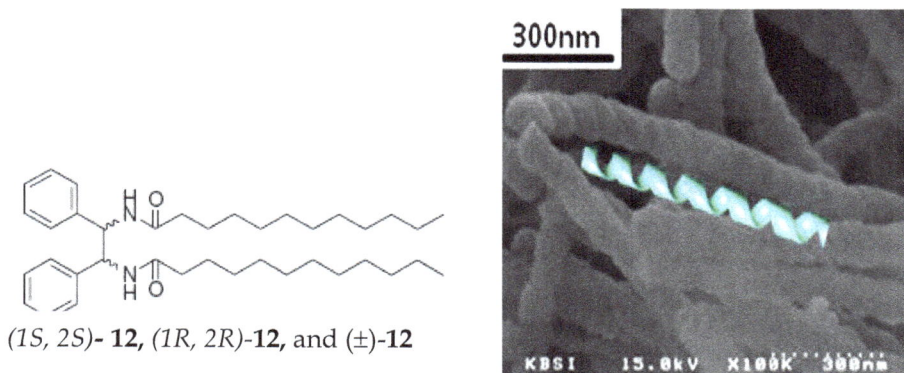

(1S, 2S)- **12**, *(1R, 2R)*-**12**, and (±)-**12**

Fig. 11. SEM image of left-handed helical silica nanotubes obtained from the 1:1 mass mixture of *(1S,2S)*-**12** and (1S,2S)-**13**
[Reprinted with permission Editor-in-chief from (Kim et al., 2011)]

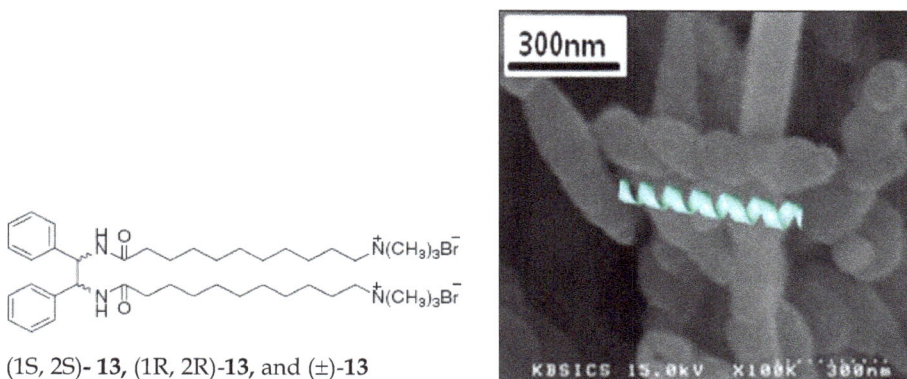

(1S, 2S)- **13**, *(1R, 2R)*-**13**, and (±)-**13**

Fig. 12. SEM image of right-handed helical silica nanotubes obtained from the 1:1 mixture of *(1S,2S)*-**12** and *(1R,2R)*-**13**
[Reprinted with permission Editor-in-chief from (Kim et al., 2011)]

In all the experiments silica nanotubes with right-handed helical structure was obtained. Similarly when the neutral gelator *(1R, 2R)*-**12** was associated with enantiomers of cationic amphiphile **13** the resulting silica nanotubes were of a left-handed helical structure. Racemic mixture of neutral gelator **12** associated with any of the stereoisomers of **13** yielded only non-helical silica tubes.

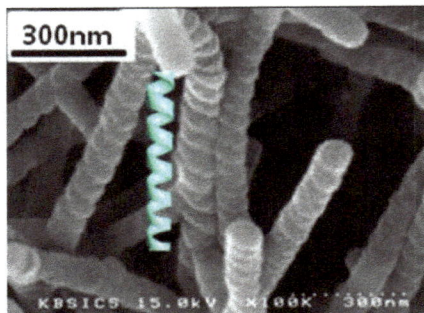

Fig. 13. SEM image of left-handed helical silica nanotubes obtained from the 1 : 1 mass mixture of (1R,2R)-**12** and (1R,2R)-**13**

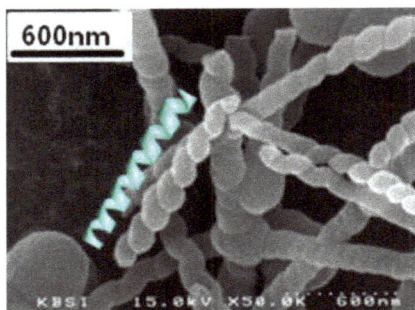

Fig. 14. SEM image of right-handed helical silica nanotubes obtained from the 1 : 1 mass mixture of (1R,2R)-**12** and (1S,2S)-**13**

The higher purity and yield of helical silica nanotubes obtained by association of diphenylethylenediamine derivatives **12** and **13** as compared with 1,2-cyclohexanediamine derivatives **8-11** may be explained by the existence of an effective π-π stacking interaction between the two phenyl groups essential to the self assembly process in the first case. As proved by the crossover experiments the association of neutral gelator and cationic amphiphile is essential for the hadedness of helical silica nanotubes which is controlled by the stereochemistry of the neutral gelator.

Double-walled silica nanotube

For certain amphiphiles with a polar head and a suitable chiral hydrophobic group self-assembly leads to aggregates with tubular structure generated by a helical ribbon. For exemple 30-crown-10-appended cholesterol gelator, **14** reported by Jung et al., (2001; 2003) with multiple binding sites (for acidic proton and cation, respectively) and two cholesterol skeletons which insure the necessary chirality of the aggregation, forms a tubular structure in the gel system probably generated by a helical ribbon as presented in the TEM image in figure 15.

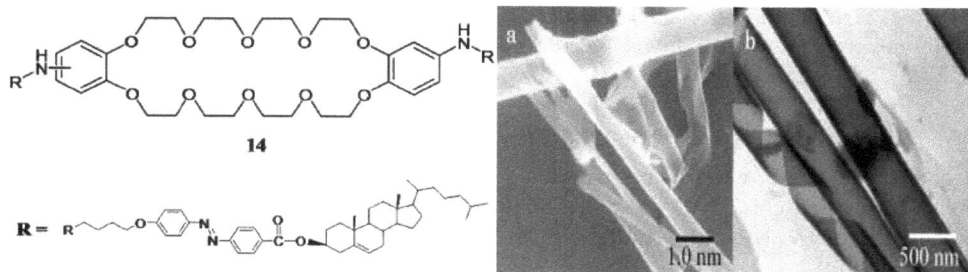

Fig. 15. SEM (a) and TEM (b) pictures of the xerogel **14** prepared from acetic acid. [Reprinted from (Jung et al., 2003) with permission from John Wiley and sons].

Sol-gel polycondensation of TEOS mediated by **14** in the presence of acetic acid yielded, after calcinations a right-handed helical ribbon structure 450-1500 nm width (Fig. 16) and a constant outside diameter of ~ 560 nm. A double layer structure was also observed with an interlayer distance of 8-9 nm (Fig. 16, c) which could be explained by the absorption of TEOS or oligomeric silica particles on both sides/surfaces of the organogelator tubules. Thus, after calcination, smaller cavities with layers of 8-9 nm were generated by the wall of the tubules formed by organogelator whereas the inner cavities with almost constant diameter were created the growth of the helical ribbon.

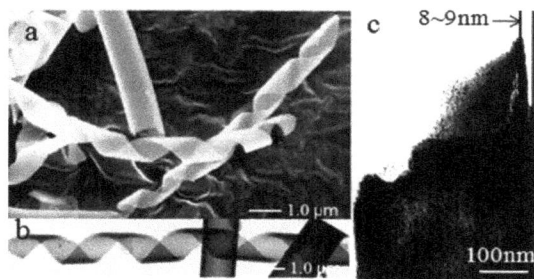

Fig. 16. (a) SEM and (b and c) TEM images of the double-walled silica nanotubes obtained from organogel 14. [Reprinted with Permission from (Jung et al., 2001), Copyright 2001, American Chemical Society]

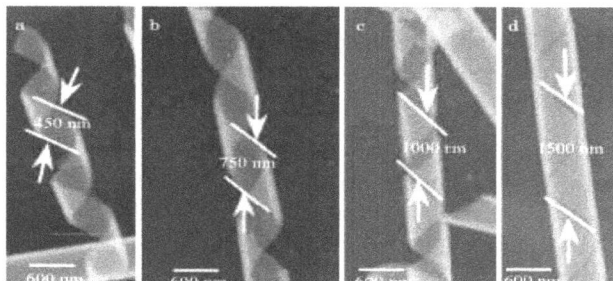

Fig. 17. TEM pictures (a, b, c, and d) of the silica obtained from the **14**- acetic acid gel after calcinations. [Reprinted from (Jung et al., 2003) with permission from John Wiley and sons]

Such materials, with helical higher-order morphology obtained by transcription of chiral assemblies created by weak intermolecular forces, are very useful as chiral catalysts (Sato et al., 2003).

Mesoporous-type helical silica nanofibers

Mesoporous – type helical silica nanofibers were fabricated by Hanabusa et al. using an amino-acid based chiral cationic hydrogelator **15** (Yang et al., 2006). Sol-gel transcription performed in acidic conditions under a shear flow yielded a material consisting of 300 nm in diameter bundles of ultrafine right-handed helical silica nanofibers with 50nm diameters (Fig. 18 a). TEM analysis revealed that the nanofibers present mesoporous pores of 2nm (Fig. 18 b).

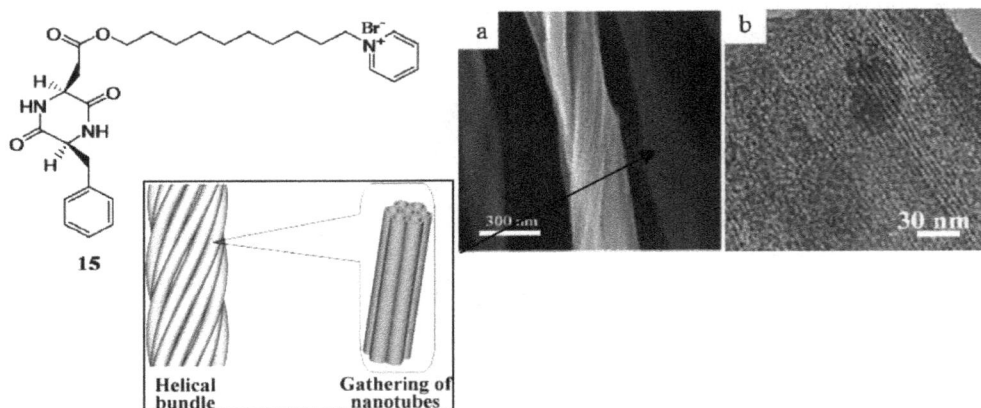

Fig. 18. (a) SEM and (b) TEM images of the right-handed helical silica nanotube with mesopores obtained from hydrogel 15. [Reprinted with Permission from (Yang et al., 2006), Copyright 2006, American Chemical Society]

Such hierarchical structure for the mesoporous silica nanofibers results from a particular formation mechanism: self- assembly of the chiral gelator **15** produces helical single strand gel fibrils on which the silica oligomers are absorbed and the sol-gel polycondensation of silica precursor begins on the surface. This process is parallel with the association of single strand gel-fibrils into multiple strand fibrils which are gathered in helical bundles. Alignment of the formed helical nanofibers is achieved under shearing, Fig. 19.

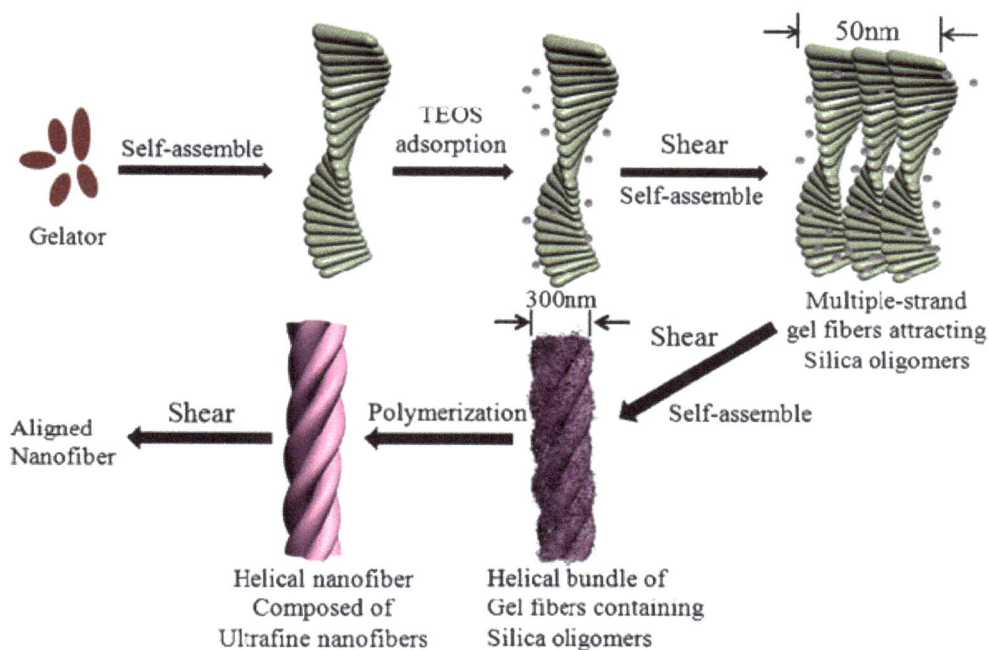

Fig. 19. Schematic representation of formation of mesoporous righthanded helical nanofibers and alignment. [Reproduced from (Jung et al., 2010) by permission of The Royal Society of Chemistry]

Mesoporous silica nanotubes

Hierarchiral silica nanotubes with radially oriented mesoporous channels perpendicular to the central axis of the tube were synthesized by sol-gel polycondensation of TEOS in the presence of the self-assembly anionic surfactant, partially neutralized carboxylate C14-L-AlaS, **16**, a co-structure directing agent, TMAPS (N-trimethoxysilylpropyl- N,N,N-trimethyl ammonium choride and acid catalysis at different molar ratios, C14-L-AlaS:TMAPS: TEOS: HCl: $H_2O\frac{1}{4}$1:x:7:y:1780, (x= 0.1–0.5 ; y =0.3–0.6) (Yu et al., 2008).

SEM and TEM images (Fig. 20) for the calcined silica nanotubes revealed that tube diameter decreased with the degree of neutralization of surfactant **16** and silica tube wall thickness increased both with de degree of neutralization of **16** and the TMAPS/ surfactant molar ratio.

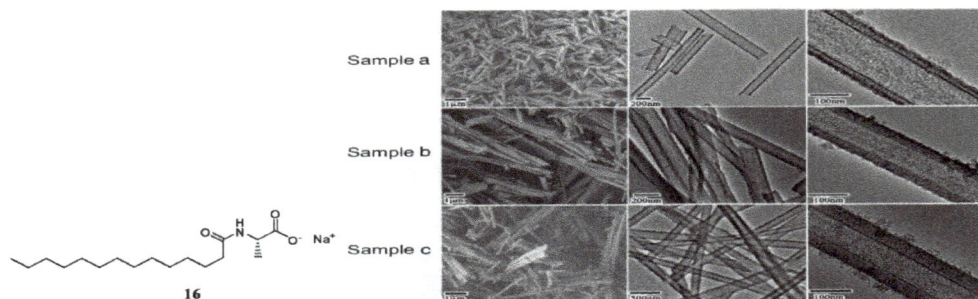

Fig. 20. SEM and (L-TEM and H-TEM images of the calcined mesoporous silica nanotubes synthesized with different degrees of neutralization of **16** (a and b) TMAPS/ surfactant molar ratio (b and c). [Reprinted from (Yu et al., 2008) with permission from John Wiley and sons]

The chiral surfactant has a helical supramolecular aggregation forming tubules with lamellar structured wall composed of several spring-like coiled bilayer structures (Fig. 21 a - c). On addition, TMAPS and TEOS can penetrate into both sides of the tubular cylinder assembly and create by re-assembly the mesoporous structure (Fig. 21 d, e).

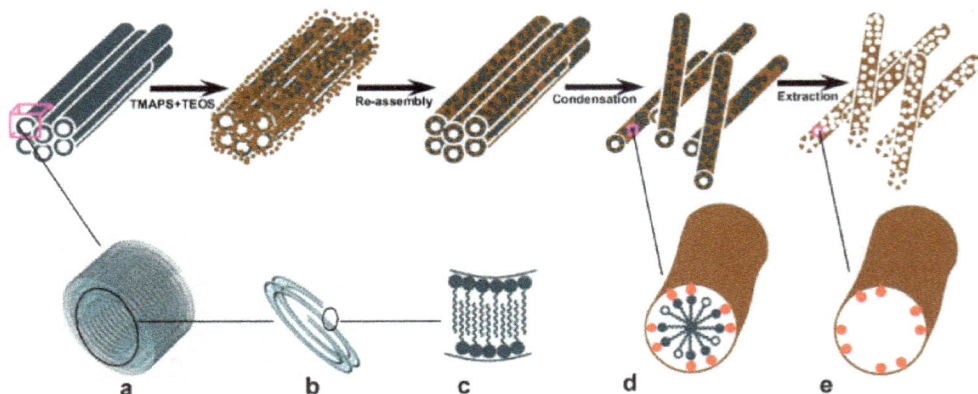

Fig. 21. Schematic illustration of the mesoporous silica nanotube formation process. [Reprinted from (Yu et al., 2008) with permission from John Wiley and sons]

5.2 Preparation of silica nanotubes using inorganic templates

Preparation of silica nanotubes using inorganic templates is based on two main classes of inorganic materials, porous inorganic membranes and inorganic materials, for example carbon nanotubes. Although the first method has limited versatility, especially for the synthesis of chiral materials, it provides the easiest way to control nanotube size and shape because benefits from the use of fabricated inorganic membrane templates with uniform in size cylindrical pores. The inorganic template synthesis of silica nanotubes was developed by Martin group (Hillebrenner et al., 2006; Kang et al., 2005; Martin, 1996; Mitchell et al., 2002) using porous alumina membranes and a sol-gel coating technique. Porous alumina templates can be prepared by electrochemical anodization on aluminum plate, procedure established by Masuda and Fukuda (1995) and developed lately (Lee et al., 2006) with pore dimension ranging from five to a few hundred nanometers and length from tens of nanometer to hundred of nanometers depending on the anodization time, potential, electrolyte, etc. A (Fig. 22).

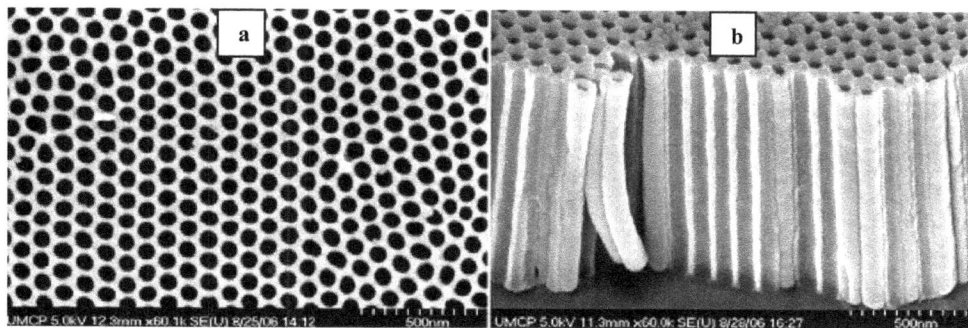

Fig. 22. Field emission scanning electron micrographs (FESEM) of home-made alumina template (60-nm diameter); (a) Top-viewed image and (b) cross-sectional viewed image [Reprinted from (Lee et al., 2006), Copyright 2007, with permission from Elsevier]

A typical template synthesis of silica nanotube using porous alumina membrane is presented in Fig. 23: in the first step thin layers of silica are formed by sol-gel chemistry onto the cylindrical walls of nanopores of the membrane, and then the top layers on both sides are removed by mechanical polishing. The alumina template is selectively dissolved in 25% phosphoric acid to liberate single silica nanotubes (free-standing nanotubes) which are collected by filtration. Thus, depending on the technical characteristics of the membrane as pore diameter and length and the extend of the sol-gel process tubules, test tubes or fibers may be obtained.

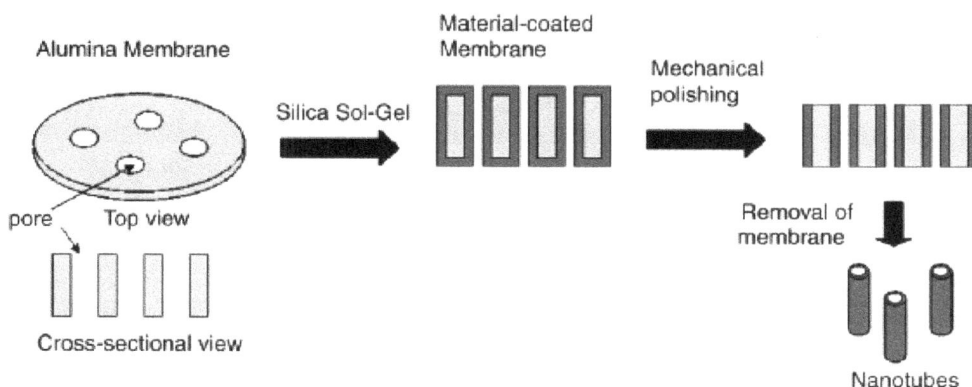

Fig. 23. Schematic illustration of template synthesis using a porous alumina membrane. [Reprinted from (Son et al., 2007), Copyright 2006, with permission from Elsevier]

Another advantage of the inorganic template synthesis of silica nanotubes is the possibility to apply, after the generation of nanotubes in the pores of the alumina template, an useful surface modification method called *diferential functionalization* of the inner and outer surfaces (Mitchell et al., 2002; Son et al., 2006; 2007) using silane chemistry (Fig. 24). The functionalization of the inner surface of the nanotube is performed selectively while still embedded within the pores of the membrane using different silane derivatives. The outer surface of the nanotube, in contact to the pore wall of the template is masked and protected. After dissolution of the template, the free outer surface is modified by a second functionalization.

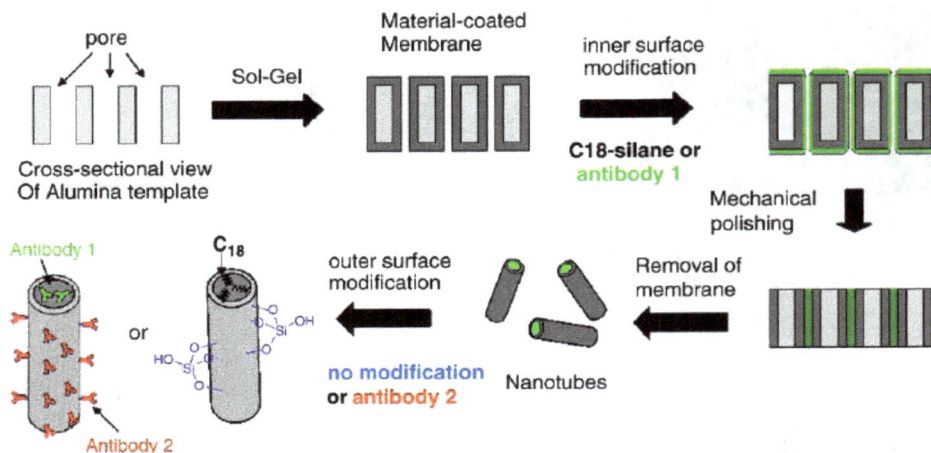

Fig. 24. Schematic of differential functionalization procedure for nanotubes obtained by inorganic template synthesis. [Reprinted from (Son et al., 2007), Copyright 2006, with permission from Elsevier]

Differentially functionalized nanotubes were used as smart nanophase extractors to remove molecules from solution: a hydrophilic outer surface and a hydrophobic inner cavity may be used for extracting lypophilic molecules from aqueous solution (Mitchell et al., 2002). A refinement of the method may lead to molecule-specific nanotubes when the surfaces are modified with enzymes, antibodies, etc. using an intermediate modification of the surfaces with aldehyde silane and attachment of the protein by reaction with free amino groups *via* Schiff base chemistry.

Magnetic silica nanotubes

Magnetic nanoparticles proved to be useful tools in drug-delivery systems, biosensors, different separation processes, enzyme encapsulation and contrast enhancement in magnetic resonance imaging (MRI). Differential functionalization of the spherical particles encounters technical difficulties because of a limited amount of reactive groups and the competition between the two reagents used to introduce different functionalities. This is why magnetic silica nanotubes can be an attractive alternative because they have all the previously described advantages of silica nanotubes-facile synthesis, differential functionalization of outer surfaces with environment friendly and/or probe molecules to identify specific targets, a geometry that allows a high degree of loading with high amounts of the desired chemical or biological species inside – and the specific applications of magnetic particles for example as targeting drug delivery with MRI capability or magnetic field assisted chemical and biochemical separations (Son et al., 2005). Magnetic silica nanotubes have been synthesized by layer-by-layer deposition methods of preformed magnetite colloidal particles or molecular precursor onto the inner surface of pre-prepared silica nanotubes embedded in a porous alumina template. Another possibility is to perform "surface-sol-gel"(SSG) methods which involves a repeat of two-step deposition cycles, in which the absorption of molecular precursor and the hydrolysis (in the case of oxide film growth) are separated by a post-absorption wash (Kovtyukhova et al., 2003). In SSG procedure, in order to achieve high reproducibility and homogeneity, TEOS is replaced with silicon tetrachloride because its reaction with active hydroxyl groups from the alumina template or silica surface is fast and quasi stoichiometric. The SSG method is composed of multiple cycles of $SiCl_4$ treatment and hydrolysis process, each cycle adding ~1nm of silica on the surface. The layer of magnetite particles was generated after several cycles of silica deposition by dip-coating with a 4:1 mixture solution of 1M $FeCl_3$ and 2M $FeCl_2$, followed by treatment with aqueous ammonia (Son et al., 2006). Magnetic silica nanotubes were prepared by classical sol-gel procedure by polycondensation of TEOS mediated by a cholesterol-based gelato pre-impregnated with nickel acetate. After thermal treatment a reducing agent was used to mineralize Ni nanocrystals on the silica nanotube (Bae et al., 2008).

5.3 The use of silica nanotubes for biological applications

Facile synthetic pathways, the ability to control size and shape, including chirality and the possibility to functionalize the inner and outer surfaces of silica nanotubes in order to specifically accommodate different molecules and finally multifunctionality given by association with magnetic particles or fluorescent molecules enables us to affirm that silica nanotubes play an important and increasing role in biological sciences as drug delivery, imaging and screening, targeting and cell recognition, etc. Several recent examples of contributions made by silica nanotubes in biological sciences are presented in the following sections.

Recognition of protein containing cysteine groups

Magnetic silica nanotubes containing Ni were functionalized with mercaptopropyl-triethoxysilane in order to dope the surface with gold nanoparticles (Fig. 25) which specifically recognized proteins with cysteine groups (Bae et al., 2008).

The biomolecular recognition of Au-doped magnetic silica nanotubes (**Au-MSNT**) was verified by binding of glutathione S-transferase (GST) a protein containing cysteine groups and examination of selective binding of protein on the surface by an immunofluorescence method: anti–GST antibodies could bind specifically to GST on the surface of **Au-MSNT** and can be visualized by confocal microscopy after subsequent binding of fluorescent secondary antibodies (Fig. 26 A). The specificity of the interaction between **Au-MSNT** and cysteine was demonstrated by repeating the experiment with ubiquitin, a protein lacking cysteine from which no fluorescence was observed due to the impossibility to bind to the functionalized nanotubes (Fig. 26 B).

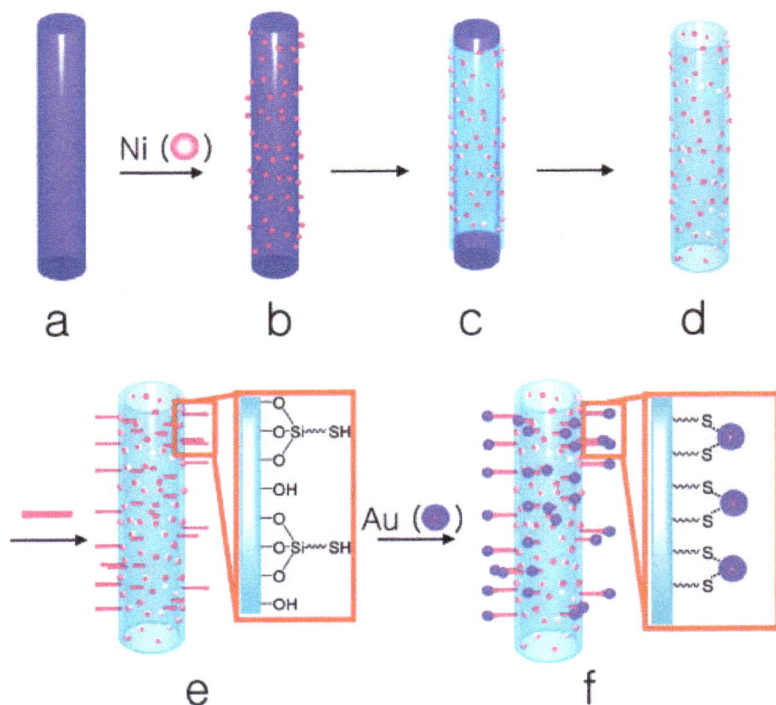

Fig. 25. Overall schemes for the synthesis of the gold-doped silica nanotube obtained from sol-gel transcription. [Reprinted with Permission from (Bae et al., 2008), Copyright 2008, American Chemical Society]

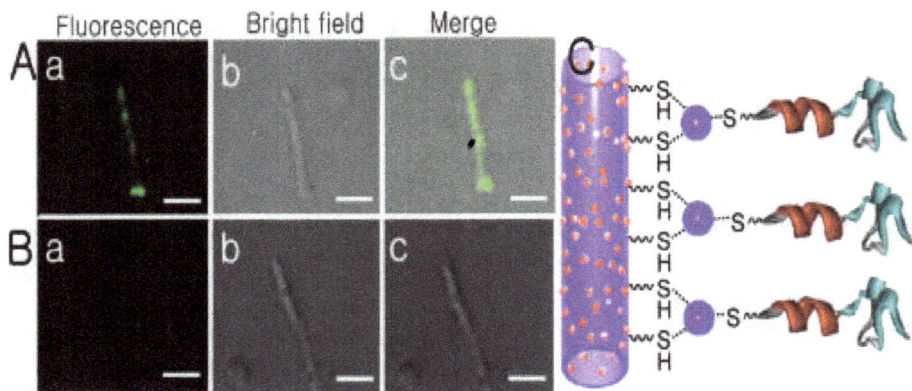

Fig. 26. Confocal images of **Au-MSNT** with (A) GST by treatment of anti-GST antibody and (B) ubiquitin by treatment of antiubiquitin antibody: (a) fluorescence, (b) bright field, and (c) merge images. (C). Illustration for binding mode of **Au-MSNT** with GST protein. Scale bar: 2 μm. [Reprinted with Permission from (Bae et al., 2008), Copyright 2008, American Chemical Society]

Specificity of Au-MSNT for cysteine may be exploited for a biosensor to detect cysteine containing proteins.

Separation of oligonucleotides

Detection of a specific messenger RNA biomolecule can serve as an indicator of the expression of its corresponding protein and as a diagnostic method for some diseases. This can be achieved, for example by specific recognition of an oligoadenosine tail by a solid support bearing covalently attached oligodeoxythymidine through the formation of specific A-T base pairs. A simple and accurate method to separate oligoadenosine derivatives was developed using nucleic acids functionalized silica nanotubes. Sol-gel template synthesis of helical silica nanotubes previously described (Jung et al., 2000) was followed by protection of outer surface with chloropropylsilane and after template removal the inner surface was functionalized with a thymidine derivative as a receptor for adenosine derivatives (Kim et al., 2010). The thymidine-immobilized silica nanotubes (**T-SNTs**) exhibited a well-defined tubular structure with 20 nm of inner diameter and 100 nm of outer diameter (Fig. 27 a, and b), which showed slight aggregation as the result of outer-surface modified by the attachment of chloropropyl silane.

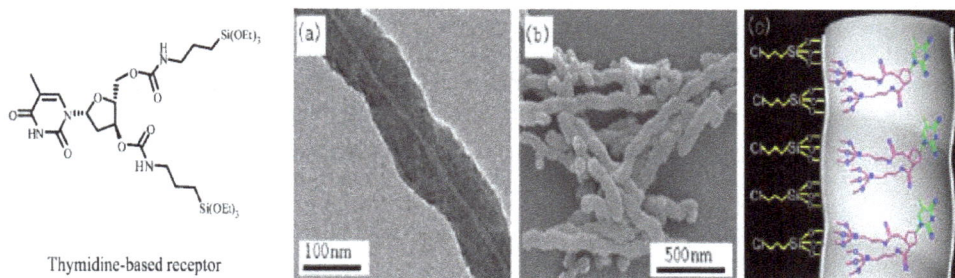

Fig. 27. (a) TEM and (b) SEM images of T-SNTs. (c) Representation of immobilization of thymidine receptor inside SNTs by covalent bonds. [Reproduced from (Kim et al., 2010) by permission of The Royal Society of Chemistry]

Adsorption capacities of **T-SNTs** for nucleic acids and oligonucleotides and specificity for oligoadenosine derivatives were tested on guest molecules containing adenosine and guanosine moieties. Fluorescence spectra of oligoadenosine derivative and oligoguanosine derivative, before and after addition of **T-SNTs**, combined with HPLC analyses demonstrated selectivity for adenosine derivatives (>95%) through the efficient formation of complementary hydrogen bonds in A-T pairs. The presence of the selectively bonded oligoadenosine derivative on **T-SNTs** was demonstrated by a strong fluorescence of the nanotubes isolated after separation (Fig. 28).

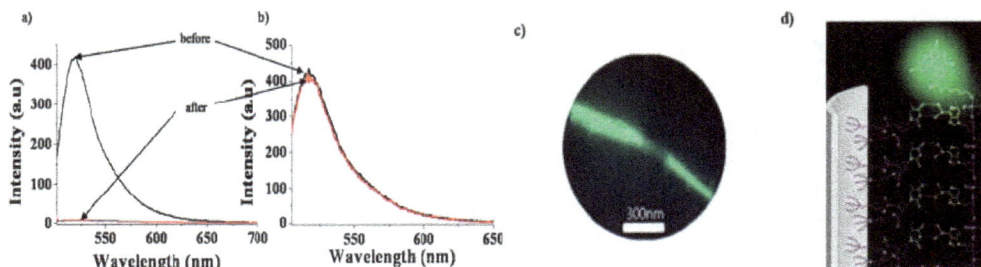

Fig. 28. (a) Fluorescence spectra of oligoadenosine and (b) oligoguanosine (λ = 519 nm) before and after addition of T-SNTs at 37 °C. (c) Fluorescence microscopic image of T-SNTs in the presence of oligoadenosine (λ = 519 nm). d) Representation for binding mode of T-SNTs with 24 by complementary hydrogen bonds between A-T base pairs. [Reproduced from (Kim et al., 2010) by permission of The Royal Society of Chemistry]

Bioseparation of a racemic mixture

Antibody-functionalized nanotubes obtained by differential functionalization were used to separate racemic mixtures. Thus inner and outer surfaces of silica nanotubes obtained with a porous alumina membrane template were functionalized with an aldehyde silane attached before or after they were liberated from the template. The RS selective Fab fragments of the antibody produced against the drug 4-[3-(4-fluorophenyl)-2-hydroxy-1-[1,2,4]-triazol-1-yl-propyl]-benzonitrile (**FTB**) were attached to the nanotubes *via* Schiff base chemistry of aldehyde moieties with free amino groups of the antibody. Racemic mixtures of **FTB** were incubated with Fab-functionalized nanotubes and the efficiency of the chiral separation (selective drug removal) was measured by chiral HPLC (Fig. 29). The efficiency of the

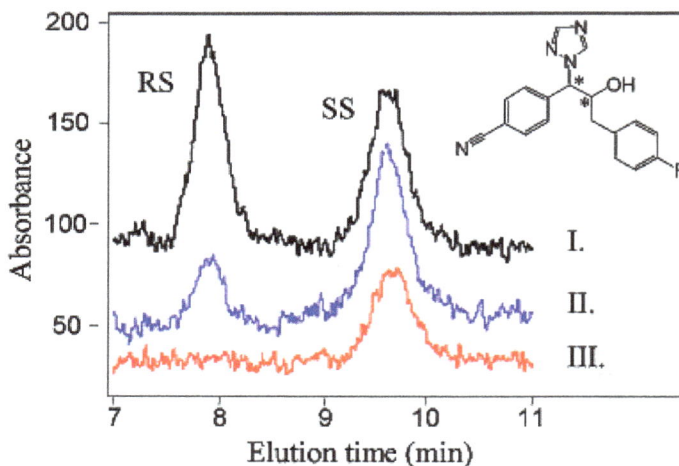

Fig. 29. Chiral HPLC chromatograms for racemic mixtures of FTB before (I) and after (II, III) extraction with 18 mg/mL of 200-nm Fab-containing nanotubes. [Reprinted with Permission from (Mitchell et al., 2002), Copyright 2002, American Chemical Society]

selective extraction of RS enantiomer of **FTB** with Fab-functionalized silica nanotubes depends on the initial concentration of the racemic mixture: 75% for 20μM (II) and 100% for 10μM (III). Non-functionalized nanotubes did not extract measurable quantities of each **FTB** enantiomer from 20μM solution.

Bioseparations by using magnetic silica nanotubes

The ability of functionalized magnetic silica nanotubes (**MNT**) to accommodate targets in the inner void was exploited in magnetic-field-assisted bioseparations by preparing materials with inner surface bonded human immunoglobulins (human-**IgG**) or Bovine Serum Albumin (BSA) using glutardialdehyde as a coupling agent (Son et al., 2005). The human **IgG-MNT** and the **BSA-MST** were tested in separation of a mixture of fluorescein-labeled anti-bovine IgG (green color) and Cy3-labelled antihuman IgG (red). After magnetic separation, the solution changed from the original pink to greenish blue only when **IgG-MNTs** were added, while the solution with **BSA-MNT** remained in its original pink color. This means that red Cy3-labeled anti-human IgG was separated specifically from the solution by human IgG-MNTs. Fluorescence spectra showed that 84% of Cy3-labeled anti-human IgG was separated by human **IgG-MNT** but only 9% by **BSA-MNT**.

Magnetic properties of **IgG-MNT** facilitate and enhance the biointeraction between the outer surface of a functionalized **MNT** and a specific target surface when a magnetic field is applied. Thus, a **MNT** with fluorescein isothiocyanate (**FITC**)-modified inner surface and Rabbit **IgG**-modified outer surface were incubated with the anti-rabbit IgG-modified glass slide for 10 min with and without magnetic field from the bottom of the glass slide (Son et

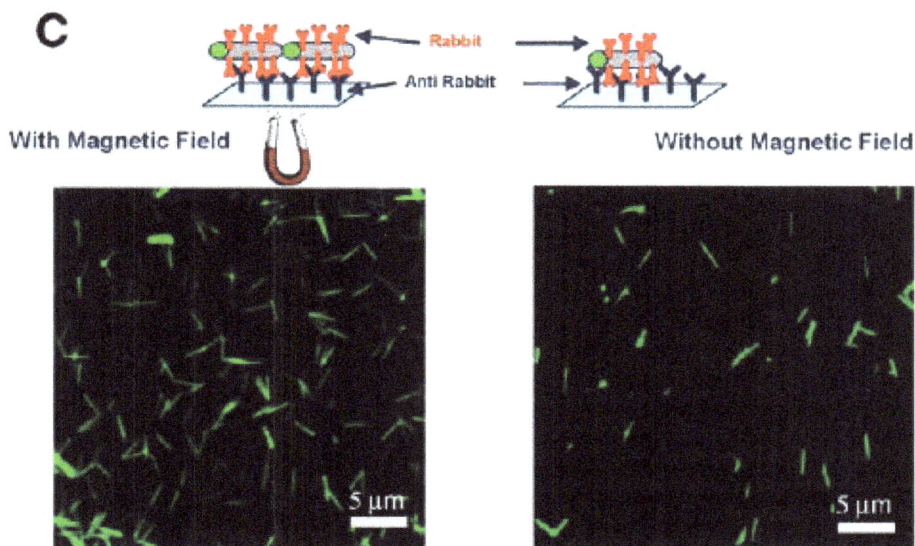

Fig. 30. Fluorescence microscope image of bound FITC-**MNT**- Rabbit **IgG** (60 nm diameter, 3 μm) after incubation with anti-rabbit IgG modified glass with and without magnetic field from the glass substrate. [Reprinted with Permission from (Son et al., 2005) Copyright 2005, American Chemical Society]

al., 2005). The efficiency of antigen–antibody interactions and the influence of magnetic field were investigated by fluorescence microscopy, revealing that binding was enhanced by 4.2 fold when the magnetic field was applied (Fig. 30). Magnetic-field assisted bio-interaction may improve significantly the drug-delivery efficiency when the carrier is a **MNT** loaded with drug inside and having probe molecules such as an antibody, on the outer surface.

Gene delivery

Green and red fluorescent silica nanotubes prepared by layer-by-layer deposition of fluorescent CdSe/ZnS core-shell semiconductor nanocrystals (of ~4nm and ~8nm diameters, respectively) onto the inner surface of pre-prepared silica nanotubes embedded in a porous alumina template (Chen et al., 2005). The inner surfaces were then functionalized with 3-(aminopropyl) trimethoxysilane (APTMS) ligands to facilitate the subsequent loading of negatively charged DNA by electrostatic interactions. The fluorescent nanotubes were loaded with plasmid DNA –carrying the GFP gene labeled with green DNA-stain SYTO-11 in order to monitor DNA localization. Monkey-kidney COS-7 cell were treated with the DNA loaded fluorescent nanotubes and confocal microscopy of the incubated samples showed that nanotubes entered in 60-70% of the cells and they are located mostly in the cytoplasm. Cytotoxicity of DNA loaded fluorescent nanotubes was investigated as well as the protection provided by the silica tube wall from environmental damage. The results indicate that the GFP gene can be loaded into silica nanotubes and successfully delivered to cells.

Immobilization of enzyme catalysts

Hollow silica nanotubes have been utilized as matrices for immobilization of enzymes in order to improve their catalytic efficiency. Glucose-oxidase (GOD) was loaded both to the inner and outer surfaces of silica nanotubes *via* the aldehyde silane route. Dispersed in a solution containing 90 nM glucose, the enzyme activity was monitored by standard dianisidine-based assay and a GOD activity of 0.5 ± 0.2 units/mg of nanotubes was obtained [88]. When the enzyme doped-nanotubes were filtered from the solution, all GOD activity ceased, an indication that immobilization *via* Schiff base chemistry is efficient, no protein leached from the support. Another example is immobilization of Penicillin G acylase (PGA) porous hollow silica nanotubes synthesized *via* a sol–gel route using nano-sized needle-like $CaCO_3$ inorganic templates (Xiao et al., 2006). PGA uptake onto the silica nanotubes was 97.20% and with an adsorption equilibrium time of the enzyme on suport of ca. 120 min, which is far faster than those previously reported supports such as pure silica SBA-15, MCM-41 and poly(vinylacetate-co-divinylbenzene) due to pore size and uniquely large entrances at two ends of the nanotubes.

One can be concluded that the template synthesis of silica nanotubes using organic and inorganic templates presents important advantages as the possibility to control lengths, diameters and wall thickness by adjusting the reaction parameters, obtaining of specific tubular nanostructures as multilayered, chiral or helical, multifunctionality by embedding magnetic particles or fluorescent molecules.

Moreover, functionalization of inner and outer surfaces improved molecular interactions with species of biological interest.

6. References

Adamczyk, Z. & Weronski, P., Unoriented adsorption of interacting spheroidal particles. J Colloid Interface Sci. (1997). 189, pp. (348-360).

Adamczyk, Z., Siwek, B., Zembala, M. & Belouschek, P. Kinetics of localized adsorption of colloid particles. Adv Colloid Interface Sci (1994). 48, pp. (151-280).

Adamczyk, Z., Weronski, P. & Musial, E. Particle adsorption under irreversible conditions: kinetics and jamming coverage. Colloids and Surfaces a-Physicochemical and Engineering Aspects (2002), 208, pp. (29-40).

Adamczyk, Z., Zembala, M., Siwek, B., Warszynski, P. Structure and ordering in localized adsorption of particles. J Colloid Interface Sci (1990), 140, pp. (123-137).

Bae, D.R., Lee, S.J., Han, S.W., Lim, J.M, Kang, D. & Jung, J.H. Au-Doped Magnetic Silica Nanotube for Binding of Cysteine-Containing Proteins. Chem. Mater. (2008). 20, pp. (3809-3813).

Bagshaw, S.A., Prouzet, E. & Pinnavaia, T.J. Templating of Mesoporous Molecular Sieves by Nonionic Polyethylene Oxide Surfactants. Science. (1995). 269, pp. (1242 –1244).

Banerjee S., Das R.K. & Maitra, U. Supramolecular gels in action. J. Mater. Chem. (2009). 19, pp. (6649–6687).

Beck, J.S., Vartuli, J.C., Roth, W.J., Leonowicz, M.E., Kresge, C.T., Schmitt, K.D., Chu C.T. W., Olson, D.H., Sheppard, E.W., McCullen, S.B., Higgins, J.B. & Schlenker, J.L. A new family of mesoporous molecular sieves prepared with liquid crystal templates. J. Am. Chem. Soc. (1992a). 114, pp. (10834–10843).

Botterhuis, N.E., Sun, Q., Magusin, P.C.M., Van Santen, R.A., Nico, A.J.M. & Sommerdijk, N.A.J.M. Hollow silica spheres with an ordered pore structure and their application in controlled release studies. Chem. Eur. J. (2006). 12, pp. (1448–1456).

Boyd, P.W. (2007). *Science,* 315, pp. (612–617).

Brusatori, M.A. & Van Tassel, P.R. A kinetic model of protein adsorption/surface-induced transition kinetics evaluated by the scaled particle theory. J Colloid Interface Sci. (1999). 219, pp. (333-338).

Brzezinski, M.A., Nelson, D.M. (1996). *Deep-Sea Res I,I* 43, pp. (437–453).

Cha, J.N., Shimizu, K., Zhou, Y., Christiansen, S.C., Chmelka, B.F., Stucky, G.D. & Morse, D.E. (1999). Silicatein filaments and subunits from a marine sponge direct the polymerization of silica and silicones in vitro, *Proc. Nat. Acad. Sci. USA,* 96, pp. (361-365).

Chen, C.C., Liu, Y.C., Wu, C.H., Yeh, C.C., Su M.T. & Wu, Y.C. Preparation of fluorescent silica nanotubes and their application in gene delivery. Adv. Mater. (2005). 17, pp. (404–407).

Courrier, H.M., Butz, N. & Vandamme, T.F. Pulmonary drug delivery systems: recent developments and prospects. Crit Rev Ther Drug Carrier Syst. (2002). 19, pp. (425-98).

Croce, G., Frache, A., Milanesio, M., Marchese, L., Causà, M., Viterbo, D., Barbaglia, A., Bolis, V., Bavestrello, G., Cerrano, C., Benatti, U., Pozzolini, M., Giovine, M. & Amenitsch, H. (2004). Structural characterization of siliceous spicules from marine sponges, *Biophys. J.,* 86, pp. (526-534).

Desai, M.P. Gastrointestinal uptake of biodegradable microparticles: effect of particle size. Pharm Res. (1996a). 13, pp. (1838- 45).

Fairhead, M., Johnson, K.A., Kowatz, T., McMahon, S.A., Carter,L.G., Oke, M., Liu, H., Naismith, J.H. & Van der Walle C.F. (2008). Crystal structure and silica condensing activities of silicatein – cathepsin L chimeras. Chemical Communications 15, pp. (1765-1767).

Fang, F. & Szleifer, I. Kinetics and thermodynamics of protein adsorption: a generalized molecular theoretical approach. Biophys J. (2001). 80, pp. (2568-2589).

Field, C.B., Behrenfeld, M.J. & Falkowski, P. (1998). *Science*, pp. (281-237).

Field, C.B., Behrenfeld, M.J., Randerson, J.T. & Falkowski, P. (1998). Primary production of the biosphere: integrating terrestrial and oceanic components, *Science*, 281, pp. (237-240).

Flodstrçm, K., Wennerstrçm, H., Teixeira, C.V., Amenitsch, H., LindRn, M. & Alfredsson, V. Time-Resolved in Situ Studies of the Formation of Cubic Mesoporous Silica Formed with Triblock Copolymers, Langmuir. (2004c). 20, pp. (10311–10316).

Fuhrmann, T., Landwehr, S., El Rharbi-Kucki, M. & Sumper, M. (2004). Diatoms as living photonic crystals, *Appl. Phys. B, 78*, (257-6028).

Fukui, K., Ueki, T., Ohya, H. & Michibata, H. Vanadium-Binding Protein in a Vanadium-Rich Ascidian Ascidia sydneiensis samea: CW and Pulsed EPR Studies, J. Am Chem. Soc., (2003). 125, pp. (6352-6353).

Hamada, T., Asanuma, M., Ueki, T., Hayashi, F., Kobayashi, N., Yokoyama, S., Michibata, H. & Hirota, H. Solution Structure of Vanabin2, a Vanadium(IV)-Binding Protein from the Vanadium-Rich Ascidian Ascidia sydneiensis samea. J. Am Chem. Soc. (2005). 127(12), pp. (4216-4222).

Hillebrenner, H., Buyukserin, F., Stewart J.D. & Martin, C.R. Template synthesized nanotubes for biomedical delivery applications. Nanomedicine. (2006). 1, pp. (39-50).

Huang, X. & Weiss, R.G. Silica Structures Templated on Fibers of Tetraalkylphosphonium Salt Gelators in Organogels, *Langmuir*. (2006). 22, pp. (8542- 8552).

Hughes, G.A. Nanostructure-mediated drug delivery, Nanomedicine: Nanotechnology, Biology, and Medicine. (2005). pp. (122–130).

Hussain, N. & Jaitley, V. Florence AT. Recent advances in the understanding of uptake of microparticulates across the gastrointestinal lymphatics. Adv Drug Delivery Rev. (2001b). 50, pp. (107- 42).

Hyun, M.-H., Shin, M.-S., Kim, T.-K., Jung, O.-S., Kim, J.-P., Jeong, E.-D., & Jin, J. S. The role of the neutral and cationic gelators from (1S,2S)-(–)-diphenylethylene diamine for the preparation of silica nano tubes. Bulletin of the Korean Chemical Society. (2009), 30, pp. (1641–1643).

John, V.T., Simmons, B., McPherson, G.L. & Bose, A. Recent developments in materials synthesis in surfactant systems. Curr. Opin. Colloid Interface Sci. (2002e). 7, pp. (288–295).

Jung, J. H., John, G., Masuda, M., Yoshida, K., Shinkai, S. & Shimizu, T. Self-Assembly of a Sugar-Based Gelator in Water: Its Remarkable Diversity in Gelation Ability and Aggregate Structure, Langmuir. (2001). 17, pp. (7229).

Jung, J. H., Lee S.-H., Yoo, J.S., Yoshida, K., Shimzu, T. & Shinkai, S. Creation of Double Silica Nanotubes by Using Crown-Appended Cholesterol Nanotubes. Chem. Eur. J. (2003). 9, pp.(5307 – 5313).

Jung, J. H., Ono, Y. & Shinkai, S. Sol-Gel Polycondensation in a Cyclohexane-Based Organogel System in Helical Silica: Creation of both Right- and Left-Handed Silica Structures by Helical Organogel Fibers, Chem.–Eur. J. (2000). 6, pp.(4552).

Jung, J. H., Yoshida, K. & Shimizu, T. Creation of Novel Double-Helical Silica Nanotubes Using Binary Gel System , Langmuir. (2002), 18, pp. (8724).

Jung, J.H. & Shinkai, S. Gels as templates for nanotubes, Top. Curr. Chem. (2004b). 58, pp. (223-260).

Jung, J.H., Amaike, M. & Shinkai, S. Sol–gel transcription of novel sugar-based superstructures composed of sugar-integrated gelators into silica: creation of a lotus-shaped silica structure, Chem. Commun. (2000). pp. (2343–2344).

Jung, J.H., Kobayashi, H., Shimzu, T., Masuda, M. & Shinkai, S. Helical Ribbon Aggregate Composed of a Crown-Appended Cholesterol Derivative Which Acts as an Amphiphilic Gelator of Organic Solvents and as a Template for Chiral Silica Transcription. J. Am. Chem. Soc. (2001c). 123, pp. (8785-8789).

Jung, J.H., Kobayashi, H., Shimzu, T., Masuda, M. & Shinkai, S. Helical Ribbon Aggregate Composed of a Crown-Appended Cholesterol Derivative Which Acts as an Amphiphilic Gelator of Organic Solvents and as a Template for Chiral Silica Transcription. J. Am. Chem. Soc. (2001). 123, pp. (8785-8789).

Jung, J.H., Ono, J. & Shinkai, S. Sol-Gel Polycondensation in a Cyclohexane-Based Organogel System in Helical Silica Chem.–Eur. J. (2000a). 6, pp. (4552- 4557).

Jung, J.H., Ono, Y., Hanabusa, K.K. & Shinkai, S. Creation of Both Right-Handed and Left-Handed Silica Structures by Sol−Gel Transcription of Organogel Fibers Comprised of Chiral Diaminocyclohexane Derivatives. J. Am.Chem. Soc. (2000b). 122, pp. (5008- 5009).

Jung, J.H., Ono, Y., Sakurai, K., Sano, M. & Shinkai, S. Novel Vesicular Aggregates of Crown-Appended Cholesterol Derivatives Which Act as Gelators of Organic Solvents and as Templates for Silica Transcription. J. Am. Chem. Soc. (2000d). 122, pp. (8648- 8653).

Jung, J.H., Parka, M. & Shinkai, S. Fabrication of silica nanotubes by using self-assembled gels and their applications in environmental and biological fields. Chem. Soc. Rev. (2010). 39, pp. (4286–4302).

Kang, M.C., Trofin, L., Mota M.O. & Martin, C.R. Protein capture in silica nanotube membrane 3-d microwell arrays. Anal. Chem. (2005). 77, pp. (6243–6249).

Ke Li, Qiao-Lian Li, Nai-Yun Ji, Bo Liu, Wei Zhang & Xu-Peng Cao, Mar. Drugs (2011), 9.

Kim, N.H., Lee, H.Y., Cho, Y., Han, W.S., Kang, D., Lee, S.S. & Jung, J.H. Thymidine-functionalized silica nanotubes for selective recognition and separation of oligoadenosines. J. Mater. Chem. (2010). 20, pp.(2139-2144).

Kim, T.K., Jeong, E.D., Oh, C.Y., Shin, M.S., Kim, J.-P., Jung, O.-S., Suh, H., Rahman, F., Khan, N., Hyun, M.H. & Jin, J.S. Helical silica nanotubes: Nanofabrication architecture, transfer of helix and chirality to silica nanotubes. Chemical Papers. (2011). 65 (6), pp. (863–872).

Klueh, U., Seery, T., Castner, D.G., Bryers, J.D. & Kreutzer, D.L. Binding and orientation of fibronectin to silanated glass surfaces using immobilized bacterial adhesin-related peptides. Biomaterials (2003), 24, pp. (3877-3884).

Kovtyukhova, N.I. , Mallouk T.E. & Mayer T.S. Templated surface sol-gel synthesis of SiO_2 nanotubes and SiO_2-insulated metal nanowires. Adv. Mater. (2003). 15, pp. (780–785).

Krasko, A., Lorenz, B., Batel, R., Schröder, H.C., Müller, I.M. & Müller, W.E.G. (2000). Expression of silicatein and collagen genes in the marine sponge Suberites domuncula is controlled by silicate and myotrophin, Eur. J. Biochem., 267, pp. (4855-4874), (4878-4887).

Krasko, A., Lorenz, B., Batel, R., Schröder, H.C., Müller, I.M. & Müller, W.E.G. (2000). Expression of silicatein and collagen genes in the marine sponge Suberites domuncula is controlled by silicate and myotrophin, Eur. J. Biochem., 267, pp. (4855-4874), (4878-4887).

Kresge, C.T., Leonowicz, M.E., Roth, W.J., Vartuli, J.C. & Beck, J.S. Ordered mesoporous molecular sieves synthesized by a liquid-crystal template mechanism. Nature. (1992b). 359, pp. (710–712).

Kröger, N., Bergsdorf, C. & Sumper, M. (1996). Frustulins: domain conservation in a protein family associated with diatom cell walls, Eur. J. Biochem., 239, pp. (259-284).

Kröger, N., Lehmann, G., Rachel, R. & Sumper, M. (1997). Characterization of a 200-kDa diatom protein that is specifically associated with a silica-based substructure of the cell wall, *Eur. J. Biochem.,* 250, pp. (99-105).

Kröger, N., Lorenz, S., Brunner, E. & Sumper, M. (2002). Self-assembly of highly phosphorylated silaffins and their function in biosilica morphogenesis, *Science, 298, pp.* (584-586).

Latour, R.A. Jr. & Rini, C.J. Theoretical analysis of adsorption thermodynamics for hydrophobic peptide residues on SAM surfaces of varying functionality. Journal of Biomedical Materials Research. (2002). 60, pp. (564-577).

Lee, W., Ji, R., Sele, U.G. & Nielsch, K. Fast fabrication of long-range ordered porous alumina membranes by hard anodization. Nature Materials. (2006). 5, pp. (741-747).

Leynaert A., Treguer P., Lancelot C. & Rodier M. (2001). *Deep-Sea Res I,* 48, pp. (639–660).

Lin, H.P., Cheng, Y.R. & Mou, C.Y. Hierarchical Order in Hollow Spheres of Mesoporous Silicates. Chem. Mater. (1998c). 10, pp. (3772– 3776).

Llusar, M. & Sanchez, C. Inorganic and Hybrid Nanofibrous Materials Templated with Organogelators. Chem. Rev. (2008). 20, pp. (782-820).

Lu, D.R. & Park, K. Protein adsorption on polymer surfaces: calculation of adsorption energies. J Biomater Sci Polym Ed. (1990). 1, pp. (243-260).

Martin, C.R. Membrane-Based Synthesis of Nanomaterials. Chem. Mater. (1996). 8, pp. (1739- 1746).

Masuda, H. & Fukuda, K. Ordered Metal Nanohole Arrays Made by a Two-Step Replication of Honeycomb Structures of Anodic Alumina. Science. (1995). 268, pp. (1466–1468).

Menzel, H., Horstmann, S., Behrens, P., Bärnreuther, P., Krueger, I. & Jahns, M. (2003). Chemical properties of polyamines with relevance to the biomineralization of silica, *Chem. Commun.,* pp. (2994-2995).

Michibata, H., Yamaguchi, N., Uyama, T. & Ueki, T. Molecular biological approaches to the accumulation and reduction of vanadium by ascidians. Coord. Chem. Rev. 237. (2002). pp. (41-51).

Mitchell, D.T., Lee, S.B, Trofin, L., Li, N.C, Nevanen, T.K., Soderlund, H. & Martin, C.R. Smart Nanotubes for Bioseparations and Biocatalysis. J. Am. Chem. Soc. (2002). 124, pp. (11864).

Mock T., Samanta M.P., Iverson V., Berthiaume C., Robison M., Holtermann K., Durkin C., BonDurant S.S., Richmond K. & Rodesch M. (2008). *Proc Natl Acad Sci USA,* 105, pp. (1579–1584).

Morse, D. E. (1999). Silicon biotechnology: harnessing biological silica production to make new materials. Trends Biotechnol. 17, pp. (230-232).

Nelson, D.M., Treguer, P., Brzezinski, M.A., Leynaert, A. & Queguiner, B. (1995). *Global Biogeochem Cycles,* 9, pp. (359–372).

Noinville, V., Vidalmadjar, C. & Sebille, B. Modeling of protein adsorption on polymer surfaces-computation of adsorption potential. J Phys Chem. (1995). 99, pp. (1516-1522).

Oberholzer, M.R., Lenhoff, A.M. Protein adsorption isotherms through colloidal energetics. Langmuir (1999), 15, pp. (3905-3914).

Oberholzer, M.R., Wagner, N.J. & Lenhoff, A.M. Grand canonical Brownian dynamics simulation of colloidal adsorption. J Chem Phys (1997), 107, pp. (9157-9167).

Ono, Y., Nakashima, K., Sano, M., Kanekiyo, K., Inoue, K., Hojo, J. & Shinkai, S. Organic gels are useful as a template for the preparation of hollow fiber silica. Chem. Commun. (1998), pp. (1477- 1478).

Ormos, P., Fabian, L., Oroszi, L., Wolff, E.K., Ramsden, J.J. & Der A. Protein-based integrated optical switching and modulation. Appl Phys Lett (2002), 80, pp. (4060-4062).

Patwardhan, S.V., Clarson, S.J. & Perry, C.C. (2005). On the role(s) of additives in bioinspired silicification, *Chem. Commun., pp.* (1113-1121).

Perry, C.C. & Keeling-Tucker, T. (2000). Biosilicification: The role of the organic matrix in structure control, *J. Bio. Inorg. Chem.,* 5, pp. (537-550).

Pouget, E., Dujardin, E., Cavalier, A., Moreac, A., Valery, C., Marchi-Artzner, V., Weiss, T., Renault, A., Paternostre, M. & Artzner, F. (2007). Hierarchical architectures by synergy between dynamical template self-assembly and biomineralization, *Nat. Mater.,* 6, pp. (434-439).

Poulsen, N., Sumper, M. & Kröger, N. (2003). Biosilica formation in diatoms: Characterization of native silaffin-2 and its role in silica morphogenesis, *Proc. Natl. Acad.Sci. U.S.A.,* 100, pp. (12075-12080).

Sangeetha, N.M. & Maitra, U. Supramolecular gels: Functions and uses. Chem. Soc. Rev. (2005). 34,pp. (821-836).

Sarikaya, M., Tamerler, C., Jen, A.K., Schulten, K. & Baneyx, F. Molecular biomimetics: nanotechnology through biology. Nat Mater (2003), 2, pp. (577-585).

Sato, I., Kadowaki, K., Urabe, H., Jung, J.H., Ono, Y., Shinkai, Se. & Soai, K. Highly enantioselective synthesis of organic compound using right- and left-handed helical silica, Tetrahedron Lett.Volume. (2003). 44, pp. (721-724).

Satulovsky, J., Carignano, M.A. & Szleifer, I. Kinetic and thermodynamic control of protein adsorption. Proc Natl Acad Sci USA. (2000), 97, pp. (9037-9041).

Sayari, A. & Hamoudi, S. Periodic Mesoporous Silica-Based Organic–Inorganic Nanocomposite Materials. Chem. Mater. (2001d). 13, pp. (3151–3168).

Sayes, C.M, Liang, F., Hudson, J.L. , Mendez, J., Guo, W., Beach, J.M., Moore, V.C., Condell, D., Doyle, C.D., West, J.L., Billups, V.E., Ausman, K.D. & Colvin, V.L. Functionalization density dependence of single-walled carbon nanotubes cytotoxicity in vitro. Toxicol. Lett. (2006). 161, pp. (135–142).

Schröder, H.C., Boreiko, A., Korzhev, M., Tahir, M.N., Tremel, W., Eckert, C., Ushijima, H., Müller, I.M. & Müller, W.E.G. (2006). Co-expression and functional interaction of silicatein with galectin: Matrix-guided formation of siliceous spicules in the marine demosponge Suberites domuncula, *J. Bio. Chem.,* 281, pp. (12001- 12009).

Schröder, H.C., Wang, X., Tremel, W., Ushijima, H. & Müller, W.E.G. (2008). Biofabrication of biosilica-glass by living organisms, *Nat. Prod. Rep.,* 25, pp. (455-474).

Sen, T., Tiddy, G.J.T., Casci, J.L. & Anderson, M.W. Synthesis and Characterization of Hierarchically Ordered Porous Silica Materials. Chem. Mater. (2004f). 16, pp. (2044–2054).

Shimizu, K., Cha, J., Stucky, G.D. & Morse, D.E. (1998). Silicatein α: Cathepsin L-like protein in sponge biosilica, *Proc. Nat. Acad. Sci. USA,* 95, pp. (6234-6238).

Son, S.J ,. Bai, X., Nan, A., Ghandehari, H. & Lee, S.B. Template synthesis of multifunctional nanotubes for controlled release. Journal of Controlled Release. (2006). 114, pp. (143–152).

Son, S.J., Bai, X. & Lee, S.B. Inorganic hollow nanoparticles and nanotubes in nanomedicine: Part 1. Drug/gene delivery applications, Drug Discovery Today. (2007). 12(15-16), pp. (650-656).

Son, S.J., Bai, X. & Lee, S.B. Inorganic nanoparticles and nantubes in nanomedicine. Part 1: Drug/gene delivery applications, Drug Discovery Today, (2007). 12, (15/16), pp. (650-655).

Son, S.J., Reichel, J., He, B., Schuchman, M. & Lee, S.B. Magnetic nanotubes for magnetic-field-assisted bioseparation, biointeraction, and drug delivery. J. Am. Chem. Soc. (2005). 127, pp. (7316–7317).

Stahlberg, J. & Jonsson, B. Influence of charge regulation in electrostatic interaction chromatography of proteins. Anal Chem. (1996). 68, pp. (1536-1544).

Sumper, M. & Kröger, N. (2004a). Silica formation in diatoms: The function of long-chain polyamines and silaffins, *J. Mater. Chem.*, 14, pp. (2059-2065).

Sumper, M. & Lehmann, G. (2006). Silica pattern formation in diatoms: Species-specific polyamine biosynthesis, *ChemBioChem*, 7, pp. (1419-1427).

Sumper, M. (2004b). Biomimetic patterning of silica by long-chain polyamines, *Angew Chem Int. Ed.*, 43, pp. (2251-2254).

Szleifer, I: Protein adsorption on surfaceswith grafted polymers: a theoretical approach. Biophys J. (1997). 72, pp. (595-612).

Tamaru, S., Takeuchi, M., Sano, M. & Shinkai, S. Sol–Gel Transcription of Sugar-Appended Porphyrin Assemblies into Fibrous Silica: Unimolecular Stacks versus Helical Bundles as Templates, Angew. Chem., Int. Ed. (2002e). 41, pp. (853-856).

Terech, P. & Weiss, R.G. Low-Molecular Mass Gelators of Organic Liquids and the Properties of their Gels. Chem. Rev. (1997). 97, p. (3133).

Vallet-Regi, M., Rámila, A., Del Real, R.P. & Pérez-Pariente, J. A new property of MCM-41: drug delivery system. Chem. Mater. (2001). 13, pp. (308–311).

Van Bommel, K.J.C, Friggeri, A. & Shinkai, S. Angew. Chem.,Int. Ed., Organic templates for the generation of inorganic materials. (2003a). 42, pp. (980-999).

Van Tassel, P.R., Guemouri, L., Ramsden, J.J., Tarjus, G., Viot, P. & Talbot, J. A particle-level model of irreversible protein adsorption with a postadsorption transition. J Colloid Interface Sci. (1998). 207, pp. (317-323).

Weiss, R.G. & Terech, P. *Molecular Gels: Materials with self-assembled Fibrillar Networks*; Springer: New York. (2006).

Wiens, M., Bausen, M., Natalio, F., Link, T., Schlossmacher, U. & Muller, E.G. Biomaterials 30 (2009) pp. (1648–1656).

Xiao, Q.-G., Tao, X., Zhang, J.-P. & Chen, J.-F. Hollow silica nanotubes for immobilization of penicillin G acylase enzyme. J. Mol. Catal. B: Enzym. (2006). 42, pp. (14-19).

Yang, X., Tang, H., Cao, K., Song, H., Sheng, W. & Wu, Q. Templated-assisted one-dimensional silica nanotubes: synthesis and applications. J. Mater. Chem. (2011). 21, pp. (6122).

Yang, Y., Suzuki, M., Fukui, H., Shirai, H. & Hanabusa, K. Preparation of Helical Mesoporous Silica and Hybrid Silica Nanofibers Using Hydrogelator. *Chem. Mater.* (2006). 18 (5), pp. (1324–1329).

Yu, Y., Qiu, H., Wu, X., Li, H., Li, Y., Sakamoto, Y., Inoue, Y., Sakamoto, K., Terasaki O. & Che, S. Synthesis and Characterization of Silica Nanotubes with Radially Oriented Mesopores. Adv. Funct. Mater. (2008). 18, pp. (541-550).

Zhang, S., Marini, D.M., Hwang, W. & Santoso, S. Design of nanostructured biological materials through self-assembly of peptides and proteins. Curr Opin Chem Biol (2002), 6, pp. (865-871).

Zhao, D., Feng, J., Huo, Q., Melosh, N., Fredrickson, G.H., Chmelka, B.F. & Stucky, G.D. Triblock Copolymer Syntheses of Mesoporous Silica with Periodic 50 to 300 Angstrom Pores. Science (1998a). 279, pp. (548 – 552).

Zhao, D., Huo, Q., Feng, J., Chmelka, B.F. & Stucky, G.D. Nonionic Triblock and Star Diblock Copolymer and Oligomeric Surfactant Syntheses of Highly Ordered, Hydrothermally Stable. Mesoporous Silica Structures. J. Am. Chem. Soc. (1998b). 120, pp. (6024–6036).

6

Impacts of Surface Functionalization on the Electrocatalytic Activity of Noble Metals and Nanoparticles

Zhi-you Zhou and Shaowei Chen
Department of Chemistry and Biochemistry,
University of California, Santa Cruz, CA,
USA

1. Introduction

Low temperature fuel cells are a promising clean energy source, which can directly convert the chemical energy stored in hydrogen and small organic molecules (e.g., methanol, ethanol, and formic acid) into electrical energy with high efficiency, and thus find extensive applications in, for instance, automobiles and portable electronic devices. However, the commercialization of fuel cells is seriously impeded by the poor performance of the electrocatalysts for oxygen reduction reactions (ORR), and oxidation of small organic molecules.[1, 2] The sluggish kinetics of these reactions, even catalyzed by high loading of Pt-based catalysts, results in significant losses in thermal efficiency. Therefore, great efforts have been dedicated to improving the electrocatalytic performance of noble metals in the past decades.

Traditionally, the studies of fuel cell electrocatalysts focus on the design and manipulation of the composition (or alloy), particle size and shape (or crystal planes exposed), where rather significant progress has been achieved.[3-9] For example, bifunctional PtRu alloy catalysts significantly decrease the self-poisoning of carbon monoxide (CO) in methanol oxidation.[10-12] Pt-monolayer catalysts and Pt–M (M = Fe, Co, Ni, Pd, etc.) alloyed catalysts have considerably improved the catalytic activity for ORR.[13, 14] The shape-controlled synthesis of nanocrystals, which can provide uniform surface sites, has also produced electrocatalysts with good activity and selectivity.[15-17]

Recently, there emerges another promising strategy to tune electrocatalytic performances, i.e., modification of noble metal surfaces by deliberate chemical functionalization with specific molecules/ions. Organic ligands (such as poly(vinylpyrolidone) (PVP), cetyl trimethylammonium bromide (CTAB), and oleylamine) are often used in solution-phase syntheses of metal nanoparticles to control particle size, structure, and stability.[18] However, for applications in fuel cell catalysis, the ligands are either removed from the nanoparticle surface prior to use, or are simply considered as part of the supporting matrix.[19, 20] It is well-known that chemically modified electrodes can greatly improve the sensitivity and selectivity for electroanalytic reactions.[21-23] This suggests that the modification of specific molecules/ions on metal surfaces may also improve the electrocatalytic properties by virtue of the interactions between reactants and modifying molecules/ions.

Generally, there are several possible interactions between reactants (and spectator molecules/ions) and modifying molecules, as depicted in Figure 1: (1) Steric blocking effect. The modifying molecules occupy part of the metal surface, which can greatly hinder the adsorption of other molecules with a relatively large size that need multiple surface sites. (2) Electrostatic effects or dipole–dipole interactions. For example, the modification of anions (cations) may enhance the co-adsorption of cations (anions). (3) Electronic structure effects. The interactions between modifying molecules and metal atoms may result in adsorbate-induced distortions of the surface lattice and partial electron transfer, which may alter the adsorption energy of a second adsorbate.

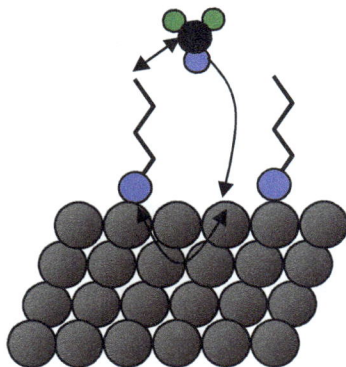

Fig. 1. Possible interactions between the reactants and modifying molecules on metal surfaces.

In this chapter, we highlight some recent key progress in surface functionalization on massive Pt single crystal planes with calixarene and CN⁻ ions, and their effects on the electrocatalytic activity and selectivity in ORR, hydrogen oxidation reactions (HOR), and methanol oxidation. We then extend the discussion to the synthesis of Pt and Pd nanoparticles capped with aryl and phosphine derivatives, and discuss the impacts of the molecule-molecule and molecule-nanoparticle interactions on the catalytic activities.

2. Surface modification of massive single crystal planes

Metal single crystal planes have well-defined surface atomic arrangements. Fundamentally, the use of single crystal planes as model electrocatalysts is anticipated to lead to a better understanding of the reaction mechanisms, surface processes, etc.[24, 25]

2.1 Modification of Pt single crystal planes with calix[4]arene

The ideal electrocatalysts for fuel cells should have both good activity and stability. It has been observed that during shutdown and startup that are inevitable for, for instance, automobile applications, the performance of cathodic Pt/C catalysts degrades quickly.[26] Under these conditions, the undesired ORR will occur on the anode side that is originally designed only for hydrogen oxidation. This reverse polarity phenomenon will push the cathode potential up to 1.5 V in some regions, and result in quick corrosion of Pt nanoparticles and carbon support. Clearly, it is very important to design selective anode catalysts that can efficiently suppress ORR while fully preserving the ability for HOR.

The Markovic group shows that chemically modified platinum with a self assembled monolayer of calix[4]arene molecules may meet this challenging requirement.[27] The structure of a calix[4]arene thiol derivative that they used is illustrated in Figure 2a. The wide rim containing sulfur atoms can firmly adsorb onto the Pt(111) terrace, leaving some edge or step sites (i.e., natural defects) and narrow terraces between the calix[4]arene

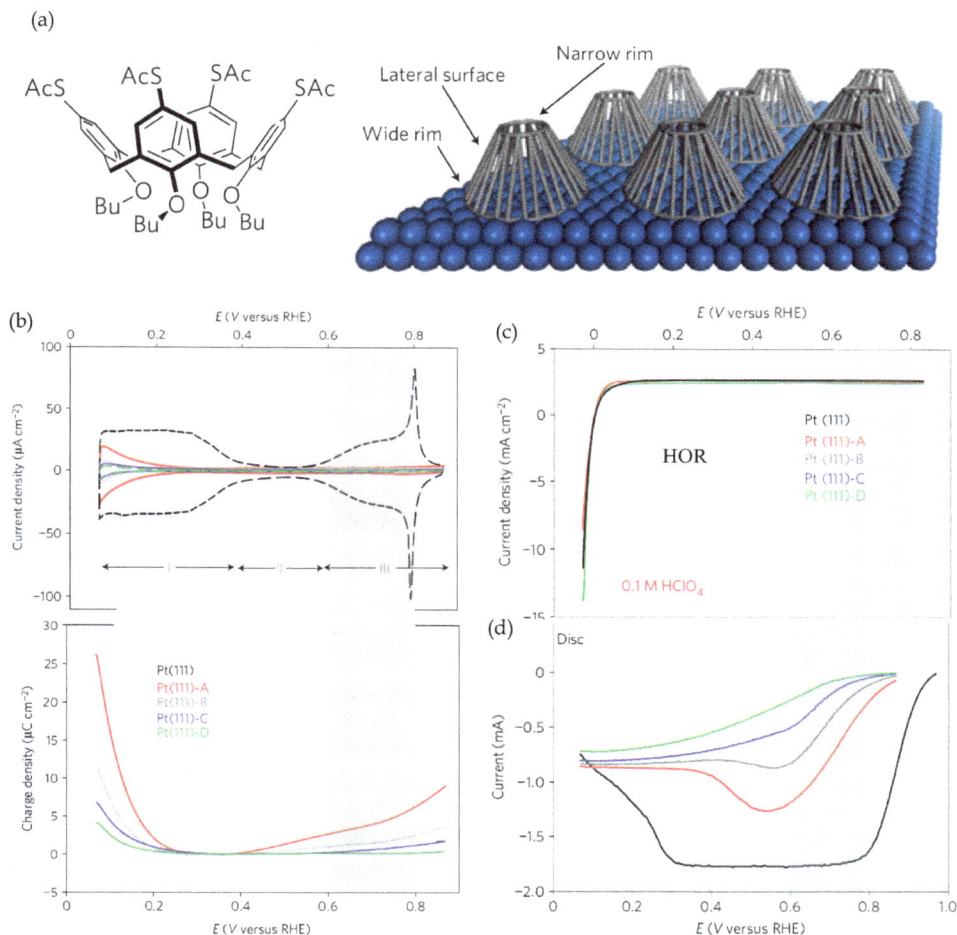

Fig. 2. (a) Structure and adsorption model for calix[4]arene molecules on Pt(111); (b) cyclic voltammograms of Pt(111) modified with different coverage (θ) of calix[4]arene molecules in 0.1 M HClO$_4$. θ increases from 0.84 to 0.98 (from curve A to D), calculated from the charge of hydrogen adsorption; HOR (c) and ORR (d) of calix[4]arene-modified Pt(111). Reprinted by permission from Macmillan Publishers Ltd: *Nature Materials* (2010, 9 (12), 998-1003), copyright 2010.

molecules. Cyclic voltammograms recorded in 0.1 M $HClO_4$ (panel b) indicate that the calix[4]arene molecules totally suppress the oxygen adsorption/desorption (+0.5 ~ +0.9 V), and most of hydrogen adsorption/desorption (+0.05 ~ +0.4 V), as their coverage increases from 0.84 to 0.98 (from curve A to D). This suggests that the ions/water from the supporting electrolyte are unable to penetrate the narrow rim of the calix[4]arene molecules, and the only sites available for the adsorption of electrolyte components are the relatively small number of unmodified Pt step-edges and/or the small ensembles of terrace sites between the anchoring groups.

The calix[4]arene-modified Pt(111) exhibits significantly different selectivity for ORR and HOR, as shown in panels (c) and (d). That is, ORR is greatly suppressed as the coverage of calix[4]arene increases, while HOR is essentially the same. It is accepted that the critical number of bare Pt atoms required for the adsorption of O_2 is higher than that required for the adsorption of H_2 molecules and a subsequent HOR. As for the ORR, the turnover frequency (TOF) of free Pt sites that are not occupied by calix[4]arene changes little (decreasing from 9 to 6 at +0.80 V), while the TOF for HOR increases sharply from 8 to 388 at +0.1 V as the coverage of calix[4]arene increases. Clearly, the decrease of ORR activity can be attributed to blocking effect, i.e., the calix[4]arene molecules block the adsorption of O_2 and its dissociation. However, the reason for enhanced HOR on the remaining Pt sites is not clear yet, and may be related to electronic effects, and/or mass-transfer limit for naked surfaces. The observed exceptional selectivity is not unique to the Pt(111)-calix system and has also been observed on Pt(100) and polycrystalline Pt electrodes, as well as Pt nanoparticles.[27, 28]

The successful preparation of oxygen-tolerant Pt catalysts for HOR indicates that surface functionalization of Pt surfaces is a promising approach to tuning the selectivity of electrocatalysts, which has great applications such as methanol-tolerant ORR catalysts.

2.2 Modification of Pt(111) with cyanide

Cyanide (CN^-) anions can adsorb strongly on metal surfaces. On the Pt(111) surface, they form a well defined $(2\sqrt3 \times 2\sqrt3)$–R30° structure (inset to Figure 3a, denoted as Pt(111)–CN_{ad}).[29, 30] The structure consists of hexagonally packed arrays, each containing six CN groups adsorbed on top of a hexagon of Pt atoms surrounding a free Pt atom. The cyanide coverage is 0.5. The cyanide adlayer on Pt(111) is remarkably stable, and can survive even after repetitive potential cycling between +0.06 and +1.10 V (vs RHE).[31] The CN- groups act as a third body, blocking Pt atoms that they occupy, but leaving other Pt atoms free onto which H, OH, or CO can still adsorb. The negative charge of the CN_{ad} dipole can also interact with other ions or polar molecules, so that the Pt(111)–CN_{ad} electrode has been considered as a chemically modified electrode.

Cuesta used the Pt(111)–CN_{ad} as a model catalyst to probe the reaction pathways of methanol electrooxidation.[32] As shown by the dashed curve in Figure 3a, there is no hysteresis of the oxidation current of methanol (+0.83 V) between the positive and the negative going sweeps on this surface. In addition, the hydrogen adsorption region (+0.05 ~ +0.5 V) remains unaffected in comparison with that recorded in a blank solution (solid curve). Electrochemical in situ FTIR studies confirm no CO formation in methanol oxidation on the Pt(111)–CN_{ad} electrode (Figure 3b). In addition to the product of CO_2 (~ 2340 cm[-1]) and the original CN adlayer (~2100 cm^{-1}), no adsorbed CO signals can be observed in the region of 1800 to 2100 cm^{-1}.

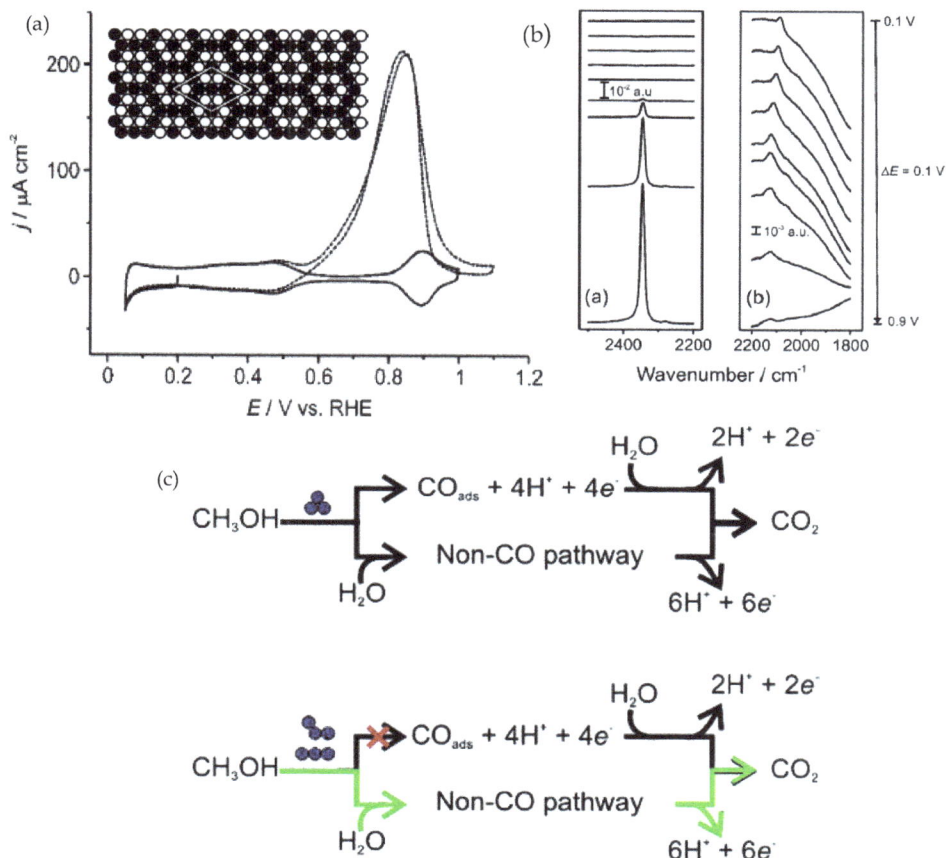

Fig. 3. (a) Cyclic voltammograms of Pt(111)–CN$_{ad}$ electrode in 0.1 M HClO$_4$ (solid line) and 0.1 M HClO$_4$ + 0.2 M CH$_3$OH (dashed line). The inset is a model of the (2√3 × 2√3)-R30° structure for a cyanide adlayer on Pt(111). Black circles correspond to Pt atoms, and white circles to linearly adsorbed CN-. (b) Electrochemical in situ FTIR spectra of methanol oxidation on cyanide-modified Pt(111) electrode. No adsorbed CO can be detected. (c) Illustration of the reaction pathways of methanol oxidation on Pt. Reprinted with permission from *the Journal of the American Chemical Society* (2006, 128 (41), 13332-13333). Copyright 2006 American Chemical Society.

There are four possible free Pt sites on the Pt(111)–CN$_{ad}$: (i) a single Pt atom; (ii) two adjacent Pt atoms; (iii) three Pt atoms arranged linearly; (iv) three Pt atoms arranged forming a chevron with a 120° angle. No adsorbed CO formation on these reactive sites suggests that the dehydrogenation reaction of methanol to CO needs at least three contiguous Pt atoms, likely as triangle assembly (Figure 3c).

In another study, the Markovic group uses the Pt(111)–CN$_{ad}$ as a model catalyst for ORR.[33] In acidic solutions, some anions such as SO$_4^{2-}$ and PO$_4^{3-}$ can adsorb strongly onto Pt surfaces (i.e., specific adsorption), and block surface sites for the adsorption of O$_2$. As a result, the ORR activity of Pt catalysts is much lower in H$_2$SO$_4$ (or H$_3$PO$_4$) than in HClO$_4$.[34] In the latter case, the adsorption of ClO$_4^-$ is negligible. It has been found that if the Pt(111) electrode is modified with cyanide, the adsorption of SO$_4^{2-}$ and PO$_4^{3-}$ is greatly suppressed (Figure 4 a and b, Region II), and the butterfly peaks as well as the specific adsorption of SO$_4^{2-}$ almost disappear. This can be simply understood in terms of through-space electrostatic repulsive interactions between the electronegative CN$_{ad}$ adlayer and the negative SO$_4^{2-}$ or PO$_4^{3-}$. However, the remaining free Pt sites on the Pt(111)–CN$_{ad}$ are still accessible for hydrogen and oxygen adsorption, as indicated by the current between +0.05 and +0.5 V and the pair of voltammetric peaks at +0.90 V. More importantly, these free metal sites are sufficient for the chemisorption of O$_2$ molecules and to break the O–O bond, as demonstrated by the ORR activity of the Pt(111)–CN$_{ad}$ electrode in 0.05 M H$_2$SO$_4$ (red solid curve, Figure 4c) similar to that of the naked Pt(111) in 0.1 M HClO$_4$ (grey dashed curve). It is determined that in comparison with naked Pt(111) (black solid curve), the ORR activity on Pt(111)–CN$_{ad}$ shows a 25-fold increase in the H$_2$SO$_4$ solution, and a ten-fold increase in H$_3$PO$_4$ solution due to less suppression of PO$_4^{3-}$ adsorption (Figure 4d).

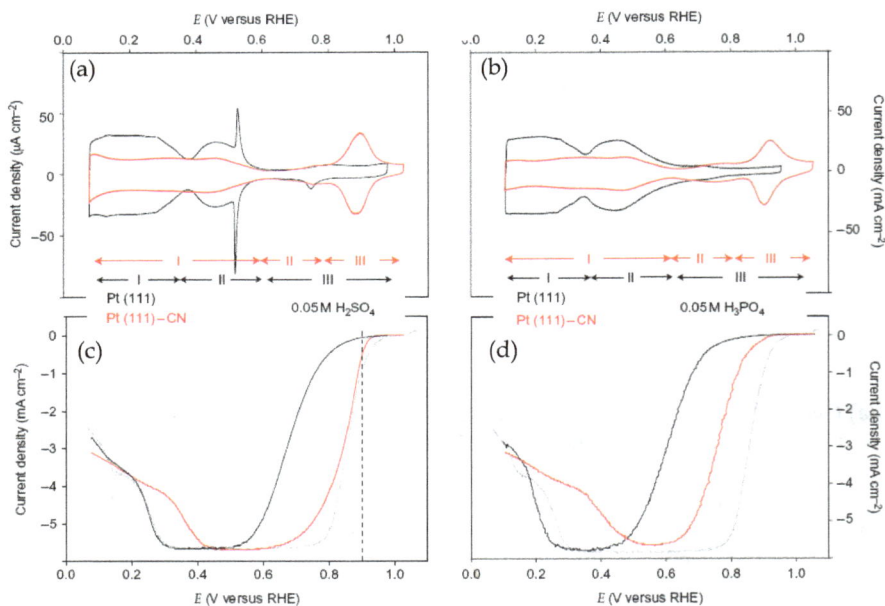

Fig. 4. Electrochemical measurements on CN-modified and unmodified Pt(111) surfaces in (a and c) 0.05 M H$_2$SO$_4$ and (b and d) 0.05 M H$_3$PO$_4$. (a and b) Cyclic voltammograms recorded in argon-saturated solutions. (c and d) ORR polarization curves in O$_2$ saturated solutions at 1600 rpm. Grey-dashed curves correspond to the ORR in 0.1 M perchloric acid. Reprinted by permission from Macmillan Publishers Ltd: *Nature Chemistry* (2010, 2 (10), 880-885), copyright 2010.

In a 0.1 M KOH solution, the behaviors change greatly. In this solution, the Pt(111)–CN$_{ad}$ electrode shows a 50-fold decrease in ORR activity as compared to the naked electrode. The interaction between the CN$_{ad}$ adlayer and the hydrated K$^+$ cations may lead to the formation of large K$^+$(H$_2$O)$_x$–CN$_{ad}$ clusters, which are 'quasi-specifically adsorbed' on Pt, hence acting as large site blockers for the adsorption of reactants such as O$_2$ and H$_2$O (see Figure 5c).[35]

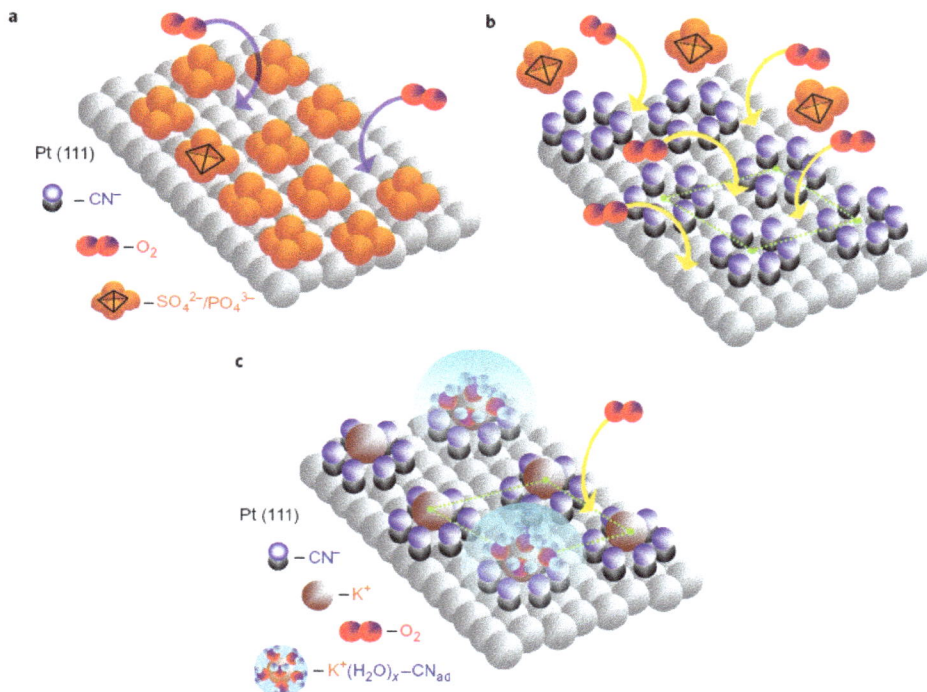

Fig. 5. Proposed models for the selective adsorption of spectator species and the available Pt surface sites for O$_2$ adsorption on CN-free and CN-covered Pt(111): (a) On Pt(111) covered by phosphoric/sulfuric acid anions O$_2$ can access the surfaces atoms only through a small number of holes in the adsorbate anion adlayer; (b) The number of holes required for O$_2$ adsorption is significantly increased on the Pt(111)–CN$_{ad}$ surface because adsorption of phosphoric/sulfuric acid anions is suppressed by the CN adlayer so that the total CN$_{ad}$/OH$_{ad}$ coverage is lower than the coverage by sulfuric or phosphoric acid anions; and (c) The adsorption of O$_2$ (and H$_2$O) is strongly suppressed on the Pt(111)–CN surface in alkaline solutions due to non-covalent interactions between K$^+$ and covalently bonded CN$_{ad}$, and the formation of spectator CN$_{ad}$–M$^+$(H$_2$O)$_x$ clusters. For clarity, the hydrated K$^+$ ions are shown as not being part of the double-layer structure. In reality, however, they are quasi-specifically adsorbed, with the locus centred between the covalently bonded CN$_{ad}$ and hydrated K$^+$. Reprinted by permission from Macmillan Publishers Ltd: *Nature Chemistry* (2010, 2 (10), 880-885), copyright 2010.

The selective adsorption of spectator species and reactants and the impacts on ORR are summarized in Figure 5, where CN$_{ad}$ repels the adsorption of SO$_4^{2-}$ and PO$_4^{3-}$ anions, resulting in more surface sites for O_2 adsorption in acidic media (panels a and b), and interacts with hydrated K^+ cations to suppress the O_2 adsorption in KOH solution (panel c).

These studies of the interactions between the CN_{ad} adlayer and H_2O, oxygen species, and spectator anions shed light on the role of covalent and non-covalent interactions in controlling the ensemble of Pt active sites required for high turn-over rates of ORR and methanol oxidation, and may be helpful in the quest of identifying new directions in designing electrochemical interfaces that can bind selectively reactive and spectator molecules.

3. Surface-functionalization of metal nanoparticles with improved electrocatalytic properties

Surface functionalization of massive single crystal electrodes offers a unique structural framework for the fundamental studies of fuel cell electrocatalysis. Yet, from the practical points of view, the preparation of surface-functionalized metal nanoparticles is necessary. In fact, most catalysts are in the form of nanosized particles dispersed on varied substrate supports. Herein, we highlight two cases, i.e., aryl- and phosphine-stabilized Pt and Pd nanoparticles, and their applications in fuel cell electrocatalysis.

3.1 Surface functionalization by aryl groups

The modification of electrode surfaces with aryl groups through electrochemical reduction of diazonium salts has been studied rather extensively.[36] A variety of electrode substrates, such as carbon, silicon, copper, iron, and gold, have been used. The covalent attachment of the aryl fragments onto the electrode surfaces is via the following mechanism[37]

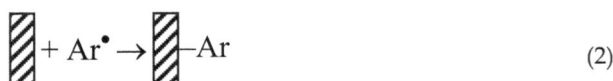

$$Ar\text{-}N\equiv N^+ + e \rightarrow Ar^\bullet + N_2 \tag{1}$$

$$\boxed{} + Ar^\bullet \rightarrow \boxed{}\text{--}Ar \tag{2}$$

At a negative enough potential, diazonium ions are electrochemically reduced to aryl radicals releasing nitrogen molecules (step (1)). The radicals are very active, and they are readily bonded onto the electrode surfaces (step (2)). Theoretical calculations predict that the adsorption energies vary from 25 to 106 kcal mol^{-1} on different substrates.[37] The strong bond strength suggests that the interaction is chemical in nature, i.e., covalent bond.

In recent years, aryl-stabilized metal nanoparticles through metal-carbon (M-C) covalent bonds, synthesized by the co-reduction of metal precursors and diazonium salts, have received great interest.[37-39] The formation mechanism is similar to that in electrochemical reduction.[36, 38] For instance, for Au nanoparticles, surface enhanced Raman scattering (SERS) have provided direct evidence for the formation of Au-C covalent bonds on the nanoparticle surface.[40]

Mirkhalaf and co-workers are the first to report the synthesis of Au and Pt nanoparticles stabilized by 4-decylphenyl fragments through the co-reduction of the corresponding metal precursors and diazonium salt.[38] The resulting Pt nanoparticles show a smaller core diameter (~ 3 nm) than the Au nanoparticles (8.1 nm), due to the stronger bonding of the aryl group on Pt than on Au. The ^{1}H NMR spectra of the Pt nanoparticles in CD_2Cl_2 solution only show some very board peaks from the 4-decylphenyl ligands. The broadening of the NMR peaks is ascribed to the fast spin-spin relaxation of atoms close to the metal core, and confirms the direct linkage of the 4-decylphenyl fragments onto Pt surfaces.

Using a similar method, our group has synthesized Pd nanoparticles stabilized by a variety of aryl moieties, including biphenyl, ethylphenyl, butylphenyl, and decylphenyl groups.[39, 41] Of these, the average core size of the Pd nanoparticles stabilized by butylphenyl group (Pd-BP) is 2.24 ± 0.35 nm (Figure 6a and b),[41] which is much smaller than that of commercial Pd black catalyst (10.4 ± 2.4 nm, Figure 6c and d). Thermogravimetric analysis (TGA) indicates that the organic component is about 19% with a main mass-loss peak at 350 °C (Figure 7a). On the basis of the weight loss and the average core size of the Pd nanoparticles, the average area occupied by one butylphenyl ligand on the Pd nanoparticle surface is estimated to be about 21 Å^2, close to the typical value (~20 Å^2) for long-chain alkanethiolate adsorbed on metal nanoparticles.[42]

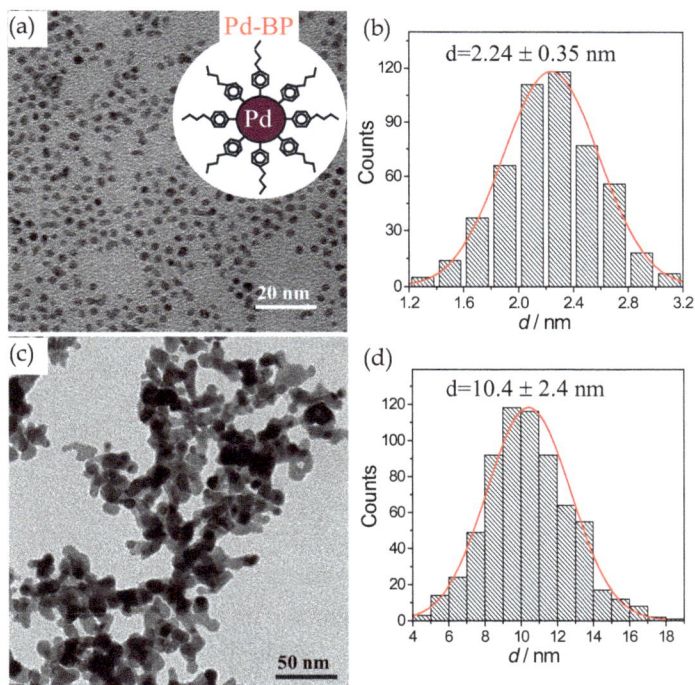

Fig. 6. TEM images and core size histograms of (a and b) Pd-BP nanoparticles and (c and d) commercial Pd black. Reprinted by permission from *Chemical Communications* (2011, 47 (21), 6075-6077). Copyright 2011 Royal Society of Chemistry.

The infrared spectrum of the Pd-BP nanoparticles (Figure 7b, curve A) shows the C-H vibrational stretch of the phenyl ring at 3027 cm[-1], and four aromatic ring skeleton vibrations at 1614, 1584, 1497, and 1463 cm[-1].[43] It is worth noting that the relative intensities of the aromatic ring skeleton vibrations of the butylphenyl groups bound onto the Pd nanoparticles are significantly different from those of monomeric n-butylbenzene and 4-butylaniline (curves B and C). For instance, the strong band observed with n-butylbenzene at 1609 cm⁻¹ and with 4-butylaniline at 1629 cm⁻¹ becomes a very weak shoulder with Pd-BP, whereas the 1584 cm⁻¹ band intensifies markedly with the ligands bound onto the nanoparticle surface. These observations suggest that the ligands on the Pd-BP surface are indeed the butylphenyl fragments by virtue of the Pd-C interfacial covalent linkage, and there are relatively strong electronic interactions between the aromatic rings and the Pd nanoparticles.

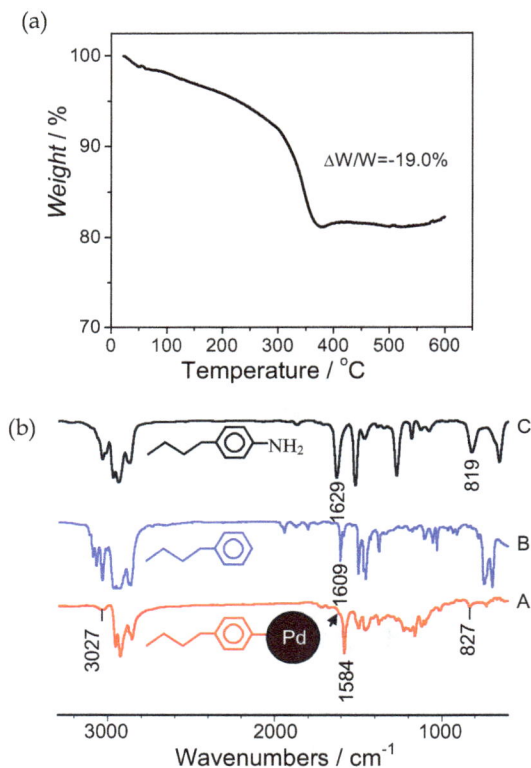

Fig. 7. (a) TGA curve of Pd-BP nanoparticles measured under a N$_2$ atmosphere at a heating rate of 10 °C min⁻¹. (b) FTIR spectra of (A) Pd-BP nanoparticles, (B) n-butylbenzene, and (C) 4-butylaniline. Reprinted by permission from *Chemical Communications* (2011, 47 (21), 6075-6077). Copyright 2011 Royal Society of Chemistry.

Cyclic voltammograms (Figure 8a) of Pd-BP nanoparticles recorded in 0.1 M H$_2$SO$_4$ indicates that the surface of the Pd-BP nanoparticles is fairly accessible for electrochemical reactions, since well-defined current peaks for the adsorption/desorption of hydrogen (– 0.25 ~ 0.0 V) and oxygen (~ +0.30 V) can be observed. The specific electrochemical surface area (ECSA) of Pd, determined through the reduction charge of monolayer Pd surface oxide, is estimated to be 122 m^2 g^{-1} for Pd-BP nanoparticles, which is about 3.6 times higher than that of commercial Pd black (33.6 m^2 g^{-1}). The theoretical ECSA of Pd-BP nanoparticles is calculated to be 223 m^2 g^{-1}, on the basis of the average core size (2.24 nm, Figure 6). This indicates that over 50% of the particle surface was accessible under electrochemical conditions.

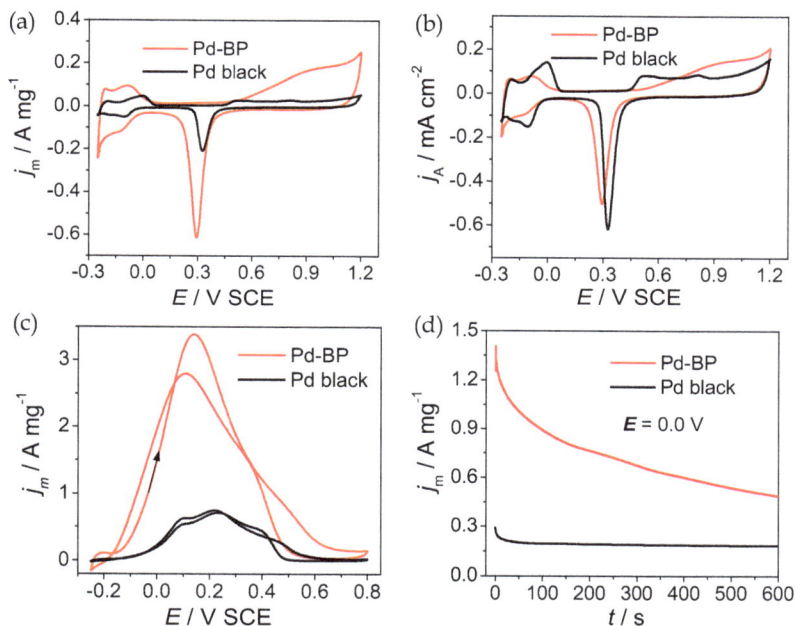

Fig. 8. Cyclic voltammograms of Pd-BP nanoparticles and commercial Pd black in 0.1 M H$_2$SO$_4$ with the currents normalized (a) by the mass loading of Pd and (b) by the effective electrochemical surface area at a potential scan rate of 100 mV s^{-1}. Panels (c) and (d) depict the cyclic voltammograms and current-time curves acquired at 0.0 V for HCOOH oxidation, respectively, at the Pd-BP nanoparticles and Pd black-modified electrode in 0.1 M HCOOH + 0.1 M H$_2$SO$_4$ at room temperature. Reprinted by permission from *Chemical Communications* (2011, 47 (21), 6075-6077). Copyright 2011 Royal Society of Chemistry.

Due to the remarkably high ECSA and surface functionalization, the Pd-BP nanoparticles exhibit a high mass activity in the electrocatalytic oxidation of formic acid. Figure 8c depicts the voltammograms of Pd-BP and Pd black catalysts recorded in a 0.1 M HCOOH + 0.1 M HClO$_4$ solution at a potential scan rate of 100 mV s^{-1}. The peak current density, normalized to the Pd mass, is as high as 3.39 A mg^{-1}$_{Pd}$ on the Pd-BP nanoparticles, which is about 4.5 times higher than that of the Pd black (0.75 A mg^{-1}). In addition, the peak potential for HCOOH oxidation is also negatively shifted by 80 mV. The Pd-BP nanoparticles also possess higher stability of the voltammetric current than Pd black for formic acid oxidation. Figure 8d shows the current-time curves recorded at 0.0 V for 600 s. The Pd-BP catalysts maintain a mass current density that is 2.7 to 4.8 times higher than that of Pd black. The high electrocatalytic activity of Pd-BP nanoparticles may be correlated with butylphenyl functionalization of the nanoparticle surface. As indicated by cyclic voltammograms where the currents are normalized to ECSA (Figure 8b), the adsorption of (bi)sulphate ions at around 0 V diminishes substantially on the Pd-BP electrode,[44] and the initial oxygen adsorption at around +0.50 V in the positive-going scan is also blocked, in comparison to Pd black. Such a discrepancy of surface accessibility and reactivity indicates that butylphenyl group may provide steric hindrance for (bi)sulphate adsorption, and also change the electronic structure of Pd.

Butylphenyl modified Pt (Pt-BP) nanoparticles have also been prepared by the co-reduction of diazonium salt and H_2PtCl_4 with $NaBH_4$ as the reducing agent.[45] TEM images (Figure 9) show that the as-prepared Pt-BP nanoparticles are well-dispersed and uniform, and well clear (111) lattice fringes can be observed. The average core size of Pt-BP nanoparticles is determined to be 2.93 ± 0.49 nm, which is comparable to the commercial Pt/C (3.1 nm) catalyst that is used in the controll experiment.

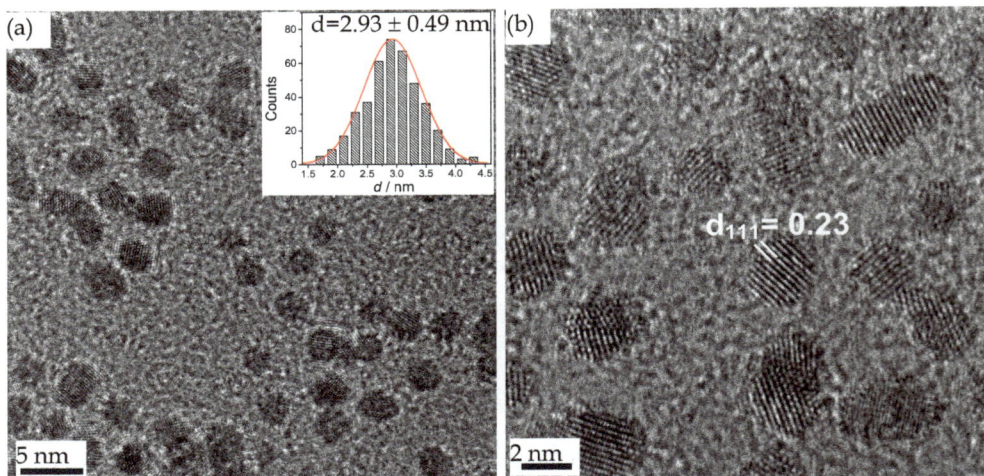

Fig. 9. TEM images of Pt-BP nanoparticles. Scale bars are 5 nm and 2 nm in panels (a) and (b), respectively. Inset to panel (a) shows the core size histogram. Reprinted by permission from *Physical Chemistry Chemical Physics* (2012, 14 (4), 1412-1417. Copyright 2012 Royal Society of Chemistry.

The IR transmission spectrum of Pt-BP nanoparticles also exhibits the characteristic peaks of the butylphenyl group, e.g., stretching vibration of the C–H moiety on the phenyl ring (3029 cm^{-1}), saturated C–H stretches in the butyl group (2955, 2929, 2858 cm^{-1}), aromatic ring skeleton vibration (1577 cm^{-1}), and out-of-plane C–H deformation vibration for para-substituted aromatic rings (827 cm^{-1}). The amount of ligands on the Pt-BP nanoparticles is quantitatively determined by TGA measurements to be about 27%. Therefore, the average area occupied by one butylphenyl ligand can be estimated to be about 5.8 Å^2, considerably lower than that observed with the aforementioned Pd nanoparticles.[41] Such a low occupied area of aryl group on Pt surfaces suggests there may exist some multilayer polyaryl structure.[46, 47]

It is well established that Pt catalysts for formic acid electrooxidation are easily poisoned by adsobed CO (CO_{ad}) intermediates.[25] Interestingly, the functionalization of Pt nanoparticles by the butylphenyl groups greatly suppresses the CO poisoning. Figure 10 compares the cyclic voltammograms of formic acid oxidation on the Pt-BP (red curve) and commercial Pt/C catalysts (black curve). On the Pt/C, along with increasing electrode potential, a weak and broad oxidation peak appears at around +0.30 V, followed by a sharp peak at +0.60 V in the forward scan. The former is attributed to the direct oxidation of formic acid to CO_2, while the latter is assigned to the oxidation of the CO_{ad} that is generated from the spontaneous dissociative adsorption of formic acid on Pt surfaces.[48] After the stripping of CO_{ad}, formic acid can then be readily oxidized on the clean Pt surface, as indicated by a very intensive current peak in the reverse scan. In contrast, on the Pt-BP nanoparticles, the hysteresis of the oxidation current of formic acid between forward and reverse scans is much less significant, and only a hump can be observed in the potential region around +0.60 V for CO_{ad} oxidation. This fact indicates that the CO poisoning pathway has been greatly suppressed on the Pt-BP nanoparticles. As a result, the peak current density of formic acid oxidation on the Pt-BP (1.03 A mg^{-1}Pt) is about 4 times larger than that on the Pt/C (0.25 A mg^{-1}Pt) in the forward scan at round +0.30 V.

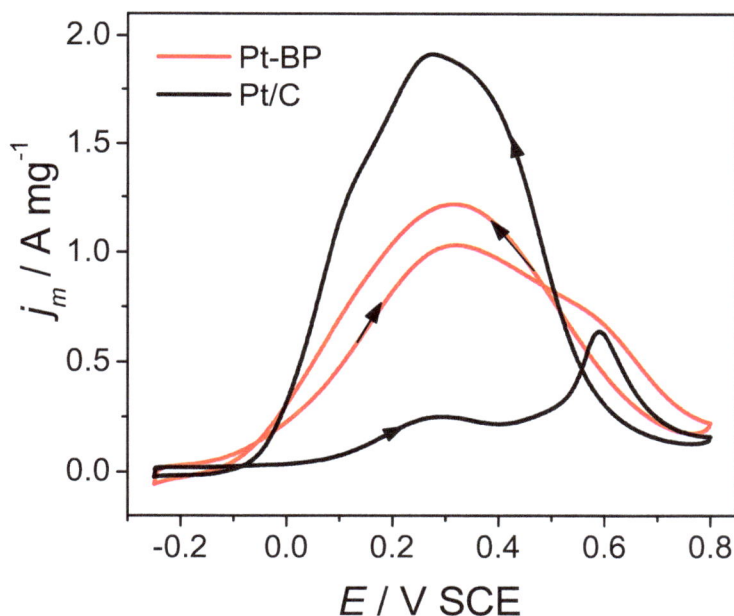

Fig. 10. Cyclic voltammograms of Pt-BP nanoparticles and commercial Pt/C catalyst in 0.1 M HCOOH + 0.1 M H_2SO_4 at 50 mV s^{-1}. Reprinted by permission from *Physical Chemistry Chemical Physics* (2012, 14 (4), 1412-1417). Copyright 2012 Royal Society of Chemistry.

The suppression of CO poisoning for formic acid on the Pt-BP nanoparticles is further confirmed by electrochemical in situ FTIR spectroscopic measurements (Figure 11). With Pt/C (panel b), the spectra show a downward band at 2343 cm^{-1} that is assigned to CO_2, the oxidation product of formic acid, and another at around 2050 cm^{-1} to linearly bonded CO

(CO$_L$), the poisoning intermediates. Clearly, unlike naked Pt, there is negligible adsorbed CO on the Pt-BP electrode (panel a), confirming that the functionalization of butylphenyl group have effectively blocked the poisoning pathway of formic acid oxidation. As a result, Pt-BP can yield more CO$_2$ than the Pt/C at a same potential.

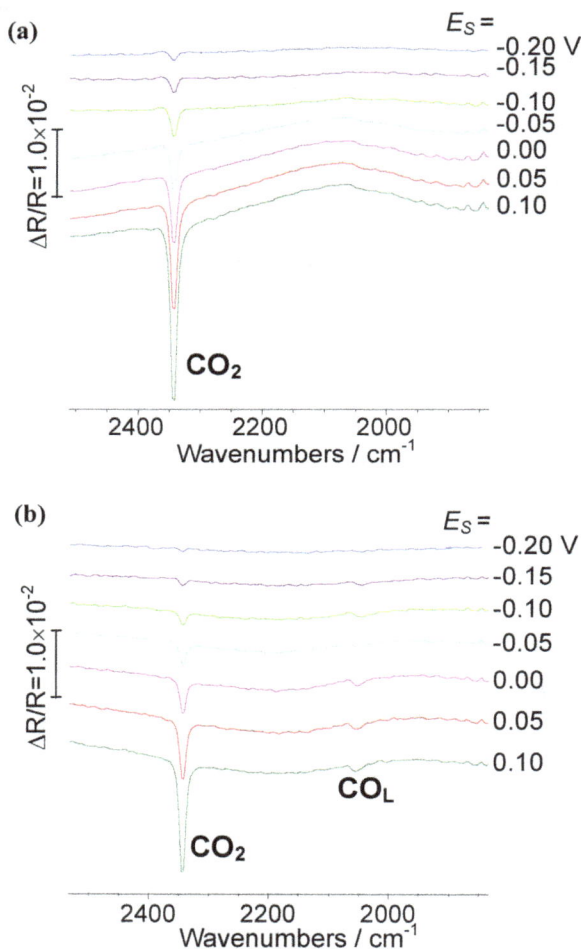

Fig. 11. Electrochemical in situ FTIR spectra of formic acid oxidation on the (a) Pt-BP nanoparticles and (b) commercial Pt/C catalyst in 0.1 M HCOOH + 0.1 M H$_2$SO$_4$ at different potentials. The sample potential (E_S) is varied from −0.20 to +0.10 V and the reference spectrum is acquired at E_R = −0.25 V. Reprinted by permission from *Physical Chemistry Chemical Physics* (2012, 14 (4), 1412-1417). Copyright 2012 Royal Society of Chemistry.

It has been known that there are at least two effects responsible for the improved electrocatalysts for formic acid oxidation: third-body effects and electronic structure effects.[49, 50] It is generally accepted that the dehydration of formic acid to form CO$_{ad}$ is a site demanding reaction, and several neighboring Pt sites are required.[50] Thus, this pathway can be blocked if Pt surfaces are covered by foreign atoms (third body). Clearly, the high catalytic activity of the Pt-BP nanoparticles may be, at least in part, reasonably attributed to third-body effects, i.e., the butylphenyl group provides steric hindrance for formic acid to form CO_{ad}.

The two examples listed above demonstrate that surface functionalization of noble metal nanoparticles with aryl groups is a promising route towards the preparation of electrocatalysts with improved performance for fuel cell electrocatalysis.

3.2 Surface functionalization by phosphine

Organophosphine ligands can strongly bind onto metal surfaces, especially for Pt group metals.[51, 52] Pietron and co-workers carry out electrochemical studies to examine the impacts of triphenylphosphine triphosphonate (TPPTP, Figure 12a) on the electrocatalytic activity of Pt nanoparticles for ORR.[53] The TPPTP-capped Pt nanoparticles are prepared through ligand exchange reactions between glycol-stabilized Pt nanoparticles and TPPTP for 2 to 4

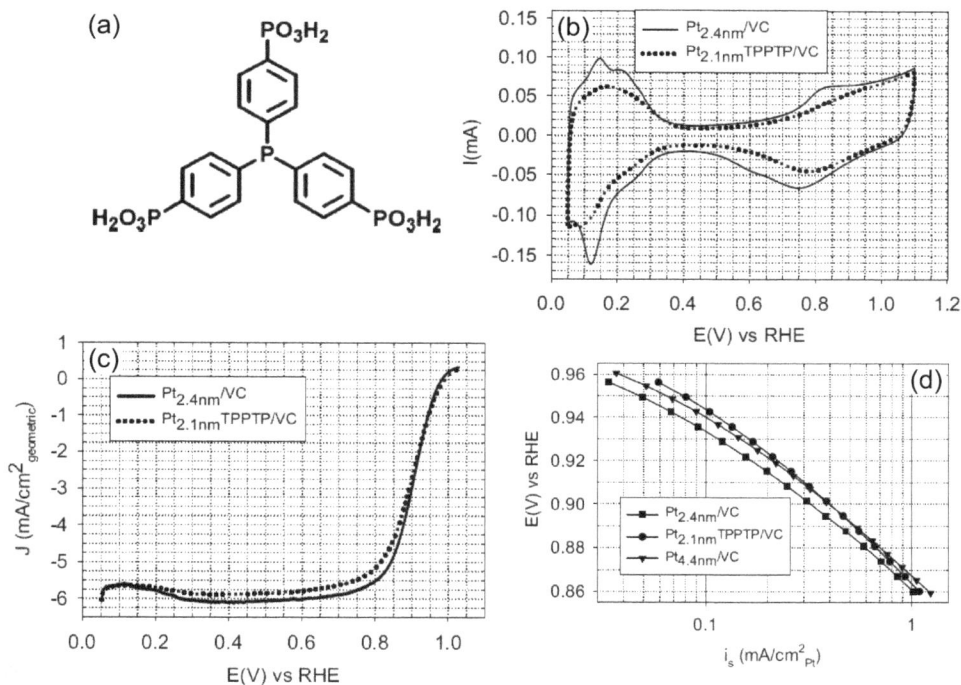

Fig. 12. (a) Molecular structure of TPPTP; (b) cyclic voltammograms of $Pt_{2.4\ nm}$/VC and $Pt_{2.2\ nm}$-TPPTP/VC in Ar-saturated 0.1 M $HClO_4$ at 25°C; (c) RDE voltammograms of the ORR; and (d) Tafel plots of the specific ORR activities at $Pt_{2.1\ nm}$-TPPTP/VC, and commercial $Pt_{2.4\ nm}$/VC and $Pt_{4.4\ nm}$/VC. Reprinted by permission from *Electrochemical and Solid State Letters* (2008, 11 (8), B161-B165). Copyright 2008 The Electrochemical Society.

days.[54] The core size of the Pt nanoparticles is about 2.1 nm. Studies based on X-ray absorption spectroscopy including XANES and EXAFS indicate that the coverage of TPPTP is about 0.3 ML, and the binding to Pt surfaces is via the P atom (Pt-P-P<tp) at +0.54 V (RHE), and approximately one-half of this converts to a Pt-O-P<tp linkage at +1.0 V.[55]

Electrochemical cyclic voltammograms (Figure 12b) show the features of weakly adsorbed hydrogen on TPPTP-caped Pt nanoparticles (dotted curve). They attribute this observation to the surface ligands that change the electronic energy of the Pt frontier orbitals and hence alter the hydrogen adsorption electrochemistry. More importantly, the TPPTP ligands weaken the oxygen adsorption on Pt, as evidenced by the positive shift of 40 mV for both onset potential and reduced desorption peak of oxygen species, as compared to that on a "bare" Pt/VC catalyst with a similar core size of 2.4 nm (solid curve). The weakening of the Pt–O bonds may improve the ORR electrocatalytic activity, since the reduced desorption of oxygen species is the rate-determined step for ORR on Pt.[7] Figure 12c and d show the polarization curves and Tafel plots of area-specific ORR activities for these Pt catalysts (along with a somewhat larger "bare" Pt/VC sample of about 4.4 nm in diameter) in O_2-saturated 0.1 M $HClO_4$ with a rotating disc electrode. The area-specific activity of TPPTP-caped Pt is enhanced by 22% in comparison with naked Pt; however, the mass-specific activity is inferior due to lower specific electrochemical area, as shown in Figure 12c. Two factors may contribute to the enhancement in area-specific activity. One is the weakening of Pt-O bonds, as mentioned above, and the other is hydrophobic effect of the aryl groups in TPPTP, which may repel water and inhibit Pt-oxide formation.

Thiol molecules have also be used to functionalize Pt nanoparticles,[56-58] but no considerable improvement of the ORR activity has been achieved.

4. Conclusion

In this chapter, we highlight some new progress in the study of the impacts of deliberate chemical functionalization of massive single crystal planes and noble metal nanoparticles with specific molecules, ions and molecular fragments on their catalytic performance in fuel cell electrochemistry. Through the interactions between the modifying groups and reactants (or spectator molecules/ions), and ligand-mediated electronic effects, the electrocatalytic activity and selectivity may be greatly tuned. This approach may have extensive applications in analytical, synthetic and materials chemistry as well as in chemical energy conversion and storage, and selective fuel production.

Further studies in this field may include the following aspects: (1) selection and optimization of functional molecules that can greatly change surface electronic structure of the metal substrates, but not occupy too many metal surface sites; (2) improvement of the stability of the modifying molecules, especially under the electrochemical conditions for ORR; (3) exploration of functional molecules that have synergistic effects for target reactions; and (4) identification of modifying molecules that possess special physicochemical properties, such as high solubility for O_2, great hydrophilic or hydrophobic characteristics, and an environment similar to enzyme, that are optimal for fuel cell electrocatalysis.

5. Acknowledgments

The authors are grateful to the National Science Foundation and the ACS-Petroleum Research Funds for financial supports of the research activities.

6. References

[1] Gasteiger, H. A.; Kocha, S. S.; Sompalli, B.; Wagner, F. T., Activity benchmarks and requirements for Pt, Pt-alloy, and non-Pt oxygen reduction catalysts for PEMFCs. *Applied Catalysis B-Environmental* 2005, 56, (1-2), 9-35.

[2] Liu, H. S.; Song, C. J.; Zhang, L.; Zhang, J. J.; Wang, H. J.; Wilkinson, D. P., A review of anode catalysis in the direct methanol fuel cell. *Journal of Power Sources* 2006, 155, (2), 95-110.

[3] Lim, B.; Jiang, M. J.; Camargo, P. H. C.; Cho, E. C.; Tao, J.; Lu, X. M.; Zhu, Y. M.; Xia, Y. A., Pd-Pt Bimetallic Nanodendrites with High Activity for Oxygen Reduction. *Science* 2009, 324, (5932), 1302-1305.

[4] Bing, Y. H.; Liu, H. S.; Zhang, L.; Ghosh, D.; Zhang, J. J., Nanostructured Pt-alloy electrocatalysts for PEM fuel cell oxygen reduction reaction. *Chemical Society Reviews* 2010, 39, (6), 2184-2202.

[5] Stamenkovic, V. R.; Fowler, B.; Mun, B. S.; Wang, G. F.; Ross, P. N.; Lucas, C. A.; Markovic, N. M., Improved oxygen reduction activity on Pt3Ni(111) via increased surface site availability. *Science* 2007, 315, (5811), 493-497.

[6] Zhang, J.; Sasaki, K.; Sutter, E.; Adzic, R. R., Stabilization of platinum oxygen-reduction electrocatalysts using gold clusters. *Science* 2007, 315, (5809), 220-222.

[7] Greeley, J.; Stephens, I. E. L.; Bondarenko, A. S.; Johansson, T. P.; Hansen, H. A.; Jaramillo, T. F.; Rossmeisl, J.; Chorkendorff, I.; Norskov, J. K., Alloys of platinum and early transition metals as oxygen reduction electrocatalysts. *Nature Chemistry* 2009, 1, (7), 552-556.

[8] Koh, S.; Strasser, P., Electrocatalysis on bimetallic surfaces: Modifying catalytic reactivity for oxygen reduction by voltammetric surface dealloying. *Journal of the American Chemical Society* 2007, 129, (42), 12624-+.

[9] Adzic, R. R.; Zhang, J.; Sasaki, K.; Vukmirovic, M. B.; Shao, M.; Wang, J. X.; Nilekar, A. U.; Mavrikakis, M.; Valerio, J. A.; Uribe, F., Platinum monolayer fuel cell electrocatalysts. *Topics in Catalysis* 2007, 46, (3-4), 249-262.

[10] Watanabe, M.; Motoo, S., Electrocatalysis by Ad-Atoms .2. Enhancement of Oxidation of Methanol on Platinum by Ruthenium Ad-Atoms. *Journal of Electroanalytical Chemistry* 1975, 60, (3), 267-273.

[11] Iwasita, T.; Hoster, H.; John-Anacker, A.; Lin, W. F.; Vielstich, W., Methanol oxidation on PtRu electrodes. Influence of surface structure and Pt-Ru atom distribution. *Langmuir* 2000, 16, (2), 522-529.

[12] Gasteiger, H. A.; Markovic, N.; Ross, P. N.; Cairns, E. J., Methanol Electrooxidation on Well-Characterized Pt-Rn Alloys. *Journal of Physical Chemistry* 1993, 97, (46), 12020-12029.

[13] Toda, T.; Igarashi, H.; Watanabe, M., Role of electronic property of Pt and Pt alloys on electrocatalytic reduction of oxygen. *Journal of the Electrochemical Society* 1998, 145, (12), 4185-4188.

[14] Paulus, U. A.; Wokaun, A.; Scherer, G. G.; Schmidt, T. J.; Stamenkovic, V.; Radmilovic, V.; Markovic, N. M.; Ross, P. N., Oxygen reduction on carbon-supported Pt-Ni and Pt-Co alloy catalysts. *Journal of Physical Chemistry B* 2002, 106, (16), 4181-4191.

[15] Tian, N.; Zhou, Z. Y.; Sun, S. G.; Ding, Y.; Wang, Z. L., Synthesis of tetrahexahedral platinum nanocrystals with high-index facets and high electro-oxidation activity. *Science* 2007, 316, (5825), 732-735.

[16] Vidal-Iglesias, F. J.; Solla-Gullon, J.; Rodriguez, P.; Herrero, E.; Montiel, V.; Feliu, J. M.; Aldaz, A., Shape-dependent electrocatalysis: ammonia oxidation on platinum nanoparticles with preferential (100) surfaces. *Electrochemistry Communications* 2004, 6, (10), 1080-1084.

[17] Zhang, J.; Yang, H. Z.; Fang, J. Y.; Zou, S. Z., Synthesis and Oxygen Reduction Activity of Shape-Controlled Pt(3)Ni Nanopolyhedra. *Nano Letters* 2010, 10, (2), 638-644.

[18] Xia, Y. N.; Xiong, Y. J.; Lim, B.; Skrabalak, S. E., Shape-Controlled Synthesis of Metal Nanocrystals: Simple Chemistry Meets Complex Physics? *Angewandte Chemie-International Edition* 2009, 48, (1), 60-103.

[19] Susut, C.; Chapman, G. B.; Samjeske, G.; Osawa, M.; Tong, Y., An unexpected enhancement in methanol electro-oxidation on an ensemble of Pt(111) nanofacets: a case of nanoscale single crystal ensemble electrocatalysis. *Physical Chemistry Chemical Physics* 2008, 10, (25), 3712-3721.

[20] Mazumder, V.; Sun, S. H., Oleylamine-Mediated Synthesis of Pd Nanoparticles for Catalytic Formic Acid Oxidation. *Journal of the American Chemical Society* 2009, 131, (13), 4588-+.

[21] Murray, R. W., Chemically Modified Electrodes. *Accounts of Chemical Research* 1980, 13, (5), 135-141.

[22] Abruna, H. D., Coordination Chemistry in 2 Dimensions - Chemically Modified Electrodes. *Coordination Chemistry Reviews* 1988, 86, 135-189.

[23] Dong, S. J.; Wang, Y. D., The Application of Chemically Modified Electrodes in Analytical-Chemistry. *Electroanalysis* 1989, 1, (2), 99-106.

[24] Clavilier, J.; Faure, R.; Guinet, G.; Durand, R., Preparation of Mono-Crystalline Pt Microelectrodes and Electrochemical Study of the Plane Surfaces Cut in the Direction of the (111) and (110) Planes. *Journal of Electroanalytical Chemistry* 1980, 107, (1), 205-209.

[25] Sun, S. G.; Clavilier, J.; Bewick, A., The Mechanism of Electrocatalytic Oxidation of Formic-Acid on Pt (100) and Pt (111) in Sulfuric-Acid Solution - An Emirs Study. *Journal of Electroanalytical Chemistry* 1988, 240, (1-2), 147-159.

[26] Borup, R.; Meyers, J.; Pivovar, B.; Kim, Y. S.; Mukundan, R.; Garland, N.; Myers, D.; Wilson, M.; Garzon, F.; Wood, D.; Zelenay, P.; More, K.; Stroh, K.; Zawodzinski, T.; Boncella, J.; McGrath, J. E.; Inaba, M.; Miyatake, K.; Hori, M.; Ota, K.; Ogumi, Z.; Miyata, S.; Nishikata, A.; Siroma, Z.; Uchimoto, Y.; Yasuda, K.; Kimijima, K. I.; Iwashita, N., Scientific aspects of polymer electrolyte fuel cell durability and degradation. *Chemical Reviews* 2007, 107, (10), 3904-3951.

[27] Genorio, B.; Strmcnik, D.; Subbaraman, R.; Tripkovic, D.; Karapetrov, G.; Stamenkovic, V. R.; Pejovnik, S.; Markovic, N. M., Selective catalysts for the hydrogen oxidation and oxygen reduction reactions by patterning of platinum with calix 4 arene molecules. *Nature Materials* 2010, 9, (12), 998-1003.

[28] Genorio, B.; Subbaraman, R.; Strmcnik, D.; Tripkovic, D.; Stamenkovic, V. R.; Markovic, N. M., Tailoring the Selectivity and Stability of Chemically Modified Platinum Nanocatalysts To Design Highly Durable Anodes for PEM Fuel Cells. *Angewandte Chemie-International Edition* 2011, 50, (24), 5468-5472.

[29] Stickney, J. L.; Rosasco, S. D.; Salaita, G. N.; Hubbard, A. T., Ordered Ionic Layers Formed on Pt(111) from Aqueous-Solutions. *Langmuir* 1985, 1, (1), 66-71.

[30] Stuhlmann, C.; Villegas, I.; Weaver, M. J., Scanning-Tunneling-Microscopy and Infrared-Spectroscopy as Combined in-Situ Probes of Electrochemical Adlayer Structure - Cyanide on Pt(111). *Chemical Physics Letters* 1994, 219, (3-4), 319-324.

[31] Huerta, F.; Morallon, E.; Quijada, C.; Vazquez, J. L.; Berlouis, L. E. A., Potential modulated reflectance spectroscopy of Pt(111) in acidic and alkaline media: cyanide adsorption. *Journal of Electroanalytical Chemistry* 1999, 463, (1), 109-115.

[32] Cuesta, A., At least three contiguous atoms are necessary for CO formation during methanol electrooxidation on platinum. *Journal of the American Chemical Society* 2006, 128, (41), 13332-13333.

[33] Strmcnik, D.; Escudero-Escribano, M.; Kodama, K.; Stamenkovic, V. R.; Cuesta, A.; Markovic, N. M., Enhanced electrocatalysis of the oxygen reduction reaction based on patterning of platinum surfaces with cyanide. *Nature Chemistry* 2010, 2, (10), 880-885.

[34] Macia, M. D.; Campina, J. M.; Herrero, E.; Feliu, J. M., On the kinetics of oxygen reduction on platinum stepped surfaces in acidic media. *Journal of Electroanalytical Chemistry* 2004, 564, (1-2), 141-150.

[35] Strmcnik, D.; Kodama, K.; van der Vliet, D.; Greeley, J.; Stamenkovic, V. R.; Markovic, N. M., The role of non-covalent interactions in electrocatalytic fuel-cell reactions on platinum. *Nature Chemistry* 2009, 1, (6), 466-472.

[36] Allongue, P.; Delamar, M.; Desbat, B.; Fagebaume, O.; Hitmi, R.; Pinson, J.; Saveant, J. M., Covalent modification of carbon surfaces by aryl radicals generated from the electrochemical reduction of diazonium salts. *Journal of the American Chemical Society* 1997, 119, (1), 201-207.

[37] Jiang, D. E.; Sumpter, B. G.; Dai, S., Structure and bonding between an aryl group and metal surfaces. *Journal of the American Chemical Society* 2006, 128, (18), 6030-6031.

[38] Mirkhalaf, F.; Paprotny, J.; Schiffrin, D. J., Synthesis of metal nanoparticles stabilized by metal-carbon bonds. *Journal of the American Chemical Society* 2006, 128, (23), 7400-7401.

[39] Ghosh, D.; Chen, S. W., Palladium nanoparticles passivated by metal-carbon covalent linkages. *Journal of Materials Chemistry* 2008, 18, (7), 755-762.

[40] Laurentius, L.; Stoyanov, S. R.; Gusarov, S.; Kovalenko, A.; Du, R. B.; Lopinski, G. P.; McDermott, M. T., Diazonium-Derived Aryl Films on Gold Nanoparticles: Evidence for a Carbon-Gold Covalent Bond. *Acs Nano* 2011, 5, (5), 4219-4227.

[41] Zhou, Z. Y.; Kang, X. W.; Song, Y.; Chen, S. W., Butylphenyl-functionalized palladium nanoparticles as effective catalysts for the electrooxidation of formic acid. *Chemical Communications* 2011, 47, (21), 6075-6077.

[42] Hostetler, M. J.; Wingate, J. E.; Zhong, C. J.; Harris, J. E.; Vachet, R. W.; Clark, M. R.; Londono, J. D.; Green, S. J.; Stokes, J. J.; Wignall, G. D.; Glish, G. L.; Porter, M. D.; Evans, N. D.; Murray, R. W., Alkanethiolate gold cluster molecules with core diameters from 1.5 to 5.2 nm: Core and monolayer properties as a function of core size. *Langmuir* 1998, 14, (1), 17-30.

[43] Lin-Vien, D.; Colthup, N. B.; Fateley, W. G.; Grasselli, J. G., *The Handbook of Infrared and Raman Characteristics Frequencies of Organic Molecules*. Academic Press: New York, 1991.

[44] Hoshi, N.; Kagaya, K.; Hori, Y., Voltammograms of the single-crystal electrodes of palladium in aqueous sulfuric acid electrolyte: Pd(S)- n(111) x (111) and Pd(S)-n(100) x (111). *Journal of Electroanalytical Chemistry* 2000, 485, (1), 55-60.

[45] Zhou, Z. Y.; Ren, J.; Kang, X. W.; Song, Y.; Sun, S. G.; Chen, S. W., Butylphenyl-Functionalized Pt Nanoparticles as CO-Resistent Electrocatalysts for Formic Acid Oxidation. *Physical Chemistry Chemical Physics* 2012, 14 (4), 1412-1417

[46] Adenier, A.; Combellas, C.; Kanoufi, F.; Pinson, J.; Podvorica, F. I., Formation of polyphenylene films on metal electrodes by electrochemical reduction of benzenediazonium salts. *Chemistry of Materials* 2006, 18, (8), 2021-2029.

[47] Kariuki, J. K.; McDermott, M. T., Nucleation and growth of functionalized aryl films on graphite electrodes. *Langmuir* 1999, 15, (19), 6534-6540.

[48] Capon, A.; Parsons, R., Oxidation of Formic-Acid at Noble-Metal Electrodes Part .3. Intermediates and Mechanism on Platinum-Electrodes. *Journal of Electroanalytical Chemistry* 1973, 45, (2), 205-231.

[49] Xia, X. H.; Iwasita, T., Influence of Underpotential Deposited Lead Upon the Oxidation of Hcooh in $HClO_4$ at Platinum-Electrodes. *Journal of the Electrochemical Society* 1993, 140, (9), 2559-2565.

[50] Leiva, E.; Iwasita, T.; Herrero, E.; Feliu, J. M., Effect of adatoms in the electrocatalysis of HCOOH oxidation. A theoretical model. *Langmuir* 1997, 13, (23), 6287-6293.

[51] Ye, E. Y.; Tan, H.; Li, S. P.; Fan, W. Y., Self-organization of spherical, core-shell palladium aggregates by laser-induced and thermal decomposition of Pd(PPh3)(4). *Angewandte Chemie-International Edition* 2006, 45, (7), 1120-1123.

[52] Son, S. U.; Jang, Y.; Yoon, K. Y.; Kang, E.; Hyeon, T., Facile synthesis of various phosphine-stabilized monodisperse palladium nanoparticles through the understanding of coordination chemistry of the nanoparticles. *Nano Letters* 2004, 4, (6), 1147-1151.

[53] Pietron, J. J.; Garsany, Y.; Baturina, O.; Swider-Lyons, K. E.; Stroud, R. M.; Ramaker, D. E.; Schull, T. L., Electrochemical observation of ligand effects on oxygen reduction at ligand-stabilized Pt nanoparticle electrocatalysts. *Electrochemical and Solid State Letters* 2008, 11, (8), B161-B165.

[54] Kostelansky, C. N.; Pietron, J. J.; Chen, M. S.; Dressick, W. J.; Swider-Lyons, K. E.; Ramaker, D. E.; Stroud, R. M.; Klug, C. A.; Zelakiewicz, B. S.; Schull, T. L., Triarylphosphine-stabilized platinum nanoparticles in three-dimensional nanostructured films as active electrocatalysts. *Journal of Physical Chemistry B* 2006, 110, (43), 21487-21496.

[55] Gatewood, D. S.; Schull, T. L.; Baturina, O.; Pietron, J. J.; Garsany, Y.; Swider-Lyons, K. E.; Ramaker, D. E., Characterization of ligand effects on water activation in triarylphosphine-stabilized Pt nanoparticle catalysts by X-ray absorption spectroscopy. *Journal of Physical Chemistry C* 2008, 112, (13), 4961-4970.

[56] Baret, B.; Aubert, P. H.; Mayne-L'Hermite, M.; Pinault, M.; Reynaud, C.; Etcheberry, A.; Perez, H., Nanocomposite electrodes based on pre-synthesized organically capped platinum nanoparticles and carbon nanotubes. Part I: Tuneable low platinum loadings, specific H upd feature and evidence for oxygen reduction. *Electrochimica Acta* 2009, 54, (23), 5421-5430.

[57] Cavaliere, S.; Raynal, F.; Etcheberry, A.; Herlem, M.; Perez, H., Direct electrocatalytic activity of capped platinum nanoparticles toward oxygen reduction. *Electrochemical and Solid State Letters* 2004, 7, (10), A358-A360.

[58] Cavaliere-Jaricot, S.; Haccoun, J.; Etcheberry, A.; Herlem, M.; Perez, H., Oxygen reduction of pre-synthesized organically capped platinum nanoparticles assembled in mixed Langmuir-Blodgett films: Evolutions with the platinum amount and leveling after fatty acid removal. *Electrochimica Acta* 2008, 53, (20), 5992-5999.

Cantilever-Based Optical Interfacial Force Microscopy

Byung I. Kim

Department of Physics, Boise State University,
USA

1. Introduction

Atomic force microscopy (AFM) is one of the most important tools that lead current nanoscience and nanotechnology in many diverse areas including physics, chemistry, material engineering, and nano-biology. The current AFM technique has been routinely applied to forced unbinding processes of biomolecular complexes such as antibody-antigen binding, ligand-receptor pairs, protein unfolding, DNA unbinding, and RNA unfolding studies (Butt et al., 2005; Fritz & Anselmetti, 1997; Schumakovitch et al., 2002). AFMs have also been applied to intermolecular friction studies (Carpick et al., 1997; Colchero et al., 1996; Fernandez-Torres et al., 2003; Goddenhenrich et al., 1994; Goertz et al., 2007; B.I. Kim et al., 2001; Major et al., 2006). These previous techniques of measuring friction employed a lateral modulation of the sample relative to the cantilever as a means to measure normal force and friction force at the same time (Burns et al., 1999a; Carpick et al., 1997; Colchero et al., 1996; Goddenhenrich et al., 1994; Goertz et al., 2007; Major et al., 2006).

However, AFM usage has been limited to passive applications (e.g., pull-off force measurement in the force-distance curve) and can only be applied to the measurement of friction while the tip is touching the sample surface because of an intrinsic mechanical instability of the tip-sample assembly near a sample surface called the "snap-to-contact problem"(Burnham, 1989; Lodge, 1983). During measurements, the mechanical instability occurs when the force derivative (i.e., dFa/dz), in respect to the tip position (z), exceeds the stiffness of the cantilever (spring constant k) (Greenwood, 1997; Israelachvili & Adams, 1978; Noy et al., 1997; Sarid, 1991), causing data points to be missed near the sample surface (Cappella & Dietler, 1999). This has been a significant barrier to understanding the nanoscopic water junction between the tip and the surface in ambient conditions, which makes it difficult, with AFM data, to directly reveal the interfacial water structure and/or analyze it with existing theories.

A decade ago for the purpose of avoiding the mechanical instability in measuring intermolecular forces, magnetic force feedback was implemented in AFM systems by attaching a magnet to the end of a cantilever (Ashby et al., 2000; Jarvis et al., 1996; A.M. Parker & J.L. Parker, 1992; Yamamoto et al., 1997). However, the magnetic force feedback requires a tedious process of attaching magnets to the backside of the cantilever using an inverted optical microscope equipped with micromanipulators (Ashby et al., 2000;

Yamamoto et al., 1997) and has poor performance in the servo system due to eddy currents (Jarvis et al., 1996; Parker & Parker, 1992; Yamamoto et al., 1997).

Interfacial force microscopy (IFM) was developed fifteen years ago, as an independent approach to avoid the mechanical instability problems related to the snap-to-contact problem associated with regular AFMs, (Chang et al., 2004; Joyce & Houston, 1991; Houston & Michalske, 1992). The IFM did not use the AFM platform, such as cantilever and optical detection schemes in measuring forces between two surfaces. Instead, the IFM uses its own force sensor with a larger tip (typical diameter around 0.1 μm to 1 μm) and an electrical detection scheme. Using force feedback, the IFM is capable of preventing the mechanical instability of the tip-sample assembly near a sample surface.

The IFM has greatly expanded applicability to the various problems at interfaces that the AFM cannot offer. IFM has been applied to diverse interfacial research including nanotribology (Burns et al., 1999b; Kiely & Houston, 1999; H.I. Kim & Houston, 2000), interfacial adhesion (Bunker et al., 2003; Huber et al., 2003) and probing of interfacial water structures (Major et al., 2006; Matthew et al., 2007). IFMs have contributed to molecular scale understanding of various surface phenomena. Kim and Houston applied IFM to measure friction force and normal force simultaneously to study the molecular nature of friction independently (H.I. Kim & Houston, 2000). The friction force is decoupled from normal force in the frequency domain. They used small vibrations to reduce the amount of wearing, getting friction information directly from the response amplitude with a lock-in amplifier. The earlier approach using AFM by Goodenhenrich et al. is similar to friction measurements but cannot be applied to friction studies at attractive regimes (Goodenhenrich et al., 1994). As an approach to study attractive regimes, friction forces have been measured using the IFM (Joyce & Houston, 1991), where the force feedback makes it possible. (Burns et al., 1999a; Goertz et al., 2007; Major et al., 2006). Burns et al. applied IFM to measure friction force and normal force simultaneously to study the molecular nature of friction to investigate the intermolecular friction along with normal forces in the attractive regime (Burns et al., 1999a).

However, the IFM technique has not been widely used due to low sensitivity and the technical complexity of the electrical-sensing method. The current IFM system uses a relatively bigger tip with the typical diameter around one micrometer for measurement of molecular interactions due to the existing low sensitivity issue in the electric force detection method of the current "teeter-totter" type of IFM force sensor (Joyce & Houston, 1991). The larger tip and the complexity of the electrical detection measurements have limited the use of the IFM as a popular tool to address the issues, especially at the single molecular level.

By combining a conventional IFM and AFM-type cantilever and its optical detection scheme, we recently developed a force microscope called the cantilever-based optical interfacial force microscope (COIFM) (Bonander & B.I. Kim, 2008). The COIFM substantively improved the force sensitivity, measurement resolution and accuracy over the conventional IFM and ordinary AFM measurements. In this chapter, the design and development of the COIFM are described along with a general description of avoiding the cantilever instability by using force feedback when measuring intermolecular forces. We derived how the interfacial force can be incorporated into the detection signal using the Euler equations for beams. Relevant calibration methods and approaches are covered for the analysis of the COIFM data. The

COIFM's unique force profiles related to ambient water structure on a surface have been demonstrated (Bonander & B.I. Kim, 2008; B.I. Kim et al., 2011). At the end of the chapter, the future applications of the COIFM system are discussed.

2. Design and development of COIFM

A schematic diagram of the overall COIFM system with the force feedback control is shown in **Figure 1(a)**. We modified a commercially available AFM system (Autoprobe LS, Park Scientific Instruments, Sunnyvale, CA), which was originally designed for general purpose use of AFM, for the base of the COIFM system to be built upon. The feedback loop was developed using an RHK SPM 1000 controller (RHK Technology, Inc., Troy, MI). The feedback control parameters, such as time-constant and gain, can be manually adjusted for the optimal feedback condition. The tip-sample distance in the z-direction was controlled by a high-voltage signal controller sent to the piezo tube. An optical beam deflection detection scheme in the AFM head of an AutoProbe LS (former Park Scientific Instruments) was used to transmit the interaction force between the tip and the surface into the electrical signal (E. Meyer et.al, 1988). The wavelength of the laser light for the optical detection is 670 nm and the position-sensitive detector is a bi-cell silicon photo-diode. The head was interfaced with an RHK SPM 1000 controller, and all data presented here were recorded through analog digital converter inputs of the controller and its software. Before experimentation, the laser beam was aligned on the backside of the cantilever and A-B was adjusted to make the laser incident zero by reflecting in the middle of the photo-diode. The zero force was set as the V_{A-B} value at large separations between the two surfaces before measurement. The tip-sample distance was controlled by moving the piezo tube in the z-direction using the high-voltage signal controller.

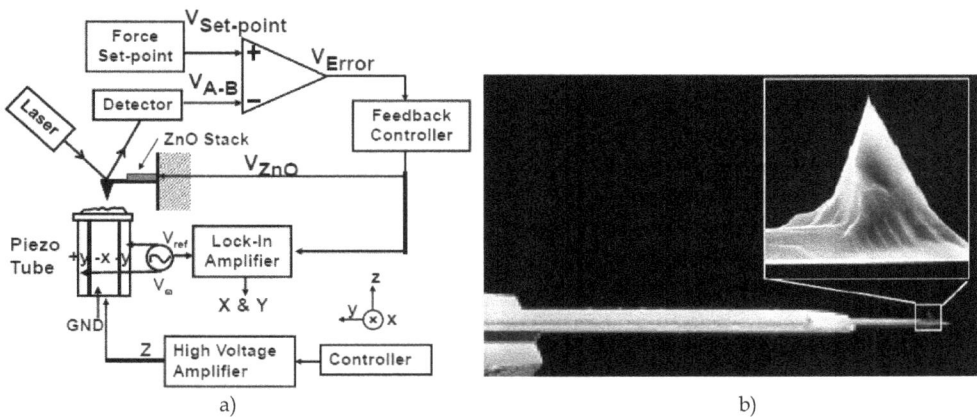

a)   b)

Fig. 1. **(a)** A schematic diagram of the COIFM with voltage-activated force feedback using an optical beam deflection detection method. The system consists of an LS AutoProbe AFM with a dimension micro-actuated silicon probe (DMASP) tip interfaced with an RHK SPM100 controller. The lock-in amplifier is used to measure the response amplitude. **(b)** Image of the DMASP tip used for COIFM. (Courtesy of Bruker Corp.)

The tip then approached the sample until touching, using the stepping motor of the piezo tube. The tip-speed was controlled by the built-in digital-to-analog converter of the RHK controller in conjunction with the high-voltage amplifier. The lateral movement was achieved by dithering the sample in the long axis direction of the cantilever, using the piezo tube with an oscillatory signal of about 1 nm amplitude at the frequency of 100 Hz (Goddenhenrich et al., 1994; Goertz et al., 2007;Major et al., 2006). A Hewlett-Packard function generator (model 33120A) was used to generate the oscillatory signal. The piezo tube sensitivities in the x and y-direction are calibrated to be 6.25nm/V and the sensitivity in the z-direction is 3.65nm/V. The amplitude of the ac-component $V_{ZnO,ac}$ was measured using a lock-in amplifier (7225 DSP, Signal Recovery, Oak Ridge, TN) with a time constant and sensitivity of 100 ms and voltage gain of 100 mV, respectively. V_{A-B}, $V_{ZnO,dc}$ and the lock-in output $V_{ZnO,ac}$ were recorded using the analog-to-digital converter of the RHK controller system. All data processing and analysis were performed with Kaleidagraph (Synergy Software, Reading, PA) after raw data acquisition.

In the present design, a cantilever with a built-in ZnO stack called a "dimension micro-actuated silicon probe" (DMASP) is employed as the COIFM sensor (DMASP, Bruker Corporation, Santa Barbara, CA). The DMASP cantilever acts as not only a detector, but also an actuator due to the ZnO stack. Voltage-activated mechanical bending of the ZnO stack serves for the force feedback in response to a displacement detection signal. The ZnO feedback loop is capable of feeding back high-frequency signals (or small forces) due to the wide frequency response of up to ~50 kHz, which is a hundred times larger than the z-bandwidth of the piezo tube feedback loop of the ordinary AFM. Thus, the feedback loop allows for more rapid, precise and accurate force measurements than ordinary commercial AFM systems in the force-distance curve. Instead of applying an opposing force on the force sensor through force-feedback, as is the case of the existing IFM, the COIFM attains zero compliance by relieving the strain built on the cantilever. This feedback mechanism protects the tip from being damaged in conjunction with the flexible spring of DMASP, thus allowing repeated use of the force sensor and improving reliability of the measurement.

Figure 1(b) shows the DMASP cantilever with the inset showing a close up of the tip. This probe is made of 1-10 Ωcm Phosphorus doped Si, with a nominal spring constant (k_z) and resonance frequency known to be 3 N/m and 50 kHz, respectively (Bruker Corp., 2011). The dimensions were measured to be L_{cant} = 485 μm, L_{tip} = 20 μm, which is in agreement with other previous measurements (Rogers et al., 2004; Vázquez et al., 2009). L_{cant} is the length of the portion of the cantilever between the base of the cantilever and the tip. A nanometer diameter tip underneath the cantilever allows for measuring the intermolecular interaction at the single molecular level between the tip and a surface. The cantilever has zero compliance during the measurement, thus preventing the snap-to-contact process associated with typical AFM force-distance measurements. Additionally, the sharp tip of the DMASP leads to probing the local structure of the interfacial water without averaging out the interfacial forces between the tip and the surface.

3. Theoretical background of COIFM with lateral modulation

3.1 Coupling of normal and friction forces through cantilever displacement

The cantilever-based optical interfacial force microscope is a combination of the AFM and IFM. The integrated COIFM employs an optical detection method of AFM and a micro-actuated silicon cantilever to self-balance the force sensor, which improves the interfacial force sensitivity by an order of magnitude and the spatial sensitivity to the sub-nanometer scale. This is enough to resolve molecular structures, such as the individual water ordering.

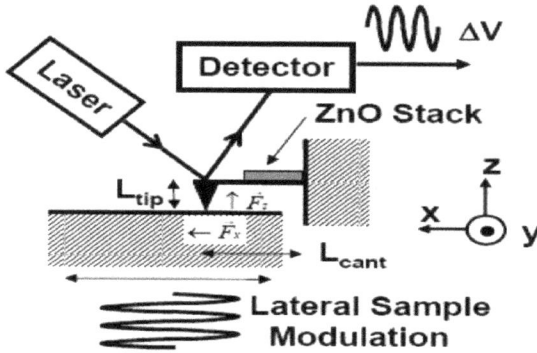

Fig. 2. The optical detection scheme of the AFM using a DMASP tip, including the applied modulation and forces. (Reprinted with permission from Rev. Sci. Instrum. **82**, 053711 (2011). Copyright 2011 American Institute of Physics.)

Figure 2 illustrates an optical beam displacement detection scheme in the AFM head that was used to transmit the interaction forces between the tip and the surface into an electrical signal (E. Meyer et al., 1988). The tip of the cantilever experiences the forces F_x and F_z by the sample surface during force measurements. A general Euler equation (Thomson, 1996) is given for the vertical displacement of the cantilever (z_z) produced by the normal force (F_z) acting at a point x = L_{cant} as follows (Chen, 1993; Sarid, 1991),

$$F_z(L_{cant} - x) = EI \frac{d^2 z_z}{dx^2} \tag{1}$$

where L_{cant} is the length of the cantilever, E is the Young's Modulus and z_z is the vertical displacement caused by F_z. The area moment of inertia (I) for a rectangular bar is given by,

$$I = \frac{t^3 w}{12} \tag{2}$$

where t is the bar thickness, and w is the width of the bar. The solution to the above equation with the boundary condition $z_z|_{z=0} = 0$, $\left. \frac{dz_z}{dx} \right|_{z=0} = 0$, is as follows,

$$z_z = \frac{3F_z}{2k_z}(\frac{x}{L_{cant}})^2(1 - \frac{x}{3L_{cant}}) \tag{3}$$

where the spring constant (k_z) is defined as the following (Sader, 2003; Sader & Green, 2004; Neumeister, 1994).

$$k_z = \frac{Et^3w}{4L_{cant}^3} \tag{4}$$

In addition to the vertical force F_z, a friction force (F_x) along the major axis of the cantilever (see **Figure 2**) also contributes to the vertical displacement of the cantilever by the following Euler equation:

$$F_xL_{tip} = EI\frac{d^2z_x}{dx^2} \tag{5}$$

where L_{tip} is the length of the cantilever tip. The vertical displacement produced by $F_x(z_x)$ can then be found in the following equation:

$$z_x = \frac{3F_x}{2k_z} \cdot \frac{L_{tip}}{L_{cant}}(\frac{x}{L_{cant}})^2 \tag{6}$$

The total displacement (z_c), the sum of both z_z and z_x, is given by the following equation:

$$z_c = \frac{3}{2k_z}(\frac{x}{L_{cant}})^2\left[\frac{L_{tip}}{L_{cant}}F_x + (1 - \frac{x}{3L_{cant}})F_z\right] \tag{7}$$

The bending motion of the cantilever due to the tip-sample interactions is detected by measuring the voltage difference (V_{A-B}) between two photodiodes, namely A and B, as shown in **Figure 1(a)**. The difference in voltage is proportional to the slope of the cantilever (according to the law of reflection) at the point $(x=L_{cant})$ where the beam is reflected as follows (Schwarz et al., 1996):

$$V_{A-B} = \alpha\vartheta_c \tag{8}$$

where α is a proportional constant. The total slope (ϑ_c) at $x=L_{cant}$ is the derivative of total displacement of the cantilever (as given in equation (7)) with respect to z in the following way:

$$\vartheta_c = \frac{dz_c}{dx}\Big|_{x=L_{cant}} = \frac{3}{2k_zL_{cant}}\left(F_z + 2\frac{L_{tip}}{L_{cant}}F_x\right) \tag{9}$$

The above relation is consistent with an earlier work (Sader, 2003). The detection signal V_{A-B} is related to the two forces as follows:

$$V_{A-B} = \frac{3\alpha}{2k_zL_{cant}}(F_z + \frac{2L_{tip}}{L_{cant}}F_x) \tag{10}$$

The above relation enables one to measure the normal force F_z and the friction force F_x simultaneously through the measurement of the optical beam displacement signal, V_{A-B}.

However, the inability of the current AFM system to control the cantilever displacement causes limitations in measuring forces using equation (10) over all distance ranges due to the snap-to-contact problem.

3.2 Voltage activated force feedback

We apply the concept of the COIFM technique to the simultaneous measurement of normal and friction forces on approach to overcome the limitations of the AFM, using the DMASP cantilever as a detector and an actuator. Here we apply this concept to the case where friction force (F_x) and normal force (F_z) exist together by making displacement zero while measuring the normal force F_z through the voltage activated force feedback (Bonander & B.I. Kim, 2008). When in feedback, the detector signal (V_{A-B}) in **Figure 1(a)** is maintained at zero by generating a feedback force $F_{feedback}$ as follows:

$$V_{A-B}(t) = \frac{3\alpha}{2k_z L_{cant}} \left(F_z + 2\frac{L_{tip}}{L_{cant}} F_x - F_{feedback} \right) = 0 \tag{11}$$

In the voltage activated force feedback, a voltage V_{ZnO} is applied to the ZnO stack of the DMASP cantilever. Because V_{ZnO} is linearly proportional to V_{A-B} with a proportional constant β as given,

$$V_{A-B} = \beta \cdot V_{ZnO} \tag{12}$$

the feedback condition in equation (11) can be solved in terms of V_{ZnO} as follows:

$$V_{ZnO} = \frac{3\alpha}{2\beta k_z L_{cant}} \left(F_z + \frac{2L_{tip}}{L_{cant}} F_x \right) \tag{13}$$

This equation suggests that the ability to obtain normal and friction forces while overcoming the snap-to-contact problem will make feedback measurements much more advantageous than non-feedback measurements. Equation (13) suggests that instead of the V_{A-B} signal, the feedback signal V_{ZnO} is used in measuring normal and friction forces.

4. Calibration of COIFM

4.1 Feedback response test

To find the time resolution of the COIFM, a square-wave voltage with amplitude 0.2 V and frequency 10 Hz was applied to the set-point voltage ($V_{set\ point}$) with the force feedback far away from the surface (B.I. Kim, 2004). **Figure 3(a)-3(d)** shows that the feedback controller preamp output (V_{A-B}) follows this square wave by applying appropriate voltages to the ZnO stack of the DMASP sensor (V_{ZnO}). The square wave causes the cantilever to try to create a torque on the cantilever so as to achieve a zero error voltage V_{Error} with the feedback on (**Figure 3(c)**). The controller is set up to optimize the transient response in order to achieve the necessary time response for a COIFM experiment. The transient feedback response test signal (**Figure 3(d)**) shows that the COIFM has a practical time resolution ~ 1.5 msec. The force resolution is less than 150 pN, which is a higher force sensitivity by two orders of magnitude than the existing IFM with electrical detection method (Joyce & Houston, 1991).

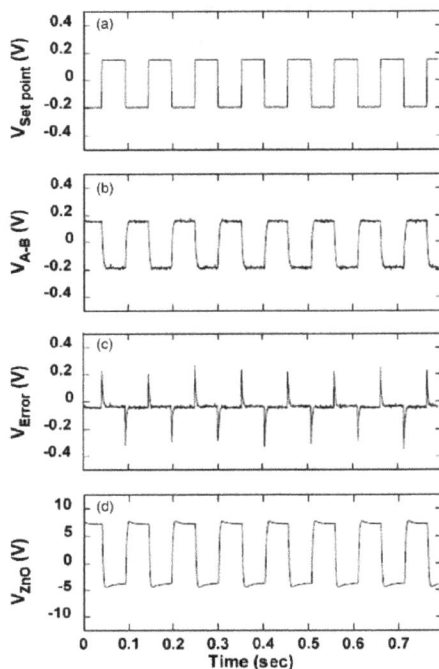

Fig. 3. **(a)** A square wave ac signal with a frequency of 10 Hz as a set-point voltage of the feedback loop. **(b)** The deflection V_{A-B} signal that follows the set-point voltage. **(c)** The error signal V_{Error} between $V_{Set-point}$ and V_{A-B}. **(d)** The signal V_{ZnO} sent from the controller to the ZnO stack material. (Reprinted with permission from Appl. Phys. Lett. **92**, 103124 (2008). Copyright 2008 American Institute of Physics.)

4.2 Normal and friction force calibration

A lock-in amplifier (7225 DSP, Signal Recovery, Oak Ridge, TN) is used to generate a sinusoidal driving signal and to detect the modulated output of the force sensor. By dithering the tip in the x-axis with amplitude 3 nm at 100 Hz and detecting the response amplitude with a lock-in amplifier, we can measure normal force and friction force simultaneously as a function of separation distance. With an amplitude of 100 mV at a load force of 10-20 nN, the friction force is typically 3-6 nN.

For qualitative understanding, the signal V_{A-B} should be converted into force using the relation between the signal and the two forces in equation (10). Equation (10) suggests how to calibrate the conversion factors from the V_{A-B} signal to both forces experimentally. To do this we need to find the proportional constant α between V_{A-B} and ϑ_c. When the tip is in contact with the substrate, the detection signal can be expressed as:

$$V_{A-B} = \frac{3\alpha}{2L_{cant}} z_c \quad (14)$$

where z_c is the cantilever displacement along z-axis. The above equation suggests that α can be found by measuring the V_{A-B} signal as a function of z_c. The cantilever displacement can be changed systematically by contacting the cantilever to the sample surface (assuming that the indentation between the tip and the surface is negligible).

If an external friction force is applied to the probe tip, the system will be balanced by generating an additional force by a change in voltage. The output is the force feedback voltage to maintain zero compliance in the cantilever and represents the interfacial friction force between tip and sample. The actual force applied to the tip is related to the voltage sent to the ZnO stack via "voltage-to-force conversion factor." In actual operation, the controller system records variations in the voltage applied to the ZnO stack as a function of relative tip/sample separation. To experimentally determine the value of the voltage-to-force conversion factor, two measurements are taken. The first measurement is done with the tip in contact with the substrate. As a voltage is applied to the piezo tube, the piezo tube moves, causing the cantilever to bend. The optical beam detection will record a change in voltage. A plot is made of the change in detection signal versus distance, and the slope of the line $\left(\dfrac{3\alpha}{2L_{cant}} \right)$ is found to be 20.4 mV/nm from the relationship between V_{A-B} and z_c shown in **Figure 4(a)**. From the figure, α is found to be 6.60×10^3 V/rad with $L_{cant} = 485\mu m$. The normal force conversion factor from V_{A-B} into F_z is determined to be 147 nN/V from $\dfrac{2k_z L_{cant}}{3\alpha}$ with the spring constant k_z of ~3 N/m.

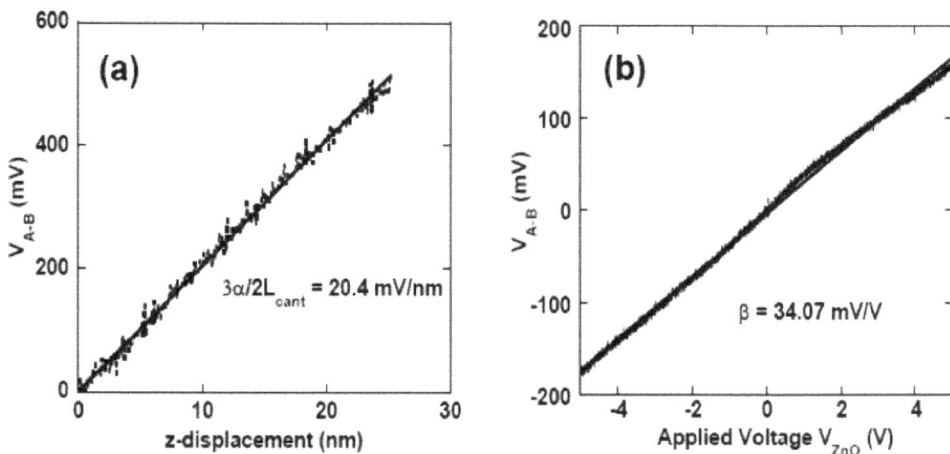

Fig. 4. **(a)** Change in V_{A-B} versus cantilever displacement (z_c) due to normal force. The slope of the line allows us to determine the conversion factor α. **(b)** The resulting change in V_{A-B} when a voltage is applied to the ZnO stack from -10 V to 10 V. The slope of this line is β and is used in the equation to find the voltage-to-force conversion factor. (Reprinted with permission from Rev. Sci. Instrum. **82**, 053711 (2011). Copyright 2011 American Institute of Physics.)

The second measurement is done with the tip not in contact with the substrate. Here a voltage is applied to the cantilever, causing it to bend. Again, the change in voltage of the optical beam detection method is recorded (see Figure 4(b)). The slope of the line is the constant β which is found to be 34.07 mV/V through the linear fitting of the data in **Figure 4(b)**. Using the obtained α value along with the measured β value, and the spring constant, the calculated conversion factor for normal forces while the system is in feedback $\left(\dfrac{2\beta k_z L_{cant}}{3\alpha} \right)$ is found to be 5.01 nN/V. Using $L_{tip} = 20$ μm (Bruker Corp., 2011), the

conversion factor for friction forces $\dfrac{\beta k_z L_{cant}^2}{3\alpha L_{tip}}$ is calculated to be 182.14 nN/V. It is important

to note that this lateral conversion factor is 12 times larger than the normal force conversion factor because of $\dfrac{L_{cant}}{2L_{tip}}$ in equation (13).

5. Demonstration of COIFM in measuring force-distance curves

The capability of this COIFM as a second generation of IFM has been demonstrated by revealing the hidden structures of the interfacial water on a silicon surface at the molecular scale. **Figure 5(a)** illustrates a typical force distance curve taken on a silicon surface (SPI Supplies) in air with feedback off as the tip approaches with the speed of 8 nm/s at a distance of 50 nm away from the surface. In the force-displacement curve, the distance zero was defined as the intersection between the contact force line and the line where the interfacial force is zero (Senden, 2001). The voltage units were converted into force units using the conversion factors found above, as shown on the right axis of each panel. A long-range repulsive force appears monotonously at the distances between 5 nm and 30 nm from the silicon surface, possibly resulting from the electrostatic dipole-dipole interaction observed by Kelvin probe measurement (Verdaguer et al., 2006). The same experiment was repeated with the feedback ON. The voltage signal to the ZnO material, V_{ZnO}, and the error signal V_{A-B} were recorded as a function of tip to sample distance, as shown in **Figure 5(b)** and **Figure 5(c)**, respectively. One of the key features in **Figure 5(c)** is that the V_{A-B} voltage remains zero during approach, indicating that all forces on the cantilever remain balanced or "zero compliance" by relieving the strain built up in the ZnO stack through force feedback. However, the sensing cantilever starts to bend as soon as the tip touches the silicon surface, indicating the breakdown of force feedback. The long- range interaction is reproducibly obtained in the force-distance curve with feedback on (**Figure 5(b)**). The background noise level (0.1~0.2 nN) is smaller than the background noise (1~2 nN) with feedback off by an order of magnitude.

Direct comparison between two force curves with feedback on and off in the distance range between 0 nm and 5 nm shows that fine periodic structures with several peaks and valleys appear from the surface in the force curve with feedback on, whereas they are not absent in the force-displacement curve with feedback off. Interestingly, the periodicity of the peaks is 0.32 ± 0.13 nm as marked with arrows in the detailed force-distance curve between 0 nm and 3 nm (inset of **Figure 5(b)**), which is comparable with the diameter of a single water molecule. In recent years, a few groups have observed similar periodic features at interfaces between solid surfaces and liquid water using amplitude modulation methods,

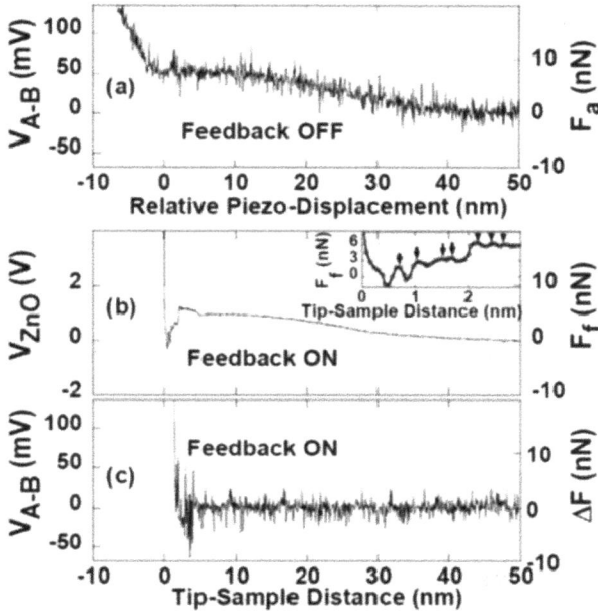

Fig. 5. **(a)** A force-displacement curve between the tip and the silicon surface obtained without a force-activated voltage feedback system. **(b)** The force applied to the ZnO stack material was graphed as a function of tip and silicon sample distance. **(c)** Force-distance curve between the tip and the silicon surface obtained with a force-activated voltage feedback system. (Reprinted with permission from Appl. Phys. Lett. **92**, 103124 (2008). Copyright 2008 American Institute of Physics.)

suggesting the possible ordering of water molecules near surfaces (Antognozzi et al., 2001; Jarvis et al., 2000; Jeffery et al., 2004; Uchihashi et al., 2005). This COIFM data on interfacial water demonstrates that the COIFM is capable of unveiling structural and mechanical information on interfacial water at the single molecular level, which has not been previously reported with the existing IFM. In contrast to the recent IFM studies of interfacial water, in which IFMs large diameter (1 µm-10 µm) tips were used (Major et al., 2006; Matthew et al., 2007), the sharp tip of the DMASP was able to probe the local structure of the interfacial water without averaging out the interfacial forces between the tip and the surface.

6. Application of COIFM to moleular interaction measurements

6.1 Background

The capability of the COIFM over both the IFM and AFM systems is demonstrated by measuring friction forces and normal forces generated by the water molecules in an ambient environment. Friction is of great importance in micromechanical systems where water is trapped between two surfaces (Komvopoulos, 2003). The trapped water has a critical effect on the performance of the systems through interfacial tribological properties (de Boer &

Mayer, 2001). This could include the roles of water in inter-molecular/inter-surface friction and in the reduction of the friction in water-based bio-materials such as artificial cartilage. Water molecules in bulk water are understood as being ordered at the liquid-solid interface, with that order decaying with each water molecule's distance from the interface (Jarvis et al., 2000).

Far less is known, comparatively, about the structure of interfacial water in ambient conditions (Verdaguer et al., 2006). In an ambient environment, in addition to interacting with the substrate surface, the interfacial water also interacts with the surrounding water vapor. More recent attempts have been made to describe the viscosity of interfacial water using the IFM (Goertz, 2007; Major et al., 2006). The structured water exists not only on crystalline hydrophilic surfaces, but even on amorphous surfaces (Asay & S.H.J. Kim, 2005, 2006; Verdaguer et al., 2006). These studies indicate that the behavior of interfacial water molecules in an ambient environment is substantially different from the behavior of water at the liquid-solid interface. Here we applied the developed COIFM to probe the structure of interfacial water.

6.2 Materials and methods

All measurements were taken on a freshly cleaned silica wafer, Si (100) (SPI Supplies, West Chester, PA), in an ambient condition with the relative humidity monitored using a thermo-hygro recorder (Control Co., Friendswood, TX). The top of the surface is expected to be covered with natural oxide in air, thereby forming the silica, the most abundant material in the Earth's crust (Iler, 1979). The wafer was attached to a 15 mm steel disk using double-stick tape and then mounted on a magnetic sample stage on top of the piezo tube. To remove all organic contaminates, the silica was cleaned using a piranha solution made from a 3:1 concentrated H_2SO_4/30% H_2O_2 (Pharmco and Fischer Scientific, respectively). It was then sonicated in acetone for 5 minutes, then in ethanol for 5 minutes, rinsed with DI water, and then dried with a dry N_2 flow. Tips were cleaned using a UV sterilizer (Bioforce Nanosciences Inc., Ames, IA) to remove the residual hydrocarbon molecules. The tip speed was chosen as 10 nm/sec. The output signals of the lock-in amplifier were converted into forces using the conversion factors found in the results section. The converted force scales are displayed on the right axis of each panel and voltage units on the left axis.

6.3 Large oscillatory force of water in an ambient condition

We also measured both normal and friction forces in the water junction between the probe and the surface with the COIFM with lateral modulation. **Figure 6(a), 6(b)** and **6(c)** show the measured V_{A-B}, $V_{ZnO,ac}$, and $V_{ZnO,dc}$ data, respectively, as a function of piezo displacement. The data clearly demonstrates that the force feedback allows for the COIFM to measure the normal force of water between the tip and the sample for all distance regimes, overcoming the snap-to-contact effect associated with the conventional AFM method. The zero distance was defined as the point where the friction force increases sharply as marked in **Figure 6(b)**. As the tip approaches, both normal and friction forces remain at zero until interaction with interfacial water occurs around 12 nm away from the substrate. Surprisingly, the data show oscillatory patterns in both normal force and friction force. The V_{A-B} signal also displays a periodic change with the tip-sample distance, as shown in an enlarged inset in **Figure 6(a)**. These periodic features are consistent with the earlier AFM-based observation of a stepwise change of the force gradient related with the thin water bridge in an ambient environment (Choe et al., 2005). The absence of modulation turns out to decrease the number of oscillations by several times, therefore suggesting that the kinetic energy due to the lateral modulation promotes layering transitions by overcoming the activation barriers between two successive layered states of the interfacial water.

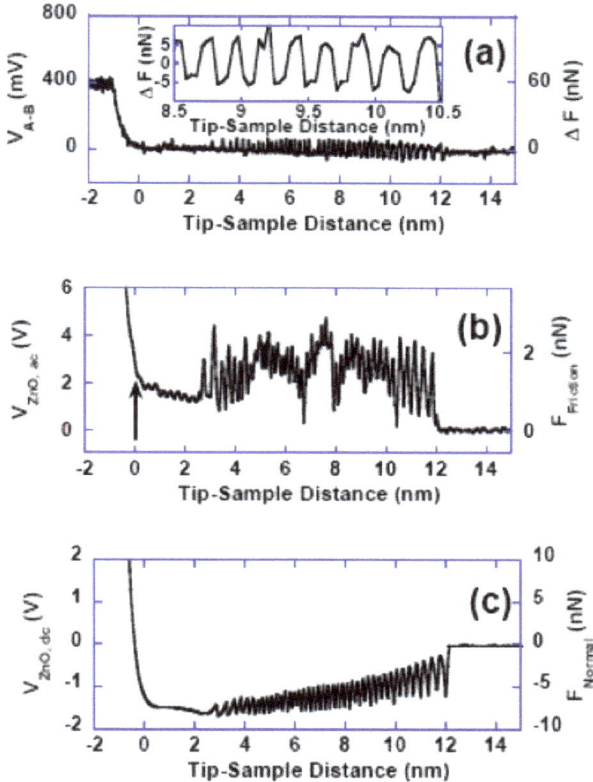

Fig. 6. Force-distance curves for V_{A-B} (a), the frictional force (b), and normal force (c) during the tip-approach towards the surface with velocities of 10 nm/s at 55% relative humidity. The inset in (a) is an expanded plot of V_{A-B} in a range of 8.5 - 10.5 nm. (Reprinted with permission from Rev. Sci. Instrum. 82, 053711 (2011). Copyright 2011 American Institute of Physics.)

The periodicity is found to be 0.227±0.056 nm for valley-valley distance analysis and 0.223±0.055 nm for peak-peak distance analysis. This periodicity matches the diameter of water, which is consistent with Antognozzi et al., who found the periodicity of water layers to be 0.24-0.29 nm even for distilled water deposited on a mica sample surface using a near-field scanning optical microscope (NSOM) (Antognozzi et al., 2001). This result is also in agreement with other earlier studies using AFM at the liquid-solid interface between a hydrophilic surface (e.g., mica) and bulk water (Higgins et al., 2006; Jarvis et al., 2000; Jarvis et al., 2001; Jeffery et al., 2004; Li et al., 2007; Uchihashi et al., 2005). The interfacial water confined between two surfaces forms water layers with periodicity of one water diameter 0.22 nm (Jarvis et al., 2000), 0.23 +0.003 nm (Jarvis et al., 2001), 0.25 + 0.05 nm (Jeffery et al., 2004), 0.23 $\pm$ 0.03 nm (Uchihashi et al., 2005), 0.29 $\pm$ 0.006 nm (Higgins et al., 2006), and 0.22-0.29 nm (Li et al., 2007). This can be understood that the ordering of confined water molecules leads to oscillatory solvation forces, which are reflections of the geometric packing experienced by the molecules due to the imposing surfaces, with the period of oscillation roughly equal to the molecular diameter of water. The oscillations occur in the molecular force due to the transition between solid (ordering) and liquid (disordering), depending on the commensuration and incommensuration between the spacing and the molecular diameter (Chaikin & Lubensky, 1995). These data on the interfacial water suggest that the COIFM is capable of providing unprecedented information on the structural and mechanical properties of molecules.

Most published data concerning interfacial water has never shown distinct oscillatory behavior (Goertz, 2007; Major et al., 2006). This is because previous methods exhibit too much noise to see the distinct oscillatory patterns or have mechanical instabilities that prevent measurements over all distance regimes. One important aspect of data collection that is visible in **Figure 6** is the lack of noise in COIFM data. This lack of noise is due to the size of the tip and the sensitivity of the system. With the lack of noise, a very distinct oscillatory pattern can be seen in both normal and friction force, which starts with onset of chain formation and continues until the tip and sample come in contact. Also, due to the lack of noise, it is evident that a peak in normal force corresponds to a peak in friction force. Once the tip and sample come in contact, there is a very sharp rise in normal force and an increase in friction force as well. So the amount of noise in IFM data makes it difficult to see the correlation that is present in COIFM data. The COIFM employs an optical detection method of AFM and a commercially available micro-actuated silicon cantilever to self-balance the force sensor, which improves the interfacial force sensitivity by an order of magnitude and the spatial sensitivity to the sub-nanometer scale, enough to resolve the individual water ordering on a silicon surface. The change in tip size and increased sensitivity for electric force detection in the COIFM in comparison to the IFM allows it to be used at the sub-nanometer range and make it a more useful technique in analyzing forces at nanoscopic ranges. While simultaneous measurement of normal and friction forces is not new, the resolution with which they have been measured by the COIFM is novel.

7. Conclusion

Here we reported the integration of the existing two scanning-probe techniques (AFM and IFM) through the development of an instrument called a "cantilever-based optical interfacial force microscope" (COIFM) (Bonander & B.I. Kim, 2008). The COIFM is a new tool developed for the study of interfacial forces, such as interfacial water structural forces. The COIFM employs a commercially available "dimension micro-actuated silicon probe" (DMASP) cantilever in its voltage-activated force feedback scheme (Aimé et al., 1994; Burnham & Colton, 1989; Lodge, 1983; G. Meyer & Amer., 1988). The diminished size of the DMASP cantilver at a radius of 10 nm, compared to that of the IFM's 0.1 μm to 1 μm probe radius, enables one to study water structures at the single molecular level. The smaller probe size of the cantilever enables improved force resolution over conventional techniques by at least an order of magnitude. Additionally, due to the optical detection scheme, the force resolution is improved by two orders of magnitude over the existing IFM with electrical detection method (Joyce & Houston, 1991). The ZnO feedback loop allows for more rapid, precise and accurate force measurements than ordinary commercial AFM systems in the force-distance curve. The COIFM attains zero compliance by relieving the strain built on the cantilever protecting the tip from being damaged in conjunction with the flexible spring of DMASP, thus allowing repeated use of the force sensor and improving reliability of the measurement.

The recently developed COIFM technique was used to measure normal and friction forces simultaneously for studies of interfacial structures and mechanical properties of nanoscale materials. We derived how the forces can be incorporated into the detection signal using the classical Euler equation for beams. A lateral modulation with the amplitude of one nanometer was applied to create the friction forces between tip and sample. The COIFM with lateral modulation allows for simultaneous measurement of normal and friction forces in the attractive regime as well as in the repulsive regime by utilizing the force feedback capability of the instrument (Bonander & B.I. Kim, 2008). We demonstrated the capability of the COIFM by measuring normal and friction forces of interfacial water at the molecular scale over all distance ranges. It also demonstrated the capability of this COIFM as a second generation of the IFM by revealing the hidden structures of the interfacial water between two silica surfaces (Bonander & B.I. Kim, 2008). The distinct oscillations observed when measuring interfacial water potentially will reveal new information about molecular water orientation with further analysis of the data. Although there have been many friction measurements using scanning probe techniques, there are few explicit relations between the detection signal, normal, and frictional forces. The ability of COIFM to measure normal and friction force simultaneously, along with the incorporation of force-feedback control, make this type of microscopy a very useful technique in analyzing thin films and interactions between two surfaces, especially when measuring large amounts of force are necessary.

8. Future applications

Due to its excellent capability, the COIFM will improve the understanding of hidden interfacial phenomena. The COIFM will reveal new information about interfacial water and also other molecules where conventional AFM and IFM systems have been used, such as DNA (Hansma et al., 1996). The usefulness and uniqueness of the COIFM in studying unprecedented structural and mechanical properties of interfacial water confined between a tip and a sample surface has already been characterized. Future applications of the COIFM will be extended to the recognition and investigation of biomolecular interactions.

Intermolecular friction forces play a fundamental role in many biological processes, such as transport along cytoskeletal filaments (Mueller et al., 2010) or inside human and animal joints (Flannery et al., 2010). Biomolecular functions are the most important phenomena to sustain our lives (Alberts, 2002). They carry out delicate structural conformation changes along the reaction coordinates during the biomolecular activation. The conformation changes are related to metastable intermediate states and energy barriers due to chemical and mechanical forces between a protein and a protein, between a protein and a ligand, and within proteins (Schramm, 2005). Correlating the metastable intermediate states and energy barriers with known structural information is extremely important in understanding each step of the biomolecular functions. A biomolecular system passes through several metastable states before it reaches a stable state with the lowest energy. Even with such importance, however, metastable intermediate states and energy barriers are difficult to observe because of a relatively short life span ($\sim 10^{-13}$ sec) and their non-equilibrium nature in a solution phase (Schramm, 2005). It is for this reason that these extremely important and challenging problems mentioned above remain largely unsolved. As an attempt to solve these problems, instead of controlling the binding time along a reaction coordinate (e.g., using kinetic isotope effects), detaching two bounded single molecules from each other by pulling both ends has been employed as an alternative method (Evans & Ritchie, 1997). The single molecular pulling measurements provide individual molecular value for a physical quantity (e.g., force) under non-equilibrium conditions because the measurement is conducted for each single molecule individually (Evans & Ritchie, 1997; Hermanson, 1995; Lee et al., 1994; Liphardt et al., 2002; Oberhauser et al., 1998; Rief et al., 1997; Ros et al., 1998; Strunz et al., 1999); whereas, the current biomolecular and biochemistry investigations measure the average value of a biochemical quantity over a huge number of molecules ($\sim 10^{23}$) under equilibrium conditions. The COIFM's sensitive distance and force control capability allows for investigating the metastable states along the reaction coordinates level.

9. Acknowledgment

The author thanks Jeremy Bonander, Jared Rasmussen, Ryan Boehm, Edward Kim, Joseph Holmes, and Thanh Tran for all their help and support on this project. This work is partly funded by NSF DMR-1126854, NSF DBI-0852886, NSF EPSCOR Startup Augmentation Funding, and the Research Corporation Single-Investigator (Cottrell College Science Award No. CC7041/7162).

10. References

Aimé, J.P., Elkaakour, Z., Odin, C., Bouhacina, T., Michel, D., Curély, J., & Dautant, A. (1994). Comments on the use of the force mode in atomic force microscopy for polymer films. *J. Appl. Phys.*, Vol. 76, No. 2, (July 1994), pp. (754-762), 0021-8979

Alberts, B. (2002). *Molecular biology of the cell* (4th Ed.), Garland Science, 978-0815332183, New York

Antognozzi, M., Humphris, A.D.L., & Miles, M.J. (2001). Observation of molecular layering in a confined water film and study of the layers viscoelastic properties. *Appl. Phys. Lett.*, Vol. 78, No. 3, (January 2001), pp. (300-302), 0003-6951

Asay, D.B., & Kim, S.H. (2005). Evolution of the adsorbed water layer structure on silicon oxide at room temperature. *J. Phys. Chem. B*, Vol. 109, No. 35, (August 2005), pp. (16760-16763), 1520-6106

Asay, D.B., & Kim, S.H. (2006). Effects of adsorbed water layer structure on adhesion force of silicon oxide nanoasperity contact in humid ambient. *J. Chem. Phys.*, Vol. 124, No. 17, (May 2006), pp. (174712/1-174712/5), 0021-9606

Ashby, P.D., Chen, L., & Lieber, C.M. (2000). Probing intermolecular forces and potentials with magnetic feedback chemical force microscopy. *J. Am. Chem. Soc.*, Vol. 122, No. 39, (September 2000), pp. (9467-9472)

Bonander, J.R., & Kim, B.I. (2008). Cantilever based optical interfacial force microscope. *Appl. Phys. Lett.*, Vol. 92, No. 10, (March 2008), pp. (103124), 0003-6951

Bruker Corporation (2011). *Probes and Accessories*, Bruker, Camarillo, CA

Bunker, B.C., Kim, B.I., Houston, J.E., Rosario, R., Garcia, A.A., Hayes, M., Gust, D., & Picraux, S.T. (2003). Direct observation of photo switching in tethered spiropyrans using the interfacial force microscope. *Nano Lett.*, Vol. 3, No. 12, (November 2003), pp. (1723-1727)

Burnham, N.A., & Colton, R.J. (1989). Measuring the nanomechanical properties and surface forces of materials using an atomic force microscope. *Journal of Vacuum Science Technology A: Vacuum, Surfaces, and Films*, Vol. 7, No. 4, (July 1989), pp. (2906 -2913), 0734-2101

Burns, A.R., Houston, J.E., Carpick, R.W., & Michalske, T.A. (1999A). Friction and molecular deformation in the tensile regime. *Phys. Rev. Lett.*, Vol. 82, No. 6, (February 1999), pp. (1181-1184), 0031-9007

Burns, A.R., Houston, J.E., Carpick, R.W., & Michalske, T.A. (1999B). Molecular Level Friction As Revealed with a Novel Scanning Probe. *Langmuir*, Vol. 15, No. 8, (April 1999), pp. (2922-2930)

Butt, H.J., Cappella, B., & Kappl, M. (2005). Force measurements with the atomic force microscope: technique, interpretation and applications. *Surf. Sci. Rep.*, Vol. 59, No. 1-6, (October 2005), pp. (1-152)

Cappella, B., & Dietler, G. (1999). Force-distance curves by atomic force microscopy. *Surface Science Reports*, Vol. 34, No. 1-3, (July 1999), pp. (1-104), 0167-5729

Carpick, R.W., Ogletree, D.F.and Salmeron, M. (1997). Lateral stiffness: a new nanomechanical measurement for the determination of shear strengths with friction force microscopy. *Appl. Phys. Lett.*, Vol. 70, No. 12, (March 1997), pp. (1548-1550), 0003-6951

Chaikin, P.and Lubensky, T. (2000).*Principles of condensed matter physics*, Cambridge University Press, 0-521-43224-3, New York

Chang, K.K., Shie, N.C., Tai, H.M., & Chen, T.L. (2004), A micro force sensor using force-balancing feedback control system and optic-fiber interferometers. *Tamkang Journal of Science and Engineering*, Vol. 7, No. 2, pp. (91-94)

Chen, C. (1993).*Introduction to scanning tunneling microscopy*, Oxford University Press, 0-19-507150-6, New York

Choe, H., Hong, M.H., Seo, Y., Lee, K., Kim, G., Cho, Y., Ihm, J., & Jhe, W. (2005). Formation, manipulation, and elasticity measurement of a nanometric column of water molecules. *Phys. Rev. Lett.*, Vol. 95, No. 18, (October 2005), pp. (187801/1-187801/4), 0031-9007

Colchero, J., Luna, M., & Baró, A.M. (1996). Lock-in technique for measuring friction on a nanometer scale. *Appl. Phys. Lettl.*,Vol. 68, No. 20, (March 1996), pp. (2896-2898), 0003-6951

De Boer, M.P., & Mayer, T.M. (2001). Tribology of MEMS. *MRS Bull.*, Vol.26, No.4, (April 2001) pp. (302-304)

Evans, E. &. Ritchie, K. (1997). Dynamic strength of molecular adhesion bonds, *Biophys. J.*, Vol. 72, No. 4, (April 1997), pp. (1541-1555), 1541-1555

Fernandez-Torres, L., Kim, B.I., & Perry, S. (2003). The frictional response of VC(100) surfaces: influence of 1-octanol and 2,2,2-trifluoroethanol adsorption. *Tribology Letters*, Vol. 15, No. 1, (June 2003), pp. (43-50), 1023-8883

Flannery, M., S. Flanagan, E. Jones and C. Birkinshaw. (2010). Compliant layer knee bearings: part I: friction and lubrication. *Wear*, Vol. 269, No. 5-6, (July 2010), pp. (325-330), 0043-1648

Goertz, M.P., Houston, J.E., & Zhu, X. (2007). Hydrophilicity and the viscosity of interfacial water. *Langmuir*, Vol. 23, No. 10, (May 2007), pp. (5491-5497), 0743-7463

Greenwood, J.A. (1997). Adhesion of elastic spheres. *Proceedings of the Royal Society of London. Series A: Mathematical, Physical and Engineering Sciences*, Vol. 453, No. 1961, (June 1997), pp. (1277-1297), 1471-2946

Göddenhenrich, T., Müller, S., & Heiden, C. (1994). A lateral modulation technique for simultaneous friction and topography measurements with the atomic force microscope. *Rev. Sci. Instrum.*, Vol. 65, No. 9, (September 1994), pp. (2870-2873), 0034-6748

Hansma, H.G., Sinsheimer, R.L., Groppe, J., Bruice, T.C., Elings, V., Gurley, G., Bezanilla, M., Mastrangelo, I.A., Hough, P.V., & Hansma, P.K. (1993). Recent advances in atomic force microscopy of dna. *Scanning*, Vol. 15, No. 5, (Sep-Oct 1993), pp. (296-299), 0161-0457

Hermanson, G.T. (1995). *Bioconjugate techniques* (1st Ed). Academic Press, 978-0123886231, San Diego

Higgins, M.J., Polcik, M., Fukuma, T., Sader, J.E., Nakayama, Y., & Jarvis, S.P. (2006). Structured water layers adjacent to biological membranes. *Biophys. J.*, Vol. 91, No. 7, (October 2006), pp. (2532-2542), 0006-3495

Houston, J.E., & Michalske, T.A. (1992). The interfacial-force microscope. *Nature*, Vol. 356, No. 6366, (March 1992), pp. (266-267), 0028-0836

Huber, D.L., Manginell, R.P., Samara, M.A., Kim, B., & Bunker, B.C. (2003). Programmed adsorption and release of proteins in a microfluidic device. *Science*, Vol. 301, No. 5631, (July 2003), pp. (352-354), 0036-8075

Iler, R.K. (1979).*The chemistry of silica : solubility, polymerization, colloid and surface properties, and biochemistry*, Wiley, New York

Israelachvili, J.N., & Adams, G.E. (1978). Measurement of forces between two mica surfaces in aqueous electrolyte solutions in the range 0-100 nm. *J. Chem. Soc., Faraday Trans. 1*, Vol. 74, No. 0, (January 1978), pp. (975-1001), 0300-9599

Jarvis, S.P., Ishida, T., Uchihashi, T., Nakayama, Y., & Tokumoto, H. (2001). Frequency modulation detection atomic force microscopy in the liquid environment. *Applied Physics A: Materials Science & Processing*, Vol. 72, No. 7, (March 2001), pp. (S129-S132), 0947-8396

Jarvis, S.P., Yamada, H., Yamamoto, S., Tokumoto, H., & Pethica, J.B. (1996). Direct mechanical measurement of interatomic potentials. *Nature*, Vol. 384, No. 6606, (November 1996), pp. (247-249), 1476-4687

Jarvis, S.P., Uchihashi, T., Ishida, T., Tokumoto, H., & Nakayama, Y. (2000). Local solvation shell measurement in water using a carbon nanotube probe. *The Journal of Physical Chemistry B*, Vol. 104, No. 26, (June 2000), pp. (6091-6094), 1520-6106

Jeffery, S., Hoffmann, P.M., Pethica, J.B., Ramanujan, C., Özer, H., & Oral, A. (2004). Direct measurement of molecular stiffness and damping in confined water layers. *Phys. Rev. B*, Vol. 70, No. 5, (August 2004), pp. (054114.1-054114.8), 1098-0121

Joyce, S.A., & Houston, J.E. (1991). A new force sensor incorporating force-feedback control for interfacial force microscopy. *Review of Scientific Instruments*, Vol. 62, No. 3, (March 1991), pp. (710-715), 0034-6748

Kiely, J.D., & Houston, J.E. (1999). Contact hysteresis and friction of alkanethiol self-assembled monolayers on gold. *Langmuir*, Vol. 15, No. 13, (June 1999), pp. (4513-4519), 0743-7463

Kim, B., Lee, S., Guenard, R., Torres, L.F., Perry, S., Frantz, P., & Didziulis, S. (2001). Chemical modification of the interfacial frictional properties of vanadium carbide through ethanol adsorption. *Surface Science*, Vol. 481, No. 1-3, (June 2001), pp. (185 - 197), 0039-6028

Kim, B.I. (2004). Direct comparison between phase locked oscillator and direct resonance oscillator in the noncontact atomic force microscopy under ultrahigh vacuum. *Review of Scientific Instruments*, Vol. 75, No. 11, (November 2004), pp. (5035-5037), 0034-6748

Kim, B.I., Bonander, J.R., & Rasmussen, J.A. (2011). Simultaneous measurement of normal and friction forces using a cantilever-based optical interfacial force microscope. *Review of Scientific Measurements*, Vol. 82, No. 5, (May 2011), pp. (053711), 0034-6748

Kim, H.I., & Houston, J.E. (2000). Separating mechanical and chemical contributions to molecular-level friction. *J. Am. Chem. Soc.*, Vol. 122, No. 48, (August 2000), pp. (12045-12046)

Komvopoulos, K. (2003). Adhesion and friction forces in microelectromechanical systems: mechanisms, measurement, surface modification techniques, and adhesion theory. *Journal of Adhesion Science and Technology*, Vol. 17, No. 4, (May 2003), pp. (477-517)

Lee, G.U., Kidwell, D.A., Colton, R. J. (1994). Sensing discrete streptavidin biotin interactions with atomic-force microscopy. *Langmuir*, Vol. 10, (February 1994), pp. (354-357)

Li, T., Gao, J., Szoszkiewicz, R., Landman, U., & Riedo, E. (2007). Structured and viscous water in subnanometer gaps. *Phys. Rev. B*, Vol. 75, No. 11, (March 2007), pp. (115415/1-115415/6), 1098-0121

Liphardt, J., Dumont, S., Smith, S.B., Tinoco, I., & Bustamante, C. (2002). Equilibrium information from nonequilibrium measurements in an experimental test of jarzynski's equality. *Science*, Vol. 296, No. 5574, (June 2002), pp. (1832-1835)

Lodge, K.B. (1983). Techniques for the measurement of forces between solids. *Advances in Colloid and Interface Science*, Vol. 19, No. 1-2, (July 1983), pp. (27 - 73), 0001-8686

Major, R., Houston, J., McGrath, M., Siepmann, J., & Zhu, X.Y. (2006). Viscous water meniscus under nanoconfinement. *Phys. Rev. Lett.*, Vol. 96, No. 17, (May 2006), pp. (5-8), 0031-9007

Meyer, E., Heinzelmann, H., Grütter, P., Jung, T., Weisskopf, T., Hidber, H., Lapka, R., Rudin, H., & Güntherodt, H. (1988). Comparative study of lithium fluoride and graphite by atomic force microscopy (afm). *Journal of Microscopy*, Vol. 152, No. 1, (October 1988), pp. (269-280), 1365-2818

Meyer, G., & Amer, N.M. (1988). Novel optical approach to atomic force microscopy. *Appl. Physics. Lett.*, Vol. 53, No. 12, (September 1988), pp. (1045-1047), 0003-6951

Müller, M.J., Klumpp, S., & Lipowsky, R. (2010). Bidirectional transport by molecular motors: enhanced processivity and response to external forces. *Biophys. J.*, Vol. 98, No. 11, (June 2010), pp. (2610 - 2618), 0006-3495

Neumeister, J.M., & Ducker, W.A. (1994). Lateral, normal, and longitudinal spring constants of atomic force microscopy cantilevers. *Review of Scientific Instruments*, Vol. 65, No. 8, (August 1994), pp. (2527-2531), 0034-6748

Noy, A., Vezenov, D.V., & Lieber, C.M. (1997). Chemical force microscopy. *Annual Review of Materials Science*, Vol. 27, No. 1, (August 1997), pp. (381-421), 0084-6600

Oberhauser, A.F., Marszalek, P.E., Erickson, H.P., & Fernandez, J.m., (1998). The molecular elasticity of the extracellular matrix protein tenascin. *Nature*, Vol. 393, (February 1998), pp. (181-185), 0028-0836

Rief, M., Gautel, M., Oesterhelt, F., Fernandez, J.M., & Gaub, H.E., (1997). Reversible unfolding of individual titin immunoglobulin domains by AFM. *Science*, Vol. 276, No. 5315, (May 1997), pp. (1109-1112)

Rogers, B., Manning, L., Sulchek, T., & Adams, J. (2004). Improving tapping mode atomic force microscopy with piezoelectric cantilevers. *Ultramicroscopy*, Vol. 100, No. 3-4, (August 2004), pp. (267 - 276), 0304-3991

Ros, R. Schwesinger, F., Anselmetti, D., Kubon, M., Schafer, R., Pluckthun, A., & Tiefenauer, L. (1998) Antigen binding forces of individually addressed single-chain Fv antibody molecules. *Proc. Natl. Acad. Sci. USA*, Vol. 95, No. 13, (June 1998), pp. (7402-7405)

Sader, J.E., & Green, C.P. (2004). In-plane deformation of cantilever plates with applications to lateral force microscopy. *Rev. Sci. Instrum.*, Vol. 75, No. 4, (April 2004), pp. (878-883), 0034-6748

Sader, J.E. (2003). Susceptibility of atomic force microscope cantilevers to lateral forces. *Rev. Sci. Instrum.*, Vol. 74, No. 4, (April 2003), pp. (2438-2443), 0034-6748

Sarid, D. (1994).*Scanning force microscopy: with applications to electric, magnetic, and atomic forces*, Oxford University Press, 0-19-509204-X, Oxford University, New York

Schramm, V.L. (2005) Enzymatic Transition States: Thermodynamics, dynamics and analogue design. *Arch. Biochem. Biophys.*, Vol. 433, No. 1, (January 2005), pp. (13-26), 0003-9861

Schumakovitch, I., Grange, W., Strunz, T., Bertoncini, P., Güntherodt, H., & Hegner, M. (2002). Temperature dependence of unbinding forces between complementary dna strands. *Biophys. J.*, Vol. 82, No. 1, (January 2002), pp. (517-521), 0006-3495

Schwarz, U.D., Köster, P., & Wiesendanger, R. (1996). Quantitative analysis of lateral force microscopy experiments. *Rev. Sci. Instrum.*, Vol. 67, No. 7, (July 1996), pp. (2560-2567), 0034-6748

Senden, T.J. (2001). Force microscopy and surface interactions. *Current Opinion in Colloid & Interface Science*, Vol. 6, No. 2, (May 2001), pp. (95 - 101), 1359-0294

Stewart, A.M., & Parker, J.L. (1992). Force feedback surface force apparatus: principles of operation. *Rev. Sci. Instrum.*, Vol. 63, No. 12, (December 1992), pp. (5626-5633), 0034-6748

Strunz, T., Oroszlan, K., Schafer, R., & Guntherodt, H.J. (1999). Dynamic force spectroscopy of single DNA molcules. *Proc. Natl. Acad. Sci. USA*, Vol. 96, (September 1999) pp. (11277-11282)

Thomson, W. (1993). *Theory of vibration with applications* (4th edition), Prentice-Hall, New York

Uchihashi, T., Higgins, M., Nakayama, Y., Sader, J.E., & Jarvis, S.P., (2005). Quantitative measurement of solvation shells using frequency modulated atomic force microscopy. *Nanotechnology*, Vol. 16, No. 3, (January 2005), pp. (S49-S53), 0957-4484

Vázquez, J., Rivera, M.A., Hernando, J., & Sánchez-Rojas, J.L. (2009). Dynamic response to low aspect ratio piezoelectric microcantilevers actuated in different liquid environments. *J. Micromech. Microeng.*, Vol. 19, No. 1, (January 2009), pp. (1-9), 0960-1317

Verdaguer, A., Sacha, G.M., Bluhm, H., & Salmeron, M. (2006). Molecular structure of water at interfaces: wetting at the nanometer scale. *Chem. Rev.*,Vol. 106, (March 2006), pp. 1478-1510

Yamamoto, S.I., Yamada, H., & Tokumoto, H. (1997). Precise force curve detection system with a cantilever controlled by magnetic force feedback. *Review of Scientific Instruments*, Vol. 68, No. 11, (August 1997), pp. (4132-4136)

8

Nano-Engineering of Molecular Interactions in Organic Electro-Optic Materials

Stephanie J. Benight, Bruce H. Robinson and Larry R. Dalton

University of Washington,
USA

1. Introduction

Integration of electronic and photonic devices, especially chip-scale integration, is dramatically impacting telecommunication, computing, and sensing technologies (Dalton et. al., 2010; Dalton & Benight, 2011; Benight et. al., 2009). For organic electronics and photonics, control of molecular order is crucial for effective device performance. For organic photonics, formation of molecular aggregates can result in unacceptable optical loss. For organic electronics, molecular aggregates can adversely influence charge mobilities. Control of intermolecular electrostatic interactions can be exploited to control molecular organization including molecular orientation. Such control can lead to optimized material homogeneity and optical transparency and also to control of molecular conductivity. Here we focus on techniques for systematically nano-engineering desired intermolecular electrostatic interactions into organic electroactive materials. While our primary focus will be on electro-optic materials (which require acentric molecular organization), the discussion is also relevant to optimizing the performance of photorefractive, electronic, photovoltaic, and opto-electronic materials.

Molecular order is known to play a pivotal role in defining bulk material processes specific to electronics and photonics (i.e. photovoltaics, electronics, light-emitting devices and electro-optics) through impact on processes such as charge mobility, nonlinear optical processes, exciton diffusion, etc. A challenge common to electronics and photonics is to develop a fundamental understanding of how intermolecular interactions introduced into functional materials can be used to influence long-range molecular order.

1.1 Organic Electro-Optic (OEO) materials

Organic electro-optic (OEO) materials have the potential to be critical components for next generation computing, telecommunication, and sensing technologies (Dalton et. al., 2010). Applications of OEO materials extend to the military and medical sectors where super lightweight aircrafts and non-invasive, powerful imaging are applications of great utility. OEO materials afford the potential for great size, weight, and power efficiency as well as greater bandwidth, lower drive voltages, and greater flexibility in manufacturing and integration (Benight et. al., 2009).

OEO molecules or "chromophores" are comprised of conjugated π-electrons that make up their core chemical structure. In the presence of time-varying electric fields, OEO chromophores exhibit ultrafast charge displacements (on the order of tens of femtoseconds) (Munn & Ironside, 1993). The response time is the phase relaxation time of the π-electron correlation. These response times are superior to that of conventional inorganic crystals (e.g. lithium niobate) typically used widespread in commercial applications; the response time of lithium niobate is defined by ion displacement and is on the order of picoseconds. Of course, other factors such as velocity mismatch of electrical and optical waves and the electrical conductivity of metal electrodes can also influence the bandwidth performance of practical devices (Dalton & Benight, 2011). OEO materials also offer other advantages including the potential for significantly higher EO activity, lower dielectric constants, and amenability to a plethora of techniques including, but not limited to crystal growth, sequential synthesis/self-assembly layer-by-layer growth, deposition from either solution or the gas phase, soft and nano-imprint lithography and gray scale and photolithography, etc. (Kwon et.al., 2010; Frattarelli et. al., 2009; Dalton et. al., 2010).

Several different types of OEO materials have been developed in previous research, some of which have been implemented into viable photonic device structures (Shi et.al., 2000; Blanchard-Desce et. al., 1999; Sullivan et. al., 2007). One of the most prevalent types of OEO systems involves a strongly dipolar ellipsoidal-shaped molecule comprised of an electron donor, π-conjugated bridge, and electron acceptor units; this type of molecule is typically denoted as a "push-pull" chromophore. Because they are comprised of electron donor, bridge, acceptor, OEO chromophores tend to have large dipole moments which translate to exhibiting strong dipole-dipole anti-parallel pairing intermolecular interactions (lowest energy configuration) in a bulk system. Fig. 1 illustrates an example of an OEO push-pull chromophore (A) and a cartoon illustrating bulk anti-parallel dipole-dipole pairing (B).

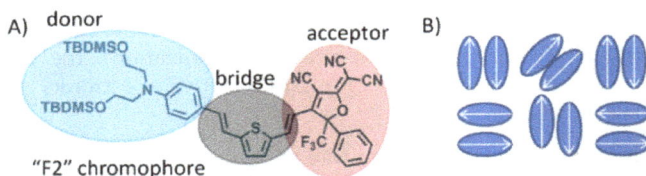

Fig. 1. A) Example of a push-pull organic electro-optic chromophore with the donor, bridge, and acceptor units highlighted. B) Cartoon illustrating dipole-dipole anti-parallel pairing with the blue ellipsoids representing chromophores and the arrows representing dipoles

In addition to dipolar chromophores, other types of chromophores such as octupolar have been explored (Blanchard-Desce et. al., 1999; Valore et.al., 2010; Ray & Leszcynski, 2004). Unfortunately, such chromophores have yet to be implemented into device structures. The reader is referred elsewhere for more information on these types of EO materials as our discussion here will focus on dipolar rod-shaped chromophores and systems based from these types of molecules (Dalton et. al., 2011).

Before proceeding further, it is important to discuss the basis for the bulk properties of OEO materials. The EO activity of materials originates from the nonlinear susceptibility or macroscopic polarization response of a material, given in Eq. 1.

$$\mathbf{P}_i = P_0 + \chi_{ij}^{(1)}\mathbf{E}_j + \chi_{ijk}^{(2)}\mathbf{E}_j\mathbf{E}_k + \chi_{ijkl}^{(3)}\mathbf{E}_j\mathbf{E}_k\mathbf{E}_l\ldots \tag{1}$$

Certain materials express higher ordered moments of susceptibility when exposed to high intensity electromagnetic radiation. In the equation above, $P_i = P_0 + \chi_{ij}^{(1)}$, is the first order (linear) polarizability, $\chi_{ijk}^{(2)}$ is second order and $\chi_{ijkl}^{(3)}$ is third order, etc.

Several physical processes can be expressed within the higher ordered moment terms (second order, third order, etc.) of the macroscopic polarization, but materials must exhibit certain properties or behavior to do so. For example, the basis for EO activity is the Pockels Effect, which is a second order nonlinear process inherent in the second ordered term, $\chi_{ijk}^{(2)} E_j E_k$ in Eq. 1. The Pockels effect is expressed when a change in the refractive index (birefringence) is induced in the material as caused by a constant or varying electric field (Munn & Ironside, 1993; Sun & Dalton, 2008).

In order for the Pockels effect or any other even ordered process to be expressed, the material must be noncentrosymmetric (bulk acentric order). In addition to a macroscopic response, a microscopic polarization response must also be demonstrated on a molecular basis. The microscopic nonlinear polarization, p_i, is represented as

$$p_i = p_0 + \alpha_{ij}^{(1)} E_j + \beta_{ijk}^{(2)} E_j E_k + \gamma_{ijkl}^{(3)} E_j E_k E_l ... \qquad (2)$$

The terms a, β, γ, etc. in the equation are second, third, and fourth rank tensors, respectively. The term, $\beta_{ijk}^{(2)} E_j E_k$, is the second order polarization response and $\beta_{ijk}^{(2)}$ is the molecular first hyperpolarizability for an individual chromophore molecule. The component of the tensor along the dipolar axis of the chromophore is denoted as β_{zzz}. Within these tensor elements lies three interrelated components: (1) vector components of the molecular electronic density distribution, (2) direction and polarization of light that is propagating through the material and (3) the direction and polarization of externally low-frequency applied fields. Optimizing β has a direct impact on the macroscopic NLO second order susceptibility, $\chi^{(2)}$ through the relationship expressed in Eq. 3.

$$\chi_{zzz}^{(2)}(\omega) = N\beta_{zzz}(\omega,\varepsilon)\langle cos^3 \theta \rangle g(\omega) \qquad (3)$$

where $\chi_{zzz}^{(2)}$ is dependent on the frequency (ω) of the incident light; N is the number density of the material or the concentration of chromophores in the material system; $\beta_{zzz}(\omega,\varepsilon)$ is dependent on both frequency and material dielectric; $\langle cos^3 \theta \rangle$ is the acentric order parameter in the material or the orientational average of the angle (θ) between the z (dipolar) axis of a chromophore and the z direction of the externally applied electric field; and $g(\omega)$ represent effects of local fields upon the chromophore molecules.

The EO coefficient, r_{33}, is the principal tensor element of the linear EO coefficient, r_{ijk}. Short for r_{333}, r_{33} is the tensor element that corresponds to the polarization of the optical beam and the direction of the externally applied electrical field both in the z direction. $\chi_{zzz}^{(2)}$ is related to r_{33} through Eq. 4 (Munn & Ironside, 1993).

$$\frac{2\chi_{zzz}^{(2)}(\omega)}{n^4} = r_{33}(\omega) \qquad (4)$$

where n in the material refractive index.

Eq. 4 demonstrates that r_{33} can be improved by targeting improvement of three parameters, $\beta_{zzz}(\omega, \varepsilon)$, N, and $\langle \cos^3 \theta \rangle$ as illustrated in Eq. 5.

$$r_{33} \propto N\beta \langle \cos^3 \theta \rangle \tag{5}$$

Research efforts in the past have focused on improving one or more of these parameters through molecular and material system design. In order to further understand the nano-engineering approach described herein, a brief history of EO systems development will be discussed.

1.2 Past approaches in improving EO activity

Many past efforts have had the goal of designing chromophores with superior hyperpolarizability, β_{zzz}. Early EO materials consisted of a chromophore as a guest in a host amorphous polymer system (Zhang et.al., 2001). The most sophisticated, best-performing chromophores with the largest hyperpolarizabilities are those that contain a heteroaromatic or polyene (isophorone-protected) bridge in the chromophore core structure. An example of both of these types of chromophores is given in Fig. 2.

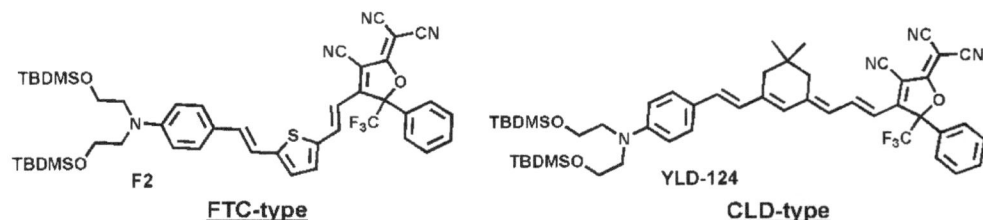

Fig. 2. Example structures of FTC-type and CLD-type chromophores are given with a heteroaromatic bridge and isophorone bridge as identifiers. The specific names for these structures, F2 and YLD-124, are also given beneath each structure. "F2" is given as a name of the structure for ease of reference and "YLD-124" is named after the inventor of the compound

Chromophores of this nature, known as FTC-type and CLD-type, respectively, are the basis for OEO chromophores used in the limited number of commercial devices today and in ongoing research for implementing OEO materials into silicon hybrid inorganic device structures (Baehr-Jones et. al., 2008; Ding et. al., 2010; Michalak et.al., 2006). "FTC" is an abbreviation that stands for "furan-thiophene chromophore" since the chemical structure contains these moieties; "CLD" is named after the authors who first introduced the molecular structure (Zhang et. al., 2001). Such acronyms are widely referred to in specialized literature and we include them here for ease of reference. While many other improvements have been made in exploring various chromophore donors, bridges, and acceptors (Davies et. al., 2008; Cheng et. al., 2007; Cheng et.al., 2008), these two base structures are the best in terms of hyperpolarizability, performance, ease of synthesis, and low absorption at the telecommunications (applications) wavelengths.

However, in examining Eq. 5, we see that r_{33} also depends on the number density or the concentration of chromophores in the system. Because of the strong intermolecular dipole-dipole antiparallel pairing interactions, such guest-host systems are often limited to number densities of ~20 % (Benight et. al., 2010). As was discussed earlier, chromophores possessing large hyperpolarizabilities also possess large dipole moments (on the order of 25 Debye), making designing chromophore with larger hyperpolarizabilities problematic. Additionally, in order to have an appreciable electro-optic effect (a second order nonlinear process), these molecules must exhibit noncentrosymmetry (be ordered in an acentric manner (i.e. uniformly or in one direction)). With increasing concentration of chromophores in the system, dipole-dipole pairing increases which negates any acentric order ($<\cos^3\theta>$) in the system. Such dipole-dipole intermolecular pairing also causes aggregation of chromophores at high number densities, leading to undesired optical loss and conductivity at poling temperatures.

To prevent or limit dipole-dipole intermolecular paring in bulk systems containing EO chromophores, bulky groups have been synthetically attached to chromophores to inhibit close approach. This method of "site (chromophore)-isolation" has been shown to limit chromophore aggregation (Liao et. al., 2005; Hammond et. al., 2008; Kim et. al., 2008). In this approach, bulky groups such as dendrons are covalently attached the chromophore core, therefore prohibiting one chromophore molecule from getting too close spatially to another chromophore molecule. In addition, multi-arm EO chromophore dendrimers have also been designed and synthesized and shown to permit high number densities of chromophores to be achieved without unwanted effects (Sullivan et. al., 2007).

Another approach that has been utilized and has yielded high EO activities on the order of 300 pm/V is preparation of binary chromophore organic glasses (BCOGs) (Kim et. al., 2006, 2008; Sullivan et. al., 2007). BCOGs are prepared by mixing chromophore guest molecules into chromophore-containing host molecules. Such systems can tolerate chromophore concentrations on the order of 60%. Both guest and host molecules are polar, resulting in favorable entropy of mixing and an absence of solvatochromic shifts with changing chromophore compositions. A dimensional restriction is also imposed because less freedom for movement of guest chromophores is able to be achieved, contributing to higher chromophore order and also higher EO activity (Benight et. al., 2010).

1.3 Improving EO activity through Increased acentric order

With the success of site-isolation and BCOG approaches, improvement of acentric order by chromophore modification, such as incorporation of additional and specific intermolecular electrostatic interactions, has remained a less pursued goal. Realization of molecular and supramolecular architectures with desired long-range acentric order in glassy materials remains a fundamental challenge.

1.3.1 Electric field poling

Most as-synthesized OEO materials are disordered, i.e. $<\cos^3\theta>$ and $r_{33} = 0$. The most commonly employed method for inducing acentric order into a chromophore system is to apply an electric field in the z direction (perpendicular to a device substrate) while heating the material to near its glass transition temperature. This electric field poling is the primary method utilized to induce acentric order into conventional poled guest-host polymer composites, EO dendrimers, and BCOG systems (Dalton et. al., 2010). For chromophore-polymer composites using solely electric field poling, the highest acentric order that is commonly achieved is on the order of 0.05 or lower (Benight et. al., 2010).

1.3.2 Other acentric ordering approaches

Approaches other than electric field poling that have also been explored have mainly consisted either of sequential synthesis of self-assembling chromophore systems or growing noncentrosymmetric crystal lattices of OEO chromophores (Kang et. al., 2004; Facchetti et. al., 2004; van der boom et. al., 2001; Frattarelli et. al., 2009; Kwon et. al., 2010). In addition and more recently, chromophores capable of being vapor deposited onto a substrate while undergoing electric field or laser-assisted electric field poling have demonstrated high acentric order (Wang et. al., 2011).

Self assembly approaches that have been successful to the present have utilized chromophores with modest β values. These chromophores are deposited through solution or vapor-phase self assembly with a functionalized monolayer tethered to the surface of the substrate. Marks and coworkers have demonstrated this concept for both OEO materials and materials in Organic Field Effect Transistors (OFETs) (Kang et. al., 2004; Facchetti et. al., 2004; van der boom et. al., 2001; Frattarelli et. al., 2009; DiBenedetto et. al., 2008). Recently, another self assembly approach has involved utilizing chromophores with weak hyperpolarizabilities and vapor depositing them on a surface under the influence of a polarized optical (laser) field as well as an electric poling field (Wang et. al., 2011). In this approach utilized by Chen and coworkers, the chromophore BNA (Benzyl-2-methyl-4-nitroaniline) has achieved r_{33} values ~ 40 pm/V with an acentric order parameter nearly unity. This approach is referred to as Laser-Assisted Poling - Matrix Assisted Poling (LAP-MAP) because the intermolecular BNA-BNA crystal forming interactions juxtaposed with the assistance of the poling field orient the chromophores in the material matrix. Albeit, BNA has been demonstrated to grow into noncentrosymmetric crystals without the presence of a poling field, laser-assisted electric field poling permits the acentric order to be achieved in a direction appropriate for waveguide device structures (Wang et. al., 2011).

Even though systems have been demonstrated in which chromophores exhibit higher acentric order as a result of self-assembly mechanisms, such systems have not utilized chromophores with large hyperpolarizabilities (β) and their fabrication methodologies are not easily adapted to chromophores with large β. High order does not adequately compensate for lower β and lower r_{33} values are the result. Moreover, defects propagate in sequential synthesis/self-assembly methods and thus film thicknesses are typically limited to 150 nm or less.

An alternative route to achieving acentric order in EO materials has been to grow noncentrosymmetric crystals of chromophore molecules (Kwon et. al., 2010; Yang et. al., 2007; Hunziker et. al., 2008; Weder et. al., 1997). A handful of EO chromophores when grown from a melt are able to crystallize in a noncentrosymmetric crystal lattice. Examples of such molecules are shown in Fig. 3.

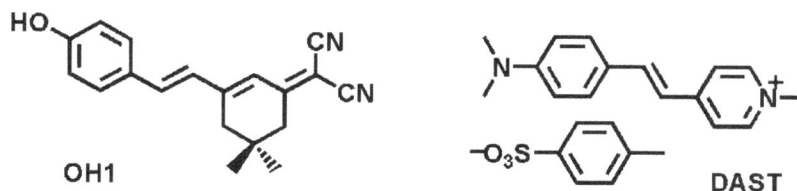

Fig. 3. The crystal growth chromophores, OH1 and DAST, shown

These chromophores, possessing low hyperpolarizabilities, yield acentric order parameters of $<\cos^3\theta>$ between 0.7-0.9 and EO activities of r_{33} = 52 pm/V at 1313 nm for OH1 and r_{11} = 53 pm/V at 1319 nm for DAST (Rainbow Photonics, 2011). However, designing chromophores that will crystallize into a noncentrosymmetric crystal is unpredictable and extremely sensitive to small structural details. For example, in attempting to improve the chromophores structure of OH1, changing chirality of one bond, extending a bond by one additional carbon, or any slight alteration in chemical structure can result in a centrosymmetric rather than a noncentrosymmetric crystal. This unpredictability makes chromophore design and improvement quite difficult and left to synthetic trial and error. It is possible that such crystal formations may be able to be modeled using theoretical methods, but such guidance is not yet available. In addition, it can be difficult and time-consuming to achieve such crystals, taking away from the efficacy in efficient processing of organic EO chromophores. Another disadvantage of crystal growth methods is the potential for light scattering from crystalline micro-domain formation.

1.3.3 Nano-engineering of directed interactions for increased acentric order

As already mentioned, the primary method for inducing acentric order is electric field poling. Recently, efforts to improve poling-induced order have involved incorporation of additional and specific intermolecular electrostatic interactions into synthesized chromophores. Dalton and Jen have employed covalent attachment of directed molecular interactions to enhance the poling-induced acentric order (Dalton et. al., 2011). Interactions such as arene-perfluoroarene (Ar-pFAr) and coumarin-coumarin have yielded some of the highest EO activities to date, ranging from 140-450 pm/V (Kim et. al., 2007, 2008; Zhou et. al., 2009; Benight et. al., 2010). Such systems embody covalent attachment of pendant groups, which interact with each other, to the chromophore cores of the molecules. In the case of Ar-pFAr interactions, an arene group is covalently attached to the chromophore donor substituent while a pFAr group is attached to the bridge substituent of the chromophore. In a bulk system the Ar and pFAr pendant groups intermolecularly interact (via quadrupolar interaction), exhibiting a self-assembly-like behavior. In the presence of an electric poling field, these pendant groups help to assist the acentric ordering of the chromophore molecules (Benight et. al., 2010; Dalton et. al., 2011; Dalton & Benight, 2011).

In addition to Ar-pFAr interactions, coumarin-based pendant groups can also be covalently attached to different portions of the chromophore core. In this case, Benight and coworkers have synthesized a series of compounds in which two coumarin-containing pendant groups are attached to the chromophore core of the molecule. Specifically, an alkoxybenzoyl-coumarin with a 6 carbon chain linker was covalently attached the chromophore core at both the donor and bridge portions to make the prototype molecule "C1" (Benight et. al., 2010). Through extensive characterization, to be recounted later in this Chapter, it was shown that the coumarin groups interact intermolecularly to restrict the rotational movement or lattice dimensionality of the chromophore cores in the bulk system. Lattice dimensionality can be further explained using theoretical methods (see Section 2.6) such as Monte Carlo theoretical modeling (see Section 2.7). As a result, the material system exhibits higher centrosymmetric and acentric order, yielding significantly enhanced EO activities. Experimental methods also demonstrate that coumarin-coumarin interactions lead to long range molecular cooperativity and that the coumarin moieties are oriented in a plane orthogonal to the poling-induced order of the chromophores (Benight, 2011; Benight et. al., 2011). These experimental results are also corroborated by molecular dynamics theoretical methods. This lattice dimensionality restriction and resulting improvement in poling-induced order is referred to as Matrix Assisted Poling (MAP). This approach can be expanded to include covalently attaching pendant groups which are molecules that differ in structure and property from the EO material and that organize in type of intermolecular interaction unique from that of the EO material (Benight et. al., 2010). Such spatially anisotropic interactions include but are not limited to dipolar, quadrupolar, and ionic interactions.

Herein, this nano-engineering of chromophores will be described in detail with an overview and explanation of the behavior of C1 through a developed toolset of comprehensive experimental characterization and theoretical methods. Instrumental methods are of the utmost importance to detect and probe pendant group molecular interactions, chromophore-chromophore interactions, and acentric/centric order of the molecular system. Several experimental methods are of great utility including optical ellipsometric methods, methods for characterizing electro-optic activity (e.g., attenuated total reflection, ATR), and methods for measuring viscoelastic behavior. Specific methods to be discussed include Attenuated Total Reflection (ATR) (Chen et. al., 1997), Variable Angle Polarization Referenced Absorption Spectroscopy (VAPRAS) (Olbricht et. al., 2011), Variable Angle Spectroscopic Ellipsometry (VASE) (Woollam, 2000), Shear-Modulation Force Microscopy (SM-FM) (Ge et. al., 2011), Intrinsic Friction Analysis (IFA) (Knorr et. al., 2009a), and Dielectric Relaxation Spectroscopy (DRS) (Dalton et. al., 2011). These methods provide complementary and synergistic understanding of the roles played by various intermolecular electrostatic interactions. Furthermore, theoretical methods such as rigid body Monte Carlo (RBMC) methods and coarse-grained Molecular Dynamics (MD) simulations are invaluable and will also be discussed

2. Recent results and discussion of nano-engineered OEO chromophore systems

2.1 The preparation of C1

The strategy of nano-engineering intermolecular interactions into organic functional materials systems involves a three pronged approach of synthesis, characterization, and theoretical simulation. As an example, we focus on the C1 molecule involving coumarin attachment to an FTC-type chromophore core. The C1 chemical structure is shown in Fig. 4. Alkoxybenzoyl-coumarins were attached to the donor and bridge portions of an EO (FTC-type) chromophore with large hyperpolarizability. The alkoxybenzoyl-coumarin pendant groups in C1 have been shown to exhibit highly planar liquid crystalline thermotropic phases (Tian et. al., 2003, 2004).

The approach in the design of C1 was to use the intermolecular interactions amongst coumarin molecules that govern liquid crystal phase formation to assist in the unidirectional (acentric) ordering of the highly dipolar chromophores in an electric field. As explained above, being able to mitigate intermolecular interactions in the EO material system is important in order to be able to induce and sustain acentric order and therefore, appreciable EO activity (r_{33}). The effect of coumarins can be interpreted as improvement of EO activity by reduction of effective lattice symmetry in the vicinity of the EO chromophore (Benight et. al., 2010).

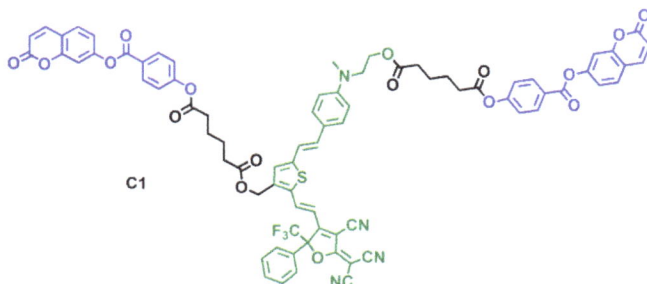

Fig. 4. The structure of C1 is shown with the EO active chromophore unit highlighted in green and the alkoxybenzoyl coumarin units highlighted in blue

C1 was synthesized in a multi-step (15 total steps) synthesis (Benight et. al, 2011; Benight, 2011). First, the chromophore core comprised of donor and bridge components was prepared according to procedures established in the literature (Sullivan et. al., 2007). The chromophore was modified to have two ethoxy groups to act as points of pendant group attachment on both the donor and bridge of the chromophore. In parallel, the coumarin pendant group was prepared through a 4-step synthesis involving a series of modified Steglich esterifications by first preparing the pendant group itself (coumarin-containing mesogen) followed by attaching an adipic acid linker to act as the connection point between chromophore and pendant group. Once the chromophore core (all components minus acceptor) and the pendant group was prepared, the pendant group was then attached to the chromophore through an esterification reaction. The final step in the synthesis of C1 was attaching the electron acceptor (also prepared separately according to a literature procedure) via a Knoevenagel condensation (Liu et. al., 2003).

2.2 EO activity of C1

To evaluate physical properties including EO activity, samples of C1 were spin coated from organic solvent into thin films on ITO coated substrates and titanium dioxide (TiO_2) coated ITO substrates. TiO_2 has been shown to be quite useful as a blocking layer in organic materials-based devices by acting as a Schottky–type barrier in blocking excess charge injection into the organic layer (Sprave et. al, 2006; Enami et. al., 2007; Huang et. al., 2010). We have found this is to be the case in our EO characterization experiments both in poling more conventional poled polymer materials as well as MAP systems like C1, achieving upwards of 30% more effective electric field poling and EO activities. r_{33} values for C1 were acquired using the widely applied technique of attenuated total reflection (ATR). The use of

ATR for characterization of EO materials is described in detail elsewhere (Benight, 2011; Davies et. al., 2008; Herminghaus et. al., 1991; Chen et. al., 1997). Although several methods for characterizing EO coefficients exist, most notably Teng-Man Simple Reflection Ellipsometry (Teng & Man, 1990; Park et. al., 2006; Verbiest et.al., 2009), ATR remains one of the most reliable techniques. Thin film samples of C1 were electric field poled at various applied voltages to obtain a linear relationship between r_{33} and applied poling field strength (E_p, in units of V/micron). C1 was demonstrated to give r_{33} values ~ 140 pm/V, three times higher than that of the same chromophore core (FTC-type) in an amorphous polymer host at optimal (20% chromophore) number density (Benight et. al., 2010). C1 is 41% chromophore and spin coated neat with no host. In addition, the poling efficiency for C1 on TiO_2-coated substrates was measured to be $r_{33}/E_p = 1.92$ (Benight et. al., 2010).

2.3 Molecular order of C1

Upon observing that C1 exhibited superior EO activity (greater than what would be expected with simply increasing the number density), methods which could ascertain the degree of poling-induced order, <P_2>, were employed. <P_2> is related to <$cos^2\theta$>, the second degree order parameter through Eq. 6

$$\langle P_2 \rangle = \frac{3\langle cos^2 \theta \rangle - 1}{2} \tag{6}$$

Currently, no direct method for measuring acentric order, <$cos^3\theta$>, in a material system is available. To obtain a quantitative measurement of acentric order in the system, other components of the r_{33} equation must be calculated and <$cos^3\theta$> back-calculated from knowledge of the other parameters. However, the <P_2> or the centric (order) of molecules can be measured using several experimental methods. The normal incidence method (NIM), in which UV-visible absorption spectra are acquired at normal to the surface of the chromophore sample before and after poling, has been employed (Kim et. al, 2009; Benight et. al., 2010). The <P_2> is able to be measured based on the assumption that as strongly absorbing chromophore molecules are poled in the z direction of the substrate, the chromophores orient more parallel to the poling axis, therefore absorbing less in the plane of the substrate. While the NIM does yield insight into the centrosymmetric order of materials, the measurement is simple and prone to error in that photo-induced chromophore degradation can lead to significant over-estimation of poling-induced order.

Recently, new and improved methods for measuring <P_2> of EO materials have been introduced. The method developed by Robinson and coworkers is known as Variable Angle Polarization Referenced Absorption Spectroscopy (VAPRAS) and measures, at various angles of incidence, absorption of s and p polarizations of light (Olbricht et. al., 2011). Utilization of the ratio of s and p absorptions negates and attenuates the effects of Fresnel reflections. That is, the s polarized absorption acts as a reference. The C1 material was studied using NIM and VAPRAS techniques to yield <P_2> values of <P_2> = 0.16 and <P_2> = 0.19, respectively for samples poled at 50 V/micron (Benight et. al., 2010).

While the absorption of the chromophore unit of the C1 molecule could be easily probed due to being in a desirable wavelength range for the VAPRAS instrument, the absorption profiles of the coumarin units after poling remained inaccessible. Utilizing another technique capable of ascertaining absorption behavior over a wide range of wavelengths (190nm -1700nm), Variable Angle Spectroscopic Ellipsometry (VASE) was employed for unpoled and poled samples of C1 (Woollam, 2000). Isotropic and anisotropic models can both be applied to analyze optical constants of samples. In our case, an anticipated anisotropy in the xz, yz (poling) axes prompted us to use an anisotropic model for samples that have been poled. Using an anisotropic model, the absorptions of the coumarin and chromophore units were able to be obtained for both the plane perpendicular to the sample surface (as studied in UV-Vis absorption spectroscopy) and for the plane within the sample (Benight et. al., 2011). A clear illustration of the absorption profiles as measured with VASE is given in Fig. 5.

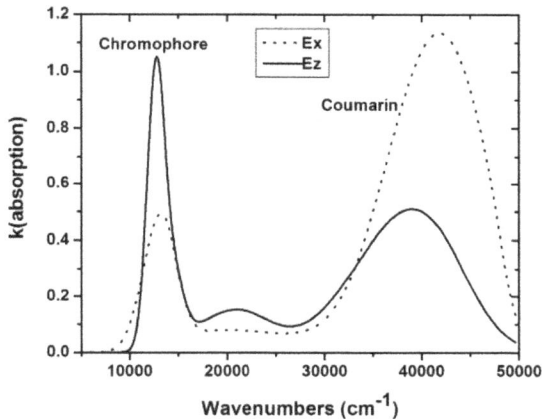

Fig. 5. The absorption component of the refractive index, k, is shown for representative samples of the C1 system. Absorption in the x-plane is shown by the dotted line while absorption containing the z-plane is shown by the solid line

These results illustrate that the coumarin units in poled samples of C1 were oriented in the x-y planes of the sample, orthogonal to that of the chromophores.

Not only can VASE be used to detect absorption behavior for a wide range of transparent and absorptive materials, but VASE can also be used to compute the $<P_2>$ of a material using Eq. 7.

$$\langle P_2 \rangle = \frac{k_P - k_\perp}{k_P + 2k_\perp} \qquad (7)$$

Here, the k_p coefficient represents the absorption parallel to the normal of the film (z axis) and $k_\perp$ represents the absorption perpendicular to the poling (z) axis (Michl & Thulstrup, 1986). The $<P_2>$ for samples of C1 poled at 50 V/micron was measured to be $<P_2>$ = 0.24, in good agreement with VAPRAS measurements of the material (Benight et. al., 2011; Benight, 2011). In addition, the $<P_2>$ of the coumarin units was found to be $<P_2>$ = -0.19. Since mathematically, the highest negative order that can be achieved is $<P_2>$ = -0.5, the coumarins are approximately ~40% centrosymetrically ordered in the xy plane (Benight et. al., 2011; Benight, 2011).

2.4 BCOG system incorporating C1

A BCOG system including the C1 material as a chromophore-containing host and a guest chromophore (polyene bridge, CLD-based) was also characterized. The guest chromophore YLD-124 (CLD-type) was doped at 25 wt % into C1 and spin coated onto TiO_2 coated ITO glass substrates. From poling experiments and measurements of EO activity using ATR technique, the poling efficiency (r_{33}/E_p) was measured to be 2.00, slightly higher than the poling efficiency for neat C1 at r_{33}/E_p = 1.92 (Benight et. al., 2010). This poling efficiency is a bit lower compared to other BCOG systems, but a small enhancement is still observed (Kim et. al., 2008). The lack of more improved enhancement in the BCOG system with C1 as a host is likely due to the addition of more chromophore molecules complicating the reduced fractional lattice dimensionality experienced by the chromophores in C1. The addition of more chromophores is likely restricting coumarin movement slightly, but the restriction in chromophore movement from fractional lattice dimensionality imposed on the chromophores is likely already fully reached with the neat C1 system.

2.5 Investigation of molecular cooperativity and mobility in C1

Intermolecular electrostatic interactions leading to improved poling-induced order should also be manifested through impact on viscoelastic properties. Traditionally, differential scanning calorimetry (DSC) is a method for ascertaining thermal transitions in a material. Upon DSC investigation of C1, it was observed that a glass transition-like change was apparent around 80 °C.

To probe the thermal transitions on a more sensitive scale, we employed methods capable of detecting molecular mobility and cooperativity on the nanoscale. Length scales and degrees of cooperativity (quantitative energetics of activations of various phases) can be deduced by nanoviscoelastic and nano-thermorheological measurements. The nanoscopic methods of Shear Modulation Force Microscopy (SM-FM), Intrinsic Friction Analysis (IFA) and Dielectric Relaxation Spectroscopy (DRS) were used to examine C1. These methods have also been applied to EO systems which incorporate Ar-pFAr pendant group interactions (Gray et. al., 2007, 2008; Zhou et. al., 2009; Knorr et. al., 2009b).

Before proceeding with the results of the C1 system, it is important to give a brief overview of each of these instruments. SM-FM is a nanoscopic analogue to dynamic mechanical analysis (DMA) widely used in studies of the nanoviscoelastic behavior of polymer relaxations and phase behavior. The SM-FM instrument is essentially an Atomic Force Microscope that measures the temperature-dependent shear force on a tip modulated parallel to the sample surface (Ge et. al., 2000). SM-FM measures the force required to move the tip of contact and the velocity of the moving tip at fixed temperatures. Experiments are typically acquired at increments of 1 K temperature. These perturbations in force and velocity can be quantified in a contact stiffness parameter (typically denoted as k_c). As the experiment is carried out over a range of temperatures, changes in the slope of contact stiffness versus temperature yield information about temperatures at which phase transitions in the material occur.

In order to ascertain specific information regarding the energetics of the detectable phase transitions as found in the SM-FM measurements, IFA was employed. IFA is a nanoscopic analogue based on the well-known lateral force microscopy technique capable of providing molecular descriptions of relaxation processes associated with transitions of structural molecular movement (Knorr et. al., 2009a). From the IFA experiment, quantitative energies of activation for structural transitions, referred to as "apparent energy" or E_a, and the molecular-scale cooperativity of these transitions can be measured. Non-cooperative processes are generally described taking into account dynamic enthalpy ΔH solely. However, if a process is cooperative, the dynamic entropy (ΔS) is also taken into account as given in the classical sense from Gibbs Free energy, $\Delta G^* = \Delta H^*-T\Delta S^*$. In addition to IFA measuring E_a, the technique also measures $T\Delta S^*$.

The IFA technique utilizes the same instrument as in SM-FM. Specifically, the force required to move the tip probe from the AFM is measured at a fixed temperature to give a force curve as a function of the log of the velocity. Data are shifted to achieve a "master" curve according to the time-temperature equivalence principle (Ward, 1971; Ferry, 1980). The vertical shifting of data is related to $T\Delta S^*$ in the Gibbs energy and depicts the degree of cooperativity as observed with respect to temperature for the material. It can be valuable to also estimate the contribution due to entropic energy from the measurement. To conduct this exercise, a method developed by Starkweather can be used to analyze the cooperativity with apparent Arrhenius activation energies which can be determined from Eq. 8 (Starkweather, 1981, 1988).

$$E_a = RT \cdot \left[1 + ln\left(\frac{kT}{2\pi h f_0} \right) \right] + T\Delta S^* \tag{8}$$

where k and h are Boltzmann's constant and Plank's constant, respectively, and f_0 is the frequency at which the relaxation peak is observed. f_0 is either deduced from DRS or estimated (Sills et. al., 2005). In IFA, the friction force typically varies with temperature and the velocity at which the tip moves is accounted for using the WLF superposition (Williams et. al., 1955).

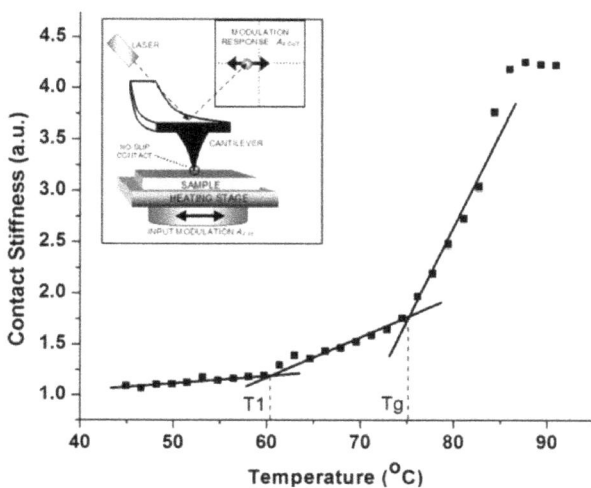

Fig. 6. SM-FM example data of the C1 system. T_1 and T_g are labeled for clarity. A schematic of the SM-FM instrument is given in the inset

In conducting SM-FM and IFA measurements for C1, two distinct thermal transitions were observed for the samples of C1. The first transition, denoted as T_1, was observed at 61 °C and a transition quite close to the T_g (at 76 °C), denoted as T_2, was also observed. An example SM-FM scan for C1 is given in Fig. 6.

The T_1 transition was not detectable clearly in the DSC scan, however was able to be probed nanoscopically using SM-FM. The results from the SM-FM and IFA for the C1 system and the molecule HD-FD (incorporating Ar-pFA interactions) is presented for comparison.

	Temperature Range	E_a [kcal/mol] [a]	$T\Delta S^*$ [kcal/mol] [b]
	$T < T_c$	16±3	-
C1	$T_c < T < T_g$	43±4	-
	$T > T_g$	72±4	54-58
	$T < T_c$	23	-
HDFD	$T_c < T < T_g$	44	-
	$T > T_g$	71	52-56

Table 1. The activation energies of HDFD and C1. [a] IFA results. [b] Estimated from Eq. 8.

Above the glass transition temperature, the mobility of the chromophore molecules in the system is the highest allowing for maximum cooperativity in the system. In fact, the entropic component of the activation energy is 75-80% of the activation energy above T_g. The vertical friction shift, related to molecular cooperativity is shown for the C1 compound in Fig. 7 below.

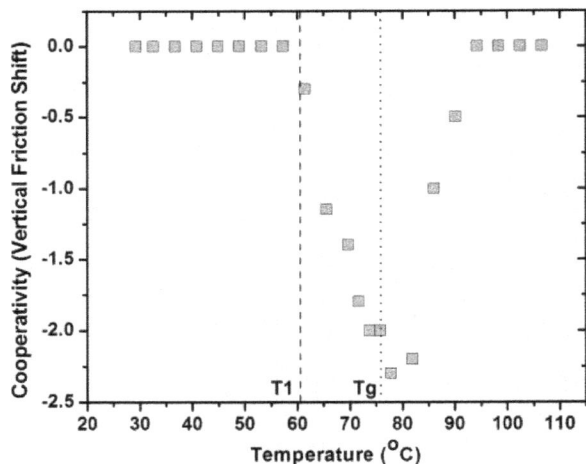

Fig. 7. The vertical friction shift, related to molecular cooperativity, is shown for C1. The plot shows that cooperativity starts at the T_1 temperature and is maximized at the glass transition or poling temperature of the material

Keeping in mind that this plot is proportional to the change in entropy and signifies the molecular cooperativity in the material, the cooperativity in C1 starts to increase at T_1 and reaching a maximum at T_g. IFA data yield an associated activation entropy change, $|\Delta S|$ ~ 160 cal/K for C1 (Benight et. al., 2011; Benight, 2011).

Another method that works in tandem with IFA in providing critical information on the length scales of molecular cooperativity is that of dielectric relaxation spectroscopy (DRS). Specifically the dissipation length, $\xi(T)$ for cooperativity or molecular mobilities (molecular interactions) can be quantified. The dissipation length is determined from Eq. 9.

$$\xi(T) = v_P(T)/f_P(T) \qquad (9)$$

where $v_P(T)$ is the peak velocity from IFA and $f_P(T)$ is the peak relaxation frequency determined from DRS (Sills et. al., 2005, Hedvig, 1977). In the actual measurement, the real and imaginary impedance values are measured and are given in Eqs. 10 and 11. These values are ascertained for a range of frequencies and temperatures. By measuring the real and imaginary values of the impedance, information of the real and imaginary portions of the dielectric data are known.

$$\varepsilon'(\omega) = \frac{-Z''}{\omega C_0 (Z''^2 + Z'^2)} \ ; \ \varepsilon''(\omega) = \frac{Z'}{\omega C_0 (Z''^2 + Z'^2)} \qquad (10,11)$$

where C_o is the vacuum capacitance of the test set-up without the sample, and ω is the angular frequency. It's important to note that in the measurement referred to here, poled samples of OEO materials are measured (in a parallel plate capacitor format).

Eq. 9 is used to acquire a dissipation length for the material, however, it can be difficult to obtain $v_p(T)$ and $f_p(T)$ at the same temperatures over large ranges of temperature. This limits the amount of information to be acquired at lower temperatures closer to T_g, however, information from both IFA and DRS can be acquired at temperatures above T_g and cooperativity can be extrapolated to lower temperatures.

It is important to note that the DRS technique is widely applied in the study of polymer relaxations, side-chain and main-chain relaxations in particular, but this technique has rarely been shown to be used on organic molecular glassy systems like C1. The dissipation length for C1 was measured to be as high as 55 ± 25 nm at 106 °C, approximately 25 °C above the T_g of the material as measured by DSC and SM-FM (Benight, 2011). To compare to other materials, typically the dissipation length for polystyrene, a conventional polymer, fades to a couple of angstroms at temperatures above T_g. As another comparison, an EO material containing Ar-pFAr interactions was analyzed using the same DRS instrument used in the investigation of C1 to give ξ = 15 ± 10 nm at T = 153 - 160 °C (Knorr, 2010). The higher dissipation length for C1 likely indicates that the interactions in C1 are more stable over longer length scales. Furthermore, this dissipation length signifies that in the C1 system, the movement of one C1 molecule is influenced by another C1 molecule up to approximately 55 nm away.

2.6 Reduced dimensionality

The C1 material system has been shown to exhibit increased centrosymmetric and acentric order, intermolecular coumarin pendant group interactions and defined optical orthogonal orientations of chromophore and coumarin units (Benight et. al., 2010). These results can be explained as the coumarin-based pendant groups imposing a lattice restriction which leads to fractional dimensional order of the chromophore molecules in the presence of an electric poling field. Inflicting a dimensional restriction upon the chromophores increases centrosymmetric order and may also increase acentric order of the bulk material system, at constant poling field strength. This lattice dimensionality restriction, as applicable to the chromophore systems presented here, can be described utilizing two different theoretical arguments both using the independent particle assumption in a low-density limit (Benight et. al., 2010). The fundamental statistical description of the order parameters is defined in Eq. 12.

$$\left\langle \cos^n \theta \right\rangle = \frac{\iint\limits_{\theta,\phi} \cos^n \theta P(\theta,\phi) e^{\frac{\mu \cdot E_0}{kT}} d\cos\theta d\phi}{\iint\limits_{\theta,\phi} P(\theta,\phi) e^{\frac{\mu \cdot E_0}{kT}} d\cos\theta d\phi} \tag{12}$$

where the probability distribution is given in Eq. 13 and is described using the many-body interactions potential (the Hamiltonian) (Benight et. al., 2010).

$$P(\theta,\phi) = \frac{e^{-\frac{H^o}{kT}}}{\iint\limits_{\theta,\phi} e^{-\frac{H^o}{kT}} d\cos\theta d\phi} \tag{13}$$

The $f = \dfrac{\mu E_0}{kT}$ term indicates the statistical weight representing the dipole interaction with the homogenous poling field, E_0. It is important to note that the E_0 given in Eq. 12 is representative of the electric field at the chromophore molecule and does not necessarily represent the overall applied poling field strength (E_p).

In the first theoretical treatment the ordering potential is considered to be uniform but the dimensionality, M, of the space is reduced. In this case the order parameter is computed using Eq. 14 (Stillinger, 1977).

$$\left\langle \cos^n \theta \right\rangle_M = \frac{\int\limits_{\theta} \cos^n \theta e^f \sin^{M-2} \theta d\theta}{\int\limits_{\theta} e^f \sin^{M-2} \theta d\theta} \tag{14}$$

This expression is easily solved in terms of Bessel functions, $I_v(f)$. For example, the lowest indexed order parameter, n=1, is computed from Eq. 15.

$$\left\langle \cos\theta \right\rangle_M = \frac{I_{\frac{M}{2}}(f)}{I_{\frac{M}{2}-1}(f)} \tag{15}$$

All higher index order parameters can be found from this one and its derivatives. In this theoretical treatment, different dimensional restrictions translate to varied degrees of rotation for the point dipoles in the system. In this model, in three dimensions, the dipolar molecule being examined can access any orientation in the Cartesian coordinate system exhibiting Langevin type behavior. By generalizing the Langevin function for any dimension, we can describe molecular order for any fractional dimension from 1 to 3 dimensions. Full derivation for the equations for 2D and 3D scenarios and 2nd degree (centrosymmetric) and 3rd degree (acentric) order parameters has been given previously (Benight et. al., 2010).

The centrosymmetric and acentric order parameters give insight into the system's dimensionality. The dimensional restrictions given for a single dipole can translate to dimensionality of molecules (chromophores in our case) in macroscopic systems including those restrictions originating from the environment. The dimensionality, M, can be approximated from a linear interpolation between $M = 2$ and $M = 3$ from the relation of $<\cos^3\theta>/<P_2>$, as shown in Eq. 16 (Benight et. al., 2010).

$$\left\langle \cos^3\theta \right\rangle_{MD} \approx \sqrt{\left(\frac{9-2M}{2+M}\right)\left(\langle P_2\rangle_{MD} - \frac{3-M}{2M}\right)} \tag{16}$$

Using this linear interpolation model, the dimensionality of C1 was found to be $M = 2.2$ D while those of the control systems utilizing more standard conventional chromophore systems with the same chromophore core were nearly 3D (Benight et. al., 2010). Such results have been corroborated by theoretical modeling of chromophores including rigid body Monte Carlo (RBMC) methods, in which the dimensional restriction was observed at experimental densities similar to that of C1 (see next section).

For the second model, called the Rigid Wall Model, a similar theoretical description of dimensionality emerges. In the rigid wall model, the order parameters of the dipoles can be defined from a simple distribution generated by a confining potential in which the dipoles may rotate unimpeded up to a stop angle, $q_o = \cos\theta_{Stop}$. The stopping angle is related to the dimensionality by Eq. 17.

$$q_0(1+q_o) = \left(\frac{3-M}{M}\right) \tag{17}$$

When this description of the confining potential is substituted into Eq. 12, the acentric order parameter for the lowest order, n=1, is given according to Eq. 18.

$$\langle \cos\theta\rangle = \frac{\cosh(f) - q_o\cosh(q_of)}{\sinh(f) - \sinh(q_of)} - \frac{1}{f} \tag{18}$$

Again, all other order parameters can be found from this one and appropriate derivatives. An illustration of the ratio of centric ($<P_2>$) to acentric ($<\cos^3\theta>$) order parameters for all of the theoretical descriptions of dimensionality that have been presented here can be of utility. Fig. 8 illustrates the relationship of $<P_2>$ and $<\cos^3\theta>$ for several different dimensionalities. Overlaid in Fig. 8 are (1) the linear interpolation of the two order parameters as derived from Bessel functions (red dotted line), (2) the dimensionality as computed from Bessel $I(z)$ functions (black solid line) and (3) the dimensionality as computed from the rigid wall model of confining potential (blue solid line).

Fig. 8 shows that for as the dimensionality is less restricted (ascending toward N = 3), the relationship between $<P_2>$ and $<\cos^3\theta>$ is the same for all of the theoretical models. All three models agree nicely for dimensionalities from N = 3 to N = 2, however, as the dimensionality is further restricted toward 1.1 D, the theoretical model which utilizes a linear interpolation of $<P_2>$ and $<\cos^3\theta>$ to calculate dimensionality is not as aligned with the other two models.

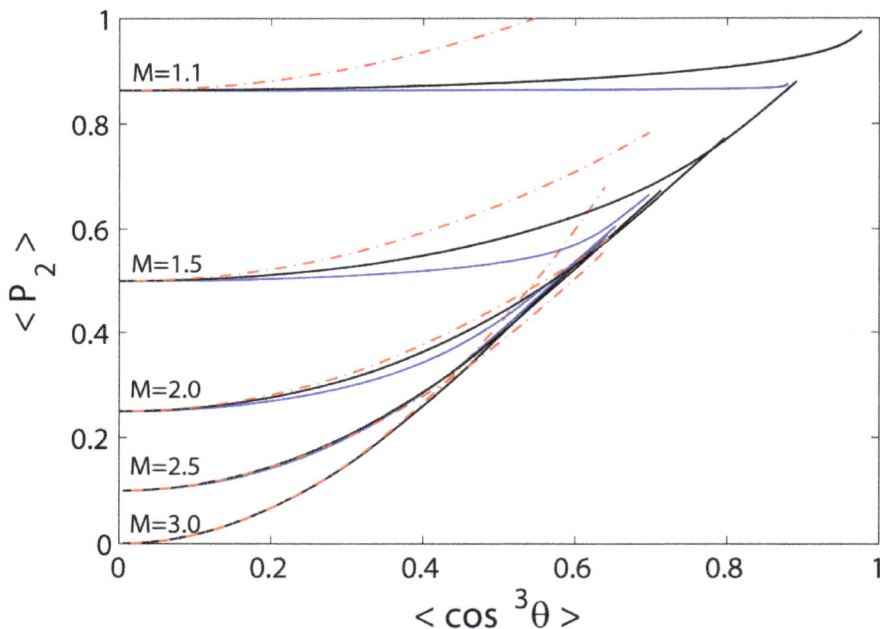

Fig. 8. The relationship of $<P_2>$ to $<\cos^3\theta>$ is presented for different dimensionalities. The methods for computing lattice dimensionality presented above are overlaid in the plot for specific quantitative dimensional systems (e.g. N = 3.0, N = 2.5, N = 2.0, N = 1.5, N = 1.1). The red dotted line illustrates the dimensionality as computed using the linear interpolation from Eq. 16. The black solid line illustrates the dimensionality as computed from Bessel $I(z)$ functions and the blue solid line represents the rigid wall model (the second method presented above)

In summary, the reduced dimensionality integrals are easily represented by Bessel functions, making the computation of the order straightforward, and are easily compared to a full 3D model with a restrictive wall barrier characterized by a designated cutoff angle. These models of dimensionality show that achieving reduced lattice dimensionality for a material system encompassing strongly dipolar chromophores will increase centrosymmetric order and may also increase acentric order at the same poling strength being applied to the system. Furthermore, one can compare <P$_2$> and <cos$^3\theta$> to determine whether the effective dimensionality of the material system has been reduced.

2.7 Theoretical methods

Theoretical methods coupled with experimental results have been integral in understanding molecular interactions in systems such as C1. The specific theoretical methods of rigid body Monte Carlo (RBMC), fully atomistic Monte Carlo (FAMC) and molecular dynamics (MD) have been crucial (Sullivan et. al., 2009; Rommel & Robinson, 2007; Robinson& Dalton, 2000; Benight et. al, 2010; Leahy-Hoppa et. al., 2006; Knorr et. al., 2009b). Such methods have been used to study EO dendrimer, chromophore guest polymer host systems, BCOGs and MAP systems. With custom code, an electric field poling experiment can be simulated with an ensemble of ellipsoids or spheres parameterized to possess the properties of the experimental chromophore systems and 1st, 2nd, and 3rd degree order parameters can be computed. Such capability allows the effect of different applied poling field strengths, different temperatures on chromophores systems and various sizes and shapes of ellipsoids to be investigated (Benight et. al., 2010).

RBMC simulations were implemented in investigation of the effect of density of chromophores systems on centrosymmetric (2nd degree) and acentric (3rd degree) order. As part of this investigation, prolate ellipsoids parameterized as the F2 chromophore (the active chromophore unit in C1, depicted in Fig. 1) were simulated under applied poling fields of 50 V/micron and 75 V/micron at low densities of chromophores in the system (~20%) ranging upwards of 60% chromophore (Benight et. al., 2010). The <P$_2$> and <cos$^3\theta$> values were computed for each simulation and the dimensionalities of the simulations from these order computations were determined. The results of these simulations illustrated that at low numbers density (~20%) in electrically-poled guest-host systems, nearly no dimensional restriction exists, yielding nearly three dimensional systems. However, upon increasing the number density of the system toward number densities of chromophore concentration similar to that of C1 (40% and beyond) the centric order of the chromophores as observed from the computed <P$_2$> values and visualizations of the simulations increased dramatically (Benight et. al., 2010). At nearly 60% chromophore concentration, the visualization of the simulation showed a system that closely resembles that of a smectic-A liquid crystal with high planar and directional order (Benight et. al., 2010). Furthermore, the dimensionality of this system was calculated to be less than 2, indicating more dimensional restriction at

E$_p$ (V/μm)	N[a]	$\langle P_2 \rangle$	$\langle cos^3 \theta \rangle$	M
50	1.26	0.015 ± 0.001	0.052 ± 0.0001	2.9
	2.52	0.017 ± 0.003	0.046 ± 0.0003	2.9
	3.78	0.017 ± 0.003	0.040 ± 0.0002	2.9
	5.04[b]	0.042 ± 0.007	0.035 ± 0.0003	2.8
	6.30[b]	0.42 ± 0.11	0.013 ± 0.001	< 2

Table 2. RBMC simulation results for ellipsoids under an electric poling field of E$_p$ = 50 V/micron parameterized as the chromophore cores in C1. T = 382 K, m = 1728 particles. [a](× 10^{20} molecules/cc); [b]Longer runs were required for convergence at these densities

increasing number densities of chromophore concentration. (Note: in RBMC simulations, the coumarin pendant groups in MAP systems are not taken into account). Results from the simulations are given in Table 2 for clarity.

In the most recent development of molecular dynamics of chromophore systems, the C1 chromophores are parameterized with the MMFF94 force field (Halgren, 1996) and MD simulations were run using the Tinker 5.1 program (Ponder, 2010) and simulated in a defined box with an Ewald boundary. Simulations can be run at room temperature or at appropriate poling temperatures near T_g (Benight et. al., 2011; Benight, 2011).

A standard statistical mechanical correlation function (McQuarrie, 2000) can be used to monitor the interactions of chromophore components in the simulation. The correlation function used in an example of recent work is given in Eq. 19 (Benight et. al., 2011; Benight, 2011).

$$g(r) = \frac{V \sum\limits_{i}^{N} \sum\limits_{j>i}^{N} n_{ij}(r, r+dr)}{2\pi N(N-1)r^2 dr} \tag{19}$$

In this equation, the volume of the system (V), the number of molecules in the system (N) and the distance between two components, r are accounted for and are indicative of the radial distribution. In the simulation of molecules specific intermolecular interactions can be probed by monitoring a given atom in the molecule of interest. For example, in the C1 system, the sulfur atom on the thiophene component of the bridge can be tagged and the interactions between intermolecular S-S can be quantified, enabling probing of the chromophore-chromophore interactions in the system. Furthermore, monitoring a specific carbonyl group of the coumarin pendant group enables monitoring of intermolecular interactions between neighboring pendant groups. The correlation function allows for monitoring at varied distances apart (Benight et. al., 2011; Benight, 2011).

As an example, MD, simulations of C1 can be run in the presence of chloroform molecules and as neat systems at experimental density. Upon viewing simulation visualizations of these simulations, the correlation functions demonstrate that in the C1 system, the chromophore-chromophore dipole-dipole interactions as monitored through the S-S bonds in the chromophore units are suppressed when compared to runs of just the chromophores in the simulation and in runs including short PMMA polymer chains in the simulation. Furthermore, distinct coumarin-coumarin interactions are evident and constant for various concentrations. In addition, visualizations of the simulations illustrate coumarin-coumarin pairs or dimers that regularly occur, further driving home the point that the intermolecular interactions of the coumarin have a strong presence in the C1 system. An example visualization of simulated neat C1 using MD is shown in Fig. 9.

In addition to MD, fully atomistic Monte Carlo (FAMC) can be employed in the study of OEO chromophore systems. It has been demonstrated that simulation under a poling field of EO chromophore dendrimers are quite useful in investigating the order parameters and poling behavior. Recent code development upgrades (done in–house) also enable the study of systems like C1 in which several levels of detail are possible in simulation the chromophores. Ellipsoids can be parameterized around the chromophore and coumarin

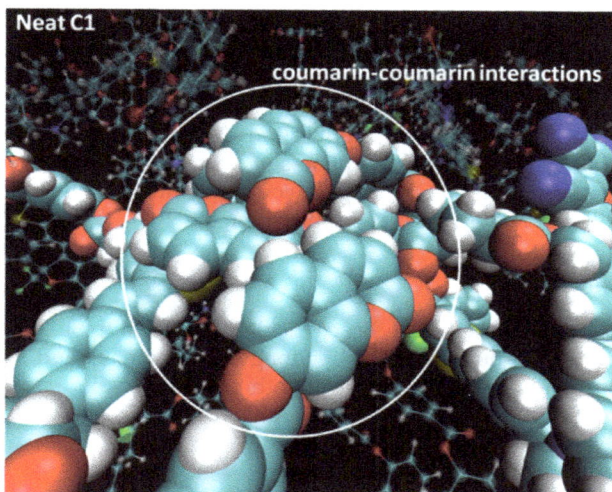

Fig. 9. A visualization of the C1 molecule as observed in the MD simulations shown

units, around specific functional units or groups within those components, or around every aromatic ring to yield further levels of detail in modeling of the chromophores. Indeed, with more levels of detail, more simulation time and expense arise. Results with updated code are still primitive, but this approach bodes well for being able to simulate more complex systems in the future.

3. Conclusions

A combination of methods for characterization of soft matter organic electro-optic materials can be utilized to yield insight into molecular order, material performance, and intermolecular interaction dynamics. The specific methods of attenuated total reflection (ATR) and *in-situ* poling are of utility for measuring the EO activity of OEO materials. Techniques such as shear-modulation force microscopy (SM-FM), intrinsic friction analysis (IFA) and dielectric relaxation spectroscopy (DRS) determine viscoelastic properties of the materials. These methods yield insight into the dynamics of intermolecular interactions in the material system, including measurement of the quantitative energies of activation for the materials discussed. In addition, optical spectroscopy methods such as Variable Angle Spectroscopic Ellipsometry (VASE) were utilized to probe the unique molecular orientations of chromophore and coumarin units in the C1 system. These experimental methods can be coupled with theoretical methods such as Monte Carlo and Molecular Dynamics to further understand the molecular ordering and intermolecular interactions in the C1 system. Utilizing all of these methods together illustrates that the coumarin-based pendant groups of the C1 molecule interact in a plane orthogonal to the chromophores to enhance the acentric order of the EO active chromophores in the presence of a poling field. It has been shown that this result can be explained through reduced dimensionality in that the coumarins are imposing a dimensional restriction on the lattice environment in which the chromophores are embedded, yielding higher centric and acentric order in the system. These results of higher order, in turn, lead to enhanced EO activities.

Utilizing this approach of nano-engineering intermolecular interactions for enhanced molecular order will aid in the development of improved OEO materials. Understanding the role and effect of intermolecular interactions yields higher EO activities. If EO performance continues to be improved, OEO materials can potentially replace the current inorganic materials being utilized in commercial applications. Such materials, for example, can be implemented into hybrid silicon-organic nano-slot waveguide device structures which, if incorporated successfully with OEO materials, could revolutionize the telecommunications industry with the ultrafast transmission of information.

4. Acknowledgments

Lewis Johnson, Prof. Antao Chen, Prof. Bruce Eichinger, Dr. Philip Sullivan, Dr. Dan Knorr, Prof. Rene Overney, Dr. Sei-hum Jang, Dr. Jingdong Luo, Su Huang and Andreas Tillack are thanked for useful discussions and assistance with experiments and calculations. Funding from the NSF [DMR-0120967] and the AFOSR [FA9550-09-0682] are gratefully acknowledged.

5. References

Baehr-Jones, T.; Penkov, B.; Huang, J.; Sullivan, P. A.; Davies, J.; Takayesu, J.; Luo, J.; Kim, T. D.; Dalton, L. R.; Jen, A. K.-Y.; Hochberg, M.; Scherer, A. (2008) Nonlinear polymer-clad silicon slot waveguide modulator with a half wave voltage of 0.25 V. *Applied Physics Letters*, Vol.92, No.7, pp. 163303-1-3.

Benight, S. J.; Bale, D. H.; Olbricht, B. C.; Dalton, L. R. (2009) Organic electro-optics: Understanding material structure/function relationships and device fabrication issues. *J. Mater. Chem.*, Vol.19, pp. 7466-7475.

Benight, S. J.; Johnson, L. E.; Barnes, R.; Olbricht, B. C.; Bale, D. H.; Reid, P. J.; Eichinger, B. E.; Dalton, L. R.; Sullivan, P. A.; Robinson, B. H. (2010) Reduced Dimensionality in Organic Electro-Optic Materials: Theory and Defined Order. *J. Phys Chem. B.*, Vol.114, No.37, pp. 11949-11956.

Benight, S. J.; Knorr, D. B., Jr.; Sullivan, P. A.; Sun, J.; Kocherlakota, L. S.; Robinson, B. H.; Overney, R. M.; Dalton, L. R. (2011) Nano-Engineering of Soft Matter Lattice Dimensionality: Measurement of Nanoscopic Order and Viscoelasticity, *Submitted*.

Benight, S. J. (2011) Nanoengineering of Soft Matter Interactions in Organic Electro-Optic Materials. University of Washington, Seattle, WA, USA.

Blanchard-Desce, M.; Baudin, J.-B.; Jullien, L.; Lorne, R.; Ruel, O.; Brasselet, S.; Zyss, J. (1999) Toward highly efficient nonlinear optical chromophores: molecular engineering of octupolar molecules. *Opt. Mater.*, Vol.12, pp. 333-338.

Chen, A.; Chuyanov, V.; Garner, S.; Steier, W. H.; Dalton, L. R. (1997) Modified attenuated total reflection for the fast and routine electro-optic measurement of nonlinear optical polymer thin films. *Organic Thin Films for Photonics Applications, 1997 OSA Technical Digest Series*, Vol.14, pp. 158-160.

Cheng, Y.-J.; Luo, J.; Hau, S.; Bale, D. H.; Kim, T.-D.; Shi, Z.; Lao, D. B.; Tucker, N. M.; Tian, Y.; Dalton, L. R.; Reid, P. J.; Jen, A. K.-Y. (2007) Large Electro-optic Activity and Enhanced Thermal Stability from Diarylaminophenyl-Containing High-β Nonlinear Optical Chromophores. *Chem. Mater.* Vol.19, pp. 1154-1163.

Cheng, Y.-J.; Luo, J.; Huang, S.; Zhou, X.; Shi, Z.; Kim, T.-D.; Bale, D. H.; Takahashi, S.; Yick, A.; Polishak, B. M.; Jang, S.-H.; Dalton, L. R.; Reid, P. J.; Steier, W. H.; Jen, A. K. Y. (2008) Donor-Acceptor Thiolated Polyenic Chromophores Exhibiting Large Optical Nonlinearity and Excellent Photostability. *Chem. Mater.*, Vol.20, No.15, pp. 5047-5054.

Dalton, L. R.; Sullivan, P. A.; Bale, D. H. (2010) Electric Field Poled Organic Electro-Optic Materials: State of the Art and Future Prospects *Chem. Rev.*, Vol.110, pp. 25-55.

Dalton, L. R.; Benight, S. J.; Johnson, L. E.; Knorr, D. B., Jr.; Kosilkin, I.; Eichinger, B. E.; Robinson, B. H.; Jen, A. K.-Y.; Overney, R. M. (2011) Systematic Nanoengineering of Soft Matter Organic Electro-optic Materials. *Chem. Mater.*, Vol.23, 3, pp. 430-445.

Dalton, L. R.; Benight, S. J. (2011) Theory-Guided Design of Organic Electro-Optic Materials & Devices. *Polymer*, Vol.3, No.3, pp. 1325.

Davies, J. A.; Elangovan, A.; Sullivan, P. A.; Olbricht, B. C.; Bale, D. H.; Ewy, T. R.; Isborn, C. M.; Eichinger, B. E.; Robinson, B. H.; Reid, P. J.; Li, X.; Dalton, L. R. (2008) Rational Enhancement of Second-Order Nonlinearity: Bis-(4-methoxyphenyl)hetero-aryl-amino Donor-Based Chromophores: Design, Synthesis, and Electrooptic Activity. *J. Am. Chem. Soc.*, Vol.130, No.32, pp. 10565-10575.

Ding, R.; Baehr-Jones, T.; Liu, Y.; Bojko, R.; Witzens, J.; Huang, S.; Luo, J.; Benight, S.; Sullivan, P.; Fedeli, J.-M.; Fournier, M.; Dalton, L.; Jen, A.; Hochberg, M. (2010) Demonstration of a low VπL modulator with GHz bandwidth based on electro-optic polymer-clad silicon slot waveguides. *Opt. Express*, Vol.18, No.15, pp. 15618-15623.

Enami, Y.; Mathine, D.; Derose, C. T.; Norwood, R. A.; Luo, J.; Jen, A. K.-Y.; Peyghambarian, N. (2007) Hybrid cross-linkable polymer/sol-gel waveguide modulators with 0.65 V half wave voltage at 1550 nm. *Appl. Phys. Lett.*, Vol.91, pp. 093505.

Facchetti, A.; Annoni, E.; Beverina, L.; Morone, M.; Zhu, P.; Marks, T. J.; Pagani, G. A. (2004) Very large electro-optic responses in H-bonded heteroaromatic films grown by physical vapour deposition. *Nature Mater.*, Vol.3, pp. 910-917.

Ferry, J. D. (1980) Ch11: Dependence of viscoelastic behavior on temperature and pressure. In *Viscoelastic Properties of Polymers*, John Wiley &Sons; pp 265-320.

Frattarelli, D.; Schiavo, M.; Facchetti, A.; Ratner, M. A.; Marks, T. J. (2009) Self-Assembly from the Gas-Phase: Design and Implementation of Small-Molecule Chromophore Precursors with Large Nonlinear Optical Responses. *J. Am. Chem. Soc.* Vol.131, No.35, pp. 12595-12612.

Ge, S.; Pu, Y.; Zhang, W.; Rafailovich, M.; Sokolov, J.; Buenviaje, C.; Buckmaster, R.; Overney, R. M. (2000) Shear modulation force microscopy study of near surface glass transition temperature. *Phys. Rev. Lett.*, Vol.85, No.11, pp. 2340-2343.

Gray, T.; Killgore, J. P.; Luo, J.; Jen, A. K.-Y.; Overney, R. M. (2007) Molecular mobility and transitions in complex organic systems studied by shear force microscopy. *Nanotechnology*, Vol.18, pp. 044009.

Halgren, T. A. (1996) Merck Molecular Force Field. I. Basis, Form, Scope, Parameterization, and Performance of MMFF94*. *J. Comput. Chem.*, Vol.17, No.5-6, pp. 490.

Hammond, S. R.; Clot, O.; Firestone, K. A.; Bale, D. H.; Lao, D.; Haller, M.; Phelan, G. D.; Carlson, B.; Jen, A. K. Y.; Reid, P. J.; Dalton, L. R. (2008) Site-Isolated Electro-optic Chromophores Based on Substituted 2,2'-Bis(3,4-propylenedioxythiophene) π-Conjugated Bridges. *Chem. Mater.* Vol.20, No.10, pp. 3425-3434.

Hedvig, P. (1977) *Dielectric spectroscopy of polymers.* 1st ed.; John Wiley & Sons: New York.

Herminghaus, S.; Smith, B. A.; Swalen, J. D. (1991) Electro-optic coefficients in electric-field-poled polymer waveguides. *J. Opt. Soc. Amer. B*, Vol.8, pp. 2311-2317.

Huang, S.; Kim, T.-D.; Luo, J.; Hau, S. K.; Shi, Z.; Zhou, X.-H.; Yip, H.-L.; Jen, A. K.-Y. (2010) Highly efficient electro-optic polymers through improved poling using a thin TiO2-modified transparent electrode. *Appl. Phys. Lett.*, Vol.96, pp. 243311.

Hunziker, C.; Kwon, S.-J.; Figi, H.; Juvalta, F.; Kwon, O.-P.; Jazbinsek, M.; Günter, P. (2008) Configurationally locked, phenolic polyene organic crystal 2-{3-(4-hydroxystyryl)-5,5-dimethylcyclohex-2-enylidene}malononitrile: linear and nonlinear optical properties. *J. Opt. Soc. Am. B*, Vol.25, No.10, pp. 1678-1683.

Kang, H.; Zhu, P.; Yang, Y.; Facchetti, A.; Marks, T. J. (2004) Self-Assembled Electrooptic Thin Films with Remarkably Blue-Shifted Optical Absorption Based on an X-Shaped Chromophore. *J. Am. Chem. Soc.*, Vol.126, pp. 15974-15975.

Kim, T.-D.; Luo, J.; Ka, J.-W.; Hau, S.; Tian, Y.; Shi, Z.; Tucker, N. M.; Jang, S.-H.; Kang, J.-W.; Jen, A. K. Y. (2006) Ultralarge and thermally stable electro-optic activities from Diels-Alder crosslinkable polymers containing binary chromophore systems. *Adv. Mater.*, Vol.18, No.22, pp. 3038-3042.

Kim, T.-D.; Kang, J.-W.; Luo, J.; Jang, S.-H.; Ka, J.-W.; Tucker, N. M.; Benedict, J. B.; Dalton, L. R.; Gray, T.; Overney, R. M.; Park, D. H.; Herman, W. N.; Jen, A. K.-Y. (2007) Ultralarge and Thermally Stable Electro-Optic Activities from Supramolecular Self-Assembled Molecular Glasses. *J. Am. Chem. Soc.*, Vol.129, pp. 488-489.

Kim, T.-D.; Luo, J.; Cheng, Y.-J.; Shi, Z.; Hau, S.; Jang, S.-H.; Zhou, X. H.; Tian, Y.; Polishak, B.; Huang, S.; Ma, H.; Dalton, L. R.; Jen, A. K.-Y. (2008) Binary Chromophore Systems in Nonlinear Optical Dendrimers and Polymers for Large Electro-Optic Activities. *J. Phys. Chem. C*, Vol.112, No.21, pp. 8091-8098.

Kim, T.-D.; Luo, J.; Jen, A. K.-Y. (2009) Quantitative determination of the chromophore alignment induced by electrode contact poling in self-assembled NLO materials. *Bull. Korean Chem. Soc.*, Vol.30, No.4, pp. 882-886.

Kwon, S.-J.; Jazbinsek, M.; Kwon, O.-P.; Gunter, P. (2010) Crystal Growth and Morphology Control of OH1 Organic Electrooptic Crystals. *Cryst. Growth Des.*, Vol.10, pp. 1552-1558.

Knorr, D. B., Jr. (2010) Molecular Relaxations in Constrained Nanoscale Systems. University of Washington, USA, PhD Dissertation.

Knorr, D. B., Jr.; Gray, T.; Overney, R. M. (2009) Intrinsic Friction Analysis - Novel Nanoscopic Access to Molecular Mobility in Constrained Organic Systems. *Ultramicroscopy*, Vol.109, No.8, pp. 991.

Knorr, D. B., Jr.; Zhou, X.-H.; Shi, Z.; Luo, J.; Jang, S.-H.; Jen, A. K.-Y.; Overney, R. M. (2009) Molecular Mobility in Self-Assembled Dendritic Chromophore Glasses. *J. Phys Chem. B.*, Vol.113, No.43, pp. 14180.

Leahy-Hoppa, M. R.; Cunningham, P. D.; French, J. A.; Hayden, L. M. (2006) Atomistic molecular modeling of the effect of chromophore concentration on the electro-optic coefficient in nonlinear optical polymers. *J. Phys. Chem. A*, Vol.110, pp. 5792-5797.

Liao, Y.; Anderson, C. A.; Sullivan, P. A.; Akelaitis, A. J. P.; Robinson, B. H.; Dalton, L. R., Electro-Optical Properties of Polymers Containing Alternating Nonlinear Optical Chromophores and Bulky Spacers. *Chem. Mater.* 2006, Vol.18, No.4, pp. 1062-1067.

Liu, S.; Haller, M.; Ma, H.; Dalton, L. R.; Jang, S.-H.; Jen, A. K.-Y. (2003) Focused microwave-assisted synthesis of 2,5-dihydrofuran derivatives as electron acceptors for highly efficient nonlinear optical chromophores. *Adv. Mater.*, Vol.15, pp. 603-607.

McQuarrie, D. A. (2000) *Statistical Mechanics*. University Science Books: Sausalito, CA.

Michalak, R. J.; Kuo, Y.-H.; Nash, F. D.; Szep, A.; Caffey, J. R.; Payson, P. M.; Haas, F.; McKeon, B. F.; Cook, P. R.; Brost, G. A.; Luo, J.; JenL, A. K.-Y.; Dalton, L. R.; Steier, W. H. (2006) High-speed AJL8/APC polymer modulator. *Photon. Tech. Lett., IEEE*, Vol.18, No.11, pp. 1207-1209.

Michl, J.; Thulstrup, E. W. (1986) *Spectroscopy with polarized light: solute alignment by photoselection, in liquid crystals, polymers, and membranes*. VCH: Deerfield Beach, FL.

Munn, R. W.; Ironside, C. N. (1993) *Principles and Applications of Nonlinear Optical Materials*. CRC Press Inc.: Boca Raton, FL.

Olbricht, B. C.; Sullivan, P. A.; Dennis, P. C.; Hurst, J. T.; Johnson, L. E.; Benight, S. J.; Davies, J. A.; Chen, A.; Eichinger, B. E.; Reid, P. J.; Dalton, L. R.; Robinson, B. H. (2011) Measuring Order in Contact-Poled Organic Electrooptic Materials with Variable-Angle Polarization-Referenced Absorption Spectroscopy (VAPRAS). *J. Phys. Chem. B*, Vol.115, No.2, pp. 231-241.

Park, D. H.; Lee, C. H.; Herman, W. N. (2006) Analysis of multiple relfection effects in reflective measurements of electro-optic coefficients of poled polymers in multilayer structures. *Optics Express*, Vol.14, No.19, pp. 8866-8884.

Ponder, J. (2010) *TINKER*, 5.1; Washington University: St. Louis MO.

Rainbow Photonics AG, (2011) In: *Terahertz Generators and Detectors*, September 28, 2011, Available from: www.rainbowphotonics.com.

Ray, P. C.; Leszcynski, J. (2004) First hyperpolarizabilities of ionic octupolar molecules: structure-function relationships and solvent effects. *Chem. Phys. Lett.*, Vol.399, pp. 162-166.

Shi, Y.; Zhang, C.; Zhang, H.; Bechtel, J. H.; Dalton, L. R.; Robinson, B. H.; Steier, W. H. (2000) Low (Sub-1-Volt) Halfwave Voltage Polymeric Electro-Optic Modulators Achieved by Controlling Chromophore Shape. *Science*, Vol.288, pp. 119-122.

Sprave, M.; Blum, R.; Eich, M. (1996) High electric field conduction mechanisms in electrode poling of electro-optic polymers. *Appl. Phys. Lett.*, Vol.69, No.20, pp. 2962.

Sills, S.; Gray, T.; Overney, R. (2005) Molecular dissipation phenomena of nanoscopic friction in the heterogeneous relaxation regime of a glass former. *J. Chem. Phys.*, Vol.123, No.13, pp. 134902.

Starkweather, H. W., Jr. (1981) Simple and complex relaxations. *Macromolecules*, Vol.14, pp. 1277.

Starkweather, H. W., Jr.(1988) Noncooperative relaxations. *Macromolecules*, Vol.21, No.6, pp. 1798-802.

Stillinger, F. H. (1977) Axiomatic basis for spaces with noninteger dimension. *Journal of Mathematical Physics*, Vol.18, No.6, pp. 1224-1234.

Sullivan, P. A.; Rommel, H.; Liao, Y.; Olbricht, B. C.; Akelaitis, A. J. P.; Firestone, K. A.; Kang, J.-W.; Luo, J.; Choi, D. H.; Eichinger, B. E.; Reid, P.; Chen, A.; Robinson, B. H.; Dalton, L. R. (2007) Theory-Guided Design and Synthesis of Multichromophore Dendrimers: An Analysis of the Electro-Opitc Effect. *J. Am. Chem. Soc.*, Vol.129, pp. 7523-7530.

Sun, S.-S.; Dalton, L. R. (2008) *Introduction to Organic Electronic and Optoelectronic Materials and Devices*. CRC Press Taylor and Francis Group: Boca Raton, FL.

Teng, C. C.; Man, H. T. (1990) Simple reflection technique for measuring the electro-optic coefficient of poling polymers. *Appl. Phys. Lett.*, Vol.56, pp. 1734-1736.

Tian, Y.; Kong, X.; Nagase, Y.; Iyoda, T. (2003) Photocrosslinkable liquid-crystalline block copolymers with coumarin units synthesized with atom transfer radical polymerization. *J. Polym. Sci., Part A: Polym. Chem.*, Vol.41, pp. 2197-2206.

Tian, Y.; Akiyama, E.; Nagase, Y.; Kanazawa, A.; Tsutsumi, O.; Ikeda, T. (2004) Synthesis and investigation of photophysical and photochemical properties of new side-group liquid crystalline polymers containing coumarin moeities. *J. Mater. Chem*, Vol.14, pp. 3524-3531.

Valore, A.; Cariati, E.; Righetto, S.; Roberto, D.; Tessore, F.; Ugo, R.; Fragala, I. L.; Fragala, M. E.; Malandrino, G.; De Angelis, F.; Belpassi, L.; Ledoux-Rak, I.; Hoang Thi, K.; Zyss, J. (2010) Fluorinated -Diketonate Diglyme Lanthanide Complexes as New Second-Order Nonlinear Optical Chromophores: The Role of f Electrons in the Dipolar and Octupolar Contribution to Quadratic Hyperpolarizability. *J. Am. Chem. Soc.*, Vol.132, pp. 4966-4970.

van der Boom, M. E.; Richter, A. G.; Malinsky, J. E.; Lee, P. A.; Armstrong, N. R.; Dutta, P.; Marks, T. J. (2001) Single Reactor Route to Polar Superlattices. Layer-by-Layer Self-Assembly of Large-Response Molecular Electrooptic Materials by Protection-Deprotection. *Chem. Mater.*, Vol.13, pp. 15-17.

Verbiest, T.; Clays, K.; Rodriguez, V. (2009) *Second-order Nonlinear Optical Characterization Techniques: an Introduction*. CRC Press: Taylor and Francis Group: Boca Raton, FL.

Wang, Z.; Sun, W.; Chen, A.; Kosilkin, I.; Bale, D.; Dalton, L. R. (2011) Organic electro-optic thin films by simultaneous vacuum deposition and laser-assisted poling. *Opt. Lett.*, Vol.36, No.15, pp. 2853–2855.

Ward, I. M. (1971) *Mechanical Properties of Solid Polymers*. Wiley-Interscience: London.

Weder, C.; Glomm, B. H.; Neuenschwander, P.; Suter, U. W.; Pretre, P.; Kaatz, P.; Gunter, P. (1997) New polyamide main-chain polymers based on 2',5'-diamino-4-(dimethylamino)-4'-nitrostilbene (DDANS). *Adv. Nonlinear Opt.*, Vol.4, pp. 63-76.

Williams, M. L.; Landel, R. E.; Ferry, F. D. (1995) The Temperature Dependence of Relaxation Mechanisms in Amorphous Polymers and Other Glass-forming Liquids. *J. Am. Chem. Soc.*, Vol.77, pp. 3701.

Woollam, J. A. (2000) Ellipsometry, Variable Angle Spectroscopic. In *Wiley Encyclopedia of Electrical and Electronics Engineering*, Wiley: New York, pp 109-117.

Yang, Z.; Worle, M.; Mutter, L.; Jazbinsek, M.; Gunter, P. (2007) Synthesis, Crystal Structure, and Second-Order Nonlinear Optical Properties of New Stilbazolium Salts. *Cryst. Growth. and Design*, Vol.7, No.1, pp. 83-86.

Zhang, C.; Dalton, L. R.; Oh, M. C.; Zhang, H.; Steier, W. H. (2001) Low V(pi) Electrooptic Modulators from CLD-1: Chromophore Design and Synthesis, Material Processing, and Chracterization. *Chem. Mater.*, Vol.13, No.9, pp. 3043-3050.

Zhou, X. H.; Luo, J.; Huang, S.; Kim, T. D.; Shi, Z.; Cheng, Y.-J.; Jang, S.-H.; Knorr, D. B., Jr.; Overney, R. M.; Jen, A. K.-Y. (2009) Supramolecular Self-Assembled Dendritic Nonlinear Optical Chromophores: Fine-Tuning of Arene-Perfluoroarene Interactions for Ultralarge Electro-Optic Activity and Enhanced Thermal Stability. *Adv. Mater.* Vol.21, No.19, pp. 1976.

Part 3

Molecular Interactions Applied in Biology and Medicine

9

New Aspect of Bone Morphogenetic Protein Signaling and Its Relationship with Wnt Signaling in Bone

Nobuhiro Kamiya
Center for Excellence in Hip Disorders,
Texas Scottish Rite Hospital for Children, Dallas, Texas,
USA

1. Introduction

Bone morphogenetic proteins (BMPs) were discovered and named in 1965 by Marshall Urist, who initially identified the ability of an unknown factor in bone to induce ectopic bones in muscle [1]. In the last 45 years, the osteogenic function of BMPs has been extensively examined, mainly using osteoblasts in culture with exogenous treatments of BMPs [2]. Based on their potent osteogenic abilities, clinical trials have been initiated to use BMP2 and BMP7 to improve fracture repair [2]. The FDA (Food and Drug Administration) has approved BMP2 and BMP7 for clinical use in long bone open-fractures, non-union fractures and spinal fusion. However, recent clinical/pre-clinical studies have shown a negative impact of BMPs on bone formation under certain physiological conditions [3-7], challenging the current dogma. This book chapter will focus on the recent findings of roles of BMP signaling in bone including its relationship with Wnt signaling through Wnt (Wngless, Int-1) receptor SOST (Sclerostin) and DKK1 (Dickkopf1). This new molecular interaction would explain the negative outcomes of BMP's therapy in orthopaedics.

2. Signaling by BMPs

Marshall Urist made the key discovery that demineralized bone matrix induced bone formation in 1965 [1]. It took another 24 years for BMPs to be discovered. The combined works of several researchers led to the isolation of BMPs and later the cloning [8-11]. BMPs belong to the transforming growth factor-β (TGF-β) gene superfamily [12]. Like other members of the TGF-β family, BMPs signal through transmembrane serine/threonine kinase receptors such as BMP type I and type II receptors. Upon ligand binding, type I and II receptors form hetero-multimers [13], and the type II receptor phosphorylates and activates a highly conserved glycine- and serine-rich domain (TTSGSGSG) called a GS box between the transmembrane and kinase domains in the type I receptor. The activated BMP type I receptors relay the signal to the cytoplasm through the Smad (Sma and Mad related protein) pathway by phosphorylating their immediate downstream targets, receptor-regulated Smads (R-Smads; Smad1, Smad5, and Smad8) proteins, which then interact with co-Smad (Smad4) protein and translocate into the nucleus [14]. It is also known that non-Smad pathways through p38 MAPK (mitogen-actiated protein kinase) and TAK1 (Transforming growth factor β–activated kinase 1) are also involved in the BMP signaling [15]. There are three type I receptors [BMPRIA (BMP receptor type IA, ALK3), BMPRIB (BMP receptor type IB, ALK6) and ACVRI (Activin receptor type I, ALK2) and three type II receptors [BMPRII (BMP receptor type II), ACVRIIA (Activin receptor type IIA) and ACVRIIB (Activin receptor type IIB)], and approximately 30 ligands are identified [16]. Type I receptor ACVRI was originally described as an activin receptor, but it is now believed to be a receptor for BMPs. In osteoblasts, BMP2, BMP4, BMP6 and BMP7 and their receptors BMPRIA and ACVRI are abundantly expressed [17]. BMPRIA is a potent receptor of BMP2 and BMP4 [18, 19], as is ACVRI for BMP7 [20]. In addition, BMP antagonists Noggin, Chordin, and Gremlin were identified in osteoblasts [21]. These antagonists fine-tune BMP signaling in osteoblasts, as BMPs upregulate expression levels of antagonists while inducing BMP signaling [22] **(Table 1, Figure 1)**.

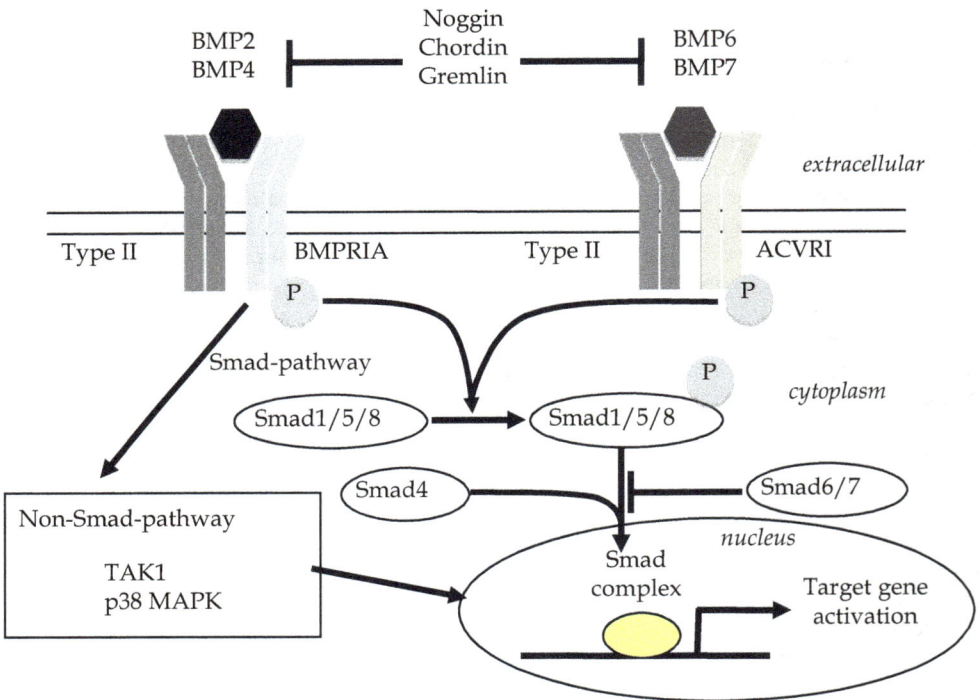

Fig. 1. Potential molecular interaction of BMP signaling in osteoblasts. BMP2, BMP4, BMP6, and BMP7 are osteoinductive and are expressed by osteoblasts. BMP2 and BMP4 are potent ligands for BMPRIA as are BMP6 and BMP7 for ACVRI. Canonical BMP signaling is through the Smad pathway via Smad1, Smad5, and Smad8 (i.e. Smad1/5/8-Smad4 complex), while non-canonical BMP signaling is through non-Smad pathways including TAK1 and p38 MAPK. Target genes are activated by these two pathways in osteoblasts.

Antagonists	Noggin, Chordin, Gremlin
Ligands	BMP2, BMP4, BMP6, BMP7
Type I Receptors	BMPRIA/ALK3, ACVRI/ALK2, (BMPRIB/ALK6)
Type II Receptors	BMPRII, ActRIIA, ActRIIB
R-Smad	Smad1, Smad5, Smad8
Co-Smad	Smad4
Non-Smad Pathways	p38 MAPK, TAK1

Table 1. Osteogenic BMPs and their signaling cascades in osteoblasts

3. Molecular interaction of BMP and Wnt

In addition to BMP signaling, Wnt signaling has been examined for a decade because of its role in bone formation and bone mass [23-27]. The physiological impact of Wnt signaling on bone mass was first reported in 2001, by showing that loss-of-function mutations in the co-receptor LRP5 (Low-density lipoprotein receptor-related protein 5) cause the autosomal recessive disorder osteoporosis-pseudoglioma syndrome (OPPG), a low bone mass phenotype in humans [28]. The importance of other Wnt ligands and receptors as bone mass effectors has been documented using genetic approaches for DKK1 [29], DKK2 [30], sFRPs (secreted frizzled-related proteins) [31], Sost/sclerostin [32], Lrp5 [33, 34] and Lrp6, all of which are expressed in osteoblasts. However, changes in BMP signaling in bone had not been reported in Wnt-related mutations in mice.

3.1 In vitro relationship

In vitro experiments using pluripotent mesenchymal cell lines or primary osteoblasts to test the interaction between BMP and Wnt signaling in osteoblasts have yielded both synergistic and antagonistic results: the treatment of C2C12 cells and primary osteoblasts with BMP2 induced Wnt3a expression and stabilized Wnt/β-catenin signaling [35-37]. The treatment of C3H10T1/2 cells with Wnt3a induced the BMP4 expression levels [38]. These suggest a positive autocrine loop [37, 39]. In contrast, inhibition of BMP signaling by treatment of primary osteoblasts with dorsomorphin, an inhibitor of BMP type I receptors, increased canonical Wnt signaling [40]. Treatment of C2C12 cells with Wnt3a repressed BMP2-dependent *Id1* (Inhibitor of DNA binding 1) expression [41]. Similarly, treatment of cultured skull bone with BMP antagonist Noggin increased canonical Wnt signaling [42]. Moreover, one study investigated intracellular cross-talk between BMP and Wnt pathways using uncommitted bone marrow stromal cells and provided a potential mechanism whereby BMP-2 antagonizes Wnt3a-induced proliferation in osteoblast progenitors by promoting an interaction between Smad1 and Dvl-1 [i.e. the human homolog of the Drosophila dishevelled gene (dsh) 1] that restricts Wnt/β-catenin activation[43]. Another interaction via Pten (phosphatase and tensin homolog)-Akt pathway has been reported in hair follicle stem/progenitor cells [44]; however, it is less likely in osteoblasts [45]. Taken together, there seems to be both positive and negative feedback loops between the two signaling pathways **(Figure 2)**.

1) Positive loop   2) Negative loop

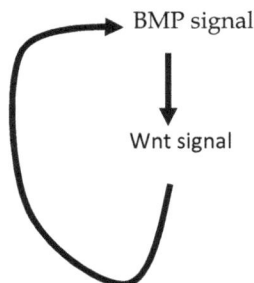

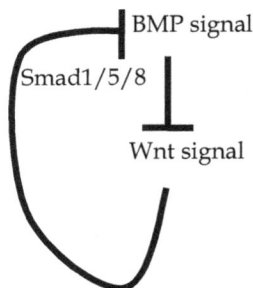

Fig. 2. A potential relationship between the two major signaling BMP and Wnt in osteoblasts based on in vitro studies. 1) Both signaling pathways function in a positive loop. 2) Both signaling pathways function in a negative loop. It is expected that these two signaling pathways may regulate each other in an age-dependent and context-dependent manner. Further studies are desired to investigate the details of each condition.

3.2 In vivo relationship

In vivo, only a few studies have revealed a link between the two signaling pathways. We recently found that loss-of-function of BMP signaling in osteoblasts via BMPRIA upregulates canonical Wnt signaling during embryonic and postnatal bone development, suggesting a negative regulation of Wnt signaling by BMP [40, 42]. In these studies, we found that upregulation of Wnt signaling is at least in part mediated by suppression of Wnt inhibitors Sost/sclerostin and Dkk1, and both Sost/sclerostin and Dkk1 are direct targets of BMP signaling. In addition, *Sost* expression was severely downregulated in *Bmpr1a*-deficient bones as assessed by microarray analysis [42]. Interestingly, both Smad-dependent and Smad-independent pathways appear to contribute to the Dkk1 expression, whereas Sost/sclerostin requires only Smad-dependent signaling, suggesting differential regulation of these genes by the BMP signaling via BMPRIA [40]. BMP and Wnt signaling regulate the development and remodeling of many tissues and interact synergistically or antagonistically in a context- and age-dependent manner *in vivo* [46, 47]. It is possible that in bone, BMP signaling inhibits Wnt signaling by upregulating the Sost/sclerostin expression in osteoblasts **(Figure 3)**.

3.3 SOST/Sclerostin and DKK1

Both SOST and DKK1 are inhibitors for canonical Wnt signaling and have been highlighted because neutralizing antibodies for SOST (AMG785) and DKK1 (BHQ880) have been developed as bone anabolic agents and these potential drugs are under clinical trial [48]. It is known that both Dkk1 and Sost/sclerostin inhibit Wnt/β-catenin signaling by binding to co-receptors. As both Dkk1 and Sost/sclerostin are secreted proteins expressed by osteoblasts, their role in regulating bone mass has been investigated using human and mouse genetic approaches.

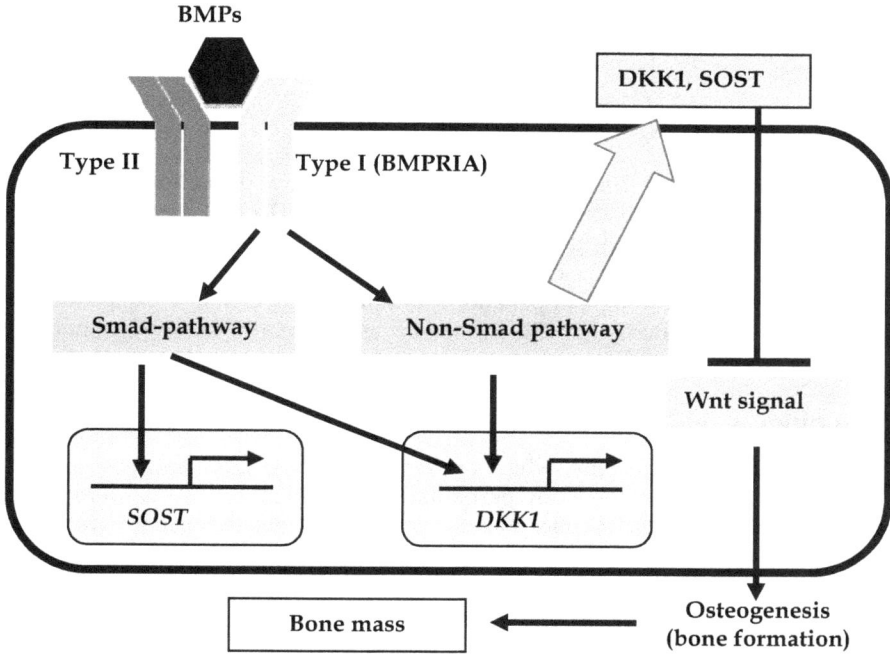

Fig. 3. Possible regulation between BMP and Wnt in Osteoblasts. A proposed model of the relationship between the BMP signaling via BMPRIA and the canonical Wnt signaling in osteoblasts. Both Dkk1 and sclerostin/Sost are downstream targets of the BMP signaling. The BMP signaling upregulates the Sost expression primarily through the Smad-dependent signaling while it upregulates the Dkk1 expression through both the Smad and non-Smad signaling (p38 MAPK). As Dkk1 and sclerostin/Sost act as Wnt signaling inhibitors, BMP signaling in osteoblasts, in turn, leads to a decrease in osteogenesis and bone mass. Dkk1 and sclerostin/Sost play an important role in regulating bone mass as downstream effectors of BMPRIA signaling in bone taking balances between BMP signaling and Wnt signaling.

3.3.1 SOST/Sclerostin

Sost/sclerostin was originally reported as a member of the BMP antagonist DAN family (i.e. the Dan gene family of BMP antagonists) [49, 50]. Although DAN family members modulate both BMP and Wnt signaling in Xenopus [51-53], recent studies suggest a primary role of Sost/sclerostin in Wnt signaling in mouse and humans: Sost/sclerostin is not a BMP antagonist [54] but rather a Wnt inhibitor [55] that binds the Wnt co-receptor low density lipoprotein receptor-related protein 5 and 6 (LRP5 and LRP6) [32, 56]. Conventional knockouts of Sost (i.e. Sost KO) are viable and exhibit increased bone mass [57]. In humans, loss-of-function and hypomorphic mutations in SOST cause sclerosteosis [58, 59] and Van Buchem disease [60, 61], respectively, with a high bone mass (HBM) phenotype. These mutants share the HBM phenotype with other gain-of-function of LRP5 mutation, due to the defect in DKK1-mediated regulation of LRP5 in humans [62-64] and overexpression of Lrp5 in mice [65]. In contrast, loss-of-function of LRP5 leads to OPPG with low bone mass [28], which is similar to the bone phenotype of mice overexpressing *Sost* [50]. In addition, recent genome-wide SNP-based analyses identified a significant association between bone mineral density and the *SOST* gene locus [66-68].

3.3.2 DKK1

Conventional knockouts of *Dkk1* die *in utero* from defective head induction and limb formation [29]. Similar to *Sost* KO mice, mice heterozygous for *Dkk1* (*Dkk1*+/− mice), however, exhibit a high bone mass (HBM) phenotype [69], while overexpression of *Dkk1* in osteoblasts causes osteopenia [70]. In addition, increased *DKK1* expression in bone marrow has also been associated with lytic bone lesions in patients with multiple myeloma [71]. Collectively, these results support the hypothesis that Dkk1 functions as a potent negative regulator of bone mass.

3.3.3 Sost/DKK1 expression in the *Bmpr1a* cKO mice

Conditional knockouts of *Bmpr1a*, which are deficient in the *Dkk1* and *Sost* expression, show a HBM phenotype [40, 42, 72]. In particular, Sost expression levels were the most dramatically reduced in the cKO mice during embryonic stages [42]. Furthermore, both Sost and Dkk1 expression levels were increased by the addition of BMP2, a potent ligand for BMPRIA, using primary osteoblasts [40]. Similarly, both Sost and Dkk1 expression levels were significantly reduced in the *Acvr1* cKO mice [73]. In addition, both Sost and Dkk1 expression levels were increased by the addition of BMP7, a potent ligand for ACVRI, using primary osteoblasts [73]. These facts support the new concept of molecular interactions between BMP signaling and Wnt signaling that Dkk1 and Sost/sclerostin act physiologically as inhibitors of canonical Wnt signaling as downstream targets of BMP receptors BMPRIA and ACVRI and that BMP signaling can negatively controls Wnt signaling in osteoblasts **(Figure 3)**.

3.4 Effects of Wnt signaling on osteoclasts

There is accumulating evidence that Wnt signaling also plays a critical role in osteoclastogenesis regulated by osteoblasts through the RANKL (Receptor activator of nuclear factor kappa-B ligand)-OPG (Osteoprotegerin) pathway. Recently, two *in vivo* studies have suggested that the canonical Wnt signaling is important in the regulation of osteoclastogenesis by osteoblasts. One study provided evidence that the Wnt pathway positively regulates the expression of *Opg* in osteoblasts [74]. Overexpression of stabilized β-catenin in osteoblasts, which results in an increase of canonical Wnt signaling level, decreases osteoclast differentiation leading to increased bone volume in mice [74]. Another study showed that an osteoblast-specific deletion of β-catenin leads to an impaired maturation and mineralization of bones in mice due to the elevated expression of *Rankl* and diminished *Opg* [75]. These facts suggest that the canonical Wnt pathway negatively regulates osteoblasts in their supporting function in osteoclastogenesis, and thus upregulation of Wnt signaling in osteoblasts can suppress osteoclast-mediated bone resorption [75]. Taken together, it is possible that the treatment of bones with BMPs can reduce Wnt activity in osteoblasts and in turn enhance osteoclast activity.

4. BMP signaling and mouse genetics

Along with the huge advancement in technologies involving mouse genetics over the last decade, many of the BMP signaling related genes have been knocked out in mice. BMP2, BMP4, BMP6 and BMP7 and their receptor BMPRIA and ACVRI are abundantly expressed in bone. However, conventional knockout mice for these genes result in an early embryonic lethality and thus, it is not possible to investigate bone development and remodeling using these models [76-82]. To avoid the embryonic lethality, a strategy of conditional knockout mice using a Cre-loxP system has been employed. A bone-specific conditional deletion of *Bmpr1a* using an *Og2-Cre* mouse, in which a Cre recombination is restricted in differentiated osteoblasts under the osteocalcin promoter, was first reported in 2004 [83]. Interestingly, this study demonstrated that the response of osteoblasts to BMP signaling is age-dependent; in the mutant mice, bone volume decreased in young mice but increased in aged mice. In addition, the activity of osteoclasts was reduced in the aged osteoblast-specific *Bmpr1a*-deficient mice, which may have lead to the complex skeletal phenotype. These facts suggest that the BMP signaling in differentiated osteoblasts can control the balance between bone formation by osteoblasts and resorption by osteoclasts, thereby affecting the final outcome of the amount of bone mass in an age-dependent manner. The increased bone mass in the *Bmpr1a*-deficient mice appeared to be in opposition to the general concept of BMPs as osteogenic inducers; however, the concept is reasonable if the target cell for BMPs as osteogenic inducers is mesenchymal cells or chondrocytes,. It is expected that BMPs have multifaceted functions *in vivo* because different cell types exhibit differing responses to BMPs. In addition, the opposite outcome in the *Bmpr1a*-deficient mice was discussed from the point of molecular interaction in the sections 3.

4.1 BMP signaling in chondrocytes, mesenchymal cells, and osteoblasts

During skeletogenesis, bones are formed via two distinct processes: intramembranous and endochondral bone formation [84]. Intramembranous bone formation occurs primarily in flat bones (*e.g.*, calvarial bones) where mesenchymal cells differentiate directly into osteoblasts [85]. Endochondral bone formation occurs primarily in long bones where condensed mesenchymal cells differentiate into chondrocytes to form cartilage templates, and then chondrocytes are replaced by osteoblasts [86]. Recently many studies have been designed to investigate the difference in the molecular mechanism by which BMP signaling regulates these cell types. Several Cre mouse lines have been used to target different cell types including osteoblast, chondrocyte, and mesenchymal cells **(Table 2)**. BMP signaling in chondrocytes and mesenchymal cells both positively control bone size and mass while BMP signaling in osteoblasts can reduce them.

4.1.1 Chondrocytes

There are several lines of evidence that show that BMP signaling in chondrocytes is required for bone size and the amount of bone mass. BMP signaling through BMPRIA is essential for postnatal maintenance of articular cartilage, using a *Gdf5*-Cre mouse line specific for chondrocytes in joints [87]. Similarly, the critical role of *Bmpr1a* together with *Bmpr1b* in chondrocytes during endochondral bone formation using a *Col2*-Cre mouse line was reported.[88]. Moreover, in chondrocytes a simultaneous deficiency in Smad 1 and Smad 5,

	Promoter Cre-mouse	BMP signal	Stage	Bone mass	Ref.
Chondrocyte					
Bmpr1a cKO	Gdf5-Cre	down	E12.5-E16.5, 7W, 9	Reduced	87
Double knockout of *Bmpr1a* and *Bmpr1b*	Col2-Cre	down	E12.5-E16.5	Reduced	88
Bmp4 overexpression	Col11a2	up	E18.5	Increased	89
Noggin overexpression	Col11a2	down	E18.5	Reduced	89
Double knockout of *Smad1* and *Smad5*	Col2-Cre	down	E12.5-newborn	Reduced	90
Mesenchymal cell					
Double knockout of B_{MP2} and B_{MP4}	Prx1-Cre	down	E10.5-newborn, 3W	Reduced	91
Bmp2 cKO	Prx1-Cre	down	5M	Reduced	92
Osteoblast					
Bmpr1a cKO	Ogl2-Cre	down	3M 10M	Reduced Increased	83
Bmp4 overexpression	2.3 kb Col1	up	E18.5	Reduced	93
Noggin overexpression	2.3 kb Col1	down	E17.5, 3W	Increased	93
Bmpr1a cKO	3.2 kb Col1-CreER	down	E18.5, 3W, 5M	Increased	40, 42, 72
Acvr1 cKO	3.2 kb Col1-CreER	down	E18.5, 3W, 5M	Increased	73
Osteoclast					
Bmpr1a cKO	Ctsk-Cre	down	8W	Increased	94

Table 2. Bone mass observed in genetically engineered mutant mice of BMP signaling

which are BMPs' downstream target molecules, reduces bone mass [90]. In parallel, studies focusing on BMP ligands and their antagonists provide further evidence that BMPs are critical for normal development of cartilage. A transgenic mouse line to overexpress *Bmp4* in mesenchymal cells/chondrocytes using a type XI collagen promoter (Col11a2) was generated, and bone mass was increased in the mutant mice [89]. Another transgenic mouse line in which *Noggin* was overexpressed in the same cells (*Col11a2-Noggin*) demonstrated a decreased bone mass. As Noggin is an antagonist for BMPs (BMP2, BMP4, BMP5, BMP6, and BMP7) with various degrees of affinity [95], these results suggest that BMP signaling positively controls proliferation and differentiation of chondrocytes.

4.1.2 Mesenchymal cells

Similar to chondrocytes, a few studies demonstrated a requirement of BMP signaling in mesenchymal cells for proper bone development and remodeling using a mesenchymal cell-specific Cre mouse line, *Prx1-Cre*, in which Cre is active in mesenchymal cells as early as embryonic day 9.5 [96]. Using the *Prx1-Cre* mouse, the simultaneously conditional deletions of *Bmp2* and *Bmp4* in mesenchymal cells resulted in an impairment of osteogenesis during late embryogenesis [91, 92]. In contrast, the conditional deletion of *Bmp2* in mesenchymal cells does not show overt developmental abnormalities; however, the resulted mice lack an initiation of fracture healing [91, 92]. Interestingly, *Bmp7*-deficiency in mesenchymal cells did not affect bone mass probably due to the compensation by Bmp4 [97]. Taken together, it is possible that the defects in the BMP signaling in chondrocytes largely contribute to the phenotypes described above because chondrocytes are derived from mesenchymal cells and play an important role in the process of fracture repair.

4.1.3 Osteoblasts

As aforementioned, a differentiated osteoblast-specific deletion of *Bmpr1a* caused an increase in bone mass in aged mice [83]. Similar to this finding, an overexpression of a BMP antagonist, Noggin, in osteoblasts increases bone volume with a reduced osteoclast number and osteoclastogenesis both at embryonic day 17.5 (E17.5) and at 3 weeks [93]. In parallel, the overexpression of *Bmp4* in osteoblasts reduced bone mass presumably due to the increase in the osteoclast number at E18.5 [93]. Recently, *Bmpr1a* was conditionally disrupted in immature osteoblasts using a tamoxifen inducible Cre driven by a 3.2-kb alpha1(I) collagen chain gene (Col1a1) promoter. In the mutant mice, bone mass was dramatically increased during the

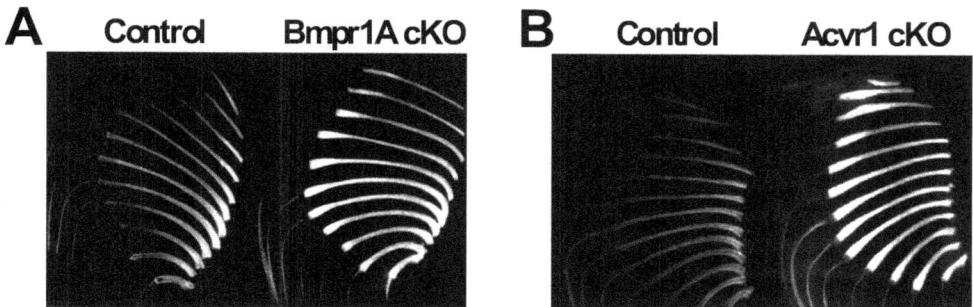

Fig. 4. Increased bone mass in the osteoblast-specific conditional knockout (cKO) mice for BMP receptors BMPRIA or ACVRI at the adult stage. *Bmpr1a* or *Acvr1* cKO mice were generated by crossing a floxed mouse line for *Bmpr1a*(*Bmpr1a*fx/fx) or *Acvr1*(*Acvr1*fx/fx) with a transgenic mouse line harboring a tamoxifen–inducible Cre driven by a 3.2 kb mouse procollagen α1(I) promoter. The Cre recombination was induced specifically in the osteoblasts by 10 weeks of tamoxifen administration from 10 weeks after birth, and bones were removed at 22 weeks. Radiodensity of rib bones was assessed by X-ray. (A) The radiodensity was dramatically increased in the *Bmpr1a* cKO mice (Cre+, *Bmpr1a*fx/fx) compared with controls (Cre–, *Bmpr1a*fx/fx). (B) The radiodensity was dramatically increased in the *Acvr1* cKO mice (Cre+, *Acvr1*fx/fx) compared with controls (Cre–, *Bmpr1a*fx/fx).

bone remodeling stage at 22 weeks as well as the bone developmental stages at E18.5 and 3 weeks [42, 72] **(Figure 4A)**. This result is an interesting contrast to previous work that disruption of *Bmpr1a* in differentiated osteoblasts results in decrease of bone mass in young adult stages (3-4 weeks). The increased bone mass in the *Bmpr1a*-deficient mice resulted from severely suppressed bone resorption due to reduced osteoclastogenesis, despite a simultaneous small reduction in the rate of bone formation [72]. Levels of RANKL and OPG are changed in the *Bmpr1a*-deficient osteoblasts and fail to support osteoclastogenesis [42, 72]. In addition, the conditional disruption of *Acvr1* in osteoblasts also demonstrated a dramatic increase in bone mass, similar to the bone phenotype of *Bmpr1a*-deficient mice **(Figure 4B)**, although osteoclastic activity is still under investigation [73]. These findings suggest that BMP signaling may have dual roles in osteoblasts; to stimulate both bone formation by osteoblasts and bone resorption supporting osteoclastogenesis. Disruption of BMP signaling in immature osteoblasts alters the balance of bone turn over to increase the bone mass, which is opposite to what people have expected for the past 4 decades.

4.1.4 Other cell type

Angiogenesis is another necessary step in new bone formation in skeletal development as well as in bone remodeling after fracture [98, 99]. Both BMP2 and BMP7 are known to induce angiogenesis by associating with other growth factors such as VEGF (vascular endothelial growth factor), bFGF (basic fibroblast growth factor), and TGF-β1 [100]. A study using an adenovirus vector in muscle demonstrated that BMP9 induces ectopic bone formation similar to BMP2 [101, 102]. As BMP9 is abundantly expressed in endothelial cells that are primarily cell types for angiogenesis [103], it is possible that BMP signaling in endothelial cells synergizes anabolic bone formation. The mechanism and origin of precursor cells for ectopic bone formation, which is physiologically observed in the patients with FOP (fibrodysplasia ossificans progressiva), is under investigation [104-106] but could be endothelial cells [107].

4.1.5 Possible interpretation

Mesenchymal cells, chondrocytes, and endothelial cells respond to BMPs by inducing bone mass and size **(Table 3)**. Recent histological findings suggest that the process of endochondral bone formation, which first forms cartilage template prior to the final bone following vessel formation (i.e. angiogenesis), plays a critical role in the process of ectopic bone formation [108]. The origin of precursor cells for the ectopic bone is under investigation [105, 106]; however, it is possible that formation of ectopic bones by BMPs [1] is largely due to the stimulation of chondrocytes, mesenchymal cells, and/or endothelial cells in soft tissue, which results in an expansion of ectopic cartilage subsequently replaced by osteoblasts. There is another possibility that the BMP signaling directly affects osteoblasts to form ectopic bone. However, this possibility is less likely based on recent evidence that reduced

Cell types that can increase bone mass	Cell types that can reduce bone mass
Mesenchymal cells	Osteoclasts
Chondrocytes	Osteoblasts
Osteoblasts	
Endothelial cells	

Table 3. A variety of cell types in bone that mediate bone mass in response to BMPs

BMP signaling in osteoblasts results in an increase in bone mass. As current methods of systemic and local treatment affect multiple cell types simultaneously in bone, it is important to evaluate the effects of BMPs on more than just osteoblasts.

4.2 Effect of BMP signaling on osteoclasts

Bone mass is determined by the balance between bone formation and bone resorption. Osteoclasts are multinuclear cells derived from hematopoietic stem cells to secrete enzymes for bone resorption [109]. Recent mouse genetic studies revealed the importance of BMP signaling for osteoclastic activity and bone resorption.

4.2.1 Regulation of osteoclast by osteoblast-dependent BMP signaling

It is expected that BMPs play roles in osteoclastogenesis and their functions, because receptors for BMPs are expressed in these cells [110]. Additionally, osteoblasts also play critical roles in bone resorption by regulating osteoclastogenesis because they produce RANK ligand (RANKL), essential to promote osteoclastogenesis, and its decoy receptor, osteoprotegerin (OPG) [111, 112]. A balance between RANKL and OPG is important to determine the degree of osteoclastogenesis, i.e. more RANKL production by osteoblasts leads to more osteoclasts; thus more bone resorption is expected. As RANKL is an osteoblastic product and BMPs induce osteoblast maturation, BMPs indirectly stimulate osteoclastogenesis and thus, osteoclastogenesis is impaired when osteoblastogenesis is blocked with BMP antagonists in culture [113]. The physiological effects of BMP signaling in osteoblasts on osteoclastogenesis were determined later using an osteoblast-specific gain-of-function or loss-of-function mouse model. For the cases of the osteoblast-specific deletion of *Bmpr1a* and osteoblast-specific over expression of *Noggin*, osteoclastogenesis is highly compromised leading to an increase of bone mass [83, 93]. In contrast, osteoblast-specific overexpression of *Bmp4* increased osteoclastogenesis [93]. The regulation of RANKL by BMPs was suggested based on an *in vitro* study [114]. This concept was recently proven in mouse studies, as *Bmpr1a*-deficient osteoblasts were not able to support osteoclastogenesis due to an imbalance between RANKL and OPG [42, 72]. It is therefore concluded that osteoblasts can respond to BMPs by inducing osteogenic (i.e. bone anabolic) action as well as osteoclastogenic (i.e. bone catabolic) action simultaneously presumably dependening on context and timing **(Table 3)**.

4.2.2 Regulation of osteoclast by osteoclast-dependent BMP signaling

BMP receptors are expressed in osteoclasts [110]. When BMP signaling through BMPRIA was deficient in osteoclasts using a Catepsin K promoter (CtsK), bone mass was increased as expected [94](Table 2). Interestingly, both bone formation rate and osteoblast number assessed by bone histomorphometry analysis were increased while osteoclast number was reduced in the mutant mice compared to their controls. It is possible that some coupling factors can control osteoblast function in an osteoclast-dependent manner in the mutant mice (i.e. osteoclast-derived coupling factors). Further studies are needed to determine whether such factors mediate BMPRIA-induced coupling from osteoclasts to osteoblasts.

5. Future direction of BMPs and Wnt

As is discussed in the former part of this review, it is important to understand that BMPs have variable and context-sensitive effects on diverse cell types in bone including chondrocytes, osteoblasts, and osteoclasts. Studies focusing on BMP receptors in chondrocytes including mesenchymal cells suggest that these cells can respond to BMP signaling by increasing bone mass during the endochondral formation process. As discussed in the latter part, BMP signals can consistently inhibit Wnt signaling and bone mass while exerting concordant effects on *Dkk1* and *Sost*. This revision of traditional understanding of the BMP signaling pathway in clinical therapeutics might suggest that in some circumstances, BMP inhibition would be desirable for promoting bone mass. More importantly, if BMP signaling reduces bone mass by inhibiting Wnt signaling through SOST/DKK1 in osteoblasts, small molecule antagonists for BMPs or BMP receptors can conversely increase bone mass and size. Therefore, development of these molecules would be a next step towards disease conditions in which bone mass is reduced such as osteoporosis and bone fracture. Although antibodies for SOST and DKk1 have been developed in order to increase bone mass, the small molecule antagonists which can be an upstream of SOST and DKK1 would be used as more potent therapeutic agents for osteoporosis. Last, the function of the BMP signaling in osteoclasts remains largely unknown in terms of coupling factors and merits future study, although the BMP signaling regulates osteoblast-dependent osteoclastogenesis via the RANKL-OPG pathway.

6. Conclusion

Understanding the complex roles of the BMP signaling pathway and its molecular interaction with other signaling pathway (i.e. Wnt) in a variety of cell-types in bone including chondrocytes, osteoblasts and osteoclasts, which contribute to normal physiological conditions (i.e. bone development, homeostasis, and remodeling) will not only help to improve current knowledge of the pathological conditions (i.e. bone fracture, osteoporosis, and other congenital and aging-related bone diseases) but may provide novel therapeutically useful strategies.

7. Acknowledgment

I would like to thank Drs. Yuji Mishina, Jian Q. Feng, Tatsuya Kobayashi, and Henry M. Kronenberg for the generation of multiple transgenic mouse lines and Harry K. W. Kim for encouragement. This work was supported by the Lilly Fellowship Foundation and TSRH Research Foundation (GL170999, GL171041).

8. References

[1] Urist, M. R., Bone: formation by autoinduction. *Science* 1965, 150, (698), 893-9.

[2] Simpson, A. H.; Mills, L.; Noble, B., The role of growth factors and related agents in accelerating fracture healing. *J Bone Joint Surg Br* 2006, 88, (6), 701-5.

[3] Aro, H. T.; Govender, S.; Patel, A. D.; Hernigou, P.; Perera de Gregorio, A.; Popescu, G. I.; Golden, J. D.; Christensen, J.; Valentin, A., Recombinant Human Bone Morphogenetic Protein-2: A Randomized Trial in Open Tibial Fractures Treated with Reamed Nail Fixation. *J Bone Joint Surg Am* 2011.

[4] Laursen, M.; Hoy, K.; Hansen, E. S.; Gelineck, J.; Christensen, F. B.; Bunger, C. E., Recombinant bone morphogenetic protein-7 as an intracorporal bone growth stimulator in unstable thoracolumbar burst fractures in humans: preliminary results. *Eur Spine J* 1999, 8, (6), 485-90.

[5] Pradhan, B. B.; Bae, H. W.; Dawson, E. G.; Patel, V. V.; Delamarter, R. B., Graft resorption with the use of bone morphogenetic protein: lessons from anterior lumbar interbody fusion using femoral ring allografts and recombinant human bone morphogenetic protein-2. *Spine (Phila Pa 1976)* 2006, 31, (10), E277-84.

[6] Seeherman, H. J.; Li, X. J.; Bouxsein, M. L.; Wozney, J. M., rhBMP-2 induces transient bone resorption followed by bone formation in a nonhuman primate core-defect model. *J Bone Joint Surg Am* 2010, 92, (2), 411-26.

[7] Vaidya, R.; Weir, R.; Sethi, A.; Meisterling, S.; Hakeos, W.; Wybo, C. D., Interbody fusion with allograft and rhBMP-2 leads to consistent fusion but early subsidence. *J Bone Joint Surg Br* 2007, 89, (3), 342-5.

[8] Sampath, T. K.; Reddi, A. H., Dissociative extraction and reconstitution of extracellular matrix components involved in local bone differentiation. *Proc Natl Acad Sci U S A* 1981, 78, (12), 7599-603.

[9] Wozney, J. M.; Rosen, V.; Celeste, A. J.; Mitsock, L. M.; Whitters, M. J.; Kriz, R. W.; Hewick, R. M.; Wang, E. A., Novel regulators of bone formation: molecular clones and activities. *Science* 1988, 242, (4885), 1528-34.

[10] Luyten, F. P.; Cunningham, N. S.; Ma, S.; Muthukumaran, N.; Hammonds, R. G.; Nevins, W. B.; Woods, W. I.; Reddi, A. H., Purification and partial amino acid sequence of osteogenin, a protein initiating bone differentiation. *J Biol Chem* 1989, 264, (23), 13377-80.

[11] Wozney, J. M., The bone morphogenetic protein family and osteogenesis. *Mol Reprod Dev* 1992, 32, (2), 160-7.

[12] Massague, J., Receptors for the TGF-beta family. *Cell* 1992, 69, (7), 1067-70.

[13] Wrana, J. L.; Attisano, L.; Wieser, R.; Ventura, F.; Massague, J., Mechanism of activation of the TGF-beta receptor. *Nature* 1994, 370, (6488), 341-7.

[14] Chen, D.; Zhao, M.; Mundy, G. R., Bone morphogenetic proteins. *Growth Factors* 2004, 22, (4), 233-41.

[15] Shim, J. H.; Greenblatt, M. B.; Xie, M.; Schneider, M. D.; Zou, W.; Zhai, B.; Gygi, S.; Glimcher, L. H., TAK1 is an essential regulator of BMP signalling in cartilage. *EMBO J* 2009, 28, (14), 2028-41.

[16] Wagner, D. O.; Sieber, C.; Bhushan, R.; Borgermann, J. H.; Graf, D.; Knaus, P., BMPs: from bone to body morphogenetic proteins. *Sci Signal* 2010, 3, (107), mr1.

[17] Lavery, K.; Swain, P.; Falb, D.; Alaoui-Ismaili, M. H., BMP-2/4 and BMP-6/7 differentially utilize cell surface receptors to induce osteoblastic differentiation of human bone marrow-derived mesenchymal stem cells. *J Biol Chem* 2008, 283, (30), 20948-58.

[18] Keller, S.; Nickel, J.; Zhang, J. L.; Sebald, W.; Mueller, T. D., Molecular recognition of BMP-2 and BMP receptor IA. *Nat Struct Mol Biol* 2004, 11, (5), 481-8.

[19] Hatta, T.; Konishi, H.; Katoh, E.; Natsume, T.; Ueno, N.; Kobayashi, Y.; Yamazaki, T., Identification of the ligand-binding site of the BMP type IA receptor for BMP-4. *Biopolymers* 2000, 55, (5), 399-406.

[20] Macias-Silva, M.; Hoodless, P. A.; Tang, S. J.; Buchwald, M.; Wrana, J. L., Specific activation of Smad1 signaling pathways by the BMP7 type I receptor, ALK2. *J Biol Chem* 1998, 273, (40), 25628-36.

[21] Rosen, V., BMP and BMP inhibitors in bone. *Ann N Y Acad Sci* 2006, 1068, 19-25.

[22] Gazzerro, E.; Gangji, V.; Canalis, E., Bone morphogenetic proteins induce the expression of noggin, which limits their activity in cultured rat osteoblasts. *J Clin Invest* 1998, 102, (12), 2106-14.

[23] Baron, R.; Rawadi, G.; Roman-Roman, S., Wnt signaling: a key regulator of bone mass. *Curr Top Dev Biol* 2006, 76, 103-27.

[24] Glass, D. A., 2nd; Karsenty, G., Molecular bases of the regulation of bone remodeling by the canonical Wnt signaling pathway. *Curr Top Dev Biol* 2006, 73, 43-84.

[25] Harada, S.; Rodan, G. A., Control of osteoblast function and regulation of bone mass. *Nature* 2003, 423, (6937), 349-55.

[26] Hartmann, C., A Wnt canon orchestrating osteoblastogenesis. *Trends Cell Biol* 2006, 16, (3), 151-8.

[27] Krishnan, V.; Bryant, H. U.; Macdougald, O. A., Regulation of bone mass by Wnt signaling. *J Clin Invest* 2006, 116, (5), 1202-9.

[28] Gong, Y.; Slee, R. B.; Fukai, N.; Rawadi, G.; Roman-Roman, S.; Reginato, A. M.; Wang, H.; Cundy, T.; Glorieux, F. H.; Lev, D.; Zacharin, M.; Oexle, K.; Marcelino, J.; Suwairi, W.; Heeger, S.; Sabatakos, G.; Apte, S.; Adkins, W. N.; Allgrove, J.; Arslan-Kirchner, M.; Batch, J. A.; Beighton, P.; Black, G. C.; Boles, R. G.; Boon, L. M.; Borrone, C.; Brunner, H. G.; Carle, G. F.; Dallapiccola, B.; De Paepe, A.; Floege, B.; Halfhide, M. L.; Hall, B.; Hennekam, R. C.; Hirose, T.; Jans, A.; Juppner, H.; Kim, C. A.; Keppler-Noreuil, K.; Kohlschuetter, A.; LaCombe, D.; Lambert, M.; Lemyre, E.; Letteboer, T.; Peltonen, L.; Ramesar, R. S.; Romanengo, M.; Somer, H.; Steichen-Gersdorf, E.; Steinmann, B.; Sullivan, B.; Superti-Furga, A.; Swoboda, W.; van den Boogaard, M. J.; Van Hul, W.; Vikkula, M.; Votruba, M.; Zabel, B.; Garcia, T.; Baron, R.; Olsen, B. R.; Warman, M. L., LDL receptor-related protein 5 (LRP5) affects bone accrual and eye development. *Cell* 2001, 107, (4), 513-23.

[29] Mukhopadhyay, M.; Shtrom, S.; Rodriguez-Esteban, C.; Chen, L.; Tsukui, T.; Gomer, L.; Dorward, D. W.; Glinka, A.; Grinberg, A.; Huang, S. P.; Niehrs, C.; Izpisua Belmonte, J. C.; Westphal, H., Dickkopf1 is required for embryonic head induction and limb morphogenesis in the mouse. *Dev Cell* 2001, 1, (3), 423-34.

[30] Li, X.; Liu, P.; Liu, W.; Maye, P.; Zhang, J.; Zhang, Y.; Hurley, M.; Guo, C.; Boskey, A.; Sun, L.; Harris, S. E.; Rowe, D. W.; Ke, H. Z.; Wu, D., Dkk2 has a role in terminal osteoblast differentiation and mineralized matrix formation. *Nat Genet* 2005, 37, (9), 945-52.

[31] Bodine, P. V.; Zhao, W.; Kharode, Y. P.; Bex, F. J.; Lambert, A. J.; Goad, M. B.; Gaur, T.; Stein, G. S.; Lian, J. B.; Komm, B. S., The Wnt antagonist secreted frizzled-related protein-1 is a negative regulator of trabecular bone formation in adult mice. *Mol Endocrinol* 2004, 18, (5), 1222-37.

[32] Li, X.; Zhang, Y.; Kang, H.; Liu, W.; Liu, P.; Zhang, J.; Harris, S. E.; Wu, D., Sclerostin binds to LRP5/6 and antagonizes canonical Wnt signaling. *J Biol Chem* 2005, 280, (20), 19883-7.

[33] Ai, M.; Holmen, S. L.; Van Hul, W.; Williams, B. O.; Warman, M. L., Reduced affinity to and inhibition by DKK1 form a common mechanism by which high bone mass-associated missense mutations in LRP5 affect canonical Wnt signaling. *Mol Cell Biol* 2005, 25, (12), 4946-55.

[34] Patel, M. S.; Karsenty, G., Regulation of bone formation and vision by LRP5. *N Engl J Med* 2002, 346, (20), 1572-4.

[35] Bain, G.; Muller, T.; Wang, X.; Papkoff, J., Activated beta-catenin induces osteoblast differentiation of C3H10T1/2 cells and participates in BMP2 mediated signal transduction. *Biochem Biophys Res Commun* 2003, 301, (1), 84-91.

[36] Mbalaviele, G.; Sheikh, S.; Stains, J. P.; Salazar, V. S.; Cheng, S. L.; Chen, D.; Civitelli, R., Beta-catenin and BMP-2 synergize to promote osteoblast differentiation and new bone formation. *J Cell Biochem* 2005, 94, (2), 403-18.

[37] Chen, Y.; Whetstone, H. C.; Youn, A.; Nadesan, P.; Chow, E. C.; Lin, A. C.; Alman, B. A., Beta-catenin signaling pathway is crucial for bone morphogenetic protein 2 to induce new bone formation. *J Biol Chem* 2007, 282, (1), 526-33.

[38] Winkler, D. G.; Sutherland, M. S.; Ojala, E.; Turcott, E.; Geoghegan, J. C.; Shpektor, D.; Skonier, J. E.; Yu, C.; Latham, J. A., Sclerostin inhibition of Wnt-3a-induced C3H10T1/2 cell differentiation is indirect and mediated by bone morphogenetic proteins. *J Biol Chem* 2005, 280, (4), 2498-502.

[39] Rawadi, G.; Vayssiere, B.; Dunn, F.; Baron, R.; Roman-Roman, S., BMP-2 controls alkaline phosphatase expression and osteoblast mineralization by a Wnt autocrine loop. *J Bone Miner Res* 2003, 18, (10), 1842-53.

[40] Kamiya, N.; Kobayashi, T.; Mochida, Y.; Yu, P. B.; Yamauchi, M.; Kronenberg, H. M.; Mishina, Y., Wnt Inhibitors Dkk1 and Sost are Downstream Targets of BMP Signaling Through the Type IA Receptor (BMPRIA) in Osteoblasts. *J Bone Miner Res* 2010, 25, (2), 200-10.

[41] Nakashima, A.; Katagiri, T.; Tamura, M., Cross-talk between Wnt and bone morphogenetic protein 2 (BMP-2) signaling in differentiation pathway of C2C12 myoblasts. *J Biol Chem* 2005, 280, (45), 37660-8.

[42] Kamiya, N.; Ye, L.; Kobayashi, T.; Mochida, Y.; Yamauchi, M.; Kronenberg, H. M.; Feng, J. Q.; Mishina, Y., BMP signaling negatively regulates bone mass through sclerostin by inhibiting the canonical Wnt pathway. *Development* 2008, 135, (22), 3801-11.

[43] Liu, Z.; Tang, Y.; Qiu, T.; Cao, X.; Clemens, T. L., A dishevelled-1/Smad1 interaction couples WNT and bone morphogenetic protein signaling pathways in uncommitted bone marrow stromal cells. *J Biol Chem* 2006, 281, (25), 17156-63.

[44] Zhang, J.; He, X. C.; Tong, W. G.; Johnson, T.; Wiedemann, L. M.; Mishina, Y.; Feng, J. Q.; Li, L., BMP signaling inhibits hair follicle anagen induction by restricting epithelial stem/progenitor cell activation and expansion. *Stem Cells* 2006.

[45] Hays, E.; Schmidt, J.; Chandar, N., Beta-catenin is not activated by downregulation of PTEN in osteoblasts. *In Vitro Cell Dev Biol Anim* 2009, 45, (7), 361-70.

[46] Huelsken, J.; Vogel, R.; Erdmann, B.; Cotsarelis, G.; Birchmeier, W., beta-Catenin controls hair follicle morphogenesis and stem cell differentiation in the skin. *Cell* 2001, 105, (4), 533-45.

[47] Barrow, J. R.; Thomas, K. R.; Boussadia-Zahui, O.; Moore, R.; Kemler, R.; Capecchi, M. R.; McMahon, A. P., Ectodermal Wnt3/beta-catenin signaling is required for the establishment and maintenance of the apical ectodermal ridge. *Genes Dev* 2003, 17, (3), 394-409.

[48] Rachner, T. D.; Khosla, S.; Hofbauer, L. C., Osteoporosis: now and the future. *Lancet* 2011, 377, (9773), 1276-87.

[49] Kusu, N.; Laurikkala, J.; Imanishi, M.; Usui, H.; Konishi, M.; Miyake, A.; Thesleff, I.; Itoh, N., Sclerostin is a novel secreted osteoclast-derived bone morphogenetic protein antagonist with unique ligand specificity. *J Biol Chem* 2003, 278, (26), 24113-7.

[50] Winkler, D. G.; Sutherland, M. K.; Geoghegan, J. C.; Yu, C.; Hayes, T.; Skonier, J. E.; Shpektor, D.; Jonas, M.; Kovacevich, B. R.; Staehling-Hampton, K.; Appleby, M.; Brunkow, M. E.; Latham, J. A., Osteocyte control of bone formation via sclerostin, a novel BMP antagonist. *Embo J* 2003, 22, (23), 6267-76.

[51] Piccolo, S.; Agius, E.; Leyns, L.; Bhattacharyya, S.; Grunz, H.; Bouwmeester, T.; De Robertis, E. M., The head inducer Cerberus is a multifunctional antagonist of Nodal, BMP and Wnt signals. *Nature* 1999, 397, (6721), 707-10.

[52] Bell, E.; Munoz-Sanjuan, I.; Altmann, C. R.; Vonica, A.; Brivanlou, A. H., Cell fate specification and competence by Coco, a maternal BMP, TGFbeta and Wnt inhibitor. *Development* 2003, 130, (7), 1381-9.

[53] Itasaki, N.; Jones, C. M.; Mercurio, S.; Rowe, A.; Domingos, P. M.; Smith, J. C.; Krumlauf, R., Wise, a context-dependent activator and inhibitor of Wnt signalling. *Development* 2003, 130, (18), 4295-305.

[54] van Bezooijen, R. L.; Roelen, B. A.; Visser, A.; van der Wee-Pals, L.; de Wilt, E.; Karperien, M.; Hamersma, H.; Papapoulos, S. E.; ten Dijke, P.; Lowik, C. W., Sclerostin is an osteocyte-expressed negative regulator of bone formation, but not a classical BMP antagonist. *J Exp Med* 2004, 199, (6), 805-14.

[55] van Bezooijen, R. L.; Svensson, J. P.; Eefting, D.; Visser, A.; van der Horst, G.; Karperien, M.; Quax, P. H.; Vrieling, H.; Papapoulos, S. E.; ten Dijke, P.; Lowik, C. W., Wnt but not BMP signaling is involved in the inhibitory action of sclerostin on BMP-stimulated bone formation. *J Bone Miner Res* 2007, 22, (1), 19-28.

[56] Semenov, M.; Tamai, K.; He, X., SOST is a ligand for LRP5/LRP6 and a Wnt signaling inhibitor. *J Biol Chem* 2005, 280, (29), 26770-5.

[57] Li, X.; Ominsky, M. S.; Niu, Q. T.; Sun, N.; Daugherty, B.; D'Agostin, D.; Kurahara, C.; Gao, Y.; Cao, J.; Gong, J.; Asuncion, F.; Barrero, M.; Warmington, K.; Dwyer, D.; Stolina, M.; Morony, S.; Sarosi, I.; Kostenuik, P. J.; Lacey, D. L.; Simonet, W. S.; Ke, H. Z.; Paszty, C., Targeted deletion of the sclerostin gene in mice results in increased bone formation and bone strength. *J Bone Miner Res* 2008, 23, (6), 860-9.

[58] Balemans, W.; Ebeling, M.; Patel, N.; Van Hul, E.; Olson, P.; Dioszegi, M.; Lacza, C.; Wuyts, W.; Van Den Ende, J.; Willems, P.; Paes-Alves, A. F.; Hill, S.; Bueno, M.; Ramos, F. J.; Tacconi, P.; Dikkers, F. G.; Stratakis, C.; Lindpaintner, K.; Vickery, B.; Foernzler, D.; Van Hul, W., Increased bone density in sclerosteosis is due to the deficiency of a novel secreted protein (SOST). *Hum Mol Genet* 2001, 10, (5), 537-43.

[59] Brunkow, M. E.; Gardner, J. C.; Van Ness, J.; Paeper, B. W.; Kovacevich, B. R.; Proll, S.; Skonier, J. E.; Zhao, L.; Sabo, P. J.; Fu, Y.; Alisch, R. S.; Gillett, L.; Colbert, T.; Tacconi, P.; Galas, D.; Hamersma, H.; Beighton, P.; Mulligan, J., Bone dysplasia sclerosteosis results from loss of the SOST gene product, a novel cystine knot-containing protein. *Am J Hum Genet* 2001, 68, (3), 577-89.

[60] Balemans, W.; Patel, N.; Ebeling, M.; Van Hul, E.; Wuyts, W.; Lacza, C.; Dioszegi, M.; Dikkers, F. G.; Hildering, P.; Willems, P. J.; Verheij, J. B.; Lindpaintner, K.; Vickery, B.; Foernzler, D.; Van Hul, W., Identification of a 52 kb deletion downstream of the SOST gene in patients with van Buchem disease. *J Med Genet* 2002, 39, (2), 91-7.

[61] Staehling-Hampton, K.; Proll, S.; Paeper, B. W.; Zhao, L.; Charmley, P.; Brown, A.; Gardner, J. C.; Galas, D.; Schatzman, R. C.; Beighton, P.; Papapoulos, S.; Hamersma, H.; Brunkow, M. E., A 52-kb deletion in the SOST-MEOX1 intergenic region on 17q12-q21 is associated with van Buchem disease in the Dutch population. *Am J Med Genet* 2002, 110, (2), 144-52.

[62] Boyden, L. M.; Mao, J.; Belsky, J.; Mitzner, L.; Farhi, A.; Mitnick, M. A.; Wu, D.; Insogna, K.; Lifton, R. P., High bone density due to a mutation in LDL-receptor-related protein 5. *N Engl J Med* 2002, 346, (20), 1513-21.

[63] Little, R. D.; Carulli, J. P.; Del Mastro, R. G.; Dupuis, J.; Osborne, M.; Folz, C.; Manning, S. P.; Swain, P. M.; Zhao, S. C.; Eustace, B.; Lappe, M. M.; Spitzer, L.; Zweier, S.; Braunschweiger, K.; Benchekroun, Y.; Hu, X.; Adair, R.; Chee, L.; FitzGerald, M. G.; Tulig, C.; Caruso, A.; Tzellas, N.; Bawa, A.; Franklin, B.; McGuire, S.; Nogues, X.; Gong, G.; Allen, K. M.; Anisowicz, A.; Morales, A. J.; Lomedico, P. T.; Recker, S. M.; Van Eerdewegh, P.; Recker, R. R.; Johnson, M. L., A mutation in the LDL receptor-related protein 5 gene results in the autosomal dominant high-bone-mass trait. *Am J Hum Genet* 2002, 70, (1), 11-9.

[64] Van Wesenbeeck, L.; Cleiren, E.; Gram, J.; Beals, R. K.; Benichou, O.; Scopelliti, D.; Key, L.; Renton, T.; Bartels, C.; Gong, Y.; Warman, M. L.; De Vernejoul, M. C.; Bollerslev, J.; Van Hul, W., Six novel missense mutations in the LDL receptor-related protein 5 (LRP5) gene in different conditions with an increased bone density. *Am J Hum Genet* 2003, 72, (3), 763-71.

[65] Babij, P.; Zhao, W.; Small, C.; Kharode, Y.; Yaworsky, P. J.; Bouxsein, M. L.; Reddy, P. S.; Bodine, P. V.; Robinson, J. A.; Bhat, B.; Marzolf, J.; Moran, R. A.; Bex, F., High bone mass in mice expressing a mutant LRP5 gene. *J Bone Miner Res* 2003, 18, (6), 960-74.

[66] Styrkarsdottir, U.; Halldorsson, B. V.; Gretarsdottir, S.; Gudbjartsson, D. F.; Walters, G. B.; Ingvarsson, T.; Jonsdottir, T.; Saemundsdottir, J.; Snorradottir, S.; Center, J. R.; Nguyen, T. V.; Alexandersen, P.; Gulcher, J. R.; Eisman, J. A.; Christiansen, C.; Sigurdsson, G.; Kong, A.; Thorsteinsdottir, U.; Stefansson, K., New sequence variants associated with bone mineral density. *Nat Genet* 2009, 41, (1), 15-7.

[67] Huang, Q. Y.; Li, G. H.; Kung, A. W., The -9247 T/C polymorphism in the SOST upstream regulatory region that potentially affects C/EBPalpha and FOXA1 binding is associated with osteoporosis. *Bone* 2009, 45, (2), 289-94.

[68] Yerges, L. M.; Klei, L.; Cauley, J. A.; Roeder, K.; Kammerer, C. M.; Moffett, S. P.; Ensrud, K. E.; Nestlerode, C. S.; Marshall, L. M.; Hoffman, A. R.; Lewis, C.; Lang, T. F.; Barrett-Connor, E.; Ferrell, R. E.; Orwoll, E. S.; Zmuda, J. M., High-density association study of 383 candidate genes for volumetric BMD at the femoral neck and lumbar spine among older men. *J Bone Miner Res* 2009, 24, (12), 2039-49.

[69] Morvan, F.; Boulukos, K.; Clement-Lacroix, P.; Roman Roman, S.; Suc-Royer, I.; Vayssiere, B.; Ammann, P.; Martin, P.; Pinho, S.; Pognonec, P.; Mollat, P.; Niehrs, C.; Baron, R.; Rawadi, G., Deletion of a single allele of the Dkk1 gene leads to an increase in bone formation and bone mass. *J Bone Miner Res* 2006, 21, (6), 934-45.

[70] Li, J.; Sarosi, I.; Cattley, R. C.; Pretorius, J.; Asuncion, F.; Grisanti, M.; Morony, S.; Adamu, S.; Geng, Z.; Qiu, W.; Kostenuik, P.; Lacey, D. L.; Simonet, W. S.; Bolon, B.; Qian, X.; Shalhoub, V.; Ominsky, M. S.; Zhu Ke, H.; Li, X.; Richards, W. G., Dkk1-mediated inhibition of Wnt signaling in bone results in osteopenia. *Bone* 2006, 39, (4), 754-66.

[71] Tian, E.; Zhan, F.; Walker, R.; Rasmussen, E.; Ma, Y.; Barlogie, B.; Shaughnessy, J. D., Jr., The role of the Wnt-signaling antagonist DKK1 in the development of osteolytic lesions in multiple myeloma. *N Engl J Med* 2003, 349, (26), 2483-94.

[72] Kamiya, N.; Ye, L.; Kobayashi, T.; Lucas, D. J.; Mochida, Y.; Yamauchi, M.; Kronenberg, H. M.; Feng, J. Q.; Mishina, Y., Disruption of BMP signaling in osteoblasts through

type IA receptor (BMPRIA) increases bone mass. *J Bone Miner Res* 2008, 23, (12), 2007-17.

[73] Kamiya, N.; Kaartinen, V.; Mishina, Y., Loss-of-function of ACVR1 in osteoblasts increases bone mass and activates canonical Wnt signaling through suppression of Wnt inhibitors SOST and DKK1. *Biochem Biophys Res Commun*, 2011, 414, (2), 326-30.

[74] Glass, D. A., 2nd; Bialek, P.; Ahn, J. D.; Starbuck, M.; Patel, M. S.; Clevers, H.; Taketo, M. M.; Long, F.; McMahon, A. P.; Lang, R. A.; Karsenty, G., Canonical Wnt signaling in differentiated osteoblasts controls osteoclast differentiation. *Dev Cell* 2005, 8, (5), 751-64.

[75] Holmen, S. L.; Zylstra, C. R.; Mukherjee, A.; Sigler, R. E.; Faugere, M. C.; Bouxsein, M. L.; Deng, L.; Clemens, T. L.; Williams, B. O., Essential role of beta-catenin in postnatal bone acquisition. *J Biol Chem* 2005, 280, (22), 21162-8.

[76] Dudley, A. T.; Lyons, K. M.; Robertson, E. J., A requirement for bone morphogenetic protein-7 during development of the mammalian kidney and eye. *Genes Dev* 1995, 9, (22), 2795-807.

[77] Gu, Z.; Reynolds, E. M.; Song, J.; Lei, H.; Feijen, A.; Yu, L.; He, W.; MacLaughlin, D. T.; van den Eijnden-van Raaij, J.; Donahoe, P. K.; Li, E., The type I serine/threonine kinase receptor ActRIA (ALK2) is required for gastrulation of the mouse embryo. *Development* 1999, 126, (11), 2551-61.

[78] Luo, G.; Hofmann, C.; Bronckers, A. L.; Sohocki, M.; Bradley, A.; Karsenty, G., BMP-7 is an inducer of nephrogenesis, and is also required for eye development and skeletal patterning. *Genes Dev* 1995, 9, (22), 2808-20.

[79] Mishina, Y.; Suzuki, A.; Ueno, N.; Behringer, R. R., Bmpr encodes a type I bone morphogenetic protein receptor that is essential for gastrulation during mouse embryogenesis. *Genes Dev* 1995, 9, (24), 3027-37.

[80] Mishina, Y.; Crombie, R.; Bradley, A.; Behringer, R. R., Multiple roles for activin-like kinase-2 signaling during mouse embryogenesis. *Dev Biol* 1999, 213, (2), 314-26.

[81] Winnier, G.; Blessing, M.; Labosky, P. A.; Hogan, B. L., Bone morphogenetic protein-4 is required for mesoderm formation and patterning in the mouse. *Genes Dev* 1995, 9, (17), 2105-16.

[82] Zhang, H.; Bradley, A., Mice deficient for BMP2 are nonviable and have defects in amnion/chorion and cardiac development. *Development* 1996, 122, (10), 2977-86.

[83] Mishina, Y.; Starbuck, M. W.; Gentile, M. A.; Fukuda, T.; Kasparcova, V.; Seedor, J. G.; Hanks, M. C.; Amling, M.; Pinero, G. J.; Harada, S.; Behringer, R. R., Bone morphogenetic protein type IA receptor signaling regulates postnatal osteoblast function and bone remodeling. *J Biol Chem* 2004, 279, (26), 27560-6.

[84] Kronenberg, H. M., Developmental regulation of the growth plate. *Nature* 2003, 423, (6937), 332-6.

[85] Nakashima, K.; de Crombrugghe, B., Transcriptional mechanisms in osteoblast differentiation and bone formation. *Trends Genet* 2003, 19, (8), 458-66.

[86] Mackie, E. J.; Ahmed, Y. A.; Tatarczuch, L.; Chen, K. S.; Mirams, M., Endochondral ossification: how cartilage is converted into bone in the developing skeleton. *Int J Biochem Cell Biol* 2008, 40, (1), 46-62.

[87] Rountree, R. B.; Schoor, M.; Chen, H.; Marks, M. E.; Harley, V.; Mishina, Y.; Kingsley, D. M., BMP receptor signaling is required for postnatal maintenance of articular cartilage. *PLoS Biol* 2004, 2, (11), e355.

[88] Yoon, B. S.; Ovchinnikov, D. A.; Yoshii, I.; Mishina, Y.; Behringer, R. R.; Lyons, K. M., Bmpr1a and Bmpr1b have overlapping functions and are essential for chondrogenesis in vivo. *Proc Natl Acad Sci U S A* 2005, 102, (14), 5062-7.

[89] Tsumaki, N.; Nakase, T.; Miyaji, T.; Kakiuchi, M.; Kimura, T.; Ochi, T.; Yoshikawa, H., Bone morphogenetic protein signals are required for cartilage formation and differently regulate joint development during skeletogenesis. *J Bone Miner Res* 2002, 17, (5), 898-906.

[90] Retting, K. N.; Song, B.; Yoon, B. S.; Lyons, K. M., BMP canonical Smad signaling through Smad1 and Smad5 is required for endochondral bone formation. *Development* 2009, 136, (7), 1093-104.

[91] Bandyopadhyay, A.; Tsuji, K.; Cox, K.; Harfe, B. D.; Rosen, V.; Tabin, C. J., Genetic Analysis of the Roles of BMP2, BMP4, and BMP7 in Limb Patterning and Skeletogenesis. *PLoS Genet* 2006, 2, (12), e216.

[92] Tsuji, K.; Bandyopadhyay, A.; Harfe, B. D.; Cox, K.; Kakar, S.; Gerstenfeld, L.; Einhorn, T.; Tabin, C. J.; Rosen, V., BMP2 activity, although dispensable for bone formation, is required for the initiation of fracture healing. *Nat Genet* 2006, 38, (12), 1424-9.

[93] Okamoto, M.; Murai, J.; Yoshikawa, H.; Tsumaki, N., Bone morphogenetic proteins in bone stimulate osteoclasts and osteoblasts during bone development. *J Bone Miner Res* 2006, 21, (7), 1022-33.

[94] Okamoto, M.; Murai, J.; Imai, Y.; Ikegami, D.; Kamiya, N.; Kato, S.; Mishina, Y.; Yoshikawa, H.; Tsumaki, N., Conditional deletion of Bmpr1a in differentiated osteoclasts increases osteoblastic bone formation, increasing volume of remodeling bone in mice. *J Bone Miner Res* 2011, 26, (10), 2511-22.

[95] Zimmerman, L. B.; De Jesus-Escobar, J. M.; Harland, R. M., The Spemann organizer signal noggin binds and inactivates bone morphogenetic protein 4. *Cell* 1996, 86, (4), 599-606.

[96] Logan, M.; Martin, J. F.; Nagy, A.; Lobe, C.; Olson, E. N.; Tabin, C. J., Expression of Cre Recombinase in the developing mouse limb bud driven by a Prxl enhancer. *Genesis* 2002, 33, (2), 77-80.

[97] Tsuji, K.; Cox, K.; Gamer, L.; Graf, D.; Economides, A.; Rosen, V., Conditional deletion of BMP7 from the limb skeleton does not affect bone formation or fracture repair. *J Orthop Res* 2010, 28, (3), 384-9.

[98] Kanczler, J. M.; Oreffo, R. O., Osteogenesis and angiogenesis: the potential for engineering bone. *Eur Cell Mater* 2008, 15, 100-14.

[99] Carano, R. A.; Filvaroff, E. H., Angiogenesis and bone repair. *Drug Discov Today* 2003, 8, (21), 980-9.

[100] Deckers, M. M.; van Bezooijen, R. L.; van der Horst, G.; Hoogendam, J.; van Der Bent, C.; Papapoulos, S. E.; Lowik, C. W., Bone morphogenetic proteins stimulate angiogenesis through osteoblast-derived vascular endothelial growth factor A. *Endocrinology* 2002, 143, (4), 1545-53.

[101] Cheng, H.; Jiang, W.; Phillips, F. M.; Haydon, R. C.; Peng, Y.; Zhou, L.; Luu, H. H.; An, N.; Breyer, B.; Vanichakarn, P.; Szatkowski, J. P.; Park, J. Y.; He, T. C., Osteogenic activity of the fourteen types of human bone morphogenetic proteins (BMPs). *J Bone Joint Surg Am* 2003, 85-A, (8), 1544-52.

[102] Phillips, F. M.; Bolt, P. M.; He, T. C.; Haydon, R. C., Gene therapy for spinal fusion. *Spine J* 2005, 5, (6 Suppl), 250S-258S.

[103] David, L.; Feige, J. J.; Bailly, S., Emerging role of bone morphogenetic proteins in angiogenesis. *Cytokine Growth Factor Rev* 2009, 20, (3), 203-12.

[104] Kan, L.; Hu, M.; Gomes, W. A.; Kessler, J. A., Transgenic mice overexpressing BMP4 develop a fibrodysplasia ossificans progressiva (FOP)-like phenotype. *Am J Pathol* 2004, 165, (4), 1107-15.

[105] Lounev, V. Y.; Ramachandran, R.; Wosczyna, M. N.; Yamamoto, M.; Maidment, A. D.; Shore, E. M.; Glaser, D. L.; Goldhamer, D. J.; Kaplan, F. S., Identification of progenitor cells that contribute to heterotopic skeletogenesis. *J Bone Joint Surg Am* 2009, 91, (3), 652-63.

[106] Yu, P. B.; Deng, D. Y.; Lai, C. S.; Hong, C. C.; Cuny, G. D.; Bouxsein, M. L.; Hong, D. W.; McManus, P. M.; Katagiri, T.; Sachidanandan, C.; Kamiya, N.; Fukuda, T.; Mishina, Y.; Peterson, R. T.; Bloch, K. D., BMP type I receptor inhibition reduces heterotopic [corrected] ossification. *Nat Med* 2008, 14, (12), 1363-9.

[107] Medici, D.; Shore, E. M.; Lounev, V. Y.; Kaplan, F. S.; Kalluri, R.; Olsen, B. R., Conversion of vascular endothelial cells into multipotent stem-like cells. *Nat Med* 2010, 16, (12), 1400-6.

[108] Chan, C. K.; Chen, C. C.; Luppen, C. A.; Kim, J. B.; DeBoer, A. T.; Wei, K.; Helms, J. A.; Kuo, C. J.; Kraft, D. L.; Weissman, I. L., Endochondral ossification is required for haematopoietic stem-cell niche formation. *Nature* 2009, 457, (7228), 490-4.

[109] Horowitz, M. C.; Lorenzo, J. A., The origins of osteoclasts. *Curr Opin Rheumatol* 2004, 16, (4), 464-8.

[110] Kaneko, H.; Arakawa, T.; Mano, H.; Kaneda, T.; Ogasawara, A.; Nakagawa, M.; Toyama, Y.; Yabe, Y.; Kumegawa, M.; Hakeda, Y., Direct stimulation of osteoclastic bone resorption by bone morphogenetic protein (BMP)-2 and expression of BMP receptors in mature osteoclasts. *Bone* 2000, 27, (4), 479-86.

[111] Simonet, W. S.; Lacey, D. L.; Dunstan, C. R.; Kelley, M.; Chang, M. S.; Luthy, R.; Nguyen, H. Q.; Wooden, S.; Bennett, L.; Boone, T.; Shimamoto, G.; DeRose, M.; Elliott, R.; Colombero, A.; Tan, H. L.; Trail, G.; Sullivan, J.; Davy, E.; Bucay, N.; Renshaw-Gegg, L.; Hughes, T. M.; Hill, D.; Pattison, W.; Campbell, P.; Sander, S.; Van, G.; Tarpley, J.; Derby, P.; Lee, R.; Boyle, W. J., Osteoprotegerin: a novel secreted protein involved in the regulation of bone density. *Cell* 1997, 89, (2), 309-19.

[112] Lacey, D. L.; Timms, E.; Tan, H. L.; Kelley, M. J.; Dunstan, C. R.; Burgess, T.; Elliott, R.; Colombero, A.; Elliott, G.; Scully, S.; Hsu, H.; Sullivan, J.; Hawkins, N.; Davy, E.; Capparelli, C.; Eli, A.; Qian, Y. X.; Kaufman, S.; Sarosi, I.; Shalhoub, V.; Senaldi, G.; Guo, J.; Delaney, J.; Boyle, W. J., Osteoprotegerin ligand is a cytokine that regulates osteoclast differentiation and activation. *Cell* 1998, 93, (2), 165-76.

[113] Abe, E.; Yamamoto, M.; Taguchi, Y.; Lecka-Czernik, B.; O'Brien, C. A.; Economides, A. N.; Stahl, N.; Jilka, R. L.; Manolagas, S. C., Essential requirement of BMPs-2/4 for both osteoblast and osteoclast formation in murine bone marrow cultures from adult mice: antagonism by noggin. *J Bone Miner Res* 2000, 15, (4), 663-73.

[114] Itoh, K.; Udagawa, N.; Katagiri, T.; Iemura, S.; Ueno, N.; Yasuda, H.; Higashio, K.; Quinn, J. M.; Gillespie, M. T.; Martin, T. J.; Suda, T.; Takahashi, N., Bone morphogenetic protein 2 stimulates osteoclast differentiation and survival supported by receptor activator of nuclear factor-kappaB ligand. *Endocrinology* 2001, 142, (8), 3656-62.

10

In Vivo Bacterial Morphogenetic Protein Interactions

René van der Ploeg and Tanneke den Blaauwen
University of Amsterdam, Swammerdam Institute for Life Sciences,
The Netherland

1. Introduction

1.1 Techniques to study protein-protein interactions in cell division

This chapter will discuss none-invasive techniques that are widely used to study protein-protein interactions. As an example, their application in exploring interactions between proteins involved in bacterial cell division will be evaluated. First, bacterial morphology and cell division of the rod-shaped bacterium *Escherichia coli* will be introduced. Next, three bacterial two-hybrid methods and three Förster resonance energy transfer detection methods that are frequently applied to detect interactions between proteins will be described and discussed in detail. The chapter concludes with a discussion about the application and results of the techniques when studying proteins involved in cell division.

2. Cell morphology

2.1 The bacterial cell wall

Bacteria have different shapes with the most common being spheres and rods. The morphology is determined by the cell wall that surrounds the cytoplasmic membrane of a bacterium. This cell wall is one large closed molecule called sacculus that keeps everything together. Breaches in the structure can be fatal causing rupture enforced by the high internal (turgor) osmotic pressure. Preserving the strength of the wall during growth and division is therefore vital. The exact architecture of the peptidoglycan network is under debate but its composition is identified (Dmitriev et al., 2005; Gan et al., 2008; Hayhurst et al., 2008; Vollmer & Seligman 2010a).

The entire structure is built from linear polymers cross-linked by short (stem) peptides. Each polymer is a repetition of a β-1,4-linked N-acetylglucosamine (GlcNAc) and N-acetylmuramic acid (MurNAc) unit, ending with a head group. The head group is different from the GlcNAc, MurNAc disaccharide unit in that the MurNAc subunit has a 1,6 intra-molecular ether-linkage from C-1 to C-6. The bridging stem peptides are synthesized as pentapeptides with an amino acid sequence of L-Ala, D-Glu, *meso*A2pm, D-Ala, D-Ala. Although, the sacculus is made of repetitions of disaccharide pentapeptides, small variations in the fine structure exist. The network is not stiff but has a remarkable flexibility and can shrink and expand with about 3 times its size (Koch & Woeste 1992;). This elastic property is functional important in a changing osmotic environment that evokes spontaneous but drastic increases and decreases in the turgor pressures. For growth and division the cell wall is continuously expanded and renewed. Proteins located in the cytoplasm, periplasm and membranes work together to make this possible. Penicillin-binding proteins (PBPs) are responsible for the synthesis and modification of the network. For the rod-shaped bacterium *E. coli* twelve different PBPs have been identified. Only a few produce clear phenotypes, most of them are not essential (Denome et al., 1999) indicating redundancy. For a broad overview of the penicillin-binding proteins from *E. coli* and several other Gram-negative and positive bacteria the reader is referred to (Sauvage et al., 2008). More information on the cell wall structure can be found in the reviews (Vollmer et al., 2008; Vollmer & Bertsche, 2008; Vollmer & Seligman, 2010).

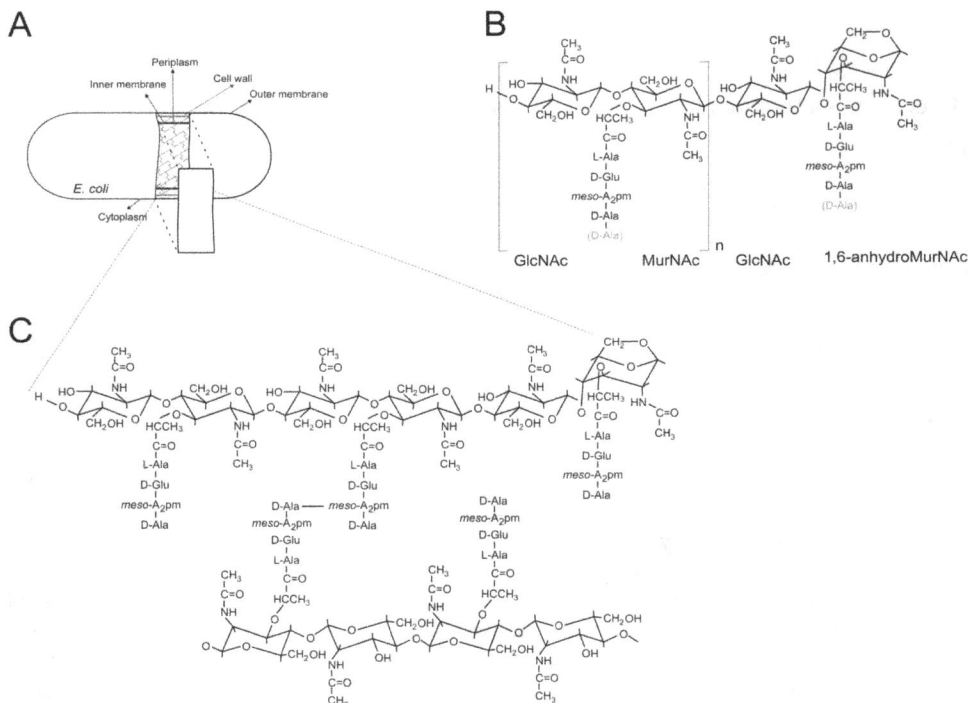

Fig. 1. A simplified presentation of the *E. coli* cell wall structure. A) The cell wall found in the periplasm of *E. coli* is build from cross-linked glycan strands. B) A glycan strand and its cross-linking in the network (C). The last D-Ala residue is enzymatically removed in the mature macromolecule and is therefore depicted in grey (B).

3. Cell division of *Escherichia coli*

3.1 The Divisome complex

Escherichia coli replicates *via* binary fission. Cells increase in length and split in two by constricting at the middle of the cell where new cell poles are synthesized. The newly created daughter cells have an identical shape in both diameter and length. An exact reproduction of the old cell poles at the constriction site is a necessity. The combined activities of hydrolases and penicillin-binding proteins make growth of the sacculus possible. The existing network is hydrolyzed at specific sites to incorporate new peptidoglycan material for elongation and cell pole synthesis.

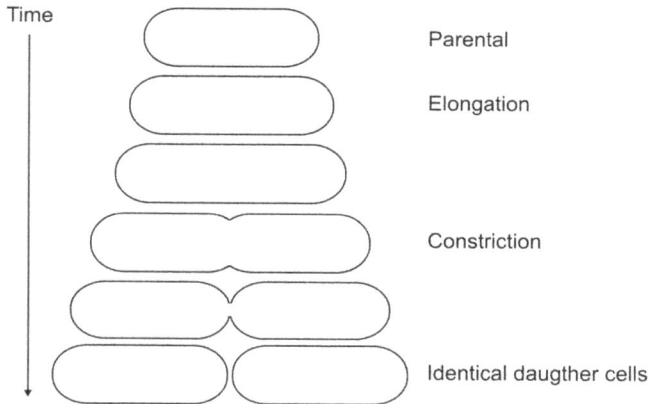

Fig. 2. *Escherichia coli* cell division. The cells elongate continuously, constriction starts in the middle of the cell producing two identical daughter cells with the same length, diameter and symmetrical poles.

The actual division process requires actions of many more different proteins at all cellular loci being cytoplasm, periplasm and inner as well as outer membrane. Before constriction starts the cell prepares itself by moving the necessary proteins to the constriction site. More than twenty proteins have been identified to be involved in the division process of which at least twelve are essential. Between the latter proteins many interactions are observed and it is therefore hypothesized that a big protein complex is formed referred to as the Divisome (Alexeeva et al., 2010; Di lallo et al., 2003; Karimova et al., 1998; Maggi et al., 2008).

The assembly of the Divisome takes place in two steps that are separated by a delay (Aarsman et al., 2005). FtsZ proteins polymerize and form a so-called Z-ring structure that marks the start of the assembly process. It localizes to the membrane by interactions with the bi-topic membrane protein ZipA and the membrane associated protein FtsA (Hale & de Boer, 1999). The ring functions as a scaffold for other proteins to attach. Several proteins have been identified that facilitate and support the Z-ring structure and its polymerization. Three of the identified proteins are called FtsZ associated proteins (Zap) A, B and C (Durand-Heredia et al., 2011; Hale et al., 2011). The first maturation step ends when FtsE and FtsX arrive. What transpires during the delay is not quite clear, possibly fine-tuning and preparation for the next group of protein that will arrive. In the second or late step a new flow of proteins move to the mid-cell position. These are the membrane proteins FtsK, FtsQ, FtsB, FtsL, FtsW, FtsI, FtsN and periplasmic protein AmiC (Aarsman et al., 2005). Four penicillin-binding proteins localize to the constriction site namely PBP1B, PBP2, PBP3 also known as FtsI and PBP5 (Bertsche et al., 2006; Den Blaauwen et al., 2003; Potluri et al., 2010; Weiss et al., 1997).

Although PBP1B and PBP3 interact, PBP1B does not play an essential role in cell constriction because it can functionally replace by PBP1A without a change in cell morphology (Bertsche et al., 2006). The function and physical presence of PBP3 is necessary to fully complete cell constriction. PBP2 on the other hand is involved in the constriction process but its function is not essential. Inhibition of PBP2's activity results an increase of the diameter of the new cell poles.

Undoubtly more proteins will be discovered that localize at mid-cell and that will have a role in fine-tuning and regulation of the constriction process. To understand the cell division process at every step, detailed information on where, when and for how long do all these proteins interact is a requisite. From localization studies it has become clear that the division proteins arrive at mid-cell in an interdependent fashion. For instance PBP1B's presence at mid-cell is completely dependent on PBP3 (Bertsche et al., 2006). An alternative view could be that some proteins only move to mid-cell as pre-complexes (Fraipont et al., 2011).

Others are only transiently there like GFP-PBP2 (Den Blaauwen et al., 2003). Because of the dynamic nature of the Divisome, biochemical techniques are not able to give a complete picture of what is going on at every step in the process. Non-invasive *in vivo* techniques are of great additional value in this field of research.

Variations of bacterial two-hybrid and FRET techniques will be described that have been used to get a better understanding of *E. coli*'s cell division interactome. After discussing the techniques results will be presented.

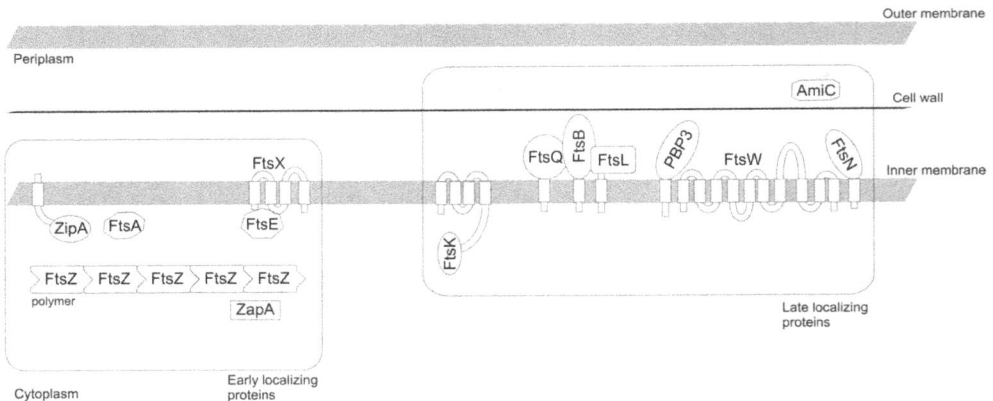

Fig. 3. Some of the cell division proteins identified in *E. coli*. On the left the early localizing proteins are found with on the right side late localizing proteins.

4. Bacterial two-hybrid

4.1 Introduction

Studies on protein-protein interactions are essential to fully specify the interactions within a cellular protein network. In the past biochemical techniques like co-immunoprecipitation, protein cross-linking and (affinity) chromatography were applied to find interacting protein partners. Although each technique has its specific advantage, often they are laborious and apply harsh wash steps to separate unspecific interactions. Moreover, an interaction in its genuine environment, a living cell is not observed. In a living cell the interaction can be weak or transient and might therefore not be observed with biochemical techniques. In 1989 Fields and Song published a new method to determine protein interactions *via* a genetic (system) method that was called yeast two-hybrid (Fields & Song, 1989). This yeast two-hybrid technique enabled the researcher to study protein interactions by screening for yeast colonies that were only able to grow on galactose, as the sole carbon source or produce a blue colony when the two proteins of interest called prey and bait interacted. In the yeast two-hybrid method the prey and bait were fused either to an N-terminal or C-terminal domain of the GAL4 protein of yeast *Saccharomyces cerevisiae*. Upon interaction of prey and bait in the nucleus, the proximity of the N- and C-terminal domain of GAL4 would be sufficiently close to form the complete GAL4 molecule. The hybrid GAL4 protein subsequently activated transcription of the upstream activating sequences for galactose genes (UASg) and the integrated *lacZ* gene, which codes for the ß-galactosidase reporter protein. The ß-galactosidase protein cuts the galactose sugar bond with 5-bromo-4-chloro-3-hydroxyindole. The indole molecule is subsequently oxidized and turns into insoluble blue product, Fig. 7A.

A year later a different genetic method was published using *E. coli* as a host. Either the ß-galactosidase protein or a bacteriophage immunity assay could be chosen as a screen. In the last case, interaction of the pray and bait leads to a complete λ repressor protein, which can inhibit λ bacteriophage production and consequently no plague is formed on the plate. The dimerization of a leucine zipper was used as a proof of principle (Hu et al., 1990). The immunity assay has been applied in many publications to investigate protein-protein interactions beside the yeast two-hybrid technique (Blackwood & Eisenman, 1995; Di lallo et al., 1999a; Longo et al., 1995; Wolfe et al., 1999). Various methods to determine protein or protein-DNA interactions have been published that prevent or allow transcription (Ladant & Karimova, 2000; Vidal & Legrain, 1999). Particularly, the switch to use *E. coli* instead of *S. cerevisiae* provided a clear advantage as *E. coli* is easier to grow, to transform and to use to make very large libraries (Joung et al., 2000). Another great improvement when working with *E. coli* is that the protein-protein interactions do not need to take place in the cell nucleus to activate transcription, which is a drawback of the yeast two-hybrid system. On the other hand, due to the difference in genetic background, yeast could still be the preferred method for proteins that need eukaryotic folding machineries that are absent in *E. coli*.

Almost a decade later two similar two-hybrid methods were published one for eukaryotes and the other for prokaryotes (Karimova et al., 1998; Rossi et al., 1997), with the latter being possibly the first generally used bacterial two-hybrid method today. What both systems have in common is that they reconstruct a catalytic site that is build from two different domains. The two systems differ in that one activates transcription of the ß-galactosidase enzym indirectly *via* cAMP whereas the other restores the catalytic site of the ß-galactosidase enzyme upon interaction of prey and bait. The prokaryotic system (*E. coli*) converts ATP to cAMP, the cAMP receptor protein (CAP) binds substrate and activates transcription of the ß-galactosidase gene, *lacZ*.

4.2 The bacterial two-hybrid technique

Today many different variants of the two-hybrid systems have been developed to study protein-protein interactions but also to investigate protein-DNA interactions. These variants are collectively called 'n'-hybrid systems. The one-hybrid system is used to study single protein-DNA interactions, the two-hybrid system to check protein-protein interactions and lastly a three-hybrid system where an RNA molecule brings two proteins together. Initially the systems were developed to screen and identify specific interactions but of course the reversal is also possible, to investigate mutations that result in a loss of interaction. This introduced the terms forward and reverse two-hybrid to discriminate screens that identify an interaction or a disruption (Vidal & Legrain, 1999). The work discussed here will focus on protein-protein interactions only. In yeast the first and most used two-hybrid method is based on transcriptional activation mediated through direct binding of the prey and bait hybrid complex to the reporter gene. Similar methods have been published for the bacterial two-hybrid systems. Variants have been developed in time for specific purposes or just to improve the system in general (Longo et al., 1995; Strauch & Georgiou, 2007). In this paragraph, three methods will be presented, two are transcriptional based and one reconstitutes the catalytic site of the reporter. The methods will be explained in more detail followed by a discussion on their pro's and contra's.

4.2.1 Bacterial two-hybrid *via* transcriptional repression

Already in 1990 Hu *et al* used the λ repressor protein to show which residues of the leucine zipper are important for an interaction (Hu et al., 1990). The λ repressor is an alpha helical protein that binds DNA upon homodimerization. The amino-terminus of each monomer contains a conserved helix-turn-helix motif that is present in many proteins involved in gene regulation. Dimerization is facilitated by the C-terminal part of the protein. As a dimer the two N-terminal DNA binding domains are sufficiently close together to allow a cooperative interaction of both termini with the nucleotide binding sequence. However, no interaction is observed with the nucleotide binding sequence when the protein is present as a monomer. Castagnoli *et al* fused the Rop protein to the λ repressor. Dimerization of the Rop protein in the cells rendered them immune to λ infections (Castagnoli et al., 1994). In 1999 and 2001 a variant of this two-hybrid method to study protein-protein interactions was published (Di lallo et al., 1999a; Di lallo et al., 2001) that combined two different phage repressor proteins, which originated from phage 434 and P22 (Di lallo et al., 2001), Fig. 4. For an increased repression of the promoter, the two operator sequences were built from four alternating half-sites with a 434 followed by P22. The λ repressor genes were genetically fused to the prey or bait genes of interest on an IPTG inducible expression plasmid. Only when the prey and bait proteins formed a hybrid protein, the N-termini of P22 and 434 were sufficiently close to merge their forces. Together the helix-turn-helix motifs bind their corresponding DNA sequences and repress the transcription of the reporter protein ß-galactosidase.

The constitutively active promoter region responsible for the transcription of the *lac*Z gene was a chimeric 434-P22 regulatory region, integrated *via* cross over in the chromosomal copy of the *glpT* gene (Di lallo et al., 2001). The integrating plasmid, called pAPA contained the *glpT* gene and crossed over after removal of the origin of replication (Di lallo et al., 1999b). Other bacterial strains can easily be converted to hosts for two-hybrid research by using this specific pAPA plasmid (Di lallo et al., 1999b).

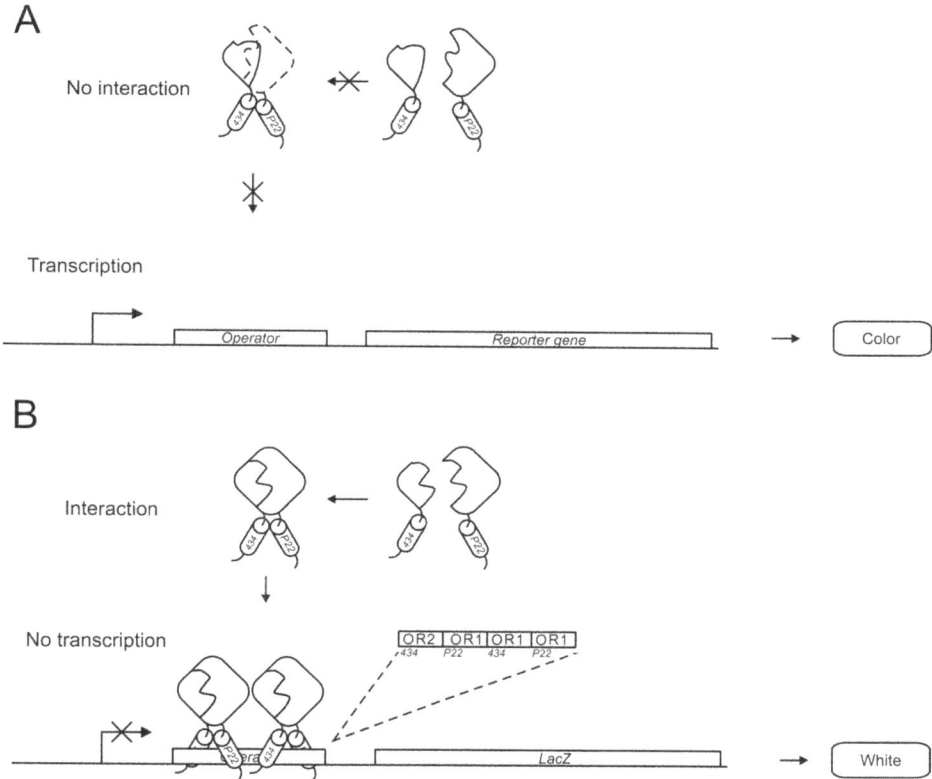

Fig. 4. Bacterial two-hybrid *via* transcriptional repression. A constitutive active promoter drives expression of the reporter protein ß-galactosidase. A) If the prey and bait proteins do not interact transcription proceeds. B) Upon interaction a complete repressor protein is formed that inhibits transcription of the reporter.

4.2.2 Bacterial two-hybrid *via* transcriptional activation through the cAMP signaling pathway

The second method described is based on a reconstitution of the E. coli cAMP signal transduction pathway during the interaction of the prey and bait proteins (Karimova et al., 1998). Daniel Ladant showed that the *Bordetella pertussis* calmodulin-dependent adenylate cyclase (*cya*) has two interaction sites that bind calmodulin (Ladant, 1988). The adenylate cyclase could be cleaved in two separate domains. The two domains are called T18 and T25 fragments. Mixed in solution these fragments do not induce the low basal calmodulin-independent acitivity that is observed for the full protein. However, the normal activity could be restored when calmodulin was administered. Calmodulin brings both domains close together and allows a reestablishment of the catalytic site. In the absence of calmodulin, the catalytic activity could also be restored when the T18 and T25 fragments were fused to two interacting proteins (Karimova et al., 1998). When these fusion proteins reconstitute adenylate cyclase activity in an E. coli *cya* deficient strain, the cAMP-signaling pathway is restored, Fig. 5. Adenylate cyclase hydrolyses ATP to cAMP, which is then bound by the cAMP receptor protein also called the catabolite activator protein (CAP). The formed cAMP/CAP complex subsequently binds to the CAP DNA binding site *via* a helix-turn-helix motif, which induces a change in DNA structure that allows binding of the polymerase to start transcription. The promoter is used to activate transcription of a reporter, like the naturally occurring bacterial genes such as *lacZ* or *mal*. Alternatively, an antibiotic resistance gene could be selected as reporter.

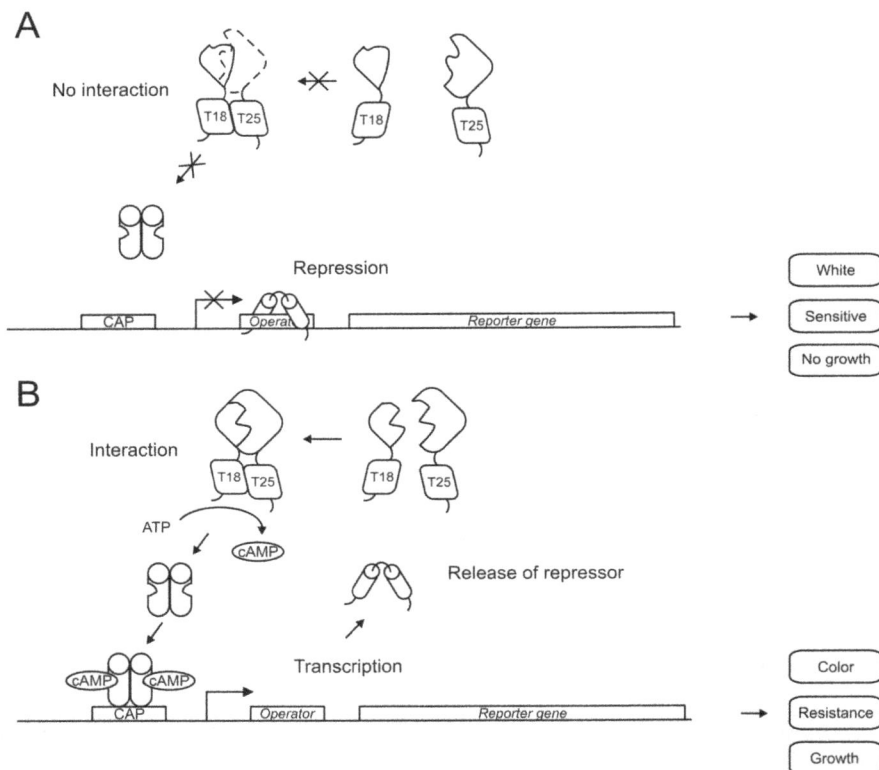

Fig. 5. Bacterial two-hybrid *via* a cAMP signaling pathway. If cAMP is present in the cell it triggers a conformational change when bound to CAP protein. In the bound state the CAP protein binds its DNA binding site, stimulates polymerase binding and release of repressor protein to promote transcription (RNA polymerase is not shown). An *E. coli* strain deficient in adenylate cyclase activity is used to study protein-protein interactions with this method. A) The prey and bait do not interact and no cAMP is produced. No reporter is produced thus no conversion of for example X-gal to indigo blue, no growth on plates when maltose is the sole carbon source and cells are sensitive to the antibiotics. B) When prey and bait proteins interact the two domains of the adenylate cyclase T18 and T25 are able to dimerize and form a functional catalytic site. In the cell cAMP is produced from ATP, which triggers transcription of a reporter gene that can be *lacZ*, the *mal* genes or the gene of antibiotic resistance marker.

4.2.3 Bacterial two-hybrid through reconstitution of a ß-galactosidase catalytic site

Bacterial two-hybrid methods often have more than one option to screen for an interaction. Frequently, a reporter protein is used to detect an interaction between the proteins of interest. With the ß-galactosidase protein possibly the most popular reporter protein today, applied in both eukaryotes and prokaryotes (Borloo et al., 2007; Wehrman et al., 2002). The method described here is unique not because it uses ß-galactosidase as a reporter protein. It is different because it lacks an 'intermediate' transcription initiation step to produce the reporter protein when the prey and bait interact (Borloo et al., 2007). In the current method the reporter is present in the cell as two separate domains. By reconstituting the active site of the ß-galactosidase protein, conversion of the reporter product can be measured directly. Possible problems in transcription of the reporter protein are thereby circumvented. In the end of the 1960's it was reported that the ß-galactosidase protein could be cleaved in two parts, an N-terminal domain and a C-terminal domain that are called α and ω, respectively (Ullmann et al., 1967). In the crystal structure the protein formed tetramers and showed how the domains interacted. An active catalytic site could be reestablished *in vitro* from the α and ω domains (Jacobson et al., 1994; Ullmann et al., 1967). A reconstitution of a functional ß-galactosidase protein was first used as a tool to study interactions in mammalian cells. Later it was successfully applied to investigate protein interactions in the cytoplasm and periplasm of prokaryotes (Borloo et al., 2007).

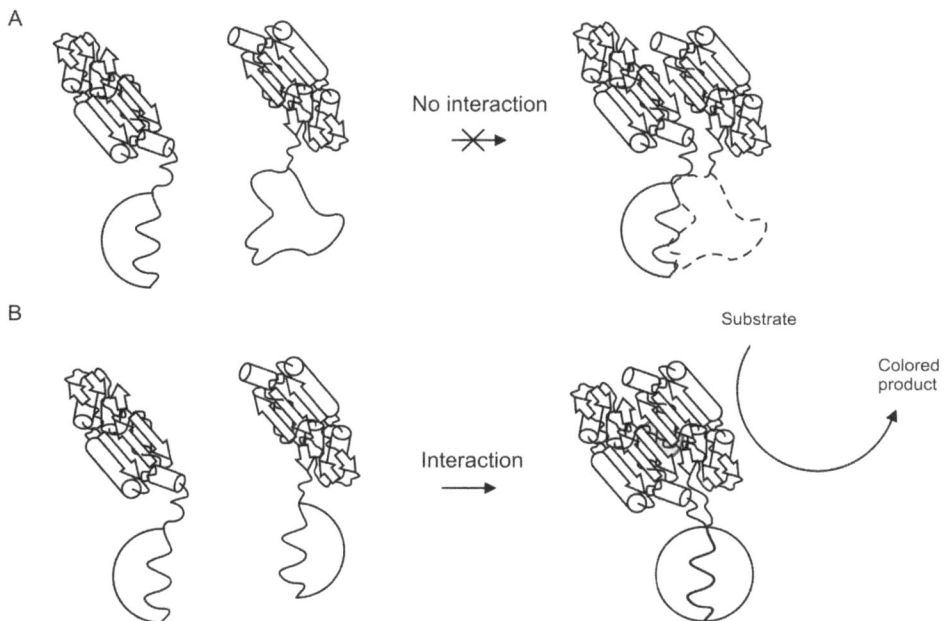

Fig. 6. Bacterial two-hybrid *via* reconstitution of a catalytic site. B) A fully functional ß-galactosidase reporter protein can be formed from two domains when the prey and bait proteins interact. An advantage is that the reporter protein is functionally active in many compartments of the cell including periplasm.

4.2.4 Pro's and contra's of the bacterial two-hybrid method

The bacterial two-hybrid method like the yeast two-hybrid method is a great accessory tool to determine protein-protein interactions. The great advantage of the bacterial two-hybrid methods is the ease by which an interaction or even better very large libraries with up to 10^8 clones can be screened (Joung et al., 2000). Upon reconstitution of a fully functional reporter protein a signal is produced that is amplified due to enzymatic activity of the reporter. Common examples are conversion of X-gal to indigo blue or ONPG (ortho-nitrophenyl-ß-galacotside) to galactose and the yellow colored ortho-nitrophenol. Plate screening assays mostly use X-gal as a substrate. To compare the interactions between different prey and bait proteins, the amount of active ß-galactosidase is measured which is proportional to the rate of ONPG hydrolysis (Miller, 1972). The more functional reporters are formed the stronger the signal. The background signal that has to be determined is the signal produced by non-interacting prey and bait proteins.

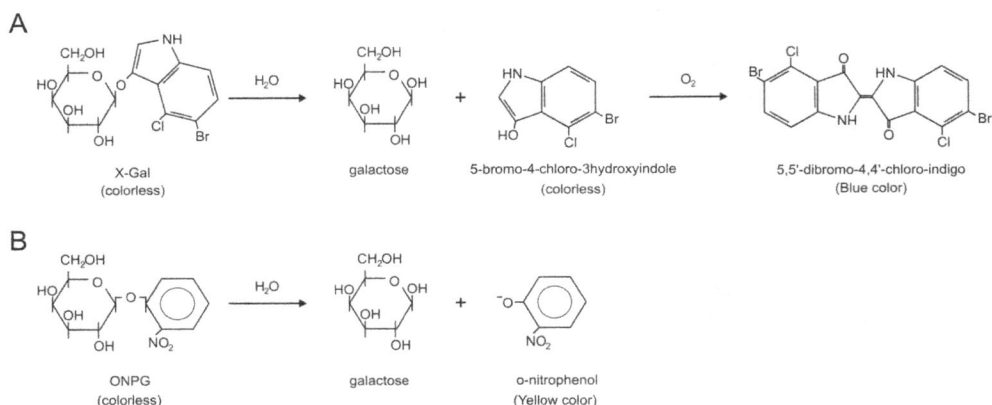

Fig. 7. Two ß-galactosidase reporter assays. A) X-gal plate assay and B) the ONPG conversion rate assay to determine the amount of active ß-galactosidase that represents the number of prey and bait interactions.

However, the results that are obtained need to be interpreted with caution because of potential false negatives and positives in the results. It is difficult to foresee all possible situations and explain them due to the fact that usually not everything is known about the proteins of interest. These false results originate from aspects in the method in combination with characteristics of the protein under investigation. Therefore, it is important to select the most suitable two-hybrid method. Depending on the proteins some methods are more fitting than others. For example, the method of transcription repression is less suitable to study the interaction of membrane proteins because the interacting proteins have to bind a DNA sequence on the chromosome that might not be easily accessible close to the membrane.

Artifacts that create the false positive and negative results can be grouped in two classes related to protein expression and folding and structure. The three methods described, use two low copy number plasmids with a p15A and ColE1 origins of replication, Table 1. The expression systems are IPTG inducible but differ in promoter strength. Nevertheless, the prey and bait fusion proteins are in many cases expressed at levels that are significantly higher than their endogenous levels. Moreover, these systems frequently use wild-type strains to express the fusion protein thus all the fusion proteins produced are extra copies in the cell. This also creates a situation wherein the fusion proteins are competing with the endogenous proteins for an interaction partner. A non-cleavable fusion protein is obligatory to prevent competition between (reporter) domain-less fusion proteins and fusion proteins for a partner. Transient interactions that need the presence of an endogenous protein will therefore be difficult to detect, a positive circumstance when measuring direct interactions.

	Method 1	Method 2	Method 3
Reference	Di lallo et al., 2001	Karimova et al., 1998	Borloo et al., 2007
Method	Repressor	Activator	Catalysis
Number of plasmids	two	two	two
Origin of replication	p15A + ColE1	p15A + ColE1	p15A + ColE1
Selection markers	Km, Ap	Cm, Ap	Cm, Cb
Strain	R721	Δcya	independent
Expression	pLac	pUV5	ptac
IPTG	100 uM	500 uM	20 mM
Induction time	90 minutes	30 hours	-
Background signal	2500 Miller units	130 ß-gal U/ mg	ß-gal activity ± 1.5 nmol /(min*mg)
Reporters	ß-gal	ß-gal, cAMP, Maltose, Lactose	ß-gal

* As presented in the cited publication. Induction levels and incubation times may be altered by the users of the particular method.
Notes: All the promoters use isopropyl ß-D-thiogalactoside (IPTG) for induction. Selection markers Cm; chloramphenicol, Ap; ampicillin, Km; kanamycin, Cb; carbenicillin. Reporters ß-gal; ß-galactosidase, cAMP; measure the cAMP concentration in the cell, Cm; resistance against chloramphenicol, *lac* and *mal*; for induction of *lac* or *mal* genes.

Table 1. Comparing of the methods*

Ideally the fusion protein is able to complement the endogenous protein but this is not always the case or even tested. Some proteins have more than one transiently interacting partner due to post-translation modifications or because they need a chaperone protein for folding. These so-called secondary third party interactions can be disrupted and as consequently lead to a false negative results. Large bulky reporter domains connected to a prey or bait protein can alter their natural movement, functioning and interaction with other proteins. The opposite result occurs when a third party protein brings the prey and bait together causing a false positive result. Therefore it is of vital importance to determine whether the prey and bait fusions are stable, functional, not harmful and do not create other unwanted phenotypes or aggregates.

The structure and the interacting sites of the prey and bait determine if an interaction can be detected with a two-hybrid method. To circumvent problems of steric hindrance the length of linker can be adjusted to introduce more freedom between the protein of interest and its reporter domain. A typical linker is for instance (Gly$_4$Ser$_3$). Linkers are constructed from amino acids with small or short side-chains, preferably without charge, which gives flexibility and freedom to move and rotate. It is noteworthy that this steric hindrance is one of the factors that make it impossible to compare the strength of the interactions between different prey and bait proteins. It needs to be kept in mind that the influence of dimerization between reporter domains on the interaction between prey and bait is unknown and can attribute to artificial results. Some interactions are transient and if the interaction of the reporter domains is strong it results in artificial high positive signal. The interaction between the adenylate cyclase domains after reconstituted does not persist when the prey and bait interaction is lost (Dautin et al., 2000). Finally, when a reporter product is formed *via* a catalytic reaction it is important to provide sufficient substrate that is readily accessible.

In conclusion, the bacterial two-hybrid system is an easy and quick method to determine whether proteins interact. The strength of the bacterial two-hybrid is most likely its simplicity. The interaction is investigated in its natural cellular environment and the amount of reporter product is a measure for how efficient a full functional reporter can be reconstituted. Furthermore, the method provides a limited number of choices. The reporter domains used are fixed, the linker length is often unaltered but can be increased. Frequently the bacterial two-hybrid methods employ the ß-galactosidase enzyme to determine the degree of interaction by looking at the conversion of substrate into colored blue or yellow product. Therefore, being more or less the standard reporter method for the bacterial two-hybrids. However, alternative reporters are sometimes offered as with the Karimova method. Additional screening options are growth on MacConkey plates with maltose or lactose as the only carbon source or antibiotic selection. Alternative reporters can be introduced but as a consequence requires some cloning. Complete independence of strain and cellular location is possible when using the Borloo bacterial two-hybrid method. The Karimova method is limited to a *cya* deficient strain. And the Di lallo method has complete freedom as long as the pAPA plasmid that contains the region for constitutive expression of the *lacZ* gene is integrated in its chromosome. It is a great advantage to being able to screen a library almost instantly using this technique. However, if little is known about the proteins being studied the results have to interpret with caution. False results can be produced because of steric hindrance, inappropriate linker length or due to third party interactions. Moreover, the expression level of the fusion proteins should be carefully chosen to obtain physiological relevant results. High overproduction conditions will eventually always produce a signal.

5. Förster Resonance Energy Transfer (FRET)

5.1 Introduction

5.1.1 Using fluorescent proteins to study movement, localization and interactions *in situ*

The first publication on green fluorescent protein (GFP) in the molecular biology had a great impact (Chalfie et al., 1994; Prasher et al., 1992). Characterization of chromophore was even performed before the potential molecular biological application in cells was recognized (Cody et al., 1993; Perozzo et al., 1988; Shimomura, 1979). The protein was functional in prokaryotic and eukaryotic cells enabling researchers to monitor protein expression, localization and interactions in living cells (Cubitt et al., 1995; Kain et al., 1995). With the publication of the crystal structure of GFP information about the chromophore environment could be used to perform more educated site-directed mutagenesis studies (Ormo et al., 1996). Improved variants and different colors were created. Additional fluorescent proteins were derived from other marine species. Nowadays, many different fluorescent proteins have been published that almost range the complete visible spectrum. Many proteins are engineered to improve their application in microbiological research. Authentic colors were changed, their tendency to oligomerize was reduced, folding and maturation times were shortened, bleach resistance and brightness increases were reported.

The development of the fluorescent proteins and improvements in the field of light microscopy allowed more quantitative and accurate localization data to be obtained. Confocal microscopy made it possible to determine the co-localization of fluorescent proteins. Unfortunately, the diffraction limit makes it impossible to very precisely determine if proteins are sufficiently close to interact. Proteins with a distance less than 250 nm were observed as one spot. Switchable fluorescent proteins are of great interest because they make it possible to go beyond this resolution barrier using super-resolution microscopy (Fu et al., 2010). By quickly turning the chromophores on and off independently, the exact mid-point of the light source can mathematically be determined from the Gaussian distribution. The application in localization microscopy introduced the term PALM for photoactivation localization microscopy. Which improved the precision up to 10 nm. Although PALM has a great potential it is an advanced microscopic technique that requires expertise. For measuring fast dynamic processes PALM is not the designated technique, at this moment. An alternative method to measure protein-protein interactions is FCCS standing for fluorescence cross-correlation spectroscopy. A requirement of the technique is that particles move through the detection volume with a minimal speed. Membrane proteins move too slowly. Therefore, interactions between these proteins cannot be determined with FCCS. Förster resonance energy transfer is a well-known applied technique to investigate interactions between proteins; furthermore FRET is not restricted per se to microscopy. For a more complete review describing the history of fluorescent proteins and their biological application within living cells the reader is referred to (Chudakov et al., 2010). This paragraph will focus on application of fluorescent proteins in studying protein-protein interactions using Förster resonance energy transfer. Three FRET techniques will be discussed, two based on an increase in acceptor fluorescence as determined by filter-based method and a spectral-based unmixing method. The third method is based on the measurement of a decrease in the donor fluorescence lifetime.

5.2 FRET methods and applications

5.2.1 Methods to study protein interactions with fluorescent proteins

In the 1940's the German physical chemist Theodor Förster published several papers describing a process were energy was transferred between two separated molecules with the same dipole-dipole moment (Förster, 1946; Förster, 1948; Förster et al., 1993). Energy was transferred *via* resonance <u>only</u> when the molecules were in very close proximity. A molecule absorbs energy and gets into an excited state. The energy can be transferred to another molecule, which depends on several factors with distance the most critical one (Lakowicz,

2006). The process was named after its discoverer Förster's resonance energy transfer (FRET). The energy transfer is not confined to chromophores therefore the name fluorescence resonance energy transfer is incomplete. When working with chromophores it is possible to detect this process by measuring changes in light intensity or in lifetime as will be discussed in the following sections. The chromophore that transfers its energy is the donor and the receiver is the acceptor. See for a description of the basic principles of FRET (Lakowicz, 2006).

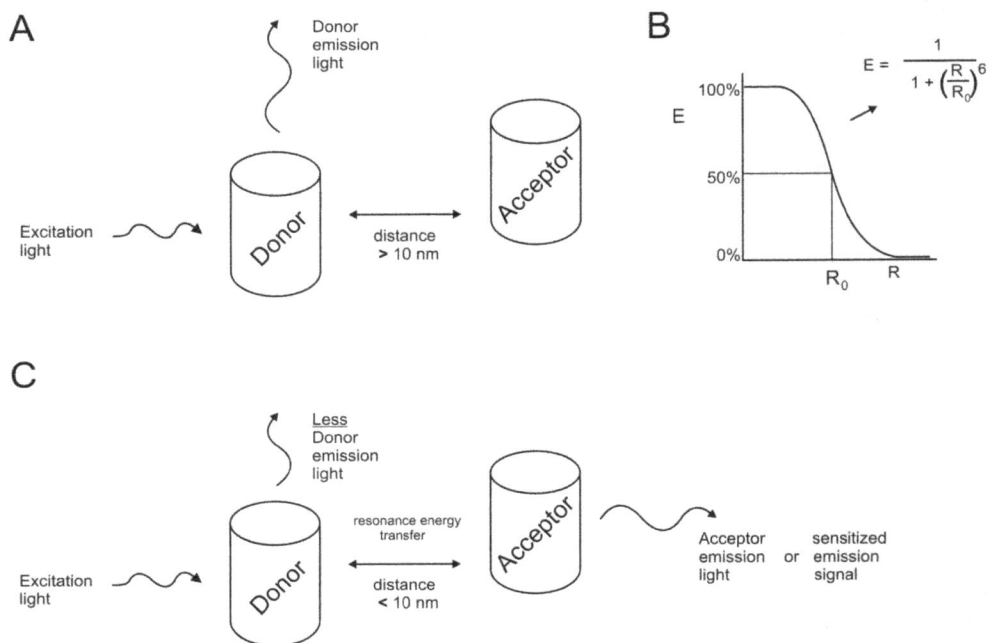

Fig. 8. Förster resonance energy transfer between donor and acceptor chromophores. A) No FRET when the chromophores are separated more than 10 nm. B) The energy transfer increases rapidly with decreasing distance. C) When FRET occurs the donor light intensity decreases and acceptor emission can be detected. R = distance between the chromophores, R_0 = Förster radius and with E = the FRET efficiency.

For localization and possibly interaction studies proteins in fixed cells can be labeled with chemical chromophores either directly or indirectly using anti-bodies conjugated to chromophores. Because this does not provide information on protein dynamics fluorescent proteins are a welcome alternative that could overcome this limitation. Without any harmful treatment the localization and interactions of proteins can be monitored. Only oxygen is essential in the cell to complete the formation of the chromophore within the protecting ß-barrel. Several different methods have been developed to determine FRET. The most common methods are based on increases or decreases in fluorescence intensity for the acceptor or donor, respectively. In the first presented method FRET is determined by quantifying the increase in acceptor signal. A more robust method is determining FRET by measuring at a decrease in fluorescence lifetime and this will be described in section 5.2.5.

5.2.2 Donor and acceptor fluorescent protein pairs

The spectral properties of a fluorescent protein are determined by the structure and conjugated pi-system of the chromophore, influenced by the surrounding environment. Fluorescent proteins have the advantage that the chromophore is well contained in a shell of ß-sheets that prevent interactions with quenching effects from molecules present in the solvent. Although extreme pH conditions still affect the protonated state of the chromophore and neighboring side chains in the ß-barrel, large changes in its properties are not expected under physiological conditions. The genes of fluorescent proteins have been cloned and proteins have been purified to obtain the exact data on their excitation and emission spectra. These data sets make it possible to estimate the application potential of each fluorescent protein, a prerequisite for setting up FRET experiments. The important parameters are the extinction coefficient, quantum yield, pKa, lifetime and bleach resistance. The extinction coefficient and quantum yield are a measure of how bright the fluorescence of the protein is. The fluorescence lifetime is the average time a chromophore is in the excited state before falling back to the ground state while emitting a photon. Each chromophore has a different sensitivity to light, known as bleach resistance. A suitable fluorescent protein combination is essential for good FRET measurements and can differ between FRET methods. For FRET to take place the so-called donor and acceptor fluorescent proteins need to overlap in excitation and emission spectra. Specifically, the higher the degree in overlap between donor emission and acceptor excitation spectra the better the energy transfer. An accurate estimation of the efficiency in energy transfer between donor and acceptor is obtained by calculating the Förster radius. This equation calculates the distance at which the donor transfers its energy with 50% efficiency to the acceptor. The larger the distance the more efficient the energy transfer is. In this calculation factors like the extinction coefficient and quantum yield are included. The angle of the fluorescent proteins, more specifically that of the chromophores determines the efficiency of energy transfer. Because the orientation is difficult to predict, for the calculation of the Förster radius a standard value of 2/3 is taken, assuming random orientations. FRET can be determined by several methods either by looking at the donor or acceptor. From the donor point of view FRET can be measured as a decrease in lifetime or in fluorescence intensity. Sensitized emission is directly measured by looking at the increase in acceptor intensity. More in-depth information about the two proposed FRET methods will be described in the next sections.

5.2.3 Protein dynamics measured with filter-based FRET

From an instrumental point of view, a filter-based FRET system is technically the easiest way to determine FRET. Filters block unwanted light thereby controlling propagation of specific excitation and emission wavelengths. An increase in acceptor signal or if possible a decrease in donor intensity is a measure for the amount of sensitized emission. The exposure time should be kept to a minimum to prevent bleaching, which would produce an artificial decrease of the donor fluorescence, mimicking FRET. However, bleach corrections are absolutely essential for calculating exact FRET percentages. A common FRET pair used in experiments is the cyan and yellow fluorescent proteins as donor and acceptor, respectively. Both proteins exist in rapidly folding and quickly mature versions. The proteins have a large spectral overlap. A large overlap in donor emission and acceptor excitation spectra raises the probability of energy transfer and indirectly the Förster radius

(4.72 nm for ECFP combined with EYFP). The disadvantage is that calculations are necessary to correct for bleed through light and direct excitation of the acceptor. Many different variants of these popular fluorescent proteins have been constructed to improve their applicability. The short folding and maturation times of these fluorescent proteins allow the measurement of FRET in living cells. A successful application for these proteins was in studying protein dynamics during chemotaxis in *E. coli* (Sourjik & Berg, 2004; Sourjik et al., 2007). The donor and acceptor proteins were fused to the chemotaxis proteins CheZ and CheY, respectively. It should be noted that an interaction between the CheZ and CheY was proven already but now FRET was used to measure the interaction dynamics of these proteins. Fluorescent proteins were fused to either N- or C-terminus of the chemotaxis proteins to test which combination would give the highest FRET signal. In accordance with the crystal structure the highest signal was detected when fusions were made to the C-terminal end bringing the fluorescent proteins in the closest proximity of each other (Volz & Matsumura, 1991; Zhao et al., 2002).

Excitation & Emission spectra

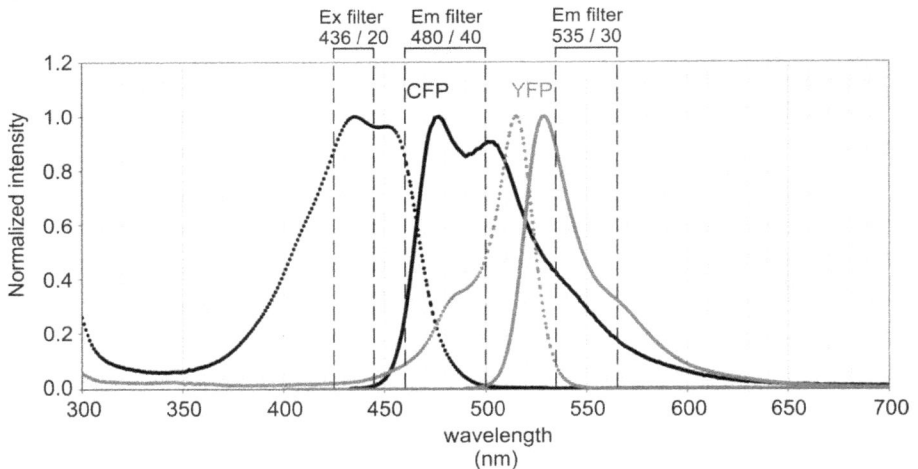

Fig. 9. CFP and YFP excitation and emission spectra. Excitation (dotted line) and emission (solid line) spectra of a CFP and YFP variant including the filters. The normalized spectra shown here have a large overlap. There is overlap in the excitation spectra but also in the emission spectra. Note that CFP covers almost the entire YFP spectrum. The bandpass filters are indicated.

The movement of *E. coli* during chemotaxis is regulated as follows. Motor proteins drive flagellas in counter clockwise rotation by default. Changes in chemoeffectors in the external environment trigger clockwise rotation of the flagella causing cells to tumble. When they resume the counter clockwise rotation they continue to swim in a new direction to find a more suitable habitat. The chemotaxis pathway of *E. coli* is well characterized; research of the last 40 years resulted in a comprehensive understanding of its mechanism (Vladimirov & Sourjik, 2009). Here a short description is given to understand the role and interaction of CheY and CheZ in chemotaxis and to be able to fully comprehend and appreciate the information FRET can give. Chemoreceptors bind repellents in the environment and promote auto-phosphorylation of CheA a histidine kinase that is bound to the chemoreceptor. The CheA phosphate group is transferred to the cytoplasmically located CheY protein that in its phosphorylated state attaches to the flagella motors to stimulate clockwise rotation. A sufficient pool of CheY is necessary for rapid motor response to changes in the chemoeffector concentrations. Therefore, CheZ dephosphorylates CheY to restore its cellular concentration. An encounter with attractants lowers CheA kinase activity and fewer CheY proteins are phosphorylated. Consequently, the number of CheY-CheZ interactions will decrease which correlates with a drop in sensitized emission signal. The FRET signal is displayed as ratio of YFP to CFP fluorescence intensity. When FRET occurs the CFP (donor) signal will decrease and the YFP (acceptor) signal will increase. By monitoring the ratios of donor and acceptor fluorescence intensity in time in the absence of effectors corrections can be included for fluctuations in the intensity of the excitation light, number of cells observed and movement of the cover slip.

A welcome advantage of this chemotaxis FRET system is the ability to regulate the level of interaction between the proteins by just adding attractant or repellents to the growth medium. Making it possible to more or less switch the interaction on and off.

5.2.4 Increased sensitivity with spectral-based FRET

The second method described is the intensity-based spectral FRET method where an increase in acceptor signal is a measure for the amount of FRET. Instead of determining the average intensity of a particular wavelength range like in a filter-based method, a complete spectrum is taken to quantify the amount of donor and acceptor. It can be applied to single data points of a fluorescence spectrum obtained by a spectrophotofluorimeter or to data points from a multiple fluorescence image. The exact contribution of donor and acceptor chromophores can be calculated using the data points for a complete reconstruction of the sample spectrum. As a consequence the FRET efficiency for both donor and acceptor can be calculated from the increase in acceptor signal. The quantification of donor and acceptor in the samples is a great advantage compared to other interaction methods. Because now not only information is obtained on whether the proteins interact but also on the contribution of each protein to the interaction. The contribution of each component can be quantified by 'linear unmixing' of the spectra (Clegg, 1992; Clegg et al., 1992; Murchie et al., 1989; Wlodarczyk et al., 2008). An elaborate unmixing description fitting to our setup can be found in the supplemental information of the corresponding publication and will be explained in less detail (Alexeeva et al., 2010).

In the presented experiments the method is used to investigate interactions at very low expression levels. The sensitivity of the method is exclusively dependent on the amount of photons that can be detected. In a microscopic setup the number of photons used for analysis is limited. For higher sample concentrations or volumes a spectrophotofluorimeter can be good alternative. By using a cuvet more sample can exposed to the excitation light and will thereby boost the photon count. When the number of fluorescent proteins is low, the exposure time is extended to obtain a more reliable spectrum. Real-time measurements will then be difficult especially when emission spectra over a broad range of wavelengths are required. To minimize the level of auto-fluorescence originating from the bacteria, red-shifted fluorescent proteins are more suitable for these studies (Alexeeva et al., 2010). The

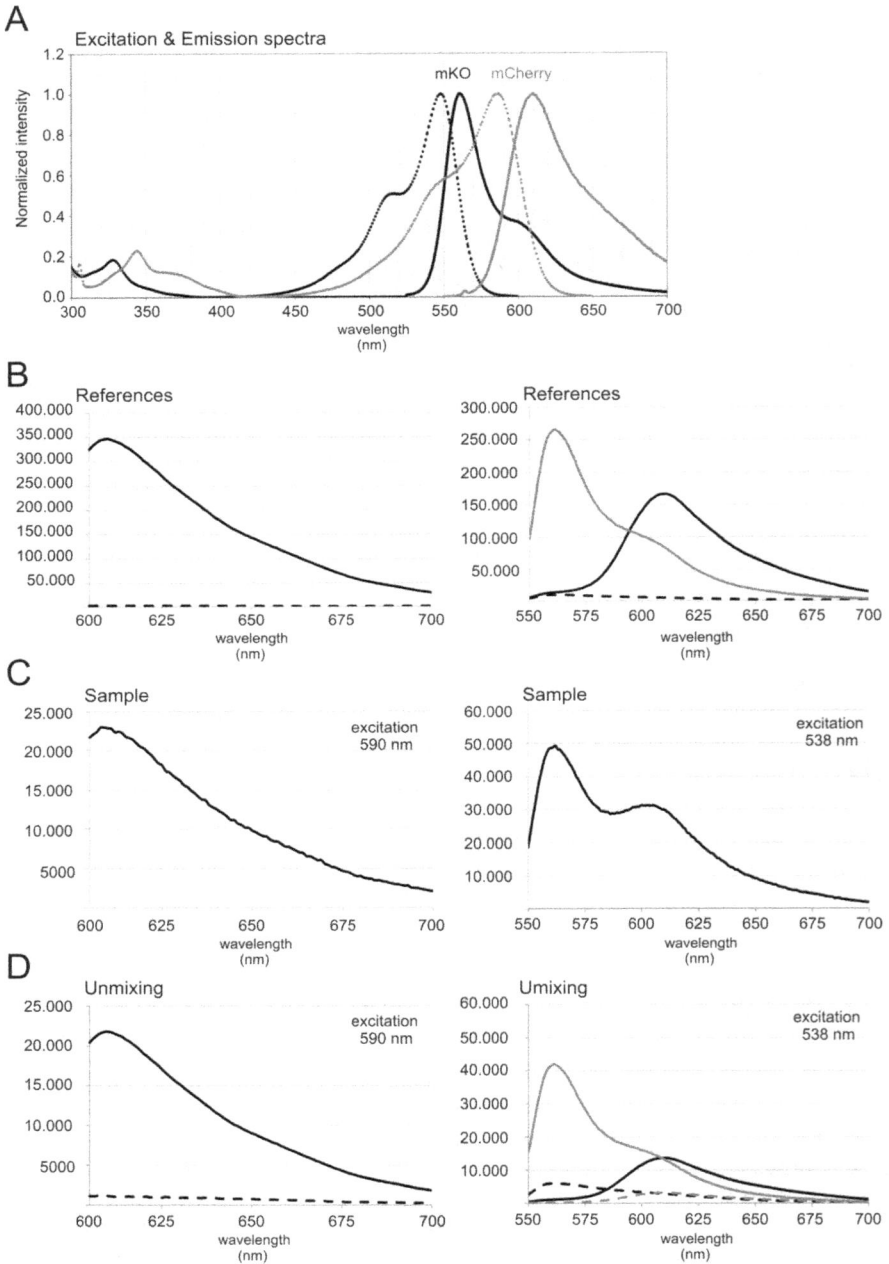

Fig. 10. Principle of spectral FRET unmixing. The excitation (dotted line) and emission (solid line) spectra of mKO and mCherry are shown in A. Cells were grown to a steady-state and fixed using a formaldehyde:gluteraldehyde solution. After washing of the cells and

maturation of mKO, two spectra are measured of the samples. Exciting first with 590 nm to obtain the mCherry spectrum (605-700 nm) alone. Followed by an excitation wavelength at 538 nm to detect the mKO – mCherry spectrum (550-700 nm). On the Y-axis of the graph the fluorescence intensity is presented with photons per second with on the X-axis the detection wavelength in nm. The sample spectrum can be reconstructed from the reference spectra when the multiplication factor for each reference is known. These factors are calculated *via* linear unmixing. For reliable mCherry and mKO reference spectra the fluorescent proteins are produced in higher numbers compared to the samples. For panels B to D the black lines represent the mCherry profile, the grey line is the mKO profile, the black dotted line is background with in the Unmixing panel (D) on the right a grey dotted line showing the sensitized emission.

auto-fluorescence signal increases when cells are exposed to more blue-shifted excitation light. A higher auto-fluorescence signal means more emission from excited biomolecules, which is less favorable because it leads to more cellular stress reactions. Therefore, red-shifted fluorescent proteins are preferable. In addition the Förster radius (R_0) is higher when red fluorescent proteins are used compared to blue-shifted proteins (Table 5.1, page 193) (Gadella, 2008). Recall that a high Förster radius means a higher sensitized emission signal upon donor excitation and thus more sensitivity. For these reasons, the red fluorescent proteins mKO and mCherry are suitable for our research on bacteria (R_0 = 6.37 nm). The orange fluorescent protein mKO has a long maturation time of about 15 hours when expressed in *E. coli* in the contrast to the 15 minutes of mCherry (Shaner et al., 2004). The high brightness of mKO and its bleach resistance in combination with its spectral profiles make it useful for sensitive measurements but its maturation time makes it useless for real-time measurements. For that reason the samples are fixed. After maturation of the chromophores the spectra of the samples are measured after administering equal amounts of bacteria to the cuvet. In the sample three different components that can produce a signal can be discriminated. The auto-fluorescence and light scattering of the cells together referred to as background signal, and the fluorescence created by the donor and acceptor. The spectral components in the sample spectra depend exclusively on the excitation wavelength that is used. First, mCherry alone is excited to determine the amount of mCherry emission in the sample. The obtained spectrum contains two components; background and mCherry. Subsequently, mKO is excited which also excites mCherry to some extend. Knowing the amount of mCherry present in the sample, the spectrum can be unmixed in spectra of the background, mKO, direct excited mCherry and sensitized emission by mCherry. The spectral overlap of the mKO and mCherry proteins makes it inevitable to excite both proteins simultaneously to directly measure FRET. In practice it means that for all the references and samples two spectra have to be measured, starting with the excitation of mCherry followed by the simultaneous excitation of both mKO and mCherry. The mCherry and mKO spectra of each sample can be reconstructed by a multiplication of the measured reference spectra and the background spectrum. If more mCherry signal is measured in the mKO spectrum compared to that in the mCherry only spectrum this is then the sensitized emission and FRET efficiencies can be calculated. Because all the samples are treated identically the method is robust to small changes.

The method has been applied successfully to study interactions between cell division proteins in *E. coli* at concentrations close to their endogenous expression level (Alexeeva et al., 2010; Fraipont et al., 2011). To reach these near physiological expression levels two compatible plasmids are used. The plasmids have a low copy number using the p15A and ColE1 origin of replication. Upstream of the multiple cloning site, a T_{RC} promoter is found that allows a lower expression level due to a mutation its promoter sequence (Weiss et al., 1999).

5.2.5 FLIM-FRET is a robust interaction and localization detection technique

A completely different method for detecting FRET between a donor and an acceptor molecule is by measuring a decrease in donor fluorescence lifetime. The structure of chromophore determines its lifetime, which is the time an electron stays in its excited state before it falls back to its ground state and emits a photon. Different techniques have been developed to determine the lifetime and both attracted a great deal of interest especially in the field of microscopic imaging. Although microscopy has a long history, the implementation of fluorescence lifetime measurements is relatively young, the publications start at the end of the 80's. Two methods have been developed simultaneously one termed time-domain and other frequency-domain. Initially frequency-domain was more applicable due to technical limitations at the time to produce short excitation pulses. At the moment both techniques are applied in wide-field and confocal microscopy. Detailed information on FLIM-FRET that handles the technical microscopic setups, the lifetime detection methods with their advantages in application can be found in (Gadella, 2008). In frequency-domain setups the excitation light is intensity-modulated. By using modulated light a shift in phase can be observed caused by the time an electron stays in the excited state, Fig. 11. And due to a loss of energy the amplitude decreases and can be measured. From these differences in phase and modulation the corresponding lifetimes are calculated. A time-domain lifetime setup is more easily explained. A pulsed laser excites a chromophore and the time that passes until a photon is measured in the detector is the lifetime of the chromophore. The time a photon needs to go from the initial excited state to emission and hitting the detector is a statistical random process. The moment a chromophore emits its energy *via* production of photon is highest just after the pulse excitation and decreases rapidly in time; therefore the lifetime is an average value. The close proximity of an acceptor offers an additional path for the excitation energy to go. In practice it means that the electron stay shorter in the excited state decreasing the average lifetime. Remember that the distance between the donor and the acceptor molecule strongly affects the likely hood of FRET, which is proportional to the decrease in lifetime of the donor chromophore.

In the research field of cell division FLIM-FRET has been used to investigate the roles of SsgA and SsgB during sporulation of *Streptomyces* aerial hyphae (Willemse et al., 2011). These proteins co-localize together with FtsZ, both SsgA and SsgB move to mid-cell in the respective order before FtsZ arrives. To find out whether the proteins interact FLIM-FRET was applied. Fluorescent protein fusions of eGFP and mCherry were constructed. The fusion proteins showed full complementation of the deficient strains. The fusion genes replaced the endogenous gene; expression was therefore under control of their authentic promoter. A decrease in the eGFP lifetime was observed with fusion protein combinations SsgA and SsgB, and SsgB with FtsZ.

In the previous section a time-domain FLIM microscopic setup was used. In the past we have applied a frequency-domain FLIM setup to proof that the spectral FRET method produces reliable data. Instead of using GFP, mKO was used as the selected donor. The interaction between FtsZ molecules was studied by fusing mKO and mCherry to its amino-terminus. The fluorescence lifetime of mKO was 2.89 ± 0.05 ns in cells coexpressing unfused mKO and mCherry. When mKO and mCherry were fused to FtsZ, the mKO fluorescence lifetime was decreased to 2.69 ± 0.03 ns corresponding to a FRET efficiency of 7% (Alexeeva et al., 2010).

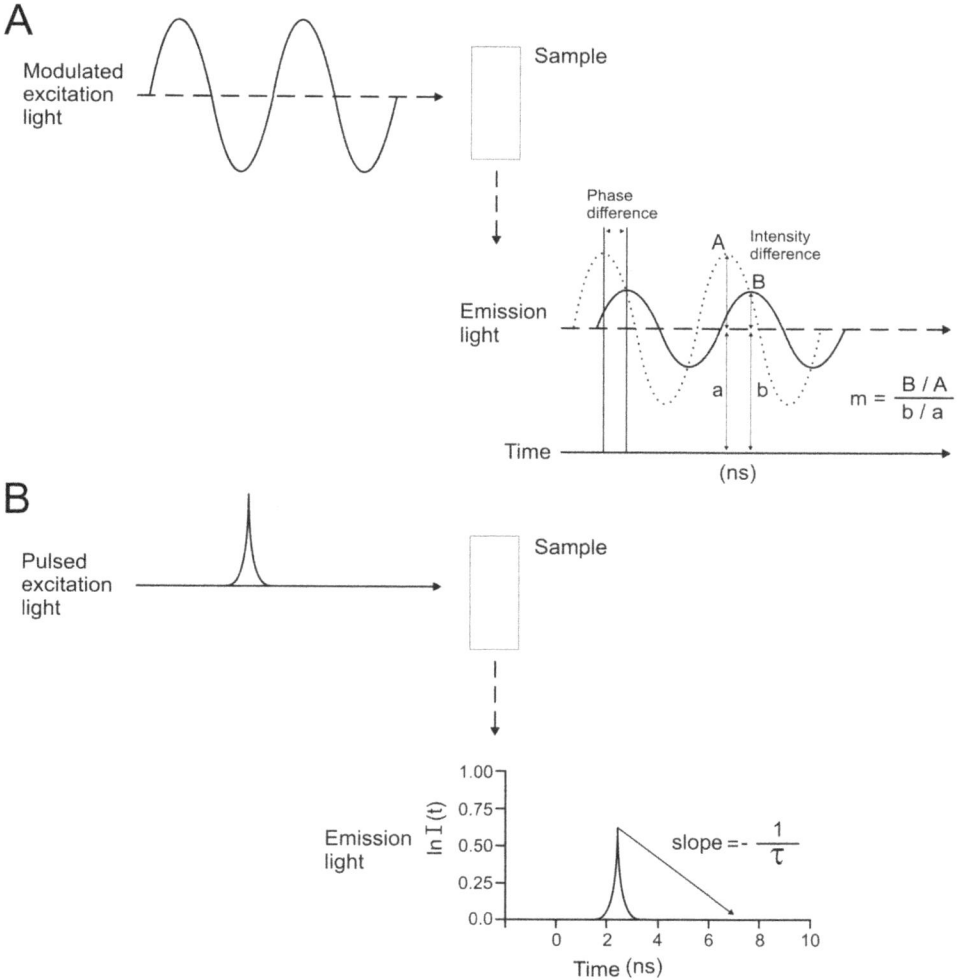

Fig. 11. Frequency-domain and time-domain FLIM. A) The frequency-domain measurements. In the picture presenting the emission light two signals are presented. The solid line is the emission signal and the dotted line the source excitation light for a comparison. From both the phase and modulation a lifetimes can be calculated. B) A schematic presentation of a time-domain measurement where the lifetime (τ) can be calculated from the slope.

5.2.6 Pro's and Contra's of lifetime and intensity based FRET methods

The application of fluorescent proteins in research allows you to observe movement and localization of proteins and study interactions with their partners. The ability to detect emission light makes it even possible to quantify data, study interactions real-time and calculate the dynamics. This provides considerable extra information compared to the bacterial two-hybrid method. However to obtain reliable FRET data, a clear understanding of the instrumental setup and a careful selection of fluorescent proteins is a necessity. Some crucial points will be discussed to provide the reader with insight in bottlenecks that can be encountered when applying a FRET method.

5.2.6.1 Instrumental settings and fluorescent protein selection

Before applying fluorescent proteins to studying protein localization and protein-protein interactions, several factors need to be taken into account. These factors have a biological, instrumental or mathematical origin when additional calculations are necessary. It is important to understand the limitations of the instrumental setup that being either a microscope or spectrophotofluorimeter. This includes knowledge on the properties and contributions of additional materials like filters. It is needless to mention but the instrumental setup should be in good conditions to reach maximum sensitivity. For a comprehensive understanding about the technical microscopic setup the reader is referred to (Gadella, 2008).

The biological factors that affect the results of the experiment originate from the use of fluorescent proteins, the linkers to fuse them to the prey and bait combined with the necessary expression tools. The barrel of a fluorescent protein is in general not interacting with other intercellular proteins and makes it therefore a very suitable protein for research. However, extended exposure of the chromophore to excitation light results in photobleaching and production of reactive oxygen species (Dixit & Cyr, 2003). Reactive oxygen arises when the energy from the (triplet) excited state of the chromophore is transferred to molecular oxygen. As a consequence the produced reactive oxygen species can lead to damage of important cellular molecules like proteins, lipids, nucleic acids etc. Additionally damage might be introduced due to a change in the redox homeostasis. Bright fluorescent proteins are desired because they decrease exposure time giving few reactive oxgen species. Therefore, the theoretically most suitable fluorescent proteins should be selected that match with the instrumental and experimental setup. Frequently the behavior of a fluorescent protein varies with organism and deviates from the reported characteristics. Often these differences arise due to dissimilar biological conditions and instrumental setups. The fluorescent protein pair should be tested in host organism using the experimental setup to learn how it behaves in the cell.

For the spectral FRET method mKO was used as a donor and was published to have a relative long maturation half time of 4.5 hours (Karasawa et al., 2004; Shaner et al., 2008). In *E. coli* the protein needed an even longer maturation time of ±15 hours to become fluorescent. By fixing the cells with a formaldehyde:gluteraldehyde solution the biological condition is frozen but it was unclear how this would affect mKO behavior. The fixation did proof not to be a problem for mKO maturation and fluorescence. Some fluorescent proteins are sensitive for fixation and lose some of their characteristic. An example is YFP that dramaticly decreases in brightness after fixation.

Many different fluorescent proteins are published with colors ranging the complete visible spectrum. The large set of fluorescent protein should give enough options for selecting the right pair for your setup.

Fluorescent proteins are fused at the amino-terminus or carboxy-terminus end but in some cases it can also be placed in a loop (Bendezu et al., 2009). Often the fluorescent protein and protein of interest (prey or bait) are fused to each other by an amino acid linker. The length of the linker needs to be adjusted to place the fluorescent protein there where it minimizes interference with the biological function of the protein of interest. To verify the functionality of the protein fusions, a deficient or depleted strain can be complemented and the normal localization of the protein can be detected. The aim should be to bring the fluorescent proteins of prey and bait in the closest proximity upon interaction. A crystal structure can help and speed up finding the correct linker length for the fusion. The risk encountered when fusing a fluorescent protein to a prey or bait protein is almost identical as described in the bacterial two-hybrid section. Competition of the fusion protein with wild-type proteins can be expected and therefore working with deficient strains is advised. Moreover, the amount of the fusion protein in the cell should be compared with the non-fused protein to detect differences in expression or biological stability. The characteristics of the linker can attribute significantly to functionality of the fusion protein thereby contributing to the stability of the fusion protein.

The three described FRET methods use completely different instrumental setups and fluorescent proteins to reach their goal, all having their own specific quality. The first method measures dynamics, the second method quantifies the level of FRET and the FLIM method is robust and semi-quantitative and suitable for single cell studies.

For measuring a dynamic process, a fast folding and bright fluorescent protein pair such as CFP and YFP is ideal. Changes in fluorescence intensity due to FRET will than easily be picked up. The application of CFP and YFP in studying the interaction between CheY and CheZ proteins shows how effective this combination is (Sourjik & Berg, 2004; Sourjik et al., 2007). A pitfall in the application of filter-based FRET is the misinterpretation of intensities for donor and acceptor levels. An overlap in the excitation and emission spectra of the donor and acceptor chromophore can lead to a process called bleed-through or leak-though. In the example with CFP and YFP, both are directly excited with the donor excitation wavelength (Sourjik et al., 2007). Fortunately, the acceptor is only excited to a minor extent and therefore hardly contributes to the FRET signal. Direct excitation of the acceptor should be kept to a minimum because an excited acceptor cannot absorb energy from an excited donor molecule. Two detection channels are used one for the donor and one for the acceptor. The FRET efficiency can be calculated when the fluorescence intensity is measured for the donor and acceptor when no interaction is taking place (FP_0). The change in fluorescence intensity for each fluorescent protein (ΔFP) by FP_0 gives the FRET efficiency assuming that the auto-fluorescence and bleaching is corrected for. However, for quantitative FRET measurements using a filter-based FRET method, many more additional factors have to be taken into account (see chapter 7 of FRET and FLIM techniques (Gadella, 2008)) when a non-interacting condition cannot be created and/ or measured. As pointed out earlier it is very important to know how the selected fluorescent proteins behave under the experimental conditions. For example photo-conversion of YFP is observed during photobleaching leading to a CFP emission signal upon excitation as reported by some but contradicted by others (Thaler et al., 2006; Valentin et al., 2005). Finally, to be able to follow a dynamic process for a very long time the fluorescent proteins should have a high bleach resistance. YFP for instance tends to be bleached relatively quickly (Griesbeck et al., 2001; Kremers et al., 2006).

For the analysis of low abundant division protein of E. coli comparable low endogenous protein expression levels were desired. The fluorescence emitted from a single cell was at the detection limit of the microscopic setup. Therefore, a spectrophotofluorimeter has been adopted instead to measure the fluorescence of many bacterial cells simultaneously. The information on localization was obtained in separate experiment. FRET calculated from a single wavelength is prone to false results. It is not always clear if the increase in fluorescence intensity of the acceptor originates from FRET. As mentioned earlier for YFP, photo-conversion induced by the excitation light might be responsible for the increase in acceptor but will not be recognized as such. More insight is obtained by the measurement of complete donor and acceptor spectra. The production of a sensitized emission signal should lead to an increase in the acceptor spectrum. If the increase in addition changes the shape of the acceptor spectrum it can be identified as an artifact. This can be especially important when working with low expression levels in a high auto-fluorescence spectral area. Because bacteria hardly give any auto-fluorescence in the red-spectral range, the orange fluorescent protein mKO and the more red-shifted mCherry were selected for the spectral FRET method. The sensitized emission is only a fraction of the emission of both proteins and therefore a low signal to noise ratio will improve the precision and reliability of the FRET signal. The bacteria behave as particles in the excitation light which causes considerable lightscattering. The unwanted light scattering polluting the emission spectrum is removed by use of specific emission filters that block the scatter light.

It is noteworthy that when working with the spectrophotofluorimeter bacterial cultures need to be grown as much as possible to a steady-state before the experiment can be started. In a steady-state culture the cells have on average a constant mass indicating that their mass increases at the same rate as their cell number. Consequently, the cells have a homogeneous metabolism and morphology that will provide a constant auto-fluorescence background in the cells. This is essential for the unmixing of the spectra, which assumes that a multiplication of the background, mKO and mCherry reference spectra determine the shape and magnitude of the measured spectrum.

Using spectra to determine the amount of each reference in a sample has advantages. By carefully selecting the fluorescent protein pair and their excitation and emission wavelengths the sensitivity can be further improved. The greater the spectral difference between the references the more easily and accurately each contribution can be calculated to reconstitute the sample fluorescence profile. Excitation at different wavelengths changes the intensity but not the shape of the emission spectra of mKO and mCherry. However, the shape of the background spectrum is dependent on the excitation wavelength and can therefore be selected. The freedom in selecting the excitation wavelength has some boundaries. When exciting the donor molecule the emission signal should be as high as possible with as little as possible direct excitation of the acceptor.

Fluctuations in the fluorescence intensity or photo-conversions of chromophore do not affect the fluorescence lifetime measurements making this method more robust. The time-domain and the frequency-domain methods have both their own advantages. Time-domain measurements can be relatively time consuming compared to frequency-domain especially when fluorescence intensity is low. On the other hand quantitative FRET percentages can more easily be retrieved from time-domain data. The detection speed of the frequency-domain method makes it more suitable to study protein dynamics. Preferably the lifetimes of the background, donor and acceptor should differ substantially. When possible, the lifetime of the background signal should be higher compared to that of the donor protein. Because it excludes the possibility that the two lifetimes overlap when the lifetime of donor is reduced due to FRET. To monitor real-time dynamics still the brightest fluorescent proteins are most desirable!

5.2.6.2 Interpretation of results

The major advantage of fluorescent proteins in cell biology is the direct representation of where a protein is in a living cell. A different but important advantage is that the proteins do not react with each or other components in the cell. The lack of affinity between fluorescent proteins makes them perfect to study protein-protein interaction *via* FRET. Fluorescent proteins have a relative large size of about 27 kDa for green fluorescent protein (GFP). A fusion of GFP to a smaller protein can alter its natural behavior significantly. And thus the ability of a fusion protein to complement a deficient strain is an important measure to test if fusion proteins can functionally replace the wild-type. In addition the stability of the fusion protein in the cell should be checked. Cleavage of fusion protein can reproduce a wild-type version product that can compete with its partner fusion protein thereby decreasing the FRET percentage. Conditions in which the fusion protein is present in the cell in levels that exceed the normal numbers should be prevented. High expression conditions can introduce artifacts due to oligomerization of the fluorescent proteins or create bystander FRET. Oligomerization of fluorescent proteins can takes place in the sub-millimolar range (0.1 mM), which is a concentration that can be easily reached when expressing membrane proteins in relative high amounts. Non-dimerizing fluorescent proteins have been created that only dimerize at very high concentrations that are almost not achievable intra-cellular. Bystander FRET is created by overproduction of the fusion proteins in the cell and as a result they come into close contact due to molecular crowding. Both situations have to be avoided. Non-interacting fusion constructs can be used as negative controls for these artificial interactions and should be added to the experiment.

Some interactions are transient or only take place in a single locus with limited space and are reasons why expression conditions should be carefully controlled and selected. Although a fusion protein can complement a deficient strain, its localization might not always be identical to the endogenous protein. A sign that there is a disturbing effect originating from the fusion with linker and fluorescent protein. Unfortunately, working with a deficient strain is not always possible. Therefore, a slight overexpression is preferred above underexpression that should be avoided at al times. In a wild-type strain fused and unfused proteins will compete for an interacting partner. To outcompete the wild-type protein a slightly higher concentration of fusion proteins is recommended.

The list of checkpoint and controls that have to be taken into account seems to be long but the advantages greatly outweigh the work that is involved. Not many other techniques can produce reliable data from a living cell in real-time. Representing interactions and dynamic process as they take place.

	Method 1	Method 2	Method 3
Reference	Sourjik & Berg, 2004	Alexeeva et al., 2010	Willemse et al., 2011
Interaction	Dynamics	Fixed	Both
Method	Filter	Spectral	Lifetime
FRET determination	Intensity	Intensity	Lifetime
Plasmids	two	two	Chromosomal
Origin of replication	p15A + ColE1	p15A + ColE1	-
Selection markers	Km, Ap	Cm, Ap	Am
Strain	independent	independent	independent
Expression	pTrc, pAra*	pTrc-down	Natural promoter
IPTG / arabinose	50 uM / 0.01 %	10 or 15 uM	-
Induction time	4 hours	6 hours	-

Notes: fluorescent proteins used cyan fluorescent protein (CFP), yellow fluorescent protein (YFP), green fluorescent protein (eGFP) and mCherry (mCh). The pTrc promoters use isopropyl ß-D-thiogalactoside (IPTG) for induction with arabinose for the promoter. Selection markers Cm; chloramphenicol, Ap; ampicillin, Km; kanamycin, Am; apramycin. * = the number of plasmids used in articles differs (Sourjik & Berg, 2002; Sourjik & Berg, 2004). The former used two and the latter one plasmid with a Trc promoter. The two plasmids were contained an arabinose and Trc promoter.

Table 2. Comparison of methods

6. Conclusion

6.1 Comparing FRET and bacterial two-hybrid

After going through the method sections it becomes immediately clear that there is a big difference between the bacterial two-hybrid methods and the FRET methods. The bacterial two-hybrid methods are verily easy and straightforward in providing an answer. Additional technical expertise is required when applying FRET in research. But the use of fluorescent proteins provides more information about the localization of the interaction and possibly also information can be gathered on the dynamics between the proteins, this in contrast to the bacterial two-hybrid. For both the bacterial two-hybrid and FRET method a wild-type behavior of the fusion protein is desired. Complementation experiments using deficient strains are effective controls to see if the fusion protein fulfills these requirements. The information obtained on the localization of the fusion proteins is lacking with the two-hybrid method. Fractionation experiments can only partially compensate for the difference between the two methods.

Another difference is the use of the linker. Linkers are used in both systems to stimulate the natural wild-type behavior of the protein in the cell. For the bacterial two-hybrid method extra freedom in movement for the reporter domain is less harmful compared to the FRET method. In a two-hybrid system, a reporter protein becomes complete upon interaction of the prey and bait. The fully functional reporter protein facilitates, direct or indirect, synthesis of a reporter product that continues to pile up and thereby amplifies the signal. With fluorescent proteins the situation is completely different. They have no affinity for each other and more freedom will result in more movement decreasing the chance that the proteins will remain or be close to another. Therefore, the sensitized emission signal will become weaker and harder to detect which make dynamic measurements impossible. The reporter methods are responsible for why the bacterial two-hybrid method produces higher false positives and the FRET method more false negatives. Endogenous protein expression levels for the fusion proteins should be maintained as much as possible. Unfortunately, high expression levels are frequently observed within use of the bacterial two-hybrid method but should be avoided at all times. Whether the bacterial two-hybrid or FRET method will be employed is completely dependent on the question that needs to be answered. A FRET experiment will gain a more comprehensive answer to what is going on but tends to be more time consuming.

6.2 Interactions and use in cell division research

In the last part of this chapter, an overview of several publications will be presented to illustrate the variation found between the interactions of E. coli cell division proteins. The proteins investigated differ in structure, function and cellular location; they are present in the cytoplasm and the inner-membrane. Some of these cell division proteins show considerable movement during the cell cycle. Several interactions are believed to be only transient, taking place only at mid-cell when new cell poles need to be synthesized. Examples of interacting cell division proteins are the FtsZ proteins that together form a polymer, which interacts with membrane proteins FtsA and ZipA. ZapA proteins bind to the polymer thereby stabilizing the structure. The membrane proteins FtsK, FtsB, FtsL and FtsQ move to the mid-cell position. Subsequently, the FtsW-PBP3 precomplex is recruited to the septation site where also FtsN arrives. More proteins are thought to be involved in the actual division process, like the periplasmic protein AmiC that is essential for the cleavage of the septum during the constriction process. Current thinking suggests that many of these proteins together form a big complex at mid-cell, called the Divisome. With multiple complexes spread across the Z-ring structure to facilitate the constriction of cell.

The division of the cell is a delicate process that needs carefully regulation. The copy number of each specific division protein in the cell is balanced. Overproduction of a cell division protein disrupts this balance which causes cell division defects like cell filamentation. Bacterial two-hybrid and FRET methods have been employed to study the interactome of the division proteins. Three bacterial two-hybrid methods are presented; two using the Di lallo method and one using the Karimova method (Di lallo et al., 2003; Karimova et al., 1998; Maggi et al., 2008). These bacterial two-hybrid results are compared with that found using spectral FRET method described in two papers, Alexeeva *et al* and Fraipont *et al*, Fig. 12. Several interactions are reproduced and are therefore most likely true. However, also contradicting results are found when applying these different methods, which can originate from the technique. In the Di lallo method an interaction between prey and bait is only observed when the dimer binds the DNA operator site. Integral membrane protein cannot easily reach DNA sequences compared to proteins dwelling in cytoplasm. False results are not unusual when applying these techniques being false positive for the bacterial two-hybrid method and false negative for the FRET method. The differences may arise due to the method but might also be caused through differences in the expression levels. Di lallo and Karimova use different promoters and induction concentration. It would be very interesting to determine if the bacterial two-hybrid and FRET experiments yield the same results when the expression and bacterial growth conditions are kept identical, preferably native protein concentrations.

Bacterial two-hybrid

A

Transcriptional repression

Di Iallo

B

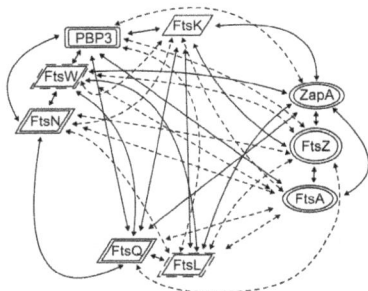
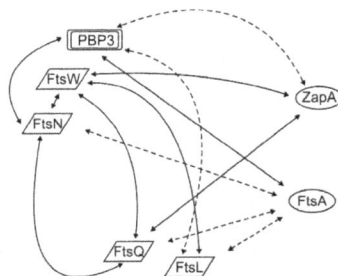

C

Transcriptional activation

Karimova

D

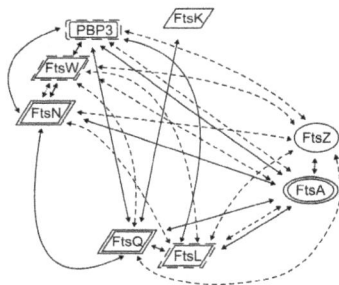
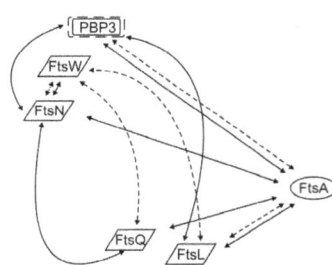

FRET

E

Spectral FRET - intensity based

Alexeeva

F

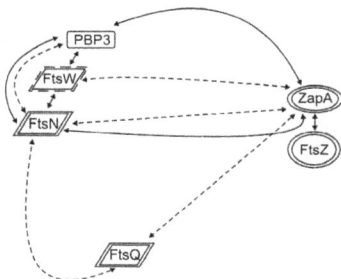
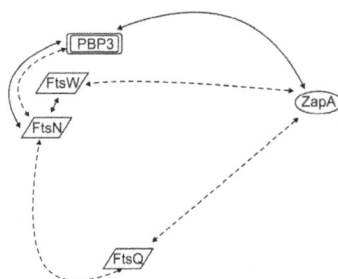

Fig. 12. A comparison of bacterial two-hybrid and spectral FRET results found for cell division proteins interactions. The solid lines represent an interaction between two proteins

and when no interaction was found the proteins are connected by a dotted line. Panels A, C and E show all the interactions studied and B, D and F represent the differences. A double line around name of the protein indicates that the protein is self-interacting. All the rectangles represent membrane proteins; the one with a straight corner is a penicillin binding protein. The circler proteins are present in the cytoplasm.

Although an interactome scheme can be constructed still many questions remain. Isolation of a complete Divisome complex has not been successful. With the current knowledge it can be speculated that isolation of a complete Divisome complex might prove to be too difficult due to weak and transient interactions. For now structural information of the Divisome complex is lacking. Fortunately more knowledge on the interactome and structure of the Divisome can be gained using bacterial two-hybrid and FRET techniques. Bacterial two-hybrid methods can be used to quickly screen many protein mutants to harvest information on interaction sites, like performed for FtsQ and PBP3 (FtsI) (D'Ulisse et al., 2007; Karimova et al., 2005). Protein-protein interactions in the periplasm can be analyzed using Borloo bacterial two-hybrid method. And the described FRET techniques can provide valuable information on dynamics, movement and localization to improve our understand about the (dis)assembly, mechanism and structure of the Divisome complex.

7. Acknowledgments

This project was supported by the Dutch NWO 'Van Molecuul tot Cel' programme grant ALW 805.47.200 and by a Vernieuwingsimpuls grant 016.001-024 (T.d.B.) and by the European Framework Programs DIVINOCELL HEALTH F3-2009-223431 (R.v. d.P). We would like to thank Eelco Hoogendoorn and Joachim Goedhart for fruitfull discussions and Mark Hink for reading the manuscript.

8. References

Aarsman, M. E.; Piette, A.; Fraipont, C.; Vinkenvleugel, T. M.; Nguyen-Disteche, M. & den Blaauwen, T. (2005). Maturation of the *Escherichia coli* divisome occurs in two steps. *Mol.Microbiol.*, Vol. 55, No. 6, pp. 1631-1645, 0950-382X; 0950-382X

Alexeeva, S.; Gadella, T. W.,Jr.; Verheul, J.; Verhoeven, G. S. & den Blaauwen, T. (2010). Direct interactions of early and late assembling division proteins in Escherichia coli cells resolved by FRET. *Mol.Microbiol.*, Vol. 77, No. 2, pp. 384-398, 1365-2958; 0950-382X

Bendezu, F. O.; Hale, C. A.; Bernhardt, T. G. & de Boer, P. A. (2009). RodZ (YfgA) is required for proper assembly of the MreB actin cytoskeleton and cell shape in E. coli. *EMBO J.*, Vol. 28, No. 3, pp. 193-204, 1460-2075; 0261-4189

Bertsche, U.; Kast, T.; Wolf, B.; Fraipont, C.; Aarsman, M. E.; Kannenberg, K.; von Rechenberg, M.; Nguyen-Disteche, M.; den Blaauwen, T.; Holtje, J. V. & Vollmer, W. (2006). Interaction between two murein (peptidoglycan) synthases, PBP3 and PBP1B, in *Escherichia coli*. *Mol.Microbiol.*, Vol. 61, No. 3, pp. 675-690, 0950-382X; 0950-382X

Blackwood, E. M. & Eisenman, R. N. (1995). Identification of protein-protein interactions by lambda gt11 expression cloning. *Methods Enzymol.*, Vol. 254, pp. 229-240, 0076-6879; 0076-6879

Borloo, J.; De Smet, L.; Vergauwen, B.; Van Beeumen, J. J. & Devreese, B. (2007). A beta-galactosidase-based bacterial two-hybrid system to assess protein-protein interactions in the correct cellular environment. *J.Proteome Res.*, Vol. 6, No. 7, pp. 2587-2595, 1535-3893; 1535-3893

Castagnoli, L.; Vetriani, C. & Cesareni, G. (1994). Linking an easily detectable phenotype to the folding of a common structural motif. Selection of rare turn mutations that prevent the folding of Rop. *J.Mol.Biol.*, Vol. 237, No. 4, pp. 378-387, 0022-2836; 0022-2836

Chalfie, M.; Tu, Y.; Euskirchen, G.; Ward, W. W. & Prasher, D. C. (1994). Green fluorescent protein as a marker for gene expression. *Science*, Vol. 263, No. 5148, pp. 802-805, 0036-8075; 0036-8075

Chudakov, D. M.; Matz, M. V.; Lukyanov, S. & Lukyanov, K. A. (2010). Fluorescent proteins and their applications in imaging living cells and tissues. *Physiol.Rev.*, Vol. 90, No. 3, pp. 1103-1163, 1522-1210; 0031-9333

Clegg, R. M. (1992). Fluorescence resonance energy transfer and nucleic acids. *Methods Enzymol.*, Vol. 211, pp. 353-388, 0076-6879; 0076-6879

Clegg, R. M.; Murchie, A. I.; Zechel, A.; Carlberg, C.; Diekmann, S. & Lilley, D. M. (1992). Fluorescence resonance energy transfer analysis of the structure of the four-way DNA junction. *Biochemistry*, Vol. 31, No. 20, pp. 4846-4856, 0006-2960; 0006-2960

Cody, C. W.; Prasher, D. C.; Westler, W. M.; Prendergast, F. G. & Ward, W. W. (1993). Chemical structure of the hexapeptide chromophore of the *Aequorea* green-fluorescent protein. *Biochemistry*, Vol. 32, No. 5, pp. 1212-1218, 0006-2960; 0006-2960

Cubitt, A. B.; Heim, R.; Adams, S. R.; Boyd, A. E.; Gross, L. A. & Tsien, R. Y. (1995). Understanding, improving and using green fluorescent proteins. *Trends Biochem.Sci.*, Vol. 20, No. 11, pp. 448-455, 0968-0004; 0968-0004

Dautin, N.; Karimova, G.; Ullmann, A. & Ladant, D. (2000). Sensitive genetic screen for protease activity based on a cyclic AMP signaling cascade in *Escherichia coli*. *J.Bacteriol.*, Vol. 182, No. 24, pp. 7060-7066, 0021-9193; 0021-9193

Den Blaauwen, T.; Aarsman, M. E.; Vischer, N. O. & Nanninga, N. (2003). Penicillin-binding protein PBP2 of *Escherichia coli* localizes preferentially in the lateral wall and at mid-cell in comparison with the old cell pole. *Mol.Microbiol.*, Vol. 47, No. 2, pp. 539-547, 0950-382X; 0950-382X

Denome, S. A.; Elf, P. K.; Henderson, T. A.; Nelson, D. E. & Young, K. D. (1999). *Escherichia coli* mutants lacking all possible combinations of eight penicillin binding proteins: viability, characteristics, and implications for peptidoglycan synthesis. *J.Bacteriol.*, Vol. 181, No. 13, pp. 3981-3993, 0021-9193; 0021-9193

Di lallo, G.; Anderluzzi, D.; Ghelardini, P. & Paolozzi, L. (1999a). FtsZ dimerization in vivo. *Mol.Microbiol.*, Vol. 32, No. 2, pp. 265-274, 0950-382X; 0950-382X

Di lallo, G.; Ghelardini, P. & Paolozzi, L. (1999b). Two-hybrid assay: construction of an Escherichia coli system to quantify homodimerization ability in vivo. *Microbiology*, Vol. 145 (Pt 6), No. Pt 6, pp. 1485-1490, 1350-0872; 1350-0872

Di lallo, G.; Castagnoli, L.; Ghelardini, P. & Paolozzi, L. (2001). A two-hybrid system based on chimeric operator recognition for studying protein homo/heterodimerization in *Escherichia coli*. *Microbiology*, Vol. 147, No. Pt 6, pp. 1651-1656, 1350-0872; 1350-0872

Di lallo, G.; Fagioli, M.; Barionovi, D.; Ghelardini, P. & Paolozzi, L. (2003). Use of a two-hybrid assay to study the assembly of a complex multicomponent protein machinery: bacterial septosome differentiation. *Microbiology*, Vol. 149, No. Pt 12, pp. 3353-3359, 1350-0872; 1350-0872

Dixit, R. & Cyr, R. (2003). Cell damage and reactive oxygen species production induced by fluorescence microscopy: effect on mitosis and guidelines for non-invasive fluorescence microscopy. *Plant J.*, Vol. 36, No. 2, pp. 280-290, 0960-7412; 0960-7412

Dmitriev, B.; Toukach, F. & Ehlers, S. (2005). Towards a comprehensive view of the bacterial cell wall. *Trends Microbiol.*, Vol. 13, No. 12, pp. 569-574, 0966-842X; 0966-842X

D'Ulisse, V.; Fagioli, M.; Ghelardini, P. & Paolozzi, L. (2007). Three functional subdomains of the *Escherichia coli* FtsQ protein are involved in its interaction with the other division proteins. *Microbiology*, Vol. 153, No. Pt 1, pp. 124-138, 1350-0872; 1350-0872

Durand-Heredia, J. M.; Yu, H. H.; De Carlo, S.; Lesser, C. F. & Janakiraman, A. (2011). Identification and characterization of ZapC, a stabilizer of the FtsZ ring in *Escherichia coli*. *J.Bacteriol.*, Vol. 193, No. 6, pp. 1405-1413, 1098-5530; 0021-9193

Fields, S. & Song, O. (1989). A novel genetic system to detect protein-protein interactions. *Nature*, Vol. 340, No. 6230, pp. 245-246, 0028-0836; 0028-0836

Förster, T. (1946). Energiewanderung und Fluoreszenz. *Naturwissenschaften.* Vol. 6, pp. 166-175

Förster, T. (1948). Zwichenmolekulare Energiewanderung und Fluoreszenz. *Ann. Phys.* Vol. 2, pp. 55-75,

Förster, T. (1993). Intermolecular energy migration an fluorescence, In: *Biological Physics,* (Mielczarek, E. V., Greenbaum, E. & Knox, R. S., Eds.), (April 1993) American Institute of Physics, 9780883188552 New York, pp. 148-160.

Fraipont, C.; Alexeeva, S.; Wolf, B.; van der Ploeg, R.; Schloesser, M.; den Blaauwen, T. & Nguyen-Disteche, M. (2011). The integral membrane FtsW protein and peptidoglycan synthase PBP3 form a subcomplex in *Escherichia coli*. *Microbiology*, Vol. 157, No. Pt 1, pp. 251-259, 1465-2080; 1350-0872

Fu, G.; Huang, T.; Buss, J.; Coltharp, C.; Hensel, Z. & Xiao, J. (2010). In vivo structure of the *E. coli* FtsZ-ring revealed by photoactivated localization microscopy (PALM). *PLoS One*, Vol. 5, No. 9, pp. e12682, 1932-6203; 1932-6203

Gadella, T. W. J. (2008). FRET and FLIM techniques, In: *Laboratory techniques in biochemistry and molecular biology*, (Pillai S. & van der Vliet, P. C., Eds.), (Dec 2008), Elsevier, 9780080549583, Amsterdam, The Netherlands

Gan, L.; Chen, S. & Jensen, G. J. (2008). Molecular organization of Gram-negative peptidoglycan. *Proc.Natl.Acad.Sci.U.S.A.*, Vol. 105, No. 48, pp. 18953-18957, 1091-6490; 0027-8424

Griesbeck, O.; Baird, G. S.; Campbell, R. E.; Zacharias, D. A. & Tsien, R. Y. (2001). Reducing the environmental sensitivity of yellow fluorescent protein. Mechanism and applications. *J.Biol.Chem.*, Vol. 276, No. 31, pp. 29188-29194, 0021-9258; 0021-9258

Hale, C. A. & de Boer, P. A. (1999). Recruitment of ZipA to the septal ring of *Escherichia coli* is dependent on FtsZ and independent of FtsA. *J.Bacteriol.*, Vol. 181, No. 1, pp. 167-176, 0021-9193; 0021-9193

Hale, C. A.; Shiomi, D.; Liu, B.; Bernhardt, T. G.; Margolin, W.; Niki, H. & de Boer, P. A. (2011). Identification of *Escherichia coli* ZapC (YcbW) as a component of the division apparatus that binds and bundles FtsZ polymers. *J.Bacteriol.*, Vol. 193, No. 6, pp. 1393-1404, 1098-5530; 0021-9193

Hayhurst, E. J.; Kailas, L.; Hobbs, J. K. & Foster, S. J. (2008). Cell wall peptidoglycan architecture in *Bacillus subtilis*. *Proc.Natl.Acad.Sci.U.S.A.*, Vol. 105, No. 38, pp. 14603-14608, 1091-6490; 0027-8424

Hu, J. C.; O'Shea, E. K.; Kim, P. S. & Sauer, R. T. (1990). Sequence requirements for coiled-coils: analysis with lambda repressor-GCN4 leucine zipper fusions. *Science*, Vol. 250, No. 4986, pp. 1400-1403, 0036-8075; 0036-8075

Jacobson, R. H.; Zhang, X. J.; DuBose, R. F. & Matthews, B. W. (1994). Three-dimensional structure of beta-galactosidase from *E. coli*. *Nature*, Vol. 369, No. 6483, pp. 761-766, 0028-0836; 0028-0836

Joung, J. K.; Ramm, E. I. & Pabo, C. O. (2000). A bacterial two-hybrid selection system for studying protein-DNA and protein-protein interactions. *Proc.Natl.Acad.Sci.U.S.A.*, Vol. 97, No. 13, pp. 7382-7387, 0027-8424; 0027-8424

Kain, S. R.; Adams, M.; Kondepudi, A.; Yang, T. T.; Ward, W. W. & Kitts, P. (1995). Green fluorescent protein as a reporter of gene expression and protein localization. *BioTechniques*, Vol. 19, No. 4, pp. 650-655, 0736-6205; 0736-6205

Karasawa, S.; Araki, T.; Nagai, T.; Mizuno, H. & Miyawaki, A. (2004). Cyan-emitting and orange-emitting fluorescent proteins as a donor/acceptor pair for fluorescence resonance energy transfer. *Biochem.J.*, Vol. 381, No. Pt 1, pp. 307-312, 1470-8728; 0264-6021

Karimova, G.; Pidoux, J.; Ullmann, A. & Ladant, D. (1998). A bacterial two-hybrid system based on a reconstituted signal transduction pathway. *Proc.Natl.Acad.Sci.U.S.A.*, Vol. 95, No. 10, pp. 5752-5756, 0027-8424; 0027-8424

Karimova, G.; Dautin, N. & Ladant, D. (2005). Interaction network among *Escherichia coli* membrane proteins involved in cell division as revealed by bacterial two-hybrid analysis. *J.Bacteriol.*, Vol. 187, No. 7, pp. 2233-2243, 0021-9193; 0021-9193

Koch, A. L. & Woeste, S. (1992). Elasticity of the sacculus of *Escherichia coli*. *J.Bacteriol.*, Vol. 174, No. 14, pp. 4811-4819, 0021-9193; 0021-9193

Kremers, G. J.; Goedhart, J.; van Munster, E. B. & Gadella, T. W.,Jr. (2006). Cyan and yellow super fluorescent proteins with improved brightness, protein folding, and FRET Forster radius. *Biochemistry*, Vol. 45, No. 21, pp. 6570-6580, 0006-2960; 0006-2960

Ladant, D. (1988). Interaction of Bordetella pertussis adenylate cyclase with calmodulin. Identification of two separated calmodulin-binding domains. *J.Biol.Chem.*, Vol. 263, No. 6, pp. 2612-2618, 0021-9258; 0021-9258

Ladant, D. & Karimova, G. (2000). Genetic systems for analyzing protein-protein interactions in bacteria. *Res.Microbiol.*, Vol. 151, No. 9, pp. 711-720, 0923-2508; 0923-2508

Lakowicz, J. R. (2006). Principles of fluorescence spectroscopy (3), Springer, 038731278-1, New York, United States of America

Longo, F.; Marchetti, M. A.; Castagnoli, L.; Battaglia, P. A. & Gigliani, F. (1995). A novel approach to protein-protein interaction: complex formation between the p53 tumor suppressor and the HIV Tat proteins. *Biochem.Biophys.Res.Commun.*, Vol. 206, No. 1, pp. 326-334, 0006-291X; 0006-291X

Maggi, S.; Massidda, O.; Luzi, G.; Fadda, D.; Paolozzi, L. & Ghelardini, P. (2008). Division protein interaction web: identification of a phylogenetically conserved common interactome between Streptococcus pneumoniae and *Escherichia coli*. *Microbiology*, Vol. 154, No. Pt 10, pp. 3042-3052, 1350-0872; 1350-0872

Miller, J. H. (1972). Experiments in molecular genetics (), Cold Spring Harbor Laboratory Press, 0879691069, New York, United States of America

Murchie, A. I.; Clegg, R. M.; von Kitzing, E.; Duckett, D. R.; Diekmann, S. & Lilley, D. M. (1989). Fluorescence energy transfer shows that the four-way DNA junction is a right-handed cross of antiparallel molecules. *Nature*, Vol. 341, No. 6244, pp. 763-766, 0028-0836; 0028-0836

Ormo, M.; Cubitt, A. B.; Kallio, K.; Gross, L. A.; Tsien, R. Y. & Remington, S. J. (1996). Crystal structure of the Aequorea victoria green fluorescent protein. *Science*, Vol. 273, No. 5280, pp. 1392-1395, 0036-8075; 0036-8075

Perozzo, M. A.; Ward, K. B.; Thompson, R. B. & Ward, W. W. (1988). X-ray diffraction and time-resolved fluorescence analyses of *Aequorea* green fluorescent protein crystals. *J.Biol.Chem.*, Vol. 263, No. 16, pp. 7713-7716, 0021-9258; 0021-9258

Potluri, L.; Karczmarek, A.; Verheul, J.; Piette, A.; Wilkin, J. M.; Werth, N.; Banzhaf, M.; Vollmer, W.; Young, K. D.; Nguyen-Disteche, M. & den Blaauwen, T. (2010). Septal and lateral wall localization of PBP5, the major D,D-carboxypeptidase of *Escherichia coli*, requires substrate recognition and membrane attachment. *Mol.Microbiol.*, Vol. 77, No. 2, pp.300-323, 1365-2958; 0950-382X

Prasher, D. C.; Eckenrode, V. K.; Ward, W. W.; Prendergast, F. G. & Cormier, M. J. (1992). Primary structure of the *Aequorea victoria* green-fluorescent protein. *Gene*, Vol. 111, No. 2, pp. 229-233, 0378-1119; 0378-1119

Rossi, F.; Charlton, C. A. & Blau, H. M. (1997). Monitoring protein-protein interactions in intact eukaryotic cells by beta-galactosidase complementation. *Proc.Natl. Acad.Sci.U.S.A.*, Vol. 94, No. 16, pp. 8405-8410, 0027-8424; 0027-8424

Sauvage, E.; Kerff, F.; Terrak, M.; Ayala, J. A. & Charlier, P. (2008). The penicillin-binding proteins: structure and role in peptidoglycan biosynthesis. *FEMS Microbiol.Rev.*, Vol. 32, No. 2, pp. 234-258, 0168-6445; 0168-6445

Shaner, N. C.; Campbell, R. E.; Steinbach, P. A.; Giepmans, B. N.; Palmer, A. E. & Tsien, R. Y. (2004). Improved monomeric red, orange and yellow fluorescent proteins derived from *Discosoma sp.* red fluorescent protein. *Nat.Biotechnol.*, Vol. 22, No. 12, pp. 1567-1572, 1087-0156; 1087-0156

Shaner, N. C.; Lin, M. Z.; McKeown, M. R.; Steinbach, P. A.; Hazelwood, K. L.; Davidson, M. W. & Tsien, R. Y. (2008). Improving the photostability of bright monomeric orange and red fluorescent proteins. *Nat.Methods*, Vol. 5, No. 6, pp. 545-551, 1548-7105; 1548-7091

Shimomura, O. (1979). Structure of the chromophore of Aequorea green fluorescent protein . *FEBS Lett*, Vol. 104, No. 2, pp. 220-222,

Sourjik, V. & Berg, H. C. (2002). Receptor sensitivity in bacterial chemotaxis. *Proc.Natl.Acad.Sci.U.S.A.*, Vol. 99, No. 1, pp.123-127, 0027-8424; 0027-8424

Sourjik, V. & Berg, H. C. (2004). Functional interactions between receptors in bacterial chemotaxis. *Nature*, Vol. 428, No. 6981, pp. 437-441, 1476-4687; 0028-0836

Sourjik, V.; Vaknin, A.; Shimizu, T. S. & Berg, H. C. (2007). In vivo measurement by FRET of pathway activity in bacterial chemotaxis. *Methods Enzymol.*, Vol. 423, pp. 365-391, 0076-6879; 0076-6879

Strauch, E. M. & Georgiou, G. (2007). A bacterial two-hybrid system based on the twin-arginine transporter pathway of *E. coli. Protein Sci.*, Vol. 16, No. 5, pp. 1001-1008, 0961-8368; 0961-8368

Thaler, C.; Vogel, S. S.; Ikeda, S. R. & Chen, H. (2006). Photobleaching of YFP does not produce a CFP-like species that affects FRET measurements. *Nat.Methods*, Vol. 3, No. 7, pp. 491; author reply 492-3, 1548-7091; 1548-7091

Ullmann, A.; Jacob, F. & Monod, J. (1967). Characterization by in vitro complementation of a peptide corresponding to an operator-proximal segment of the beta-galactosidase structural gene of *Escherichia coli*. *J.Mol.Biol.*, Vol. 24, No. 2, pp. 339-343, 0022-2836; 0022-2836

Valentin, G.; Verheggen, C.; Piolot, T.; Neel, H.; Coppey-Moisan, M. & Bertrand, E. (2005). Photoconversion of YFP into a CFP-like species during acceptor photobleaching FRET experiments. *Nat.Methods*, Vol. 2, No. 11, pp. 801, 1548-7091; 1548-7091

Vidal, M. & Legrain, P. (1999). Yeast forward and reverse 'n'-hybrid systems. *Nucleic Acids Res.*, Vol. 27, No. 4, pp. 919-929, 0305-1048; 0305-1048

Vladimirov, N. & Sourjik, V. (2009). Chemotaxis: how bacteria use memory. *Biol.Chem.*, Vol. 390, No. 11, pp. 1097-1104, 1437-4315; 1431-6730

Vollmer, W. & Bertsche, U. (2008). Murein (peptidoglycan) structure, architecture and biosynthesis in *Escherichia coli*. *Biochim.Biophys.Acta*, Vol. 1778, No. 9, pp. 1714-1734, 0006-3002; 0006-3002

Vollmer, W.; Blanot, D. & de Pedro, M. A. (2008). Peptidoglycan structure and architecture. *FEMS Microbiol.Rev.*, Vol. 32, No. 2, pp. 149-167, 0168-6445; 0168-6445

Vollmer, W. & Seligman, S. J. (2010). Architecture of peptidoglycan: more data and more models. *Trends Microbiol.*, Vol. 18, No. 2, pp. 59-66, 1878-4380; 0966-842X

Volz, K. & Matsumura, P. (1991). Crystal structure of *Escherichia coli* CheY refined at 1.7-A resolution. *J.Biol.Chem.*, Vol. 266, No. 23, pp.15511-15519, 0021-9258; 0021-9258

Wehrman, T.; Kleaveland, B.; Her, J. H.; Balint, R. F. & Blau, H. M. (2002). Protein-protein interactions monitored in mammalian cells *via* complementation of beta -lactamase enzyme fragments. *Proc.Natl.Acad.Sci.U.S.A.*, Vol. 99, No. 6, pp. 3469-3474, 0027-8424; 0027-8424

Weiss, D. S.; Pogliano, K.; Carson, M.; Guzman, L. M.; Fraipont, C.; Nguyen-Disteche, M.; Losick, R. & Beckwith, J. (1997). Localization of the *Escherichia coli* cell division protein FtsI (PBP3) to the division site and cell pole. *Mol.Microbiol.*, Vol. 25, No. 4, pp. 671-681, 0950-382X; 0950-382X

Weiss, D. S.; Chen, J. C.; Ghigo, J. M.; Boyd, D. & Beckwith, J. (1999). Localization of FtsI (PBP3) to the septal ring requires its membrane anchor, the Z ring, FtsA, FtsQ, and FtsL. *J.Bacteriol.*, Vol. 181, No. 2, pp. 508-520, 0021-9193; 0021-9193

Willemse, J.; Borst, J. W.; de Waal, E.; Bisseling, T. & van Wezel, G. P. (2011). Positive control of cell division: FtsZ is recruited by SsgB during sporulation of *Streptomyces*. *Genes Dev.*, Vol. 25, No. 1, pp. 89-99, 1549-5477; 0890-9369

Wlodarczyk, J.; Woehler, A.; Kobe, F.; Ponimaskin, E.; Zeug, A. & Neher, E. (2008). Analysis of FRET signals in the presence of free donors and acceptors. *Biophys.J.*, Vol. 94, No. 3, pp. 986-1000, 1542-0086; 0006-3495

Wolfe, S. A.; Greisman, H. A.; Ramm, E. I. & Pabo, C. O. (1999). Analysis of zinc fingers optimized *via* phage display: evaluating the utility of a recognition code. *J.Mol.Biol.*, Vol. 285, No. 5, pp. 1917-1934, 0022-2836; 0022-2836

Zhao, R.; Collins, E. J.; Bourret, R. B. & Silversmith, R. E. (2002). Structure and catalytic mechanism of the *E. coli* chemotaxis phosphatase CheZ. *Nat.Struct.Biol.*, Vol. 9, No. 8, pp. 570-575, 1072-8368; 1072-8368

11

DNA-DNA Recognition:
From Tight Contact to Fatal Attraction

Youri Timsit

Genomic and Structural Information,
CNRS - UPR2589, Institute of Microbiology of the Mediterranean,
Aix-Marseille University,
Science Park Luminy, Marseille,
France

1. Introduction

The close approach of double helices plays essential roles in the architecture and catalysis of nucleic acids. Transient or long-term DNA-DNA interactions occur in the cell and participate in various genetic functions. For example, bringing DNA sites into proximity is required for DNA recombination, chromatin packaging and building architectural complexes that control transcription and replication (Echols, 1990; Grosschedl, 1995; Segal and Widom, 2009). Thus, the juxtaposition of DNA double helices in a crossover arrangement represents a ubiquitous motif in higher-order DNA structures and is known to be implicated *in* the genetic functions. The detailed knowledge of the structure and energetics of close DNA-DNA interactions is therefore indispensable for a complete understanding of these functions at the molecular level. Since the backbone of nucleic acids is negatively charged, the close approach of double helices requires cations or polyamines that are present in the cell to reduce their electrostatic repulsion (Bloomfield, 1996). Although similar electrostatic rules govern the assembly of RNA and DNA, the helical packing modes of the two molecular cousins differ as a consequence of their distinct secondary structures. Indeed, in addition to adopt a regular A-conformation, the RNA molecules most often fold into more complex structural motifs due to the presence of the 2'-hydroxyl group and extended base pairing rules (Leontis and Westhof, 2001). Thus, RNA structures are characterized by flourishing modes of tertiary interactions, from simple inter-helical interactions to the docking of a wide repertoire of sequence-dependent 3D motifs (Leontis et al., 2006; Batey et al., 1999). In contrast, DNA molecules mainly form a regular B-DNA double helix stabilized by canonical Watson-Crick base pairing. Consequently, following the evolutionary choice of austerity, the building of higher-order DNA structures is mainly directed by the B-DNA double helix geometry and chirality (Timsit and Moras, 1994).

Although short-range contacts between double helices have been considered to be strongly repulsive, arrays of parallel stacks of helices are formed under various conditions of condensing agents and may form organized phases or DNA liquid crystals (Bloomfield, 1996; Strey et al., 1998). In such arrangements, the inter-axial distance between double helical segments is about 25-32 Å and thus DNA duplexes do not form direct intermolecular interactions (Schellman and Parthasarathy, 1984; Raspaud et al., 2005). In these conditions, the parallel packing of helices is only moderately influenced by the helical nature of DNA (Kornyshev and Leikin, 1998; Minsky, 2004; Kornyshev et al., 2005). However, recent theoretical and experimental studies have indicated that close DNA-DNA interactions can occur in the presence of divalent cations (Qiu et al., 2006; Tan and Chen, 2006; Inoue et al., 2007; Qiu et al., 2007). Indeed, crystallographic studies have shown that both DNA geometry and chirality has a more profound effect on the association of closely-packed DNA helices, as in tight crossover arrangements. Early theoretical studies of helical packing based on simple geometric rules (Srinivasan and Olson, 1992) correctly predicted many of the helical packing modes observed in the crystal structures of nucleic acids (Timsit and Moras, 1992, Murthy and Rose, 2000). This remarkable fit between theoretical and experimental packing without taking into account electrostatic forces therefore indicates that helical interactions in crystals not only minimize the electrostatic repulsion between the negatively charged sugar-phosphate backbones but also optimise the docking between the complementary surfaces of the double helices. Indeed, DNA crystals in which electrostatic repulsion between double helices is naturally minimized have unveiled helical packing rules that are dictated by the chirality of the DNA double helix (Timsit and Moras, 1994). Although each type of double helix exposes different accessible surfaces and charged groups that predispose them to interact with themselves in a different manner (Lu et al., 2000), the handedness of A- and B-double helices exerts a common constraint on their packing. Thus, the mutual fit of the backbone into the groove specifically generates right-handed crossovers that display particularly interesting structural and functional properties.

2. From DNA chirality to stable right-handed crossovers

B-DNA helices can assemble into right-handed DNA crosses by the mutual fit of their sugar-phosphate backbone into the major groove (fig. 1a). These right-handed crossovers are characterized by positive values of the crossing angle. The phosphate group penetrates into the major groove to form hydrogen bonds with the amino groups of cytosines (see below) (Timsit et al., 1989). Remarkably, most of the right-handed crosses examined to date are assembled by the major groove-backbone interaction, involve cytosine-phosphate group interaction (fig. 2a) and is stabilized by divalent cations. Although less frequent in DNA crystals, minor groove-backbone interaction has also been observed (Wood et al., 1997). This mode of interaction has also been observed in a molecular dynamics study of mini-circles of DNA (Mitchell et al., 2011). Left-handed B-DNA crossovers that are characterized by negative values (-40 and -80°) of the crossing angles prevent the self-fitting of the double helices. Within this geometry, the helices are juxtaposed by groove-groove interactions to minimize their electrostatic repulsion (Timsit et al., 1999) (fig. 1b). However, their mode of interaction is neither stabilized by direct sequence-specific contacts between DNA segments, nor by intermolecular divalent cation bridges.

In contrast, the A-form double helices preferentially self-assemble into right-handed crossovers formed by minor-groove backbone interactions. In A-form double helices, this is the shallow minor groove of the A-form that is devoted to intermolecular interactions. One of the most common elements of the ribosome structure is the interaction of RNA double helices via minor grooves (Nissen et al., 2001). Inter-helical packing involving minor-groove

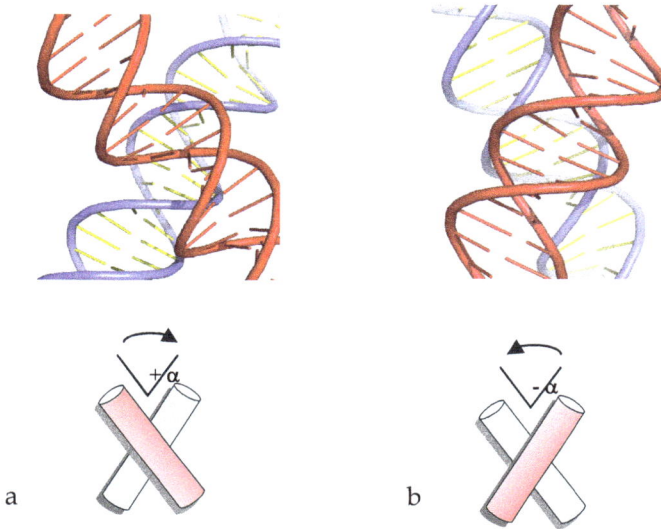

a: Right-handed crossover assembled by the mutual fit of the backbones into the major-groove; b: Left-handed crossover assembled by major groove-major groove interaction

Fig. 1. Chiral B-DNA crossovers

backbone interactions have been observed in the crystal packing of many RNA oligonucleotides (Schindelin et al., 1995; Baeyens et al., 1995). Moreover, this so called "along-groove" packing motif that has been also observed within the structure of the 23S RNA of the large ribosomal subunit is thought to play a role in ribosomal function such as tRNA translocation (Gagnon and Steinberg, 2002; Gagnon et al, 2006). The role of the DNA sequences is also different in the packing of A- and B-DNA helices. Indeed, a comparison of DNA crystal packing modes revealed that the interactions between A-DNA helices are much less dependent from the DNA sequence than the B-DNA ones (Timsit and Moras, 1992). Probably because the shallow minor groove of the A-form provides the opportunity to form many van de Waals and hydrophobic interactions, their stable association has been found less dependent from the formation of specific hydrogen bonds. Thus, a constant geometry of the A-form assemblies can be maintained for a large variety of sequences (Timsit, 1999). In contrast, the tight association of B-DNA helices is greatly influenced by the DNA sequence.

3. Cytosine and DNA self-assembly

Right-handed crossovers of B-form DNA double helices are therefore unique in that they are assembled by a sequence-dependent interaction. While the B-DNA double helix dictates the geometry of inter-helical assembly, cytosines play a key role for controlling the interaction through specific interaction of their N4 amino groups with phosphate groups (Timsit et al., 1989; Timsit and Moras, 1991; 1994). Moreover, a recent survey of the Nucleic Acids Database shows that, without exception, cytosine-phosphate interactions are strictly required for stabilising right-handed DNA crossovers (fig. 2a). Probably due to the vicinity of the N7 group that displays a negative potential, the N6 amino group of adenine has been not found to substitute the N4 amino group of cytosine for this type of interaction (Timsit and Varnai, 2011).

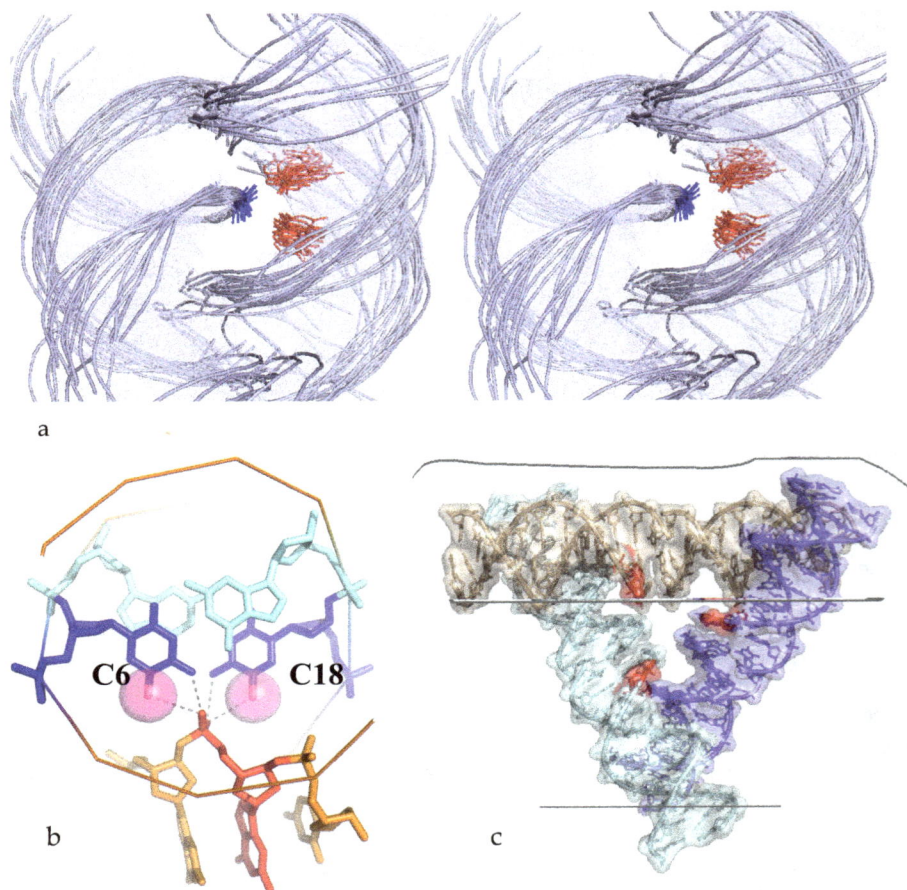

a: Stereo-view of superimposed self-fitted right-handed DNA crossovers found in the Nucleic Acids Database. The cytosines involved in the interaction are represented in red; b: 5-Methyl cytosine stabilizes the groove-backbone interaction through the formation of C-H...O interactions with the phosphate group; c: The role of the sequence in DNA self-assembly: the cytosines (represented in red) dictate the organisation of the triangular motifs

Fig. 2. Cytosine and DNA-DNA recognition

Crystallographic studies of methylated DNA duplexes showed that C5-methyl cytosines also promote and stabilize the formation of DNA crossovers at the modified C5-mpG sequences (Mayer-Jung et al, 1997). The two methyl groups form a hydrophobic clamp which traps the incoming phosphate through C-H...O interactions that further stabilize the helical assembly (fig. 2b).

Overall, these works have put in light the particular role of cytosine bases for controlling spatial organisation and the stability of tertiary DNA assemblies. The formations of inter or intramolecular H-bond between the N4 amino group of cytosine and a phosphate group play a key role for controlling DNA-DNA interactions in a sequence dependent manner. These studies provided the basic principles for designing DNA sequence that control the precise organisation of double helices into a 3D lattice. This method for designing DNA crystals has been used successfully for the systematic crystallisation of DNA molecules of various sequences and sizes (Timsit and Moras, 1992; 1996) as for example, the spectacular DNA triangular motifs of DNA dodecamer and decamer duplexes (Timsit et al., 1989; 1991; 1994) (fig. 2c). The DNA triangles recently designed by Zheng et al. (2009) fit remarkably well into this conceptual and methodological framework. Indeed, the overall architecture of these triangles is also dictated by both the cytosine positioning and the double helix geometry (Timsit and Varnai, 2011). The general applicability of these principles has subsequently been supported by the crystal structure of 4-way junctions whose stability also depends on cytosines placed at specific position (Ortiz-Lombardia et al., 1999; Khuu et al., 2006).

4. Differential stability of chiral crossovers

The free energy of interactions of DNA duplexes in right and left-handed crossovers as a function of divalent cation concentration in solution has been investigated using molecular dynamics simulations (Varnai and Timsit, 2010). This study showed that right-handed DNA crossovers (fig. 1a) are thermodynamically stable in solution in the presence of divalent cations. Consistent with recent theoretical and experimental observations of close DNA-DNA interactions in the presence of divalent cations (Qiu et al., 2007; Tan and Chen, 2006; Inoue et al., 2007), a short-range attraction of about -4 kcal mol^{-1} between the self-fitted duplexes was observed in the presence of divalent cations (Varnai and Timsit, 2010). Attractive forces at short-range stabilize the DNA-DNA association with inter-axial separation of helices less than 20 Å. Right-handed crossovers, however, dissociate in the presence of monovalent ions only. In solution, the acute angle by which the two B-DNA duplexes cross one another in the right-handed geometry fluctuates around an average value of 84 ± 6°, a value close to that observed in the R3 crystal packing (Timsit et al., 1989). The tight spread around this angle indicates that the major groove induces a strict geometric constraint on the mutually fitted structures. Thus, the crossing angle is mainly influenced by the helical geometry of the B-form duplex that effectively constrains the structure of the assembly by steric interactions in the major groove (Timsit and Moras, 1994). Consistent with the crystallographic studies, molecular dynamics simulation showed that two helices remain assembled by specific cytosine-phosphate interactions and bridging Mg^{2+} ions at the duplex interface. The repulsion of the negatively charged backbone is circumvented both by the specific relative orientation of helices and by the presence of Mg^{2+}. Therefore, similar structural features of the right-handed crossover are present in solution and in the crystal environment. Simulated DNA triangles constructed from 20-mer sequences are also stabilized by similar interactions in solution. In contrast, left-handed crossovers are unstable at similar ionic conditions and resulted in a swift dissociation of the helices. Without specific intermolecular interaction, left-handed helix juxtapositions by major groove-major groove interaction (fig. 1b) are stable only in the crystallographic environment but appeared to be unstable in solution.

5. Role of divalent cations in DNA assembly

Molecular dynamic simulation studies have also shown that the stabilisation of right-handed crossover increases as the Mg^{2+}/duplex stoichiometric ratio increases. A minimum of 8 Mg^{2+} per duplex is required to keep the duplexes anchored together with an associated binding free energy of about -4 kcal mol[-1]. Higher Mg^{2+} concentrations (16 Mg^{2+}/duplex) strengthen the helical interaction further and increase the associated binding free energy to -7 kcal mol[-1]. At lower Mg^{2+} concentrations (4 Mg^{2+}/duplex) no net attraction was visible. However, monovalent ions cannot replace the effect of Mg^{2+} to induce attraction between DNA helices even at high Na^+ concentration (56 Na^+/duplex). Importantly, during the simulation, Mg^{2+} ions occupied the divalent cation binding sites observed in the crystal structure of self-fitted duplexes. In contrast, no specific Mg^{2+} binding site was observed in the control simulations of isolated duplexes and hence we suggest that specific binding sites are formed simultaneously with the formation of the crossover structure. These data fit well with the crystallographic studies that showed that the diffraction power of crystals of duplexes assembled via groove-backbone interactions was strictly correlated with the Mg^{2+}/duplex stoichiometric ratio. Best diffracting crystals were obtained with 16 Mg^{2+} per oligonucleotide while very large crystals obtained with 1 Mg^{2+} per oligonucleotide did not diffract at all (Timsit et al., 1989).

The strict requirement for Mg^{2+} to stabilize tight DNA-DNA interactions is also consistent with recent experimental and theoretical data. For example, SAXS and light scattering experiment indicated DNA-DNA repulsion in the presence of monovalent ions (up to [Na^+] of 600 mM) but increasing attraction above [Mg^{2+}] of 50 mM (Qiu et al., 2007). Also, recent theoretical work that used the tightly bound ion model found that helices repel one other in the presence of monovalent ions while divalent cations are able to induce attraction between two DNA helices (Tan and Chen, 2006). Regions of very tight contacts between DNA segments have been observed in cryo-EM images of supercoiled DNA vitrified from a solution containing 10 mM Mg^{2+} (Adrian et al., 1990). Since such close DNA-DNA interactions were considered repulsive until recently, these observations were regarded as technical artefacts induced by cryo-congelation. These striking results were, however, supported more recently by AFM studies on supercoiled DNA (Lyubchenko and Shlyakhtenko, 1997; Shlyakhtenko, 2003). In addition, a recent study has also reported that DNA duplexes can self-assemble at nanomolar DNA concentrations in the presence of Mg^{2+} (Inoue et al., 2007). Divalent cations and in particular Mg^{2+} ions are also required for the folding of both DNA and RNA molecules. They mediate the folding of Holliday junctions from a planar open structure into a compact stacked conformation (Lilley, 2000). Indeed, four-way junctions and right-handed crosses share an analogous geometry that is *stabilized* by similar tertiary interactions involving cytosines and Mg^{2+} (Timsit and Moras, 1991; Ortiz-Lombardia et al., 1999). The folding of particular RNA motifs found in many functional RNA molecules also requires specific divalent cations (van Buuren et al., 2002; Tinoco et al., 1997; Klein et al., 2004; Woodson, 2005). A common feature in most of these structures is the anchoring of a phosphate group to a guanine base through a divalent cation bridge. Thus, among all cations available in physiological conditions, divalent cations have the unique property of stabilising specific and tight intra- and intermolecular interactions between nucleic acid segments by forming guanine-phosphate bridges. In contrast, monovalent ions that are more diffuse around DNA and RNA may have an important role in the long-range steering of duplexes, as for example in the parallel alignment of double helices found in liquid crystals (Strey et al., 1998; Murthy and Rose, 2000).

6. DNA sequence and higher-order structures

These findings revealed that like a directing piece for a "supramolecular construction set", the B-DNA double helix dictates the overall geometry of DNA self-fitted assemblies. Its periodic structure imposes elementary geometric constraints which restrict the spatial organization of DNA segments. They have also demonstrated that the DNA sequence encodes specific signals for positioning intra- or intermolecular DNA-DNA interactions. Cytosine and cluster of guanine bases that constitute preferential divalent cation binding sites act conjointly to define the emplacement of right-handed crossovers. Conversely, AT rich regions are less suitable for tight DNA-DNA interactions. These data may be useful for understanding the organisation of DNA higher-order structures such as, for example, the 30 nm chromatin fibre (Robinson and Rhodes, 2006; Wu et al., 2007). It is well established that electrostatic forces govern primarily the folding of the chromatin fibre (Robinson et al., 2006; Dorigo et al., 2004). The strong dependence of chromatin compaction on cations is reminiscent of a process that involves DNA-DNA interactions. Although recent experimental data support a compact interdigitated solenoidal structure (Robinson et al., 2006), the exact mode of organisation of nucleosomes and linker DNA within the chromatin is still a matter of controversy (Schalch et al., 2005; Dorigo et al., 2004). Consequently, whatever its exact mode of assembly is, chromatin folding is expected to involve close interactions between the linker DNA and/or between the nucleosomal DNA (fig. 3).

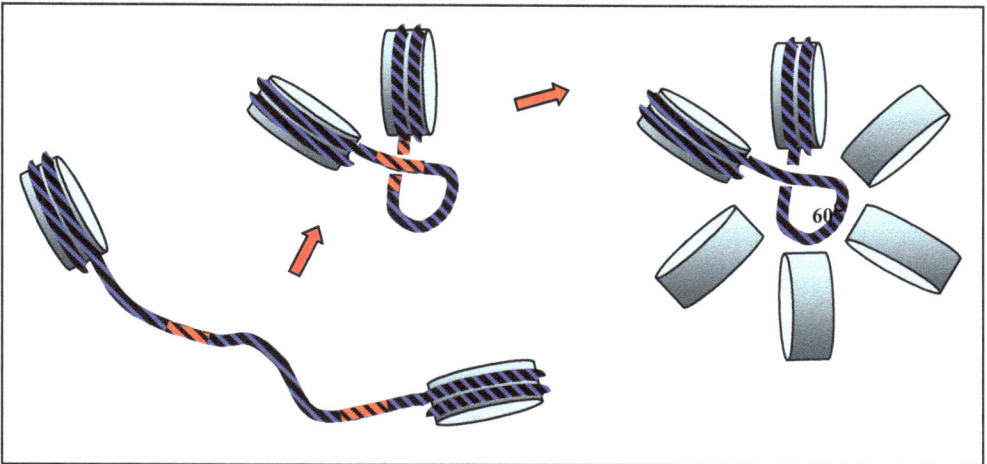

Fig. 3. DNA-DNA recognition may contribute to the assembly of the chromatin fibre.

The GC sequences suitable to form tight DNA-DNA interaction are represented in red. The model proposes that they could contribute to organise tighly packed region in chromatin. The geometric constraints imposed by the groove-backbone interaction may also help to organise the fibre into a symmetric array.

Further, the stability of right-handed crossovers at close to physiological conditions supports earlier hypotheses that groove-backbone fitting organises the nucleosomal or linker assembly within the chromatin fibre (Timsit and Moras, 1991; 1994). Moreover, groove-backbone interactions have been observed in the crystal packing of nucleosomes (Davey et al, 2002; Schalch et al., 2005) (fig. 4) and close DNA-DNA interactions are seen in the recent all-atom model of the chromatin fiber (Wong et al., 2007). Therefore, the cell may dispose of a collection of direct DNA-DNA interactions with varying degree of stability that can be exploited for tuning chromatin compaction. In addition, the finding that 5-methyl cytosine promotes tight DNA-DNA interaction may help to understand the role of DNA methylation on the compaction of the chromatin.

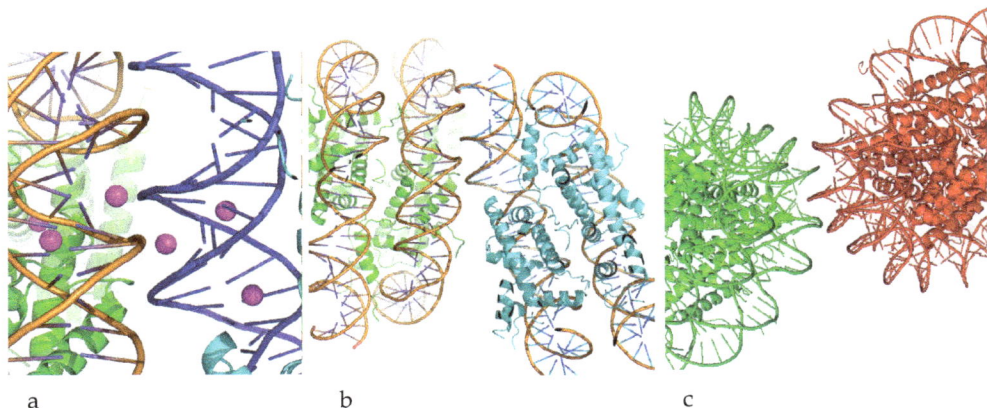

a   b   c

a: Detail of the interaction between two nucleosomes in their crystal packing; the divalent cations that *stabilize* the interaction are represented in violet sphere (Davey et al., 2002); b: global view of the packing of two nucleosomes intreracting by groove-backbone interaction; c: another mode of self-fitting of two nucleosomes in the crystal structure of Shalch et al. 2005.

Fig. 4. Tight right-handed crossovers between nucleosomal DNA observed in their crystal structure

7. From crossover geometry to DNA topology

DNA-DNA interactions have been also found to play a critical role in DNA topology. Indeed, within the interwound plectonemic supercoiled DNA, closely packed regions with intersegmental contacts occur in the presence of divalent cations, under physiological conditions (Vologodskii & Cozzarelli, 1994; Lyubchenko & Shlyakhtenko, 1997; Shlyakhtenko et al., 2003). The importance of such close contacts has been also noted for the knotting of supercoiled DNA (Schlick & Olson, 1992). DNA supercoiling participates in many cellular processes in both prokaryotes and eukaryotes, such as remote gene regulation and site-specific recombination (Wang, 2002; Travers & Muskhelishvili, 2005). Although DNA is mainly negatively supercoiled in mesophilic cells, transcription and DNA replication may generate domains of positively supercoiled DNA *in vivo* (Stupina and Wang, 2004; Postow et al., 2001; Liu & Wang, 1987). Type II topoisomerases can modulate the topological state of DNA by catalyzing the double strand passage reaction. These enzymes not only play a major role in disentangling sister chromatids during replication but also are crucial in maintaining the fine balance of superhelical density and regulating the topological state of genomic DNA (Wang, 2002).

A remarkable property of type II topoisomerases is their sense of global DNA topology. For example, when a negatively supercoiled ring is singly linked to a nicked ring, these enzymes preferentially unlink the ring rather to remove the supercoils (Roca and Wang, 1996). Topoisomerases II can also simplify DNA topology and reduce the fraction of knotted or catenated circular DNA molecules well below thermodynamic equilibrium values (Rybenkov et al., 1997). They are also capable of chiral discriminiation between knots of opposite sign (Shaw and Wang, 1997) and some of them such as DNA gyrase, topoisomerase IV and human topoisomerase IIa can discern the sign of supercoiled DNA in acting preferentially on positive supercoiled DNA (Crisona et al., 2000, Charvin et al., 2003, Stones et al., 2004; Nöllmann et al., 2007; McClendon et al., 2005). Due to this striking ability to sense the global properties of DNA from local interactions, they have been compared to Maxwell's demons (Pulleyblank, 1997). How topoisomerases can discriminate between the different global topologies of a much larger DNA? Several hypotheses have been postulated for explaining this phenomenon. For example, the protein induces a sharp bend in DNA at the binding site that provides unidirectional strand passage (Rybenkov et al., 2001; Dong and Berger, 2007). A kinetic proofreading model that requires two separate topoisomerase-DNA collisions for segment passage and a model of interaction of the enzyme with three segment-interactions have been also proposed (Yan et al., 1999; Trigueros et al., 2004). Alternatively, it has been suggested that the topological information may be embodied in the local geometry of DNA crossings and that topoisomerases act at the hooked juxtapositions of the strands (Buck and Zechiedrich, 2004; Randall et al., 2006). Indeed, recent experimental studies support this view in showing that topoisomerase IV and DNA gyrase can discriminate the sign of supercoiled DNA on the basis of the geometry of the DNA crossovers (Crisona et al., 2000, Charvin et al., 2003, Stones et al., 2004; Nöllmann et al., 2007; Corbett et al., 2005). Knowing that type II topoisomerases recognise and act on DNA crossovers (Zechiedrich and Osheroff, 1990), the question can be formulated as: how type II topoismerases are able to distinguish the global DNA topology from the local information provided by the juxtaposition of DNA segments?

Applying crystallographic lessons to DNA topology brought useful insights to solve this question. Indeed, the interplay of local and global properties constitutes a key element in the cellular function of DNA and local intra- or intermolecular DNA-DNA interactions play a central role by establishing a link between the two hierarchical levels of structural organisation in DNA (Minsky, 2004). Thus, it has been demonstrated that the various topological states of the cell are associated with different inter-segmental interactions (Timsit and Varnai, 2010). Knowing that right- and left-handed crossovers not only differ by their geometry but also by their stability helped to understand the mechanism of chiral discrimination by type II topoisomerases. Indeed, the stable right-handed DNA crossovers constitute the most probable structure of site juxtaposition in physiological conditions. Consequently, right-handed crosses that occur preferentially in (+) supercoiled DNA for geometrical reasons, is also preferentially formed in the absence of superhelical stress, as in relaxed DNA, catenanes or loose knots for electrostatic reasons. Thus, whereas the unstable left-handed crossovers are exclusively formed in negatively supercoiled DNA, stable right-handed crossovers constitute the local signature of an unusual topological state in the cell, such as the positively supercoiled or relaxed DNA. The differential stability of crossovers may be therefore exploited for sensing the global topology of DNA from local interactions providing a simple mechanism for the local discrimination of global DNA topology. This suggested that type II topoisomerases may discriminate (-) supercoiling from other topological states in preferentially acting on stable right-handed crossovers. In addition, a recent study has demonstrated how binding right-handed crossovers across their large angle imposes a different topological link between the type II topoisomerase and the plectonemes of opposite sign. The different topological links affect the enzyme freedom of motion and processivity and provide an explanation for the chiral discrimination (Timsit, 2011).

8. Right-handed double helix and the evolutionary choice of DNA topology

The asymmetrical behaviour of supercoiled DNA of opposite signs may have contributed to orient early choices for DNA topology in the nascent DNA world. DNA topology has been the subject of adaptive pressure in organisms that live at different temperatures for maintaining the balance between the melting potential and functional stability (Forterre and Gadelle, 2009). In addition, it is well known that the dynamics of plectonemic DNA supercoiling plays a critical role in promoting interactions between remote sites in processes such as transcription initiation and site-specific recombination (Kanaar and Cozzarelli, 1992; Travers and Muskhelishvili, 2007). Such dynamic should be also finely tuned in function of temperature and the topological state of the supercoiled DNA. Indeed, several studies have shown that some particular local inter-segmental contacts alter the functional dynamic of supercoiled DNA (Minsky, 2004). Similarly, divalent cations that promote formation of stacked 4-way junctions (Lilley, 2000) considerably slow down the kinetics of spontaneous branch migration (Panyutin and Hsieh, 1994).

It is therefore likely that, among other physical properties of DNA, such as its anisotropic flexibility (Olson and Zhurkin, 2000; Travers, 2004), or the fact that DNA is more easily untwisted than overtwisted, the differential stability of chiral crossovers has influenced the choice of DNA topology in mesophilic cells. In particular, the formation of stable right-handed crossovers in relaxed or (+) supercoiled DNA may have posed challenges to mesophilic cells: in the presence of divalent cations, the stable inter-segmental interactions should make (+) supercoiled DNA significantly more "sticky" than (-) supercoiled DNA, along GC rich sequences. Indeed, from a functional point of view, right-handed DNA crosses can be viewed as a Janus-like DNA structure. While the stable inter-segmental interactions can be useful for closely packaging DNA into higher-order DNA structures, they may have a detrimental effect by impeding the global dynamics of the genome, if they occur without control within a plectonemic supercoiled DNA. It is therefore possible that these two opposite features may have lead to different evolutionary strategies to adapt to mesophilic conditions where weak interactions that occur within right-handed crossovers can be expected to be stable.

Indeed, in contrast to life at high temperature that can tolerate various topological states of DNA –from negative supercoiled DNA to slightly positively supercoiled DNA- (Brochier-Armanet and Forterre, 2006; Charbonnier and Forterre, 1994; Guipaud et al., 1997; Lopez-Garcia et al., 2000; Marguet and Forterre, 1994), adaptation to mesophilic life is much more contraining on the topology of DNA: the genome of mesophilic organisms, including bacteria, archaea and eukarya, is systematically (-) supercoiled. All mesophilic bacteria have a DNA gyrase that introduce (-) supercoiling in a plectonemic form (Forterre and Gadelle, 2009). Particularly interesting is the case of mesophilic archaea. They have either acquired a gyrase that introduce negative supercoiling, or histones that wrap DNA into toroidal supercoils (Forterre and Gadelle, 2009). In other words, mesophilic organisms appear to have evolved to strictly avoid the presence of permanently relaxed or (+) supercoiled DNA in their genome. As mentioned above these topological states are expected to impede the dynamics of supercoiled DNA and affect functions. Maintaining permanent (-) supercoiling could therefore be viewed as preventing sticky interactions and promoting the "fluidity" required for various functions. Our model can also account for the observation that hyperthermophilic archaea tolerate other topological states of DNA, such as the relaxed or slightly (+) supercoiled states. Indeed, higher temperatures would decrease the stability of right-handed crossovers and restore the relative mobility of DNA segments.

Second, wrapping DNA around histones in mesophilic archaea and eukarya can be viewed as an alternative mode of adaptation to the presence of sticky DNA-DNA interactions in their genome. It can be speculated that this regular mode of DNA packaging allows the organism to precisely control the position of right-handed crosses and to exploit their physical properties. For example, it has been proposed that DNA self-fitting may contribute to stabilise the interactions between nucleosomes or DNA linkers within the chromatin fibre (Robinson and Rhodes, 2006).

9. Is DNA a helicase?

In another hand, DNA melting and strand-separation is indispensable for the initiation of the replication and transcription. However, the opening of base pairs in mesophilic conditions in which the double helix is expected to be stable represents another challenging problem. For example, it is commonly thought that the underwound DNA in (-) supercoiled DNA facilitates the strand separation required for transcription or DNA recombination in mesophilic bacteria (Travers and Muskhelishvili, 2007; Benham and Mielke, 2005; Vologodskii and Cozzarelli, 1994). Although it is commonly thought that the less stable AT base pairs are the preferential sites of initiation of DNA melting, experimental data have shown that the question is more complex. For example, it has been shown that the large scale opening of A+T rich sequences within supercoiled DNA molecule is suppressed by salts. The authors concluded that the spontaneous opening is highly unlikely under physiological conditions of salt, temperature and superhelicity and that proteins would be required to facilitate opening transitions (Bowater et al., 1994). On the basis of a particular behaviour of specific DNA sequences submitted to tight DNA-DNA interaction, the present study proposes an alternative model for the induction of strand separation in physiological conditions.

Indeed, crystallographic studies have revealed that the groove-backbone interaction can profoundly modify the secondary structure of the double helix in specific sequences (Timsit and Moras, 1995). For example, in the *tet* d(ACCGGCGCCACA) crystals, the groove-backbone interaction has triggered pre-melting of the double helix (Timsit et al., 1991). This study showed indeed that, in pulling-out the interacting cytosine, the phosphate group inserted into the major groove has broken the corresponding GC base pair in the middle of the duplex. This event has been then propagated on both sides of the anchoring points, to the neighbouring base pairs, thus leading to the concerted shift of the base pairing along the major groove (fig. 5). These observations were correlated with the great thermal sensitivity of the *tet* crystals and suggested an unstable state of the double helix induced by the tight

DNA-DNA interactions has been trapped within the crystals at low temperature. Indeed, the disruption of canonical Watson-Crick pairing to form less stable shifted non Watson-Crick pairing is thought to reduce the stability of the double helix.

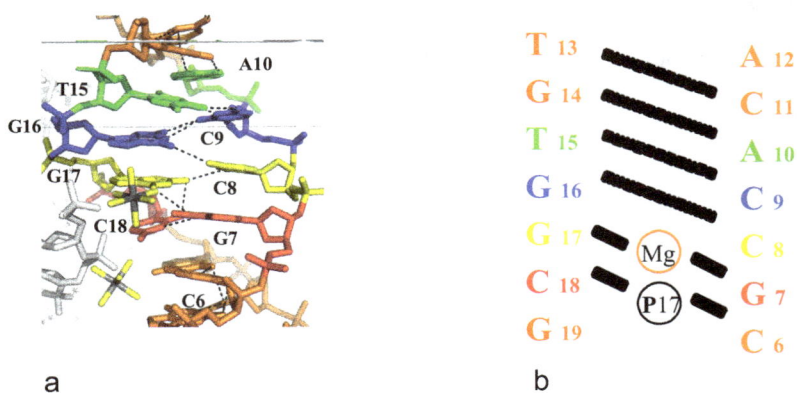

a: Shift of the base pairing in the *tet* dodecamer. The backbone of the fitted molecule is represented in grey. The bases that should be normaly paired are represented with the same color ; b : Schematic view of the shift of the base pairing. The phosphate group inserted into the major groove (P17) has induced the breaking of the G7-C18 base pair

Fig. 5. Base pair opening and helix destabisation induced by DNA-DNA interactions

It has been therefore proposed that this temperature sensitive structure represented an example of "premelted" state induced by DNA-DNA interaction within the crystal (Timsit et al., 1991; Timsit and Moras, 1995). This unique property was found to be related to special features of *tet* sequence. Indeed, although submitted to identical intermolecular contacts, the structure of the duplex *ras* d(ACCGCCGGCGCC) remained nearly unaffected and its crystals were stable up to 37°C. It was therefore concluded that due to its ability to form a rearranged pairing in the major groove and to the presence of highly unstable CA steps, the CCACA 5'-end of *tet* was responsible of its unusual properties (Timsit and Moras, 1995 ; 1996). This particular electrostatic partition of the strands, consisting of one strand with a set of major groove donor bases (amino groups of A or C) and another one with a set of major groove acceptor bases (carbonyl groups of G or T), is suitable for the propagation in a domino-like motion, of the alterations induced locally by the DNA-DNA intermolecular contacts.

A similar opening of a GC base pair induced by the insertion of phosphate group into the major groove has been also observed in the high resolution structure of a decamer duplex (Van Aalten et al., 1999). However, in that structure, the disruption of the base pairing remained located to the central base-pair and was not propagated since the sequence is not appropriate. In addition, many other examples have shown that the close approach of the helices contributes to destabilise the base pairs (Tarri and Secco, 1995). The analysis of these structures showed that both the duplex sequences and the tightness of the interaction modulate the structural response to DNA-DNA interactions, from the disruption of a single base-pair to the induction of a premelted structure. $(C/A)_n$ or $(T/G)_n$ sequences were proposed to be particularly sensitive to intermolecular interactions.

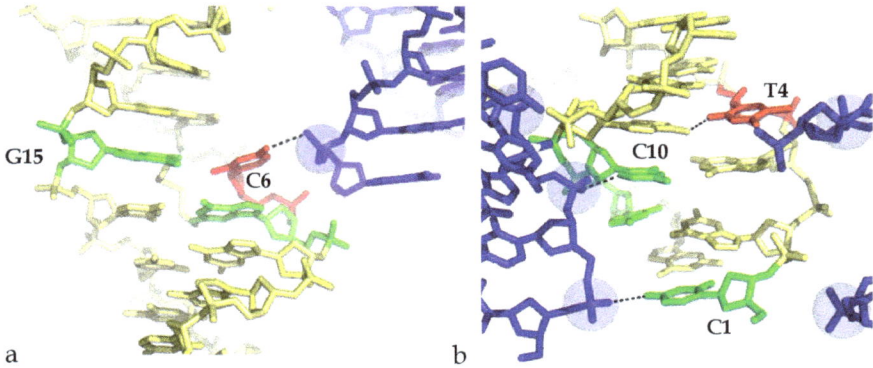

The backbone of the double helix that trigger the disruption of the base pair is represented in blue. a : Van Aalten et al. (1999) ; b : Tari and Secco (1995)

Fig. 6. Inter-helical interaction and base-pair opening

Overall, these findings suggested a molecular mechanism in which the melting of the double helix can be triggered by specific DNA-DNA intermolecular contacts. The close and specific approach of DNA segments occuring in genome packaging, DNA looping, (+) supercoiled DNA or synapsis pairing can trigger, in a sequence-dependent manner, the melting of the double helix and thus initiate the strand separation. In this model, both DNA-DNA

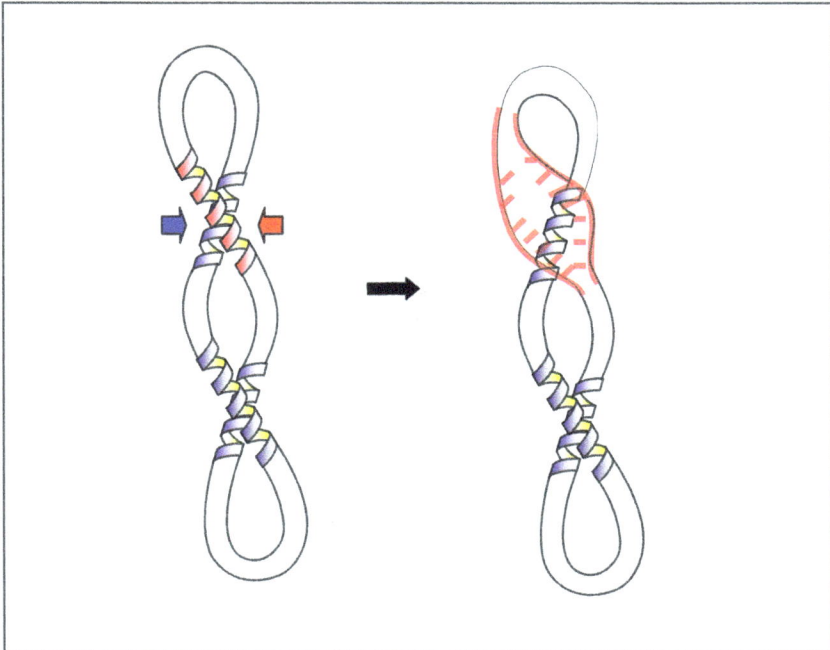

Fig. 7. DNA may be a helicase

interaction and specific sequences act together as a "catalyst" for triggering and stabilising higher energetic premelted states (transition state) of the DNA. These *findings* therefore suggest that DNA may realy be a helicase (fig. 7).

The close approach of two DNA segments in supercoiled DNA or higher-order structure can induce the initiation of strand separation in specific sequences such as $(C/A)_n$ or $(T/G)_r$ represented in red

10. Conclusion

This chapter shows firstly, that the double helix geometry and handedness play a critical role in the building rules of DNA higher-order structures. For example, due to the structural and electrostatic complementarity between the grooves and the sugar-phosphate backbone, the double helices can be mutually fitted by major groove-backbone interaction into stable right-handed crossovers. This interaction is specifically stabilised by cytosine bases of one helix that form hydrogen bonds with the phosphate group of another helix. This property has in turn important repercussions on DNA topology and on the mechanism of action of type II topoisomerases. Secondly, on the basis of crystallographic studies, it is shown here that the close DNA-DNA interactions can also trigger strand separation in a specific sequence context, thus raising the provocative question: is DNA a helicase? These observations provide alternative models for understanding mechanism DNA unwinding required in numerous genetic functions such as replication, transcription and recombination.

11. References

Adrian, M., ten Heggeler-Bordier, B., Wahli, W., Stasiak, A.Z., Stasiak, A. and Dubochet, J. (1990) Direct visualization of supercoiled DNA molecules in solution. *EMBO .J* 9, 4551-4554.

Baeyens, K., De Bondt, H.L. and Holbrook, S.R. (1996) Structure of an RNA double helix including uracil-uracil base pairs in an internal loop. *Nat. Struct. Biol.* 2, 56-62.

Batey, R.T., Rambo, R.P. and Doudna, J.A. (1999) Tertiary motifs in RNA structure and folding. *Angew. Chem. Int. Ed.* 38, 2326-2343.

Benham, C.J. and Mielke, S.P. (2005) DNA mechanics. *Annu. Rev. Biomed. Eng.* 7, 21–53.

Bowater, R.P., Aboul-ela, F. and Lilley, D.M. (1994) Large-scale opening of A + T rich regions within supercoiled DNA molecules is suppressed by salt. *Nucleic Acids Res.* 22, 2042-50.

Brochier-Armanet, C. and Forterre, P. (2006) Widespread archael reverse gyrase in thermophilic bacteria suggest a complex history of vertical inheritance and lateral gene transfers. *Archaea* 2, 83-93.

Bloomfield, V.A. (1996) DNA condensation. *Curr. Opin. Struct. Biol.* 6, 334-341.

Buck, G.R. and Zechiedrich, E.L (2004) DNA Disentangling by Type-2 Topoisomerases. *J. Mol. Biol.* 340, 933-939.

Charbonnier F, Forterre P (1994) Comparison of plasmid DNA topology among mesophilic and thermophilic eubacteria and archaebacteria. *J. Bacteriol.* 176, 1251-1259.

Charvin, G., Bensimon, D. and Croquette, V. (2003) Single-molecule study of DNA unlinking by eukaryotic and prokaryotic type-II topoisomerases. *Proc. Natl. Acad. Sci. USA* 100, 9820-9825.

Corbett, K.D, Schoeffler, A.J, Thomsen, N.D. and Berger, J.B. (2005) The structural basis for substrate specificity in DNA topoisomerase IV. *J. Mol. Biol.* 351, 545-561.

Crisona, N.J., Strick, T.R., Bensimon, D., Croquette, V. and Cozzarelli, N.R. (2000) Preferential relaxation of positively supercoiled DNA by E. coli topoisomerase IV in single-molecule and ensemble measurements. *Gene & Dev.* 14, 2881-2892.

Dong, K.C. and Berger, J.M. (2007) Structural basis for gate-DNA recognition and bending by type IIA topoisomerases. *Nature* 450, 1201-1206.

Echols, H. (1990) Nucleoprotein Structures Initiating DNA-Replication, Transcription, and Site-Specific Recombination. *J. Biol. Chem.* 265, 14697-14700.

Forterre, P. and Gadelle, D. (2009) Phylogenomics of DNA topoisomerases: their origin and putative roles in the emergence of modern organisms. *Nucleic Acids Res.* 37, 679-692.

Guipaud, O., Marguet, E., Noll, K., Bouthier de la Tour, C. and Forterre, P. (1997) Both gyrase and reverse gyrase are present in the hyperthermophilic bacterium Thermogata maritima. *Proc. Natl. Acad. Sci. USA* 94, 10606-10611.

Grosschedl, R. (1995) Higher-Order Nucleoprotein Complexes in Transcription - Analogies with Site-Specific Recombination. *Curr. Opin. Cell Biol.* 7, 362-370.

Inoue, S., Sugiyama, S., Travers, A.A. and Ohyama, T. (2007) Self-assembly of double-stranded DNA molecules at nanomolar concentrations. *Biochemistry* 46, 164-171.

Khuu, P., Regier Voth, A., Hays, F. and Ho, P.S. The stacked-X DNA Holliday junction and protein recognition *J. Mol. Recogn.* 19, 234-242 (2006).

Klein, D.J., Moore, P.B. and Steitz, T.A. (2004) The contribution of metal ions to the structural stability of the large ribosomal subunit. *RNA* 10, 1366-1379.

Kornyshev, A.A. and Leikin, S. (1998) Electrostatic interaction between helical macromolecules in dense aggregates: an impetus for DNA poly- and meso-morphism. *Proc. Natl. Acad. Sci. USA* 95, 13579-13584.

Kornyshev, A.A., Lee, D.J., Leikin, S., Wynveen, A. and Zimmerman, S.B. (2005) Direct observation of azimuthal correlations between DNA in hydrated aggregates. *Phys. Rev. Lett.* 95, 2537-2540.

Leontis, N.B. and Westhof, E. (2001). Geometric nomenclature and classification of RNA base pairs. *RNA* 7, 499-512.

Leontis, N.B., Lescoute, A., and Westhof, E. (2006). The buidling blocks and motifs of RNA architecture. *Curr. Opin. Struct. Biol.* 16, 279-287.

Lilley, D.M.J. (2000) Structures of helical junctions in nucleic acids. *Q. Rev. Biophys.* 33, 109-159.

Liu, L.F. and Wang, J.C. (1987) Supercoiling of the DNA template during transcription. *Proc. Natl. Acad. Sci. USA.* 84, 7024-7027.

Liu, Z., Deibler, R.W., Chan, H.S. and Zechiedrich, L. (2009) The why and how DNA unlinking. *Nucl. Acid Res.* 37 661-671.

Liu, Z., Mann, J.K., Zechiedrich, E.L. and Chan, H.S. (2006) Topological information embodied in local juxtaposition geometry provides a statistical mechanical basis for unknotting by type-2 DNA topoisomerases. *J. Mol. Biol.* 361, 268-285.

Lopez-Garcia, P., Forterre, P., Van der Oost, J. and Erauso, G. (2000) Plasmid pGS5 from hyperthermophilic archaeon archaeoglobus profundus is negatively supercoiled. *J. Bacteriol.* 182, 4998-5000.

Lu, X.-J., Shakked, Z. and Olson, W. (2000). A-form conformational motifs in ligand-bound DNA structures. *J. Mol. Biol.* 300, 819-840.

Lyubchenko, Y.L. and Shlyakhtenko, L.S. (1997) Visualization of supercoiled DNA with atomic force microscopy in situ. *Proc. Natl. Acad. Sci. USA* 94, 496-501.

Marguet, E. and Forterre, P. (1994) DNA stability at temperatures typical for hyperthermophiles *Nucleic Acids Res.* 22, 1681-1686.

Mayer-Jung, C., Moras, D. and Timsit, Y. (1997). Effect of cytosine methylation on DNA-DNA recognition at CpG steps *J. Mol. Biol.* 270, 328-335.

McClendon, A.K., Rodriguez, A.C. and Osheroff, N. (2005) Human topoisomerase IIα rapidly relaxes positively supercoiled DNA. Implications for enzyme action ahead of replication forks. *J. Biol. Chem.* 280, 39337-39345.

Minsky, A. (2004) Information content and complexity in the high-order organization of DNA. *Annu. Rev. Biophys. Biomol. Struct.* 33, 317-342.

Mitchell, J.S., Laughton, C.A. and Harris, S.A. Atomistic simulations reveal bubbles, kinks and wrinkles in supercoiled DNA. *Nucl. Acids Res.,* 1-11 doi:10.1093/nar/gkq1312

Murthy, V.L. and Rose, G.D. (2000) Is counterion delocalization responsible for collapse in RNA folding? *Biochemistry* 39, 14365-14370.

Nissen, P., Ippolito, J.A., Ban, N., Moore, P. and Steitz, T. (2001) RNA tertiary interactions in the large ribosomal subunit: the A-minor motif. *Proc. Natl. Acad. Sci. USA* 98, 4899-4903

Nöllmann, N., Stone, M.D., Bryant, Z., Gore, J., Crisona, N.J., Hong, S.C., Mitelheiser, S., Maxwell, A., Bustamante, C. and Cozzarelli, N.R. (2007) Multiple modes of Escherichia coli DNA gyrase activity revealed by force and torque. *Nat. Struct. Mol. Biol.* 14, 264-271.

Olson, W.K., and Zhurkin, V.B. (2000) Modeling DNA deformations. *Curr.Opin. Struc. Biol.* 10, 286-297.

Ortiz-Lombardia, M., Gonzalez, A., Eritja, R., Aymami, J., Azorin, F. and Coll, M. Crystal structure of a DNA Holliday junction. *Nat. Struct. Biol.* 6, 913-917 (1999).

Postow, L., Crisona, N.J., Peter, B.J., Hardy, C.D. and Cozzarelli, N.R. (2001) Topological challenges to DNA replication: conformations at the fork. *Proc. Natl. Acad. Sci. USA* 98, 8219-8226.

Pulleyblank, D.E. (1997) Of Topo and Maxwell's dream. *Science* 277, 648-649.

Qiu, X.Y., Kwok, L.W., Park, H.Y., Lamb, J.S., Andresen, K. and Pollack, L. (2006) Measuring inter-DNA potentials in solution. *Phys. Rev. Lett.* 96, 138101-138104.

Qiu, X., Andresen, K., Kwok, L.W., Lamb, J.S., Park, H.Y. and Pollack, L. (2007) Inter-DNA attraction mediated by divalent counterions. *Phys. Rev. Lett.* 99, 038104-038107.

Randall, G.L., Pettitt, B.M., Buck, G.R. and Zechiedriech, E.L. (2006) Electrostatics of DNA-DNA juxtapositions: consequences for type II topoisomerase function. *J. Phys. Condens. Matter* 18, S173-S185.

Raspaud, E., Durand, D. and Livolant, F. (2005) Interhelical spacing in liquid crystalline spermine and spermidine-DNA precipitates. *Biophys. J.* 88, 392-403.

Roca, J. and Wang, J.C. (1996) The probabilities of supercoil removal and decatenation by yeast DNA topoisomerase II. *Genes Cells* 1, 17-27.

Robinson, P.J.J. and Rhodes, D. (2006) Structure of the '30 nm' chromatin fibre: A key role for the linker histone. *Curr. Opin. Struct. Biol.* 16, 336-343.

Rybenkov, V.V., Ullsperger, C., Vologodskii, A.V. and Cozzarelli, N.R. (1997) Simplification of DNA topology below equilibrium values by type II topoisomerases. *Science* 277, 690-693.

Schalch, T., Duda, S., Sargent, D.F. and Richmond, T.J. (2005) X-ray structure of a tetranucleosome and its implications for the chromatin fibre. *Nature* 436, 138-141.

Schellman, J.A. and Parthasarathy, N. (1984) X-Ray-diffraction studies on cation-collapsed DNA. *J. Mol. Biol.*, 175, 313-329.

Shaw, S. and Wang, J.C. (1997) Chirality of DNA trefoils: implications in intramolecular synapsis of distant DNA segments. *Proc. Natl. Acad. Sci. USA* 94, 1692-1697.

Segal, E. and Widom, J. (2009) What controls nucleosome positions? *Trends Genet.* 25, 335-343.

Schlick, T. and Olson, W.K. (1992) Trefoil Knotting Revealed by Molecular-Dynamics Simulations of Supercoiled DNA. *Science* 257, 1110-1115.

Schindelin, H., Zhang, M., Bald, R., Fürste, J.-P., Erdmann, V.A. and Heinemann, U. (1995) Crystal structure of an RNA dodecamer containing the Escherichia coli Shine-Dalgarno sequence. *J. Mol. Biol.* 249, 595-603.

Shlyakhtenko, L.S., Miloseska, L., Potaman, V.N., Sinden, R.R. and Lyubchenko, Y.L. (2003) Intersegmental interactions in supercoiled DNA: atomic force microscope study. *Ultramicroscopy* 97, 263-270.

Stone, M.D., Bryant, Z., Crisona, N.J., Smith, S.B., Vologodskii, A., Bustamente, C. and Cozzarelli, N.R. (2003) Chirality sensing by Escherichia coli topoisomerase IV and the mechanism of type II topoisomerases. *Proc. Natl. Acad. Sci. USA* 100, 8654-8659.

Strey, H.H., Podgornik, R., Rau, D.C. and Parsegian, V.A. (1998) DNA--DNA interactions. *Curr. Opin. Struct. Biol.* 8, 309-313.

Stupina, V.A and Wang, J.C. (2004) DNA axial rotation and the merge of oppositely supercoiled DNA domains in Escherichia coli: effects of DNA bends. *Proc. Natl. Acad. Sci. USA* 101, 8608-8613.

Tan, Z.J. and Chen, S.J. (2006) Electrostatic free energy landscapes for nucleic acid helix assembly. *Nucleic Acids Res.* 34, 6629-6639.

Tari, L.W. and Secco, A.S. (1995) Base-pair opening and spermine binding--B-DNA features displayed in the crystal structure of a gal operon fragment: implications for protein-DNA recognition. *Nucleic Acids Res.* 23, 2065-2073.

Timsit, Y., Westhof, E., Fuchs, R and Moras. D. Unusual helical packing in crystals of DNA bearing a mutation hot spot. *Nature* 341, 459-462 (1989).

Timsit, Y., Vilbois, E. and Moras, D. Base pairing shift in the major groove of $(CA)_n$ tracts by B-DNA crystal structures. *Nature* 354, 167-170 (1991).

Timsit, Y. and Moras, D. (1991) Groove-Backbone Interaction in B-DNA - Implication for DNA condensation and recombination. *J. Mol. Biol.* 221, 919-940.

Timsit, Y. and Moras, D. (1992) Crystallization of DNA. *Meth. in Enzymology* 211, 409-429.

Timsit, Y. and Moras, D. (1994) DNA self-fitting: the double helix directs the geometry of its supramolecular assemblies. *EMBO J.* 13, 2737-2746.

Timsit, Y. and Moras, D. (1995). Self-fitting and self-modifying properties of the B-DNA molecule. *J.Mol. Biol.* 251, 629-647.

Timsit Y and Moras D. (1996). Cruciform structures and functions. *Q. Rev. Biophys.* 29, 279-307.

Timsit, Y., Shatzky-Schwartz, M. and Shakked, Z. (1999) Left-handed DNA crossovers. Implications for DNA-DNA recognition and structural alterations. *J. Biomol. Struct. Dyn.* 16, 775-785.

Timsit, Y. (1999) DNA structure and polymerase fidelity. *J. Mol. Biol.* 293, 835-853.

Timsit, Y. and Varnai P. (2010) Helical chirality: a link between local interactions and global topology in DNA. *Plos One* 5:e9326.

Timsit, Y. and Varnai, P. (2011) Cytosine, the double helix and DNA self-assembly. *J. Mol. Recognit.* 24, 137-138.

Timsit, Y. (2011) Local sensing of global DNA topology: from crossover geometry to type II topoisomerase processivity. Nucleic Acids Res. doi: 10.1093/nar/gkr556 First published online: July 15, 2011

Tinoco, I., Jr. and Kieft, J.S. (1997) The ion core in RNA folding. *Nat. Struct. Biol.* 4, 509-512.

Travers, A., Muskhelishvili, G. (2007) A common topology for bacterial and eukaryotic transcription initiation? *EMBO Rep.* 8, 147-151.

Travers, A.A. (2004) The structural basis of DNA flexibility. *Philos. T. Roy. Soc. A* 362, 1423-1438.

Trigueros, S., Salceda, J., Bermudez, I., Fernandez, X. and Roca, J. (2004) Asymmetric removal of supercoils suggests how topoisomerase II simplifies DNA topology. *J. Mol. Biol.* 335, 723-731.

van Aalten D.M., Erlanson D.A., Verdine G.L., Joshua-Tor L. (1999) A structural snapshot of base-pair opening in DNA. *Proc Natl Acad Sci USA.* 96, 11809-11814.

van Buuren, B.N.M., Hermann, T., Wijmenga, S.S. and Westhof, E. (2002) Brownian-dynamics simulations of metal-ion binding to four-way junctions. *Nucleic Acids Res.* 30, 507-514.

Varnai, P. and Timsit, Y. (2010) Differential stability of chiral DNA crossovers mediated by divalent cations. *Nucl. Acid Res.*, 38 4163-4172.

Vologodskii AV, Cozzarelli NR (1994) Conformational and thermodynamic properties of supercoiled DNA. Annu Rev Biophys Biomol Struct 23: 609-643.

Vologodskii, A. (2009) Theoretical models of topology simplification by type IIA DNA topoisomerases. *Nucl. Acid Res.* 37, 3125-3133.

Wang, J.C. (2002) Cellular roles of DNA topoisomerases: a molecular perspective. *Nature Rev.*, 3, 430-440.

Wong, H., Victor, J.M. and Mozziconacci, J. (2007) An all-atom model of the chromatin fiber containing linker histones reveals a versatile structure tuned by the nucleosomal repeat length. *PloS one*, 2, e877.

Wood, A.A., Nunn, C.M., Trent, J.O. and Neidle, S. (1997) Sequence-dependent crossed helix packing in the crystal structure of a B-DNA decamer yields a detailed model for the Holliday junction. *J. Mol. Biol.* 269, 827-841.

Woodson, S.A. (2005) Metal ions and RNA folding: a highly charged topic with a dynamic future. *Curr. Opin. Chem. Biol.*, 9, 104-109.

Wu, C.Y., Bassett, A. and Travers, A. (2007) A variable topology for the 30-nm chromatin fibre. *Embo Rep.* 8, 1129-1134.

Yan, J., Magnasco, M.O. and Marko J.F. (1999) A kinetic proofreading mechanism for disentanglement of DNA by topoisomerases. *Nature* 401, 932-935.

Zechiedrich, E.L. and Osheroff, N. (1990) Eukaryotic topoisomerases recognize nucleic acid topology by preferentially interacting with DNA crossovers. *EMBO J.* 9, 4555-4562.

Zheng, J., Birktoft, J.J., Chen, Y., Wang, T., Sha, R., Constantinou, P.E., Ginell, S.L., Mao, C., & Seeman, N.C. From molecular to macroscopic via the rational design of self-assembled 3D DNA crystal. *Nature* 461, 74-77 (2009).

12

Towards an *In Silico* Approach to Personalized Pharmacokinetics

Akihiko Konagaya
Tokyo Institute of Technology,
Japan

1. Introduction

The human genome sequence project has made a great impact on medical science and drug discovery (Collins et al., 2003). The rapid progress of genome sequencing technologies enables us to study personal genome sequences with reasonable costs (Mitchelson, 2007). It is now widely believed that personal genome information will be one of the most important biomedical contributions to personalized medicine, that is, medical and health care based on individual genetics (Angrist, 2007). Personalized medicine has opened the doors to new and emerging technologies in genome drug discovery, including pharmacogenetics, pharmacokinetics, and pharmacodynamics, to name but a few. Pharmacogenetics investigates genetic effects in drug metabolic enzymes and drug transporters for drug efficacy (Pirmohamed, 2011). Pharmacokinetics and dynamics (PKPD) focus on the area under plasma concentration time curve (AUC) of drugs, one of the important indices to check the drug effects in the human body, especially for preventing adverse side effects (Gabrielsson et al., 2009).

Although both pharmacogenetics and PKPD have revealed the association between genetic mutations and drug efficacy, their accomplishments are not sufficient yet for clinical purpose, especially with respect to prediction performance (Pirmohamed, 2011). Why are the associations between genetic mutations and drug efficacy so vague? Why can pharmacokinetic models not predict drug metabolism correctly? Why does model parameter fitting not work well in prediction tasks? These questions motivated us to initiate the study of personalized pharmacokinetics, the opposite side of conventional pharmacokinetics, that is, population pharmacokinetics (Willmann et al., 1994).

The basic approach to personalized pharmacokinetics is the breakdown of the pharmacokinetic problem into the dynamics of molecular interactions between drug metabolic enzymes and drug metabolites. In this approach, the molecular interactions and the drug metabolites are represented by ontologies. The key concepts of pharmacokinetics, such as drug metabolic pathways and drug-drug interactions are also represented by the aggregation of molecular interactions and the conflicts of molecular interactions on the same enzymes, respectively. The severities of drug-drug interactions are measured by mathematical simulation models represented by ordinal differential equations corresponding to molecular interactions.

Personalized pharmacokinetics is one of the *in silico* studies and trial simulations but differs from population pharmacokinetics in the following sense. The objective is the analysis of the genetic traits of an individual patient rather than the average and distribution of these traits in the patient population. A pharmacokinetic model is constructed from the knowledge base of drug metabolic reactions according to an individual regimen, not from a statistical analysis or machine learning applied to population data. The model is used for the estimation of parameter distribution reproducing the observed clinical data from the viewpoint of an inverse problem.

In order to achieve the above goals, we have developed a prototype system for personalized pharmacokinetics, which consists of the Drug Interaction Ontology, the inference programs for drug-drug interaction detection and generation of metabolic pathways and models, and the pharmacokinetic numerical simulation engine with virtual patient population convergence facility. The Drug Interaction Ontology is a kind of knowledge representation with regards to drug metabolism (Yoshikawa et al., 2004). It consists of two vocabulary hierarchies in terms of process and continuants like the SNAP-SPAN ontology, that is, a combination of a purely spatial ontology supporting snapshot views of the world at successive instants of time and a purely spatiotemporal ontology of change and process (Grenon&Smith, 2004a). The process vocabulary hierarchy defines the dynamics of drug metabolism. The continuant vocabulary hierarchy defines categories of bio-chemical molecules related to drug metabolism. A drug metabolism knowledge base is developed on the Drug Interaction Ontology as a collection of anonymous objects representing drug metabolic reactions and their aggregation, that is, drug metabolic pathways. The Drug Interaction Ontology is represented by OWL-DL. However, the logic programming language Prolog is used for drug-drug interaction detection and generation of metabolic pathways and pharmacokinetic numerical simulation models.

The inference programs compensate the lack of knowledge in the form of anonymous objects in the drug metabolic pathway knowledge base. The inference programs dynamically generate anonymous instances, such as drug metabolic pathways and detect the occurrences of drug-drug interactions, when a patient's regimens are given. The automatic generation of drug metabolic pathways is indispensable for multiple drug regimens since the total number of drug metabolic pathways becomes the number of all possible combinations of drugs, which are meaningless to be provided in the drug metabolic knowledge base in advance. Instead, the knowledge base consists of a collection of primitive drug metabolic reactions and inference programs to generate drug metabolic pathways of multiple-drug regimens. The drug-drug interaction detector detects the occurrences of drug-drug interactions in the generated drug metabolic pathways and then dynamically adds the occurrences as hypothetical assertions in the drug metabolic pathway knowledge base, mapping the assertions on process and continuant vocabulary hierarchies. The mapping gives useful background information to validate the assertions. In addition, the mapping is also helpful to generate a pharmacokinetic numerical simulation model when choosing an appropriate mathematical equation corresponding to competitive and non-competitive enzymatic inhibition.

The virtual patient population convergence is a concept for solving an inverse problem. The objective is to estimate multiple sets of parameters reproducing the personal clinical datum of an individual patient. The inverse problem approach is very different from conventional population pharmacokinetics that tries to estimate the average and diversity of a patient population. In our approach, only the clinical data of an individual patient are necessary. This enables us to analyze an "outlier" of the population. We strongly believe that the analysis of such outliers is more important than the analysis of average behaviors for practical personalized pharmacokinetics.

In order to demonstrate the effectiveness of our approach, we applied the prototype system to the pharmacokinetic studies of anti-cancer drug irinotecan with a whole body pharmacokinetic model with regards to the hepatic and renal excretions of five major irinotecan metabolites: CPT-11 (irinotecan), APC[1], NPC[2], SN-38 (active metabolite of irinotecan), and SN-38G (SN-38glucuronide). Firstly, we investigated how ketoconazole affects the irinotecan metabolite blood concentrations with the pharmacokinetic model in (Arikuma et al., 2008). Ketoconazole inhibits CYP3A4, one of the drug metabolic enzymes of irinotecan. The numerical simulation analysis revealed an interesting behavior of the drug metabolism, which is difficult to expect from pathway-level analysis. Then, we investigated how the UGT1A1*28/*28 mutation affects SN-38 blood concentration in the Arikuma model. UGT1A1*28/*28, also known as Gilbert's syndrome, is one of the important mutations that may cause severe side-effects when using irinotecan (Tukey et al, 2002). The simulation result suggests how the expression level of UGT1A1 affects the metabolite concentration of SN-38. Lastly, we investigated kinetic parameters reproducing a bile-duct cancer patient who showed a metabolite excretion profile that was completely different from other cancer patients. This is a typical example of an inverse problem. The result indicates that the analysis of the solution space of the inverse problem is a key to understanding the peculiar behavior of an outlier like this bile-duct cancer patient.

The organization of this chapter is as follows. Firstly, we describe our motivation and background of studying personal pharmacokinetics in Section 2. Sections 3 describes the prototype system which consists of the Drug Interaction Ontology, inference programs for pathway (model) generation and drug-drug interaction detection, and numerical simulation engine with the facility of virtual patient population convergence. Section 4 introduces three case studies of irinotecan pharmacokinetics, drug-drug interaction of irinotecan and ketoconazole, mutation analysis of drug metabolic enzyme UGT1A1, and an inverse problem analysis of a bile-duct cancer patient with an external bile drain. Section 5 discusses the controversial points of this work. Finally, section 6 gives the conclusions of our study.

2. Personalized pharmacokinetics

The role of *in silico* prediction of drug interactions at the pathway level is becoming more and more important for personalized medicine. Multiple-drug regimens exemplify the need for the computer-assisted prediction of drug interactions, which may be different from one patient to another. Multiple-drug regimens are commonly prescribed for elderly patients suffering from more than one disease. However, these regimens sometimes cause unexpectedly severe side effects because of the drug interactions or individual differences concerning response to the drugs (Okuda et al., 1998). Therefore, the prediction of drug

---

[1] 7-ethyl-10-[4-N-(5-aminopentanoic acid)-1-piperidino] carbonyloxycamptothecin
[2] 7-ethyl-10-(4-amino-1-piperidino) carbonyloxycamptothecin

interactions for preventing the side effects is an important issue for these regimens. On the other hand, the information useful for *in silico* drug interaction prediction has increased very rapidly in recent years. Technological innovations in genomic sciences have produced an enormous amount of bio-molecular information including sequences, structures, and pathways. In order to integrate the bio-molecular information, ontologies are attracting a lot of attention (Baker & Cheung, 2007; Konagaya, 2006a, 2006b). In addition, pharmacokinetics modeling and simulation are emerging, promising techniques to understand the dynamic behavior of drug metabolic pathways (Tsukamoto et al., 2001;Vossen et al., 2007). To develop personalized pharmacokinetics that can deal with individual drug administration including multiple-drug regimens by integrating the above information and techniques, the following issues must be solved:

-    Context dependency of drug-metabolic pathways,
-    Treatment of multi-scale events,
-    Quantitative evaluation of interactions,
-    Automatic generation of simulation models,
-    Inverse problem solving of an outlier,
     and so on.

Drug-metabolic pathways do not exist *a priori*. They strongly depend on contexts and situations including the administration route, single nucleotide polymorphism (SNP) of drug-response genes, and the administration of multiple drugs and foods. Therefore, a dynamic reconstruction of drug metabolic pathways from primitive molecular events is necessary for drug interaction prediction at the pathway level. Such a reconstruction requires the formal definition of molecular events and the relations among them, i.e., the Drug Interaction Ontology (DIO) (Yoshikawa et al, 2004).

Pathways triggered by drug administration consist of multi-scale events: from the molecular level to the body level, ranging from nanoseconds to hours or days in terms of drug response. For example, drug administration and drug excretion are body-level events, while drug transport and enzymatic reactions are molecular-level events. A comprehensive view from the molecular level to the body level is necessary in order to understand multi-scale events.

Quantitative evaluation plays an essential role to estimate the degree of side effects caused by drug interactions. More than one drug interaction may occur in drug-metabolic pathways from the qualitative reasoning viewpoint. However, not all drug interactions cause side effects because of differences in binding affinity and molecular population. Quantitative simulation models with an *in silico* drug interaction prediction system must be incorporated to discriminate serious drug interactions from negligible ones. It should also be noted that the total drug metabolism depends on not only kinetic parameters but also physiological parameters such as organ volumes and blood flows. The incorporation of kinetic and physiological parameters is necessary for a realistic simulation model to recapture experimental data.

The automatic generation of mathematical models is necessary to avoid a combinatorial problem of drug metabolic pathways caused by multi-drug regimens and the occurrences of drug-drug interactions. On-demand automatic generation of simulation models also liberates modelers from the tedious editing of complex differential equations.

In the case of pharmacokinetics, we often encounter an outlier caused by a peculiar patient who shows different clinical data from other patients. In order to analyze such an outlier, inverse problem solving is necessary. Since most pharmacokinetic models are underdetermined, that is, they have dozens of parameters which are difficult to determine by biological experiments or clinical observation, the solution space of inverse problems becomes a multiple-set or a manifold. Finding multiple sets that reproduce the observed clinical data on a pharmacokinetic model is one of the major challenges in numerical simulation. We solved this issue by restricting the search space within physiologically reasonable ranges, and by utilizing intelligent sampling techniques with a Markov-Chain Monte Carlo (MCMC) method and Support Vector Machine (SVM) method in the prototype system.

3. Prototype system

The prototype system for personalized pharmacokinetics consists of the Drug Interaction Ontology, the inference programs for drug-drug interaction detection and generation of metabolic pathways and models, and the pharmacokinetic numerical simulation engine with virtual patient population convergence facility (Fig.1).

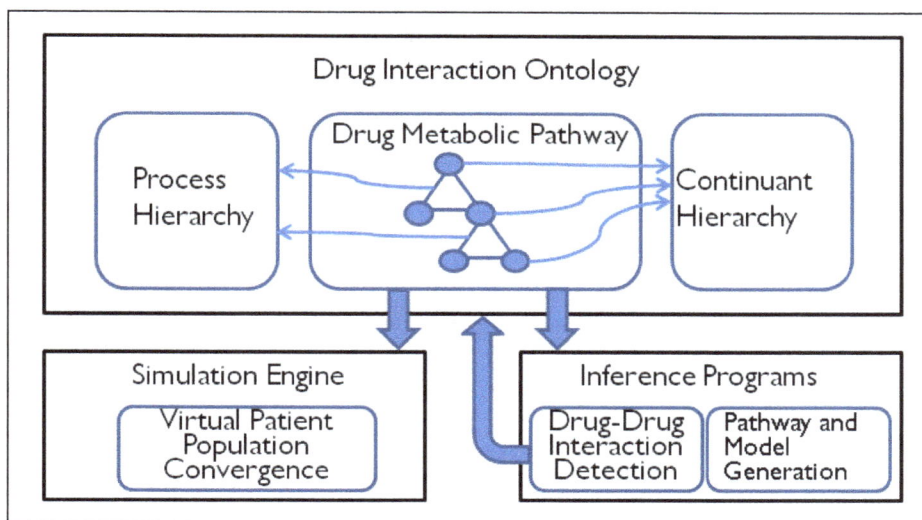

Fig. 1. The Organization of the Prototype System
The prototype system consists of the Drug Interaction Ontology, inference programs and simulation engine. A drug metabolic pathway is represented by an anonymous object whose compounds are linked to terms in the continuant hierarchy, and whose reactions are linked to terms in the process hierarchy. The inference programs generate a metabolic pathway by aggregating primitive reaction objects. After detecting the occurrences of the drug-drug interaction, the generated pathway is translated to a model, that is, a list of ordinal differential equations. The simulation engine estimates a virtual patient population reproducing an observed clinical datum by means of solving an inverse problem with the equations.

3.1 Drug Interaction Ontology (DIO)

The ontological approach in knowledge base design is adopted for resource sharing and the semantic description of molecular events and pathways. Ontology is necessary to define molecular events and pathways in a form that can be shared among computers and human beings. This enables the full use of powerful computational intelligence for dynamic pathway reconstruction in a way that human intelligence can follow and understand. Ontology is also important for establishing interoperability among web resources and thereby to make use of the latest drug reaction information published in the semantic web (Baker&Cheung, 2007; Berners-Lee&Hendler, 2001). Public biological ontologies, especially in the field of chemical biology, are now dramatically increasing and have a great potential to develop sustainable knowledge bases for molecular reaction and pathways (Ceusters&Smith, 2010).

The Drug Interaction Ontology is designed to share the knowledge of drug-drug interaction in both machine and human understandable form. The controlled vocabularies of the Drug Interaction Ontology consist of "process" (Fig. 2) and "continuant" (Fig. 3) as proposed in Basic Formal Ontology (BFO) (Grenon et al., 2004b). Molecular events are asserted in the knowledge base referring the terms of the controlled vocabularies. A drug metabolic pathway is represented by the aggregation of molecular events. This is because an infinite number of terms or classes are required to express all combinations of molecular events. We avoid this problem by treating pathways as anonymous objects deduced from prototype molecular event objects rather than treating them as instances of pathway classes.

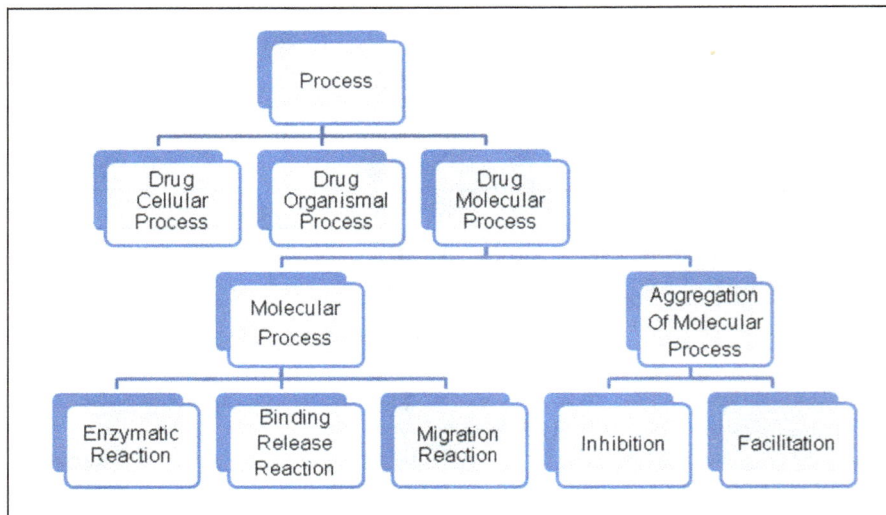

Fig. 2. A Part of Process Vocabulary Hierarchy
Process vocabulary hierarchy defines the terminology of metabolic reactions from molecular level to body level. Note that each term in the process hierarchy may have more than two kinds of anonymous reaction objects or their aggregations categorized into the same term. This facility enables to avoid combinatorial expansion of terms to deal with aggregation of anonymous reaction objects.

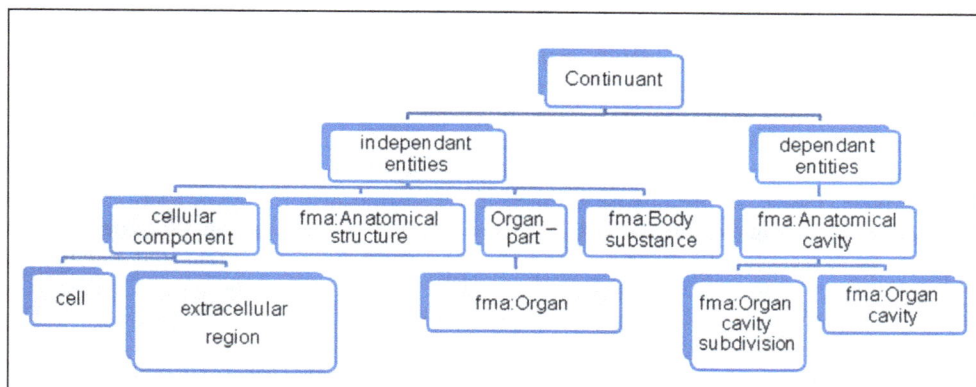

Fig. 3. A Part of Continuant Vocabulary Hierarchy
Continuant vocabulary hierarchy defines the terminology of compounds, cells and organs which contribute the human body. The terminologies imported from reference ontologies are prefixed by identifies such as "fma" and "umls" for the Foundational Model of Anatomy and Unified Medical Language System, respectively.

Ontology provides not only a class hierarchy of controlled vocabulary but also property and relations among terms useful for inferences. The inferences may include qualitative reasoning and numerical simulation, i.e., pathway generation, drug-drug interaction detection, differential equation generation, and numerical simulation as well as reasoning on OWL-DL. Obtained results such as drug interaction candidates can be dynamically mapped on the ontology as hypothetical assertions. This approach is attractive since it can use ontology as background knowledge to interpret the inference results.

The Drug Interaction Ontology (DIO) was written in OWL-DL; the controlled vocabularies of process and continuant were implemented as OWL class hierarchy, and part of attributes, including molecules and organs, were implemented as OWL instances. The molecular event objects were represented by OWL instances and OWL properties. The ontology referred to other taxonomies and ontologies for well-established vocabularies of biochemical terms, anatomical entities, and properties. This enabled the reduction of our ontology construction cost and to concentrate our efforts on the information specific to drug interaction.

3.2 Automatic generation of drug metabolic pathways

The drug metabolic pathway, due to its dynamic nature, is difficult to define *a priori* in the manner as seen in bio-molecular metabolic pathways in the Kyoto Encyclopedia of Genes and Genomics (KEGG) (Kanehisa&Goto, 2000). Therefore, modularization is necessary for the dynamic reconstruction of pathways that depend on dose conditions. Careful selection of primitive modules is the key to ensuring the soundness of pathway reconstruction. Molecular events, such as molecular transport and enzymatic reactions, are well-formed primitive modules for this purpose. In this chapter, we refer to the primitive modules as "molecular events", and the aggregation of molecular events as "pathways". To avoid redundant pathway branch constructions, which are non-essential for the target drug interactions, we adopt causality-based modularization in which each molecular event is defined by the unique relationship between key molecules before and after the event. The triadic relationship <trigger, situator, resultant> is one such causality that can be commonly found in molecular reactions (Yoshikawa et al., 2004). For example, in case of enzymatic reactions, substrates, enzymes and other products correspond to trigger, situator, and resultant, respectively. In the case of molecular transport, extra (intra) cellular molecules, transporters, and intra (extra) cellular molecules correspond to the participants of the triadic relation, respectively. The triadic relationship can be applicable to higher level events like drug dosage and drug excretion, as long as its causality is unique and clear. Figure 4 shows a simple example of pathway reconstruction with two primitive molecular events: an enzymatic reaction in which carboxylesterase (CE) metabolizes irinotecan into SN-38 (7-ethyl-10-hydroxycamptothecin) in the liver, and molecular transport in which SN-38 in the liver is transported to the bile by MRP2 (Multidrug resistance-associated protein 2). Two molecular events are connected at the resultant of the enzymatic reaction and the trigger of the molecular transport for passing SN-38 in the liver (SN-38@liver) to the bile.

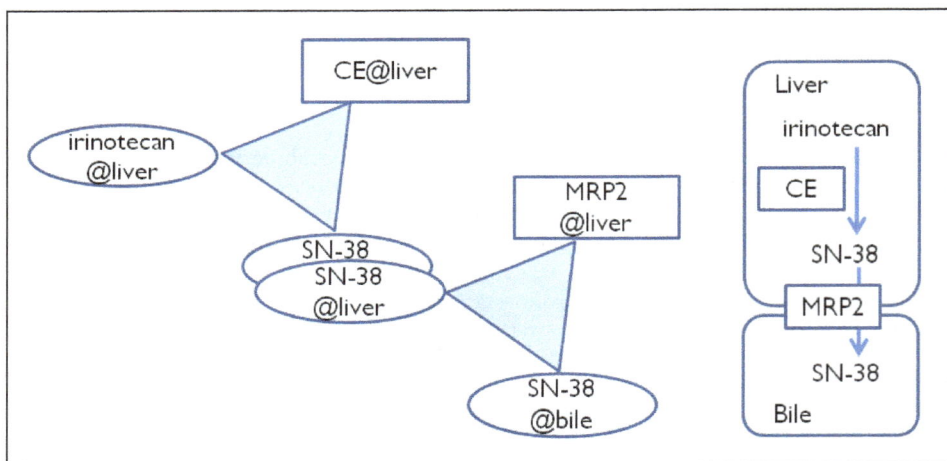

Fig. 4. Reconstruction of Drug Metabolic Pathway from Primitive Reactions
Each primitive metabolic reaction is represented by a triplet of <trigger, situator, resultant>. Metabolic pathway is constructed by aggregation of primitive reactions where a resultant becomes a trigger of the consecutive reaction as like SN-38 for CE and MRP2. Each compound has location information indicated by the @ symbol to deal with migration process across organs such as MRP2.

3.3 Automatic generation of drug metabolic pharmacokinetic models

In order to incorporate a quantitative numerical simulation into the prototype system, the following two aspects are considered: a methodology for the automatic conversion from a generated pathway to a quantitative simulation model, and a methodology to solve an inverse problem, that is, virtual patient population convergence. These two methodologies enable us to apply the prototype system to the *in silico* prediction of individual drug interactions for multiple-drug regimens, assuming that kinetic parameters and the initial enzyme concentration are roughly estimated by individual genetic variations and health indices of bio-markers. A simulation model is automatically translated from a drug metabolic pathway generated by the inference programs. The generated pathway is converted to an intermediate model by merging organs and molecular events, respectively, to fit a given simulation model such as a compartment model. Then, a list of ordinal differential equations for the simulation model is generated from the intermediate model by converting the merged events to mathematical expressions.

Figure 5 shows a simulation model automatically generated for the co-administration of irinotecan and ketoconazole. The organs and tissues are integrated into 8 compartments, i.e., blood (including rapidly equilibrating tissues: artery, heart, kidneys, lung, and veins), liver, GI (gastrointestinal consists of the large intestine, small intestine, portal vein, and stomach), adipose tissue, NET (non-eliminating tissue such as skin and muscle), GI lumen, bile lumen,

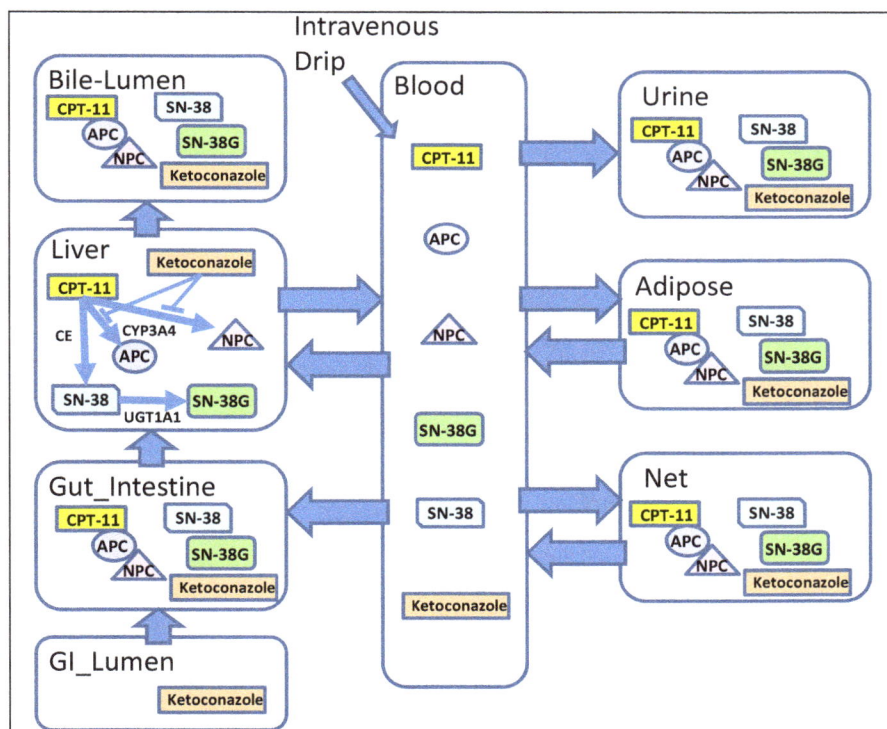

Fig. 5. Generated Pathway Model for Irinotecan and Ketoconazole Co-administration
The model consists of 8 compartments: Blood, Liver, Adipose, Gut-Intestine (GI), GI-Lumen, Urine, Bile-Lumen, and Net. Irinotecan is injected directly into the Blood compartment in the way of intravenous drip. The Blood compartment plays a central role to circulate drug metabolites to other compartments. The GI-Lumen and the GI compartments are provided for the oral administration of ketoconazole. The Urine and the Bile-Lumen compartments are provided for the renal and hepatic excretions of irinotecan metabolites. Drug metabolic reactions are occurred in the Liver compartment. The Adipose and the Net compartments are provided for the difference of blood circulation speeds of organs.

and urine. Michaelis-Menten equations are used for all enzymatic reactions. A competitive Michaelis-Menten inhibition model is used for this simulation as used for midazolam and ketoconazole inhibition by (Chien et al., 2006). In order to increase the predictive performance, a simplified pathway is used for the generation of simulation models from the viewpoint of the trade-off between model complexity and data availability. For example, in the case of irinotecan and ketoconazole metabolisms, reabsorption through small intestine and reactions concerning albumin are omitted due to the lack of information.

The generated simulation models and pharmacokinetic moment parameter values are mapped onto the Drug Interaction Ontology as hypothetical assertions. The simulation models are asserted as aggregations of objects representing terms and parameters in differential equations. Those objects having references to the components of the pathway objects from which the simulation models are generated. The moment parameter values are asserted with the drug interaction objects and the corresponding simulation model for the further analysis. See (Arikuma et al., 2008) for the details of ontology, mathematical models, and their implementations of irinotecan and ketoconazole drug metabolic pathways.

3.4 Virtual patient population convergence

Virtual patient population convergence is a concept to find a partial set of virtual patients reproducing an observed clinical datum in some error range with regards to a given mathematical model by means of iteration of numerical simulation and population selection starting from an initial virtual patient population (Fig.6). In general, the solution space of a mathematical model, that is, the virtual patient population reproducing an observed clinical data on the model, might be infinite. In addition, there is a trade-off between the virtual patient sampling time and the precision of the virtual patient population converged to the observed clinical data. Therefore, some kind of criteria with regards to the "goodness" of convergence for virtual patient population must be defined.

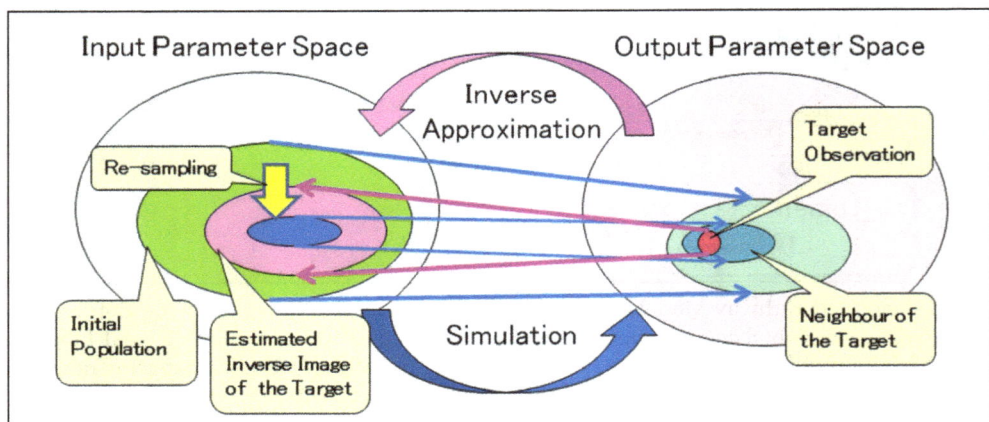

Fig. 6. Schematic View of Virtual Patient Population Convergence
Virtual Patient Population Convergence estimates a virtual population mapping onto the neighbourhood of the target by iterating simulation and re-sampling of virtual patients. Intelligent sampling method is required to reduce total simulation costs for the convergence.

First of all, each virtual patient should hold reasonable parameters from the viewpoints of physiology and pharmacokinetics. This requires that the search space should be restricted within or around the initial virtual patient population. Virtual patients with unrealistic parameters must be eliminated during the convergent process. Note that the search space is still wide enough since some pharmacokinetic parameters such as maximum velocity (Vmax) of enzymatic reaction range over the order of 10 to the power of 5 or 6 in an initial population.

Next, care must be taken for the handling of errors in the observed clinical data. The clinical data may have five or more percent of errors. Therefore, it is more important to preserve the diversity of virtual population as much as possible rather than to find a precise virtual patient reproducing the observed clinical data.

Lastly, convergence speed does matter. A mathematical model usually has dozens of unknown parameters. The virtual patient population convergence tries to find a "good" population by iteration of simulation and sampling by assigning new values to the unknown parameters of a virtual patient. Random sampling does not work well for this case due to the large search space. Sophisticated sampling algorithms must be developed to solve practical inverse problems of pharmacokinetics.

For example, Arikuma's irinotecan pharmacokinetic model has 61 input parameters and 10 output parameters with regards to irinotecan metabolism (Arikuma et al., 2008). Suppose that the goal of convergence is to find a virtual population whose error is within the ranges of 1/10 of the target output on each output parameter. This may correspond to finding a rare event occurring one-tenth for each ten parameters, that is, once every 10 billion if we assume a uniform distribution.

Intelligent sampling such as Gibbs sampling and Markov Chain Monte Carlo (MCMC) methods are helpful to avoid redundant sampling. We accelerated the sampling performance of the MCMC method by using a support vector machine (SVM) to estimate the boundary of virtual patients reproducing the neighbour of the target output values. Intelligent sampling with regards to the boundary is much faster than the sampling with simulation, although some candidates inside the boundary may fall out of the neighbourhood of the target data due to the non-linear behaviour of simulation model.

4. Case studies

In order to demonstrate the effectiveness of the prototype system, we applied it to the irinotecan pharmacokinetic studies including multi-drug administration of irinotecan and ketokonazole, mutation effects of UGT1A1*28/*28 and the hepatic and renal excretion analysis of the bile-duct cancer patient with an external bile-drain. Irinotecan is an anti-cancer drug that is commonly used for colon and breast cancers (Mathijssen et al., 2001). Irinotecan is a prodrug of SN-38, anti-neoplastic topoisomerase I inhibitor, and is bio-activated by carboxyl esterase (CE). About 60% of irinotecan is excreted as unchanged drug from bile and kidney (Slatter et al., 2000). Irinotecan is also metabolized by CYP3A4 to form APC and NPC. NPC is further metabolized by CE to form SN-38. SN-38 undergoes glucuronate conjugation by UGT1A1 to form the inactive glucronide, SN-38G.

Ketoconazole (KCZ) is an anti-fungal drug and a well-known inhibitor of CYP3A4. Ketoconazole undergoes extensive metabolism in the liver to form several metabolites (Whitehouse et al., 1994). About 2 to 4% of urinary radioactivity represents unchanged drug (Heel et al., 1982). It has been reported that the inhibition of CYP3A4 by ketoconazole influences the metabolism of irinotecan, which results in a 6% SN-38 increase (Kehrer et al., 2002). In addition, it is known that the mutations on UGT1A1, UGT1A1*28/*28 which decreases the expression of UGT1A1 enzyme down to 30%, has a strong relationship with some side effects of irinotecan (Sai et al., 2004; Ando et al., 2000).

Slatter et al. reported that the bile cancer patient with an external bile-drain showed completely different renal and hepatic metabolite excretion profiles from other cancer patients in their pharmacokinetic studies (Slatter et al., 2000). Why was the bile cancer patient so different from the other cancer patients? Which parameter caused the difference? This case is a typical example that requires personalized pharmacokinetics, in other words, the analysis of an outlier.

4.1 Drug-drug Interaction between Irinotecan and ketoconazole

The pathways of intravenously administered irinotecan and orally administered ketoconazole were inferred as aggregation of molecular events by the Pathway Object Constructor. The generated object included a metabolic pathway where irinotecan and its derivatives circulate through the veins, liver, bile, intestines, and portal vein, namely, the

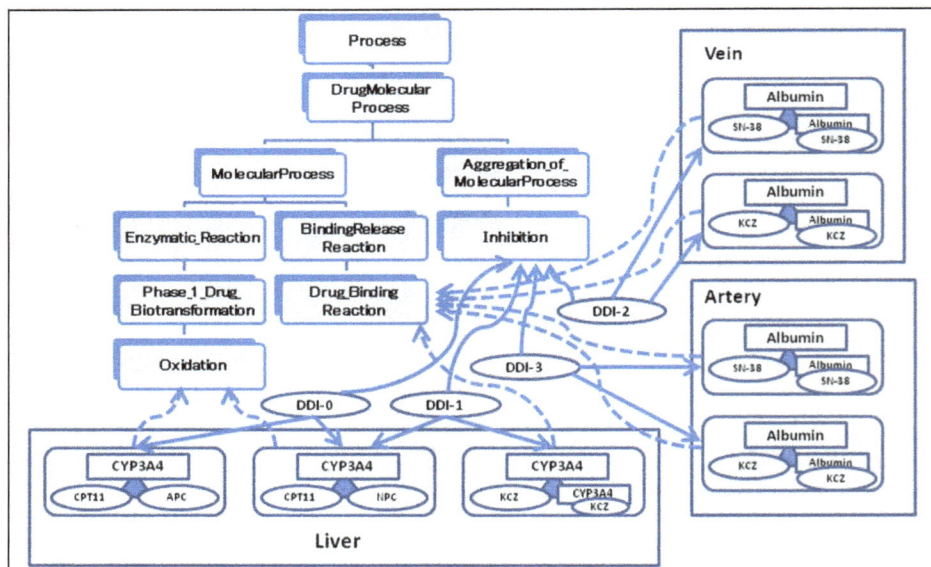

Fig. 7. Detected Drug-Drug Interactions mapped onto the Drug Interaction Ontology
Four drug-drug interaction events are detected on the irinotecan and ketoconazole coadministrated metabolic pathway, which are mapped onto the Drug Interaction Ontology. Seven primitive reactions are involved in the events. Three of them and the rest are mapped onto the oxidation process and the drug_binding process, respectively.

enterohepatic circulation, and are excreted through the kidneys or through the bile. These generated pathways were consistent with *in vivo* studies (Mathijssen et al., 2001). Interactions between intravenously administered irinotecan and orally administered ketoconazole were detected and asserted by the Drug Interaction Detector. The detected drug interactions and the hypothetic assertions are shown in Fig. 7. The assertion contains four drug interactions; two of them concern "drug binding reaction" to albumin in veins (ddi2) and arteries (ddi3), and the rest of them concern "oxidation" by CYP3A4 (ddi0 and ddi1). The detected drug interaction concerning CYP3A4 (ddi0 and ddi1) has been confirmed by the literature on *in vivo* studies (Kehrer et al., 2002).

We evaluated the effects of drug interactions concerning CYP3A4 quantitatively with numerical simulations. Intravenous drip infusion (125 mg/m^2, 90 min) was assumed for irinotecan, and oral administration (200 mg) was assumed for ketoconazole. Two simulations were performed: sole administration of irinotecan for a patient having UGT1A1*1/*1 (wild type) and co-administration of irinotecan and ketoconazole for a patient having UGT1A1*1/*1 (wild type). Fig. 8 (a) and (b) show the simulated concentration/time profiles of irinotecan, SN-38, APC, NPC, SN-38G in blood for the simulation of sole administration of irinotecan and co-administration of irinotecan and

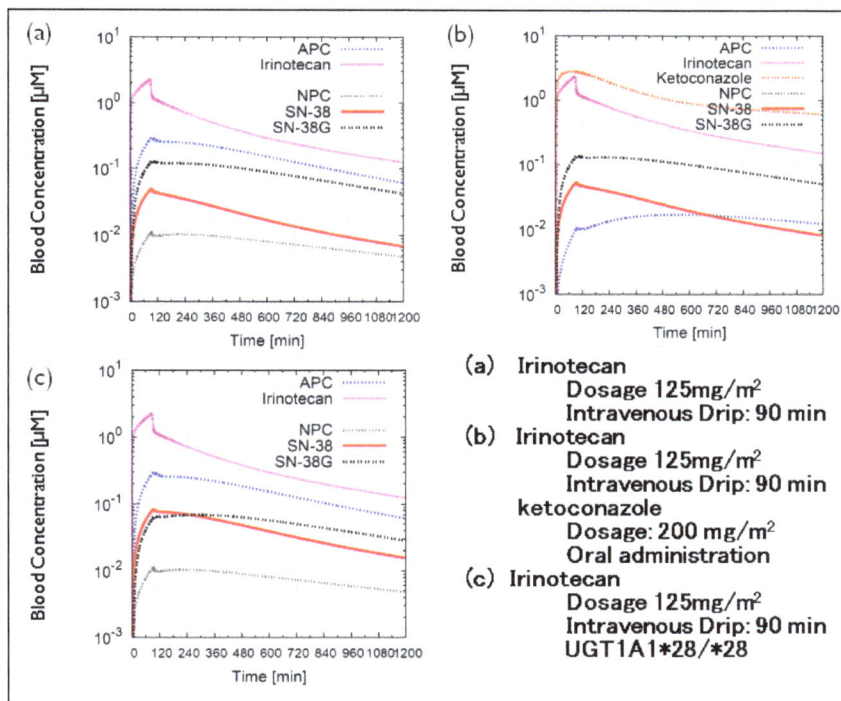

Fig. 8. Simulation Results of Time Course Concentration:
(a) Sole Irinotecan Administration (wild type)
(b)Multiple-administration of Irinotecan and Ketoconazole (wilde type)
(c) Sole Irinotecan Administration (UGT1A1*28/*28 mutation)

ketoconazole. By the ketoconazole administration, the area under the plasma concentration-time curve (AUC) of APC and NPC were decreased to 48.1% and 35.3%, respectively. The AUC of SN-38 was increased only to 108% by the ketoconazole administration. Similarly, the maximum drug concentration (Cmax) of APC and NPC were decreased to 25.6% and 20.2%, respectively, whereas the Cmax of SN-38 was increased to 105% by the ketoconazole administration. This implies that effects of co-administration of ketoconazole and irinotecan are mild for the blood concentration of SN-38 in spite of inhibition of CYP3A4 in the drug metabolic pathway.

4.2 Effects of UGT1A*28/*28 mutation

It is reported that a patient with UGT1A1*28/*28 mutation decreased the expression of UGT1A1 down to 30% of patients with UGT1A1*1/*1 (wild type) in average. According to our quantitative evaluation with numerical simulation, the patient with UGT1A1*28/*28 mutation significantly increased the AUC and Cmax of SN-38: the AUC was increased to 208% and the Cmax was increased to 165% (Fig. 8 (c)). This implies that patients with UGT1A1*28/*28 may suffer severe side effects when the doses are the same as those for patients with UGT1A1*1/*1. These results agree with previously published experimental papers (Kehrer et al., 2002; Sai et al., 2004; Ando et al., 2000).

4.3 Analysis of a bile-duct cancer patient with an external bile-drain

In their pharmacokinetic study, Slatter et al. reported that the bile-duct cancer patient with an external bile-drain showed completely different renal and hepatic metabolite excretion profiles from other cancer patients (Slatter et al., 2000). Since the bile-duct cancer patient excreted her bile acid through an external bile-drain, it is possible to observe the hepatic clearances directly. As seen in Table 1, the ratio of hepatic and renal metabolite excretions is completely different between the bile-duct cancer patient and the other cancer patients; the ratio is almost the same for the bile-duct cancer patient while the ratio is 1-to-2 for the other cancer patients. Slatter et al. explained that the difference might result from the inhibition of canalicular multiple organic anion transporter (cMOAT/MRP2/ABCC2) in the bile-duct cancer patient. This is one of the typical examples that require the analysis of an outlier, that is, personalized pharmacokinetics.

	Bile-duct Cancer Patient		Other Cancer Patients (N=7)	
	Urine Excretion	Faeces Excretion	Urine Excretion	Faeces Excretion
CPT-11	21.8	24.7	22.4(5.5)	32.3(4.5)
APC	7.7	5.6	2.2(1.5)	8.3(3.0)
SN-38G	12.0	2.7	3.0(0.8)	0.3(0.2)
SN-38	0.9	3.2	0.4(0.1)	8.2(2.5)
NPC	0.09	0.62	0.1(0.1)	1.4(0.9)
Total	45.8	43.4	30.2(6.6)	62.0(7.6)

Table 1. Percentages of Administrated Dose Excreted from Urine and Faeces in the Bile-duct Cancer Patient and the Average of Other Cancer Patients summarized from the Pharmacokinetic Study reported by (Slatter et al. 2000).

Bile acid excretion and faeces excretion are added for the bile-duct cancer patient. The total percentage does not recover all administrated dose due to the experimental limitations and drug metabolisms other than the CPT-11, APC, SN-38G, SN-38, and NPC.

Arikuma et al. have developed an irinotecan metabolic pathway model with five major irinotecan metabolites: CPT-11(irinotecan), APC, NPC, SN-38 and SN-38G (Arikuma et al., 2008). Since those components occupy about 90 percent of irinotecan metabolites, it is sufficient to consider the dynamics of blood concentration of the five compounds with renal and hepatic excretions as a pharmacokinetic model. The challenge for personalized pharmacokinetics is whether Arikuma's model can reproduce the bile-duct patient data or not.

Figure 9 shows the set of virtual patients whose simulation results fall into the neighbourhood of the bile-duct patient within a 5 percent error margin in hepatic and renal excretions of CPT-11, APC, NPC, SN-38 and SN-38G after virtual patient population convergence. The distribution shows clear dependency in hepatic and renal excretions in CPT-11, but no such dependencies in other components.

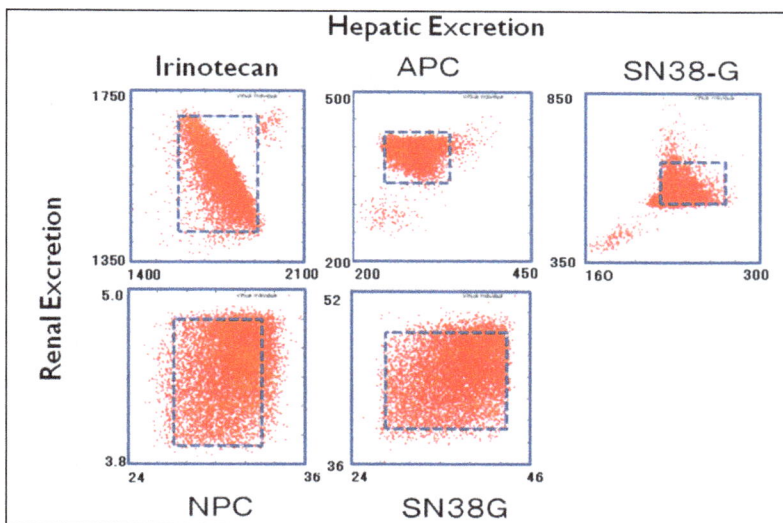

Fig. 9. Distribution of Virtual Patient Population reproducing the Hepatic and Renal Irinotecan Metabolite Excretions of the Bile-duct Cancer Patient
Horizontal and vertical axes represent hepatic and renal excretions, respectively. The centre of each square in the graphs is the target excretion of the bile-duct cancer patient estimated from the pharmacokinetic study reported by Slatter et al. 2000. Each dot represents a virtual patient whoes hepatic and renal excretions are within the five percent error ranges on the pharmacokinetic simulation.

In the case of an undetermined inverse problem, what we can obtain is just a multi-set of solution candidates. We cannot determine the true solution because of the lack of conditions. However, it would be reasonable to expect that the solution candidates may have common properties reproducing the target data in the inverse problem. Figure 10 shows the averages and distributions of the hepatic clearances converged to metabolic excretions of the bile-duct cancer patient and the averages of other cancer patients. Intriguingly, there is a clear down-regulation in APC, SN-38, and SN-38G between the two, suggesting coincidence with the pharmacogenetic study of ABCC2 (Innocenti et al., 2009).

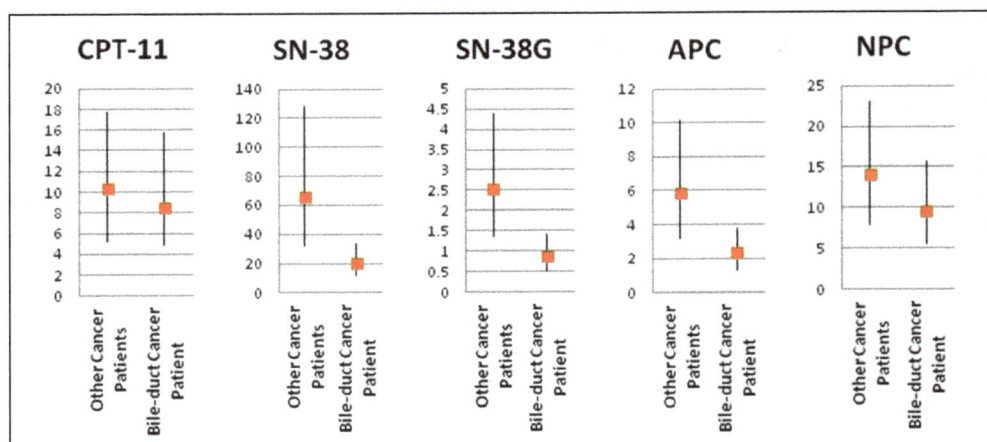

Fig. 10. Comparison of the bile-duct cancer patient and the other cancer patients with regards to the Average and Distribution of Hepatic Clearances in Virtual Patient Population after Convergence
Horizontal axis represents estimated hepatic clearances in numerical simulation (unit $l/h/m^2$).

5. Discussion

Our study on personalized pharmacokinetics is still in its infancy. Ontology and mathematical simulation are powerful information technologies but need more research for practical use, especially from the viewpoints of knowledge representation and unknown parameter estimation. In this section, we discuss these issues.

5.1 Knowledge representation and inference

It is widely known that there is a trade-off between the amount of knowledge base and deepness of inference in knowledge representation, that is, shallow inference on huge knowledge base and deep inference on small knowledge base. Both approaches have their own merits. Therefore, it is important to choose the right size of knowledge base and the right level of inference in knowledge representation.

As for biomedical ontologies, OWL becomes one of *de facto* standards, especially in the world of Semantic Web, the network of knowledge with inference in the Internet (Timizi et al, 2011). It adopts a unique representation form of knowledge with UNICODE, URI (Uniform Resource Identifier) and RDF (Resource Description Form), and provides an inference engine, named reasoning based on description logic (Zhang et al., 2006).

We adopted OWL-DL for the implementation of the Drug Interaction Ontology to make it consistent with reference ontologies such as Foundation of Model of Anatomy (FMA)

(Rosse&Mejino, 2003) and Unified Medical Language System (UMLS) (Bodenreider, 2004). The reference ontologies enable us to focus on the development of a control vocabulary specific to pharmacokinetics while allowing the usage of general technical terms as defined in the reference ontologies. In addition, the OWL reasoner helps indentify unsatisfiable classes and consistency checking in Ontology (Zhang et al., 2006).

However, it is apparent that OWL-DL is not suitable for the detection of drug-drug interactions and the generation of metabolic pathways. We strongly believe that inference programs on ontology should not be restricted to the level of reasoning. Therefore, we developed our inference programs in Prolog while using OWL-DL for consistency checking of the Drug Interaction Ontology.

The inference programs may infer new assertions that cannot be deduced from the original ontology in reasoning. In such a case, it is impossible for computers to validate the assertions automatically. In order to compensate for the lack of automatic validation, we introduced hypothetical links that map the assertions on the Drug Interaction Ontology. The hypothetical links give useful background information for human beings to validate the assertions. For example, in the case of the drug-drug interaction detection of irinotecan and ketoconazole, four drug-drug interaction events (from ddi0 to ddi3 in Fig. 6) were found by our inference program. Mapping the events onto the process vocabulary hierarchy enables us to interpret ddi0 and ddi1 as enzymatic inhibition of CYP3A4 while ddi2 and ddi3 are binding confliction on albumin.

5.2 Distribution estimation

Mathematical models expressed by ordinal differential equations define the mapping from parameter space to accumulation data space when integrating the equations from zero to a specific time point. The mapping is mostly nonlinear, and multiple sets of parameters may reproduce the same accumulation data. The virtual patient population convergence is a technique to estimate the distribution of virtual patients reproducing the same or similar accumulation data when the initial population is given in an inverse problem.

From the viewpoint of parameter distribution estimation algorithms, the virtual patient population convergence raises several issues such as initial population dependency and early convergence to local minima as well as convergence performance. As for the convergence performance, we have developed a very fast deterministic algorithm, the details of which will be published elsewhere (Aoki et al., 2011). Further studies should be carried out for the rest.

Another important issue related to the virtual patient population convergence is the interpretation of solution space, that is, the obtained population after convergence. Mathematical models often have parameter-parameter dependencies that compensate for the effect of certain parameters by means of adjusting other parameters to reproduce the same output (Azuma et al., 2007). This suggests focusing on the analysis of parameter diversities of the solution space, rather than the analysis of an individual virtual patient of the solution space. In case of the bile-duct cancer patient of irinotecan pharmacokinetics, the virtual patients in solution space falls into specific parameter ranges on hepatic clearances. However, care must be taken when the ranges seem to be strange from the viewpoint of biology and medical science. In such a case, the specific ranges may result from a deficiency of the model to explain the behavior of the outlier.

6. Conclusions

The effectiveness of personalized pharmacokinetic is demonstrated by the *in silico* analysis of an irinotecan pharmacokinetic study with the Drug Interaction Ontology and automatic drug metabolic pathway generation followed by numerical simulation.

As for drug-drug interaction detection, the prototype system detected four drug interactions for an irinotecan plus ketoconazole regimen. Two of them concerned cytochrome p450 (CYP3A4) and were consistent with known drug interactions. The numerical simulation indicates that the effect of the drug-drug interactions is mild for the increase of SN-38 blood concentration although APC and NPC blood concentrations are reduced considerably.

We then quantitatively examined the effect of genetic variation UGT1A1*28/*28 using numerical simulations. The genetic variation on UGT1A1 showed a two-fold increase of SN-38's AUC as suggested by the literature (Gagne et al., 2002).

Finally, we analyzed the pharmacokinetic parameters reproducing the bile-duct cancer patient with an external bile-drain in terms of hepatic and renal metabolic excretions of CPT-11, APC, NPC, SN-38 and SN-38G. The obtained virtual patients suggests that the difference of hepatic clearances in APC, SN-38 and SN-38G may be the major reason that causes the clinical differences between the bile-duct cancer patient and the other cancer patients in irinotecan pharmacokinetics.

7. Acknowledgments

The author expresses his thanks to Mr. Takeshi Arikuma and Mr. Takashi Watanabe for the development of the prototype system including the Drug Interaction Ontology; the inference programs of automatic generation of pathways and models, and the detection of drug-drug interactions, the numerical simulation engine with virtual patient population convergence facility.

8. References

Aoki, Y.; Hayami, K.; Hans, DS. & Konagaya, A. (2011). Cluster Newton Method for Sampling Multiple Solutions of an Underdetermined Inverse Problem: Parameter Identification for Pharmacokinetics, NII Technical Report (NII-2011-002E)

Ando, Y.; Saka, H.; Ando, M.; Sawa, T.; Muro, K.; Ueoka, H.; Yokoyama, A.; Saitoh, S.; Shimokata, K. & Hasegawa, Y. (2000). Polymorphisms of UDP-Glucuronosy ltransferase Gene and Irinotecan Toxicity: A Pharmacogenetic Analysis, *Cancer Research*, Vol. 60, pp.6921-6926

Angrist, M. (2007). *Here is a Human Being at the Dawn of Personal Genomics*, Harper

Arikuma, T.; Yoshikawa, S.; Watanabe, K.; Matsumura, K. & Konagaya, A. (2008). Drug Interaction Prediction using Ontology-driven Hypothetical Assertion Framework for Pathway Generation followed by Numerical Simulation, BMC Bioinformatics, vol. 9 (Suppl 6), S11

Azuma, R.; Umetsu, R.; Ohki, S.; Konishi, F.; Yoshikawa, S., Matsumura, K. & Konagaya, A. (2007). Discovering Dynamic Characteristics of Biochemical Pathways using Geometric Patterns among Parameter-Parameter Dependencies in Differential Equations, *New Generation Computing*, Vol.25, pp.425-441

Baker, CJ. & Cheung, KH. (2007) *Semantic Web: Revolutionizing Knowledge Discovery in the Life Sciences*, Springer

Berners-Lee, T. & Hendler J. (2001). The Semantic Web, *Nature*, Vol.410, No.6832, pp.1023-1024

Bodenreider, O. (2004). The Unified Medical Language System (UMLS): Integrating Biomedical Terminology, Nucleic Acids Research, Vol.32, D267-D270

Ceusters,W. & Smith,B. (2010). A Unified Framework for Biomedical Terminologies and Ontologies, Studies in Health Technology and Informatics, Vol.160(Pt 2),pp.1050-1054

Chien, JY. ; Lucksiri, A.: Charles, S.; Ernest, I.; Gorski, JC.; Wrighton, SA. & Hall SD. (2006). Stochastic Prediction of CYP3A-Mediated Inhibition of Midazolam Clearance by Ketoconazole, *Drug Metab Dispos*, Vol.34, No.7, pp.1208-1219

Collins, FS.; Green, ED.; Guttmacher, AE. & Guyer, MS. (2003). A Vision for the Future of Genomic Research, Nature, vol. 422, pp.835-847

Gabrielsson, J.; Dolgos, H.; Gillberg, PG.; Bredberg, U.; Benthem, B. & Duker, G. (2009). Early Integration of Pharmacokinetic and Dynamic Reasoning is Essential for Optimal Development of Lead Compounds: Strategic Consideration, Drug Discovery Today, Vol.14, No.7, pp.358-372

Gagne, JF.; Montminy, V.; Belanger, P.; Journault, K.; Gaucher, G. & Guillemette C. (2002). Common Human UGT1A Polymorphisms and the Altered Metabolism of Irinotecan Active Metabolite 7-ethyl-10-hydroxycamptothecin (SN-38), *Molecular Pharmacology*, Vol.62, No.3, pp.608-617

Grenon, P. & Smith, B. (2004a). SNAP and SPAN: Towards Dynamic Spatial Ontology. *Spatial Cognition and Computation*, Vol.1, pp.69-103

Grenon, P.; Smith, B. & Goldberg, L. (2004b). Biodynamic Ontology:Applying BFO in the Biomedical Domain, In: *Ontologies in Medicine Amsterdam*, IOS Press, pp.20-38

Heel, R.; Brogden, RA.; Carmine, PM.; Speight, T. & Avery, G. (1982). Ketoconazole: A review of its Therapeutic Efficacy in Superficial and Systemic Funfal Infections, *Drugs*, Vol.23, pp.1-36

Innocenti,F.; Kroetz, DL.; Schuetz, E.; Dolan, ME.; Ramirez, J.; Relling, M.; Chen,P.; Das,S.; Rosner,GL. & Ratain,MJ. (2009). Comprehensive Pharmacogenetic Analysis of Irinotecan Neutropenia and Pharmacokinetics, Journal of Clinical Oncology, Vol.27, No.16, pp.2601-2614

Kanehisa, M. & Goto, S. (2000). KEGG: Kyoto Encyclopedia of Genes and Genomes, *Nucleic Acids Research*, Vol.28, pp.27-30

Kehrer, DF.; Mathijssen, RH.; Verweij, J; de Bruijn, P. & Sparreboom, A. (2002). Modulation of Irinotecan Metabolism by Ketoconazole, *Journal of Clinical Oncology*, Vol.20, No.14, pp.3122-3129

Konagaya, A. (2006a). OBIGrid: Towards the Ba for Sharing Resources, Services and Knowledge for Bioinformatics, *Proceedings of the 4th International Workshop on Biomedical Computations on the Grid (BioGrid'06)*, Singapore

Konagaya, A. (2006b). Trends in Life Science Grid: from Computing Grid to Knowledge Grid, *BMC Bioinformatics*, Vol.7 (Suppl 5):S10

Mathijssen, RHJ.; van Alphen, RJ.; Verweij ,J.; Loos, WJ.; Nooter, K.; Stoter, G. & Sparreboom, A. (2001). Clinical Pharmacokinetics and Metabolism of Irinotecan (CPT-11), *Clinical Cancer Research*, Vol.7, pp.2182-2194

Mitchelson, KR. (Ed.) (2007). New High Throughput Technologies for DNA Sequencing and Genomics, Elsevier, UK

Okuda, H.; Ogura, K.; Kato, A.; Takubo, H. & Watanabe, T. (1998). A Possible Mechanism of Eighteen Patient Deaths Caused by Interactions of Sorivudine, a New Antiviral Drug, with Oral 5-fluorouracil Prodrugs. *J Pharmacol Exp Ther*, Vol. 287, No.2, pp.791-799

Pirmohamed, M. (2011). Pharmacogenetics: past, present and future, Drug Discovery Today, Aug. 22nd,Epub ahead of print

Rosse, C. & Mejino, JLV. (2003). A Reference Ontology for Biomedical Informatics: the Foundational Model of Anatomy, Journal of Biomedical Informatics, Vol.36, pp.478-500

Sai, K.; Saeki, M.; Saito, Y.; Ozawa, S.; Katori, N.; Jinno, H.; Hasegawa, R.; Kaniwa, N.; Sawada, J.; Komamura, K., Ueno, K.; Kamakura, S.; Kitakaze, M., Kitamura, Y.; Kamatani, N.; Minami, H; Ohtsu, A; Shirao, K.; Yoshida, T. & Saijo, N. (2004). UGT1A1 Haplotypes Associated with Reduced Glucuronidation and Increased Serum Bilirubin in Irinotecanadministered Japanese Patients with Cancer. Clin Pharmacol, *Ther*, Vol. 75, pp.501-515

Slatter, JG.; Schaaf, LJ.; Sams, JP.; Feenstra, KL.; Johnson, MG.; Bombardt, PA.; Cathcart, KS; Verburg, MT.; Pearson, LK.; Compton, LD.; Miller, LL.; Baker, DS.; Pesheck, CV.; Raymond, S. & Lord, I. (2000). Pharmacokinetics, Metabolism, and Excretion of Irinotecan (CPT-11) Following I.V. Infusion of [14C]CPT-11 in Cancer Patients, *Drug Metabolism and Disposition*, Vol.28, No.4, pp.423-433

Timizi,SH.; Aitken,S.; Moreira, DA.; Mungall, C.; Seueda, J.; Shah, NH. & Miranker, DP. (2011). Mapping between the OBO and OWL Ontology Languages, Journal of Biomedical Semantics 2011, Vol.2 (Suppl 1):S3

Tsukamoto, Y.; Kato Y; Ura. M.; Horii, I.; Ishitsuka, H.; Kusuhara, H. & Sugiyama, Y. (2001). A Physiologically based Pharmacokinetic Analysis of Capecitabine, a Triple Prodrug of 5-FU, in Humans, the Mechanism for Tumor-selective Accumulation of 5-FU, *Pharm Res*, Vol.18, No.8, pp.1190-1202

Tukey, RH.; Strassburg, CP. & Mackenzie, PI. (2002). Pharmacogenomics of Human UDP-Glucuronosyltransferases and Irinotecan Toxicity, *Molecular Pharmacology*, Vol.62, No.3, pp.446-450

Vossen, M.; Sevestre, M.; Niederalt. C.; Jang, IJ.; Willmann. S. & Edginton, AN. (2007). Dynamically Simulating the Interaction of Midazolam and the CYP3A4 Inhibitor Itraconazole using Individual Coupled Whole-Body Physiologically-based Pharmacokinetic (WBPBPK) Models. *Theoretical Biology and Medical Modelling*, Vol. 4, No. 13

Whitehouse, LW.; Menzies, A.; Dawson, B.; Cyr, TD.; By AW.; Black, DB. & Zamecnik, J. (1994). Mouse Hepatic Metabolites of Ketoconazole: Isolation and Structure Elucidation, *J Pharm Biomed Anal 1994*, Vol. 2, No.11, pp.1425-1441

Willmann, S.; Hohn, K.; Edginton, A.; Sevestre, M.; Solodenko, J.; Weiss, W.; Lippert, J. & Schmitt ,W. (2007). Development of a Physiology-based Whole-Body Population Model for Assessing the Influence of Individual Variability on the Pharmacokinetics of Drugs, *Pharmacokinet Pharmacodyn*, Vol. 34, No.3, pp.401-431

Yoshikawa, S.; Kenji, S. & Konagaya, A. (2004). Drug Interaction Ontology (DIO) for Inferences of Possible Drug-Drug Interactions, *Medinfo*, Vol.11, pp.454-458

Zhang, S.; Bodenreider, O. & Golbreich, C. (2006). Experiences in Reasoning with the Foundational Model of Anatomy in OWL DL, *Pacific Symposium on Biocomputing 2006*, pp.200-211

Part 4

Molecular Dynamics Simulations

13

General Index and Its
Application in MD Simulations

Guangcai Zhang[*], Aiguo Xu and Guo Lu
Institute of Applied Physics and Computational Mathematics,
The People's Republic of China

1. Introduction

In the long-term practice, it is recognized that the properties of materials are not uniquely determined by their average chemical composition but also, to a large extent, influenced by their structures. The impurities and defects in metal will hinder the movement of free electrons and reduce their conduction, therefore, the thermal conductivity of alloy is significantly smaller than that of pure metal. The yield strength, fracture strength, fatigue toughness and other mechanical properties of metal are influenced by defects, such as dislocations, grain boundaries, micro voids and cracks. In weak external magnetic field, due to the existence of spontaneous magnetization within a small area, e.g. magnetic domains, ferromagnet shows strong magnetism. The bonding strength and density of crystalline phases dramatically influence the strength of ceramic. Due to the existence of independent molecules, linear structure (including the branched-chain structure) polymers are flexible, malleable, less hard and brittle, and can be dissolved in a solvent or be heated to melt. However, in three-dimensional polymers, as there are no independent molecules, they are hard and brittle, can swell but cannot be dissolved or melt, and are less flexible. In nematic liquid crystal, the rod-like molecules are arranged parallelly to each other, but their centres of gravity are in disorder. Under external force, molecules can flow easily along the longitudinal direction, and consequently have a considerable mobility. In smectic liquid crystal, the molecules align in a layered structure via lateral interaction of molecule and interaction of functional groups contained by molecules. Two-dimensional layers can slide between each other, but the flow perpendicular to layers is difficult.

With the change of external conditions, the microscopic structures of material may change and consequently alter the material macroscopic properties. For example, the hardness of eutectoid steel with 0.77% carbon content is about HRC15 after annealing, but up to HRC62 after quenching, because the structure of carbon steel differs after different heat treatments. Electric field can change the order of liquid crystal molecules; LCD is made by using this feature, and now has been widely used in everyday life. It is one major goal of material sciences and urgent need of engineering applications to quantitatively clarify the relationship between macroscopic behaviour and microscopic structure. This goal imposes

---

[*] Corresponding Author

the task of identifying and describing those lattice defects, including their collective static and dynamic behaviour, which are responsible for specific macroscopic properties.

In a narrow sense, the structure of materials refers to microstructure. Haaseen defined it as the totality of all thermodynamic non-equilibrium lattice defects on a space scale that ranges from angstroms (e.g. non-equilibrium foreign atoms) to meters (e.g. sample surface). Its temporal evolution ranges from picoseconds (dynamics of atoms) to years (corrosion, creep, fatigue) (Dierk, 1998). The generalized structure also includes the electronic structure of atoms and chemical structure of molecules in equilibrium states. According to length scale, the various levels of material structure can be roughly divided into nanoscopic, microscopic, mesoscopic, and macroscopic regimes, where the term nanoscopic refers to the atomic level, microscopic to the level smaller than the grain scale, mesoscopic to the grain scale, and macroscopic to the sample geometry. Of course, this subdivision is to a certain extent arbitrary, various alternative subdivisions are conceivable.

Due to the multilevel nature of material structures and the correlation between them, macroscopic properties of materials are overall performance of structures of each level. Take polymer material for example, small molecules with chemical reactivity and fixed structure are made of different atoms, and can gather into macro molecules by the polymerization reaction (addition polymerization or condensation). The polymer chains consist of several units of the same structure linked in a special sequence and spatial configuration under certain formation mechanism. As the single bond can rotate, polymer conformations with certain potential energy distribution can be formed by chains. The chains with certain conformation gather into polymer body through the second-force or hydrogen bonds. Under certain physical conditions, polymer itself or with other added substances (such as fillers, plasticizers, stabilizers, etc.) can form macroscopic aggregation structure made of a number of microstructures to become a performance polymer material by means of molding.

Currently, all kinds of microscopy techniques are used to observe, trace and analyze material structures. The microscopy techniques with matching resolution are selected according to the characteristic scale of various structures. For instance, optical microscopy is commonly used for micro-scale structures, and electron microscopy for nano-scale structures. Advances in experimental observation methods represented by all kinds of microscopy technologies make possible getting more and more accurate material structure information. With the rapid development of computer technology and materials science, computer simulation is widely used in the study on the relationship between microstructures and macroscopic properties of various materials, and greatly promotes the development of material science. As different simulation methods are only suitable for their own corresponding spatial and temporal scales, we need choose appropriate simulation schemes according to the length scales of structures involved in specific issues.

Molecular Dynamic (MD) simulation is a very powerful tool to solve many-body problems. As being able to provide detailed evolution images of microscopic particles over time, it is particularly suitable for studies on the evolution law and characteristics of material microstructure. Since it was firstly used to study phase transition for hard sphere regimes in 1957, MD simulation technique has been greatly improved. Now it has become an important method for studies on characteristics and evolution law of structure of metallic materials, inorganic non-metallic materials and polymer materials. In the fields of nano-materials and bio-pharmaceuticals, molecular dynamics simulation is becoming an indispensable basic tool. In the studies on the large deformation of metal, molecular dynamic simulation plays a crucial role in explaining inverse Hall-Petch behaviour, nucleation and annihilation of dislocation involved in grain boundary and the relationship between shear band and dimple crack surface(Kumar et al., 2003). In the studies on protein folding, molecular dynamics is one of the most effective ways to simulate protein folding and unfolding. Via molecular dynamics simulation the transition state and change of energy in folding (or unfolding) process can be clearly analysed.

Structure analysis is the core issue of material simulation. Over the years, many defect identification methods have been proposed. For metal defects distinction, there are the excess energy method, coordination number method, centro-symmetry parameter method (Kelchner et al., 1998), Ackland's bond-angle method (Ackland & Jones, 2006), and bond-pair method (Faken & Jonsson, 1994), etc.

As we know, the complex computations between particles cannot be avoided in any elaborate defect identification methods. Since the computation complexity of traditional methods are of high order, the computation time needed will dramatically increase with growing the system size. Therefore, it is necessary to design new data structure and indexing algorithm to reduce computation complexity. The computation complexity of defects identification methods will be greatly reduced by using background grid and linked list. The background grid index, together with the linked list data structure, is suitable for management of uniform distributed points, and it has been widely applied in computation and analysis of many simulations. The analysis of complex structure in non-uniform system refers not only to points, but also to lines, surfaces and bodies, and their distributions are usually non-uniform. The background grid index cannot meet the needs for managing these objects, but the multi-level division of space is effective way to manage complex data. The SHT (space hierarchy tree), a newly proposed data structure, is a powerful dynamical management framework of any complex object in any dimensional space. Index of objects with complex structure can be created based on SHT, and corresponding fast searching methods could be designed to meet various search needs.

As the elements managed by SHT are abstract objects, such as geometry objects (points, lines, surfaces and bodies) in three-dimensional space and characteristic regions in phase space, the general index based on SHT data structure can be used in various application areas. In this chapter, index methods in MD simulation will be introduced, including the background grid index and general index based on SHT, and then with the emphasis on several applications of SHT general index in computational geometry (shortest path problem and Delaunay division) and characteristic structure analysis (cluster construction, defects identification, and interface construction).

2. General index

In many programs, such as post processing of MD, discrete elements simulation, computational geometry, smart recognition, spatial objects (i.e., points, lines, surfaces, bodies, structures, etc.) are used to construct complex structures in the system. Without index of spatial objects, the quantity of computation for searching object is very high. If a system contains N objects, the computation complexity related to two objects is N^2 and N^3 for three objects. If the total number of objects is more than 10^4, the computation complexity is not acceptable. Therefore, effective storage and fast search of objects are necessary, and consequently the index of spatial objects should be established. Currently, there are three kinds of spatial indexes: background grid, tree and linked list. These indexing methods mainly focus on points and neighbour search. In this chapter, an SHT general index is presented, which is suitable for management of any objects in any dimensional space. Fast searching algorithms meeting any given requirements are proposed based on SHT. In traditional programming, intuitive idea does not lead to intuitive algorithm. However, if following the idea of SHT to write codes, intuitive idea can directly lead to intuitive algorithm, so that it's beneficial to programming. Taking into account the integrity of this description, background grid index is firstly introduced, and then we introduce the SHT data structure and fast searching methods.

2.1 Background grid index

For points and small objects with uniform sizes, if their spatial distribution is not extremely scatted, the background grid index is suitable for managing. This indexing method is widely used in molecular dynamics, Monte Carlo method, smooth particle hydrodynamics, material point method and discrete element method to search neighbour objects. For this case, small object can be viewed as a point located at the centre of the object or its feature point.

The idea of background grid index (Michael, 2007) is as follows: set a box to contain all objects according to their locations; divide the box into smaller grids with a certain size, then put objects into appropriate grids according to their locations. Hence, an index for spatial objects is created. When searching for objects, several grids are firstly selected according to searching conditions, and then search in selected grids to get the needed objects. In figure 1, small black rectangles stand for objects and thin lines are for background grid. When search for objects in a given circle, what we need to search through is not all the grids but only those covered by the circle.

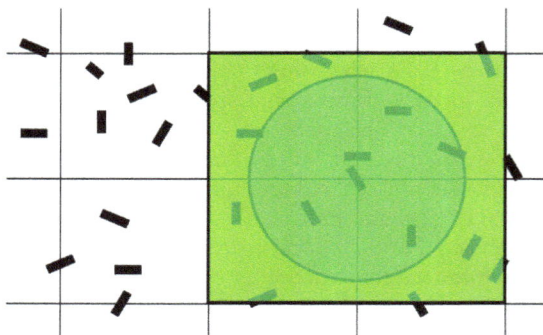

Fig. 1. Scheme for the background grid index of objects.

For creating background grid index, one way to store the objects within each grid is to use array. Thus an array with possible maximum size must be firstly defined. In this way, a large quantity of memory may be wasted. So, the array method is not suitable for the case with objects nonuniformly distributed. A good alternative is to use the linked list to store such kinds of objects. Figure 2 is the scheme for linked list of objects in background grid (two-dimensional space). The indexing steps are as follows:

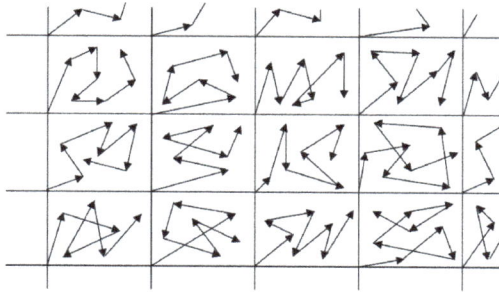

Fig. 2. Scheme for the linked list of objects in background grid.

1. Set the grid size Δ according to the search requirements;
2. Check all objects to get system size $(x_0, x_1) \times (y_0, y_1)$;
3. Calculate the grid array size (I, J), where $I = \left\lceil \dfrac{x_1 - x_0}{\Delta} \right\rceil + 1, J = \left\lceil \dfrac{y_1 - y_0}{\Delta} \right\rceil + 1$, allocate the grid array dynamically;
4. For each object A, according to its location (x_A, y_A) to get the grid (i, j) containing it, where $i = \left\lceil \dfrac{x_A - x_0}{\Delta} \right\rceil + 1$ and $j = \left\lceil \dfrac{y_A - y_0}{\Delta} \right\rceil + 1$. Insert object A into the top of linked list of grid (i, j). In figure 3, the red point stands for grid $g(i, j)$, blue point stands for object A, $g(i,j) \to obj$ stands for the object pointed to by grid $g(i, j)$. The two steps are as follows:

$$A \to next = g(i, j) \to obj$$
$$g(i, j) \to obj = A \tag{1}$$

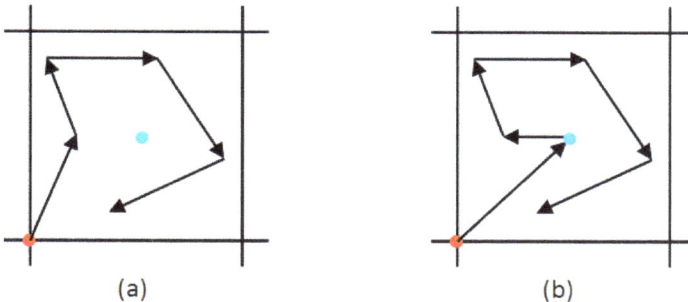

(a)   (b)

Fig. 3. Scheme for inserting an object into a linked list: (a) before inserting; (b) after inserting.

The algorithm for searching objects in a given region is as follows: (i) search for grids crossing with the given region; (ii) search for objects within those grids.

In background grid index, *correlated grid* refers to the grid crossing with the given region, *number of object included* is the number of objects within each grid. The computational complexity of background grid index is related to the number of correlated grid (N_{gird}) and the average number of object included (N_{obj}), it is of order $O(N_{gird}N_{obj})$. The larger the grid size, the less the number of correlated grid and the more the average number of objects included, and vice versa. The memory needed increases with decreasing grid size. A proper grid size has to be chosen to reduce computational complexity. Usually, the selected average number α of objects included ranges from 3 to 5. The grid size can be gained by calculating the average density of object. Such as, in two-dimensional space, the grid

size is $\Delta = \left(\dfrac{\alpha\left(x_1 - x_0\right)\left(y_1 - y_0\right)}{N} \right)^{1/2}$, and it is $\Delta = \left(\dfrac{\alpha\left(x_1 - x_0\right)\left(y_1 - y_0\right)\left(z_1 - z_0\right)}{N} \right)^{1/3}$ in three-

dimensional space.

The data structure of background grid index is very simple. If the size and spatial distributions of objects are uniform, the calculation based on this index is fast. A few disadvantages of this index scheme should also be pointed out: (i) the background grid cannot be dynamically created, it must be recreated with local dynamical adjustment; (ii) for the size of object is not contained by the index, if the size of objects is not uniform, it is hard to support index related to sizes. (iii) When the dimension is high or the spatial distribution is not uniform, the needed memory is too large.

2.2 SHT (Space Hierarchy Tree) and general fast search

Currently, there is no general scheme for managing different types of objects, there is also no universal way to quickly search for objects under flexible requirements. Different problems require different designs. A large number of complex operations are unavoidable in the corresponding coding procedure. This leads to lots of drawbacks, such as design difficulties, without university, long codes and hard to maintain.

When indexing discrete spatial points, background grid linked list structure can be used. For spatial data, tree structure can provide convenient index. Examples are referred to BSP (Binary space partitioning) tree, K-D tree (short for k-dimensional tree) and octree (de Berg, 2008; Donald, 1998). The BSP and K-D trees have their own specific index methods and are not suitable for general index. For example, the BSP tree is suitable for identification of inner or outer region of polyhedra.

The data structure of object determines its indexing, for instance, the index of discrete points by using background grid is limited to neighbor point index. A better data structure design can implement general index. A data structure meeting any indexing requirements must be appropriate to describe the locations and sizes of objects, adapted to the dynamic changes in the data, and its structure has to be standardized. The octree can manage three-dimensional points. It can also manage other kinds of spatial objects.

In this section, a new data structure, SHT, extended from octree is proposed to unify the management of any objects in any-dimensional space, and two general search methods (conditional search and minimum search) are presented to implement search for object meeting any given requirement. The computation complexities of two search schemes are both logN. By using SHT and corresponding fast searching methods, programming becomes easy.

2.2.1 SHT management structure

In three-dimensional space, SHT data structure is similar to octree. Go a further step, for a system in n-dimensional space, a n-dimensional cube (a line segment in one-dimensional space, a rectangle in two-dimensional space, and a cube in three-dimensional space) is designed to contain the system; then divide this cube in each dimension into two parts to form 2^n sub-cubes; only retain the cubes with objects inside; continue to decompose each cube until the required resolution is reached; put the objects (points, lines, surfaces, bodies) into the appropriate cube according to their locations and sizes; retained cubes are connected together, according to their belonging relationships, to form a 'spatial hierarchy tree'. Up to now, an index for a set of spatial objects has been established, needed objects can be quickly found via the index. Figure 4-a (Figure 4-b) is a scheme for the SHT management structure of two-dimensional (three-dimensional) discrete points.

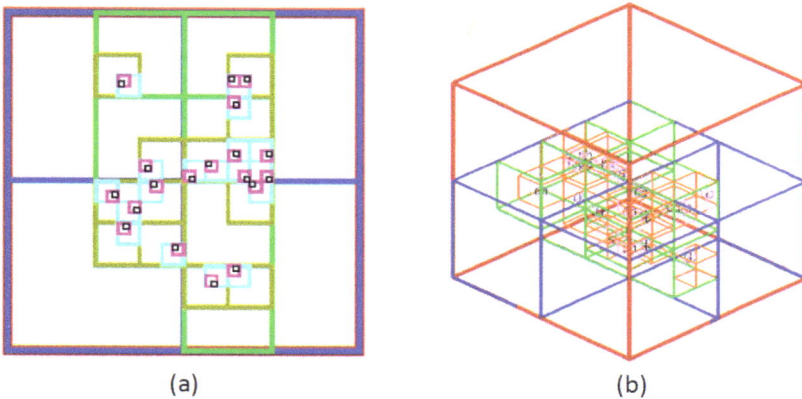

(a)   (b)

Fig. 4. Scheme for management region of SHT of discrete points. (a) Two-dimensional points; (b) three-dimensional points.

This SHT data structure contains the properties of directed tree. Each cube is named a 'branch'. Its child-cubes are named 'sub-branches' and its parent-cube is called the 'trunk'. The largest or the top-level cube is named the 'root' and the smallest or bottom-level cubes are called 'leaves'. The process for linking an object with a branch is named 'putting into' and the opposite process is called 'cutting down'. The SHT is different from the octree in two aspects. Firstly, it can be used in any-dimensional space and the number of sub-branches is arbitrary. Secondly, objects can be put into not only leaves but also any other branches, and the number of objects is arbitrary. Due to the uncertainty of the number of 'sub-branches' in a 'branch', 'branches' sharing the same 'trunk' are linked as a list; Similarly, 'objects' belonging to the same 'branch' are also linked as a list. Figure 5 is the scheme for object management by SHT. For the convenience of description, from now on, unless specifically pointed out, we do not distinct between 'branch' and its corresponding cube region, such as, the 'center of branch' is also referred to the geometrical center of the cube, the 'size of branch' is also the cube size, and the 'branch corner' is also the corner of the cube, etc. Meanwhile, tree are named according to the managed objects, such as, tree created to manage points is called 'point tree', and to triangle is 'triangle tree', etc.

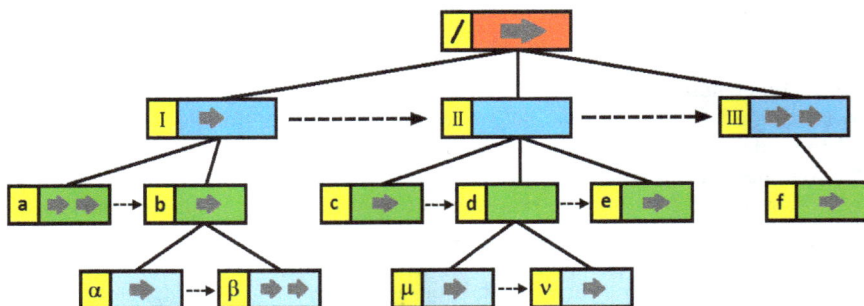

Fig. 5. Scheme for object management by SHT. The rectangle in each row stands for branch, horizontal grey arrow stands for object, and vertical black arrow stands for the list which connects the sub-branches belonging to a same branch.

In practical applications, the number of objects may be variable. Therefore, the SHT is constructed dynamically. The dynamic management procedure of SHT includes three basic operations: (I) establishment of a tree, (II) adding a new object to a tree, (III) removing an object from a tree.

(I) The establishment of a 'tree' from an object is to establish a minimum 'branch' to contain the object. The idea of algorithm is thus: Enlarge the size of the given minimum cube until the object can be put into. The typical procedure consists of two steps: (i) Get the known minimum resolution 'σ', i.e. the smallest edge length of cubes. Use the center of an object as the geometrical center of the cube to create a 'branch'; (ii) Check whether or not the branch can contain the object. If yes, the branch size is proper, and put the object into it; if not, double the branch scale and go back to step (ii) to continue. Up to this step, a 'tree' with only one 'branch' which contains one 'object' has just been established. Figure 6 shows the construction process of the original tree, where a triangular object in two-dimensional space is used as an example.

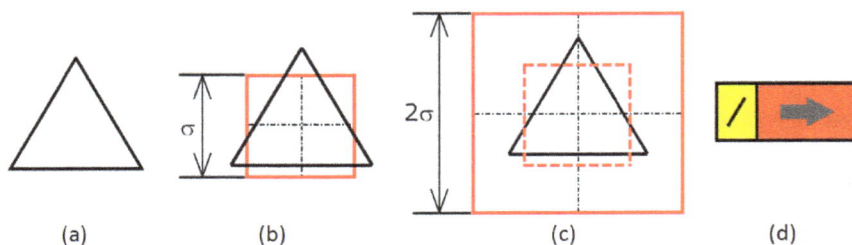

Fig. 6. Scheme for the establishment of the original tree. (a) triangle object; (b) create an original branch with minimum resolution 'σ'; (c) double the scale of original branch until the triangle can be contained; (d) established triangle tree, where '/' is root ID, grey arrow stand for the triangle.

(II) The idea of adding an object to a tree is thus: Enlarge the tree to contain the object and then add the object to an appropriate branch. The algorithm for adding a new object B to the tree is as below: (i) Adjust the root. (a) Check whether or not the object B can be contained by the current root r. If yes, end this step and go to step (ii). If not, take the center of root r as a reference point to calculate the quadrant where the center of object B is located, and then create trunk s of root r, link root r to trunk s and set trunk s as root. (b) Go back to step (a). (ii) Place object B. (c) Set the root as current branch; (d) Check whether or not the sub-branch of current branch, with respect to quadrant B, contains object B. If not, place the object into current branch and break; if yes, enter the sub-branch (if the sub-branch does not exist, create it) and take it as the current branch. (e) Go back to step (d).

Figure 7 is the scheme for adding a triangle to an existing tree, where A and B stand for triangle objects. O1 is the geometrical center of root of the original tree and O2 is the corresponding center of the enlarged root. The root is '/', the second-level sub-branches are 'I', 'II', 'III' and 'IV', 'a' is the third-level sub-branch. B0 is the geometrical center of object B. In figure 7(c), the rectangle in each row stands for the corresponding branch (symbol in the left part is the branch, and in right part is the object).

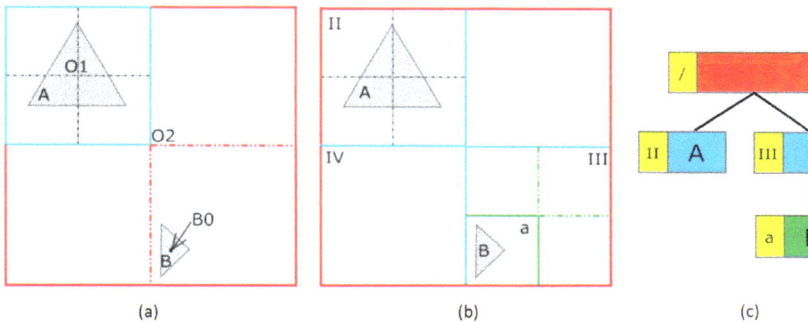

Fig. 7. Scheme for adding objects to the tree. (a) Generation of new root; (b) placing an object; (c) SHT corresponding to (b).

(III) The algorithm for removing an object from the tree is as follows: (i) According to the link, pick up the branch containing the object. Remove the object from the corresponding object-list, and then set this branch as current branch; (ii) Check whether or not current branch contains other objects and sub-branches. If not, remove this branch, enter its trunk and set the trunk as new current branch, and then go back to step (ii) to continue this process until all the useless branches are eliminated. Figure 8 is the scheme for removing an object from a tree.

In the dynamical algorithm of SHT, except for adding a sub-branch or trunk of a branch, other operations have nothing to do with space dimension. We can use four functions to describe the operations related to space dimension and to the location and size of object in this process: (i) Create an appropriate branch to contain the current object, (ii) Check whether or not a branch contains the current object, (iii) Take the current branch as reference point, and then create a trunk of current branch in the quadrant where the object is located, (iv) Take the current branch as reference point, and then create sub-branch of current branch in the quadrant where the object is located. The computer memory required by the SHT is approximately equal to kNlogN, where N is the number of objects. It is independent of the spatial dimension. When the spatial dimension is higher or spatial objects are have a highly scattered distribution, the SHT can save a large quantity of memory compared with the background grid method. In addition, because SHT is dynamically constructed, the size of the system can dynamically increase or decrease with the addition or deletion of objects. This is a second obvious advantage over the traditional background grid method.

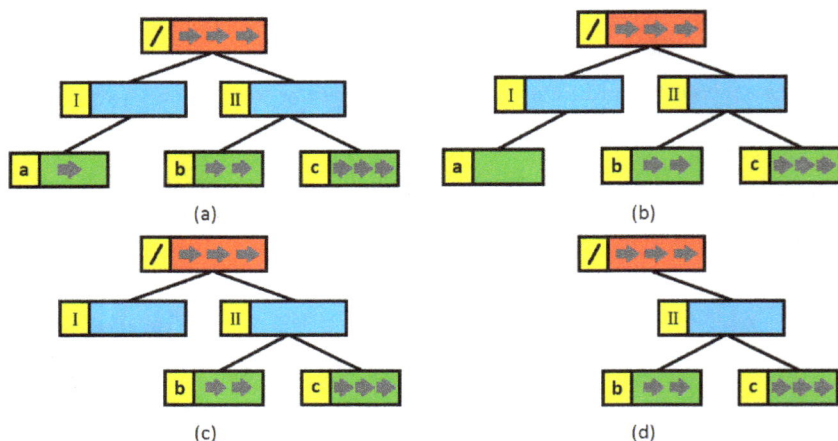

Fig. 8. Scheme for removing an object from a tree. (a) pick up the branch containing the object, in this figure, it is branch 'a'; (b) remove the object from branch 'a'; (c) remove branch 'a'; (d) remove branch 'I', which contains no more objects and sub-branches.

2.2.2 Fast searching algorithms based on SHT

In practical application, we generally need a fast search of objects satisfying certain requirements. The computational complexity of an ergodic search is N. It is clearly not practical when dealing with a huge number of objects. For such cases, we need to develop fast searching algorithms. As the objects managed by SHT can be located, fast searchers with computational complexity logN can be easily created. The basic idea is as below: It is unnecessary to search the objects directly, but rather check branches. Skip those branches without objects under consideration. Thus, searching is limited to a substantially smaller range. Depending on requirements of applications, two fast searching algorithms are presented: conditional search and minimum search. The goal of conditional search is to search for objects meeting certain conditions. For example, find objects in a given area. The goal of minimum search is to search for an object whose function value is minimum. For example, find the nearest object to a fixed point.

2.2.2.1 Conditional search

Conditional search is to search for objects meeting given conditions among the objects hanged up to the SHT. The idea of algorithm is as follows: Check whether or not a branch contains objects meeting given conditions. If not, do not search the branch (including all sub-branches of it and corresponding objects). For example, if searching for all objects in a given circle, check whether or not a branch crosses with the given circle. If not, skip this branch. In this way, the searching is limited to branches crossed with the given circle.

For the convenience of description, we define *candidate branch* as the branch which may contain the objects meeting the given condition during search procedure. Conditional search is implemented using the stack structure. The steps are as follows: (i) Push the root into a stack A; (2) Check whether or not the stack A is empty. If yes, end the search; if not, pop out a branch b from the stack A; (ii) Check whether or not each sub-branch of b is a candidate branch. If yes, push it into stack A; (iv) Check whether or not the object in b satisfies the given conditions; pick out the required objects; (v) Go back to step (ii).

Figure 9 shows the given circle and spatial division for managing planar triangles. Figure 10 is the scheme for the SHT corresponding to Figure 9, where '/' stands for the root. The process to find out objects in the given circle is shown in Figure 11. In the example, apply the procedure given above, we will finally find object A and B is in the given circle.

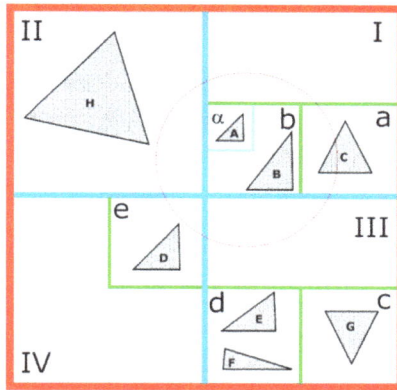

Fig. 9. Distribution of planar triangles and the corresponding spatial division.

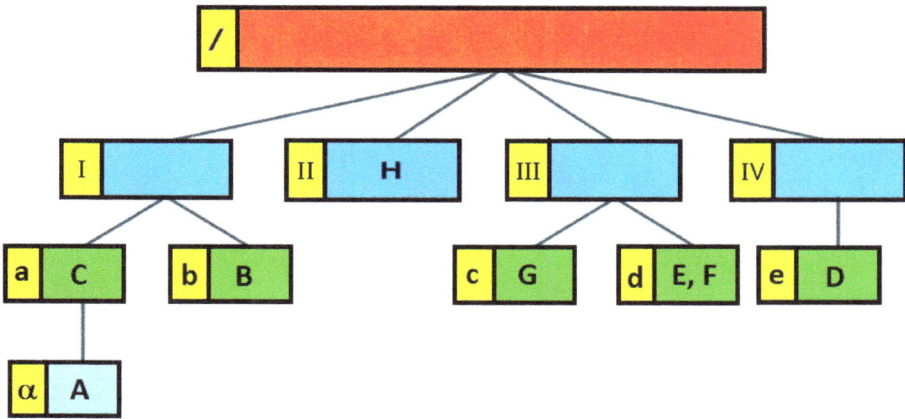

Fig. 10. SHT corresponding to Figure 9.

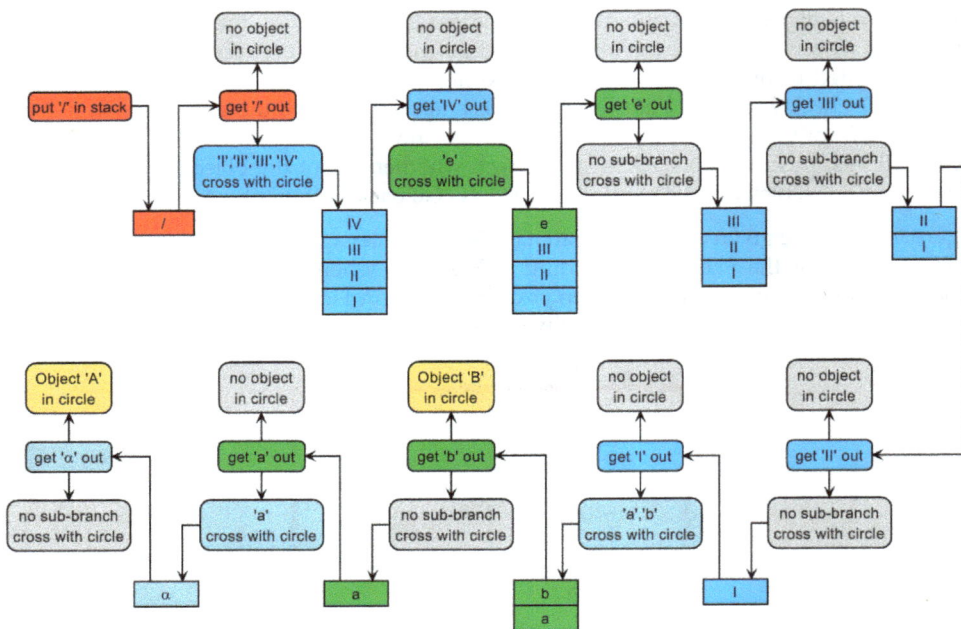

Fig. 11. Flowchart for the fast search of objects in a circular area.

In the searching, other operations have nothing to do with space dimension and type of object except for two operations, (i)checking whether or not an object is the needed one, or (ii) checking whether or not an branch is a candidate branch. So, the algorithm can be built on the abstract level. The conditional search is implemented by providing a *conditional function* and an *identification function*. The *conditional function* is used to check whether or not an object is needed. Assuming *condition(o)* is the conditional function, the argument o is object and the function value is bool number. The *identification function* is used to assess whether or not a branch is a candidate branch. Assuming *maycontain(b)* is identification function, the argument b is branch and the function value is bool number. With defining the above two functions, conditional searching meeting any given conditions can be easily implemented.

The conditional search can search for a collection of objects. With the search marks, the actual search process is done step by step. In each step, only one object can be gained. The computation complexity of conditional search is kL, where L~ln N is total level of SHT. The efficiency of the conditional searching algorithm depends on the identification function. The more efficient the identification function, the smaller the k and the faster the searching procedure. If identification function value is always true, k tends to infinity and this searching algorithm goes back to an ergodic browser. However, the computation complexity of identification function influences the k value. Since the computation for sphere regions is more efficient than for cube ones, in complex conditional searching algorithms, circum-spheres of a cube are extensively used to reduce the quantity of computations.

2.2.2.2 Minimum search

In programming related to spatial objects, we often need to find objects meeting some given extreme condition. For example, to find a point with the largest z component from a set of three-dimensional points, or to search a point with the nearest distance to a given point, or to search a sphere closest to a plane, etc. Such searches can be attributed to the minimum searching problem. For spatial objects, each one can be assigned a function value related to its location and size such that the minimum search is to find the object with the minimum function value.

Fast searching can be created based on the SHT. The idea is as follows: Design a function to assess the range of the function value of all objects. Some branches can be excluded by comparing the ranges of the function value of different branches. For example, in a region, $\sum$ is a set of discrete points, we need to search the nearest points to a given point A. The fast searching is not to compute the distance between each point in $\sum$ and point A, but assess the range of distance between 'branches' and point A to excluding unnecessary searching in branches (including the point in it and its sub-branches) with longer distance.

For convenience of description, we define a few concepts.(i) Range of a branch: It means the range covering all the values of the given function for objects in this branch. (ii) B-R-branch: It is a new branch data structure composed of the branch itself and the range of it. (iii) Candidate B-R-branch: It is the B-R-branch which may be checked in the following procedure. It may contain objects whose functional values are minimum. In the minimum searching procedure, we must keep enough candidate B-R-branches. Some of them may be dynamically added or removed according to the need. In order to accelerate the searching speed, the candidate B-R-branches should be linked as a list. According to the above definition, each B-R-branch has a range. So, each B-R-branch has a lower limit to its range. The B-R-branches in the list are arranged in such a way that their lower limits subsequently increase. Obviously, the B-R-branch with the smallest lower limit is placed at the head of the list. (iv) Candidate object: the object with the minimum values during the searching procedure. It is initialized as null, and at the end of searching, it is the object with the minimum value. (v) Candidate value: function value of candidate object.

The minimum searching algorithm is as follows: (i) The root and its range are merged as a B-R-branch; Add the B-R-branch to a candidate list named L; The candidate value V is set as positive infinity; The candidate object is set as null. (ii) Pick out a B-R- branch, for example, B, from the head of candidate list L; Check the values of its objects. If the value of an object O is smaller than V, then replace V with this value; in the meantime, set object O as a candidate object. Remove the B-R-branches whose lower limit values are greater than V from the list L. (iii) Construct a B-R-branch Z for each sub-branch of B. If the minimum value of Z is larger than V, cancel Z; If the maximum value of Z is smaller than the minimum value of the B-R-branch C in L, then all the B-R-branches behind C are removed from L; If the minimum value of Z is larger than the maximum value of a B-R-branch in L, cancel Z; Otherwise, insert Z into list L according to its lower limit. (iv) Repeat steps (ii) and (iii) until the candidate list L is empty. The final candidate object is the required one.

As an application example of the proposed minimum search algorithm, we consider a case to find the nearest point to a fixed point A. Figure 12 shows the distribution of planar points and spatial division. Figure 13 is the corresponding SHT. Suppose point A in Figure 12 is the given fixed point. To seek the nearest point to it, the range of the branch is calculated by a sphere evaluation method. Figure 14 shows the flowchat.

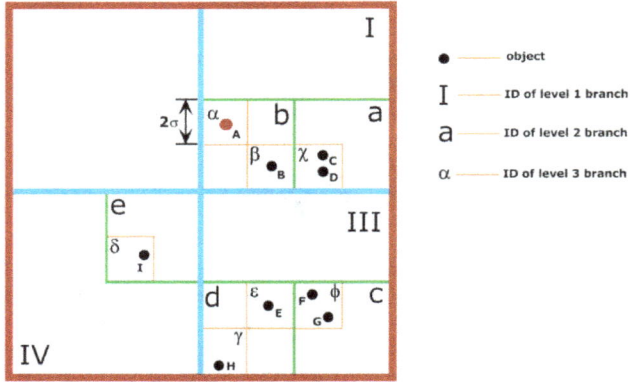

Fig. 12. Planar point distribution and corresponding spatial division.

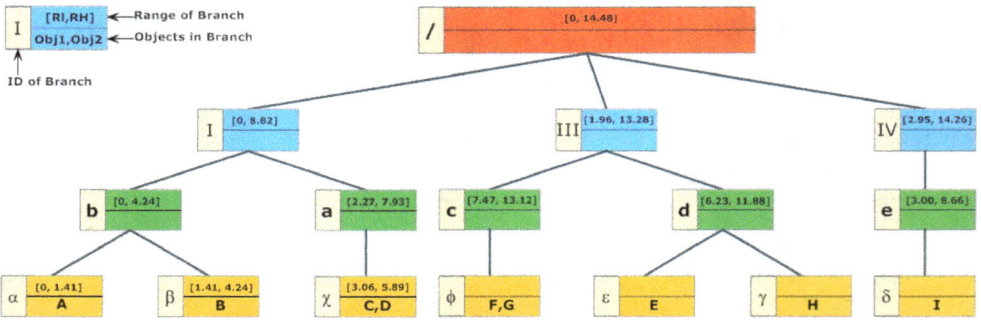

Fig. 13. SHT corresponding to Figure 12.

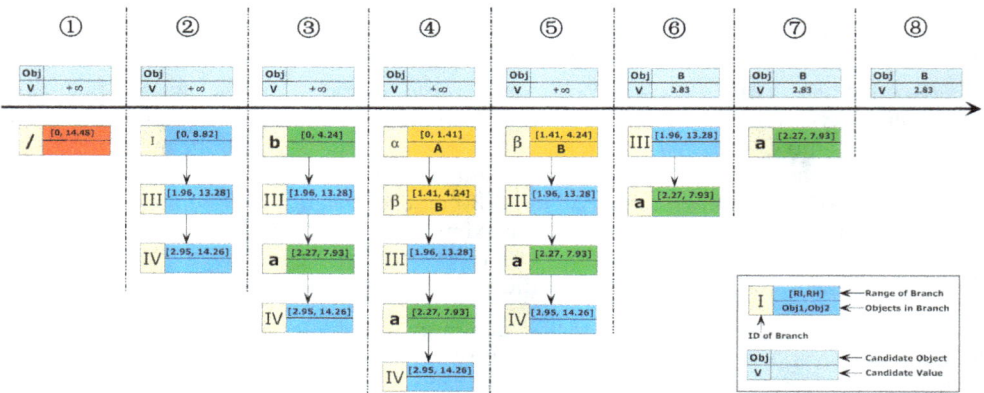

Fig. 14. Scheme for the fast search of the nearest point to a given fixed one.

In minimum search, except for computing the function value of an object and the range of a branch, other operations have nothing to do with space dimension and type of object, so that the algorithm can be built on abstract level. Similar to conditional search, the minimum search is implemented by providing a *value-finding function* and *range-evaluation function*. The *value-finding function* is used to calculate the function value of an object. Assuming *value(o)* is *value-finding function*, the argument *o* is an object and the function value is a real number. The *range-evaluation function* is used to compute the range of a branch. Assuming $M(b)$ is the upper limit and $m(b)$ is the lower limit of the range, the argument of function is branch b and the function value is a real number. With defining the above two functions, various minimum searches can be easily implemented.

The computation complexity of minimum searching is kL, where L~ln N is total level of SHT. The efficiency of the minimum searching algorithm depends on the range-valuation function. The smaller the range given by the range-evaluation function, the smaller the k and the faster the searching procedure. The worst range-evaluation function gives a range from -∞ to +∞. In such a case, the searching algorithm goes back to the ergodic browser. In the case with a large quantity of objects, one should use a good range-evaluation function to reduce the number of objects to be searched. However, a precise range-evaluation generally needs a large quantity of computations, which also decreases the global efficiency. We should find a balance between the two sides. Since the computation for sphere regions is more efficient than for cube ones, in complex minimum searching algorithms, circum-spheres of a cube are extensively used to evaluate the range of a branch.

During conditional searching and minimum searching, we often need to screen some objects. For example, in the searching for the nearest point to a point A, if A is not screened, the searching result is just A and it's not what we need. In the algorithm implementation, a screen function is designed by breaking the corresponding connections in the tree, each call of screen function can screen an object, and a recover function is been design to recover all screened objects.

3. Applications

SHT framework provides a general index of set of spatial objects and effective management of data. Most problems in specific applications can be solved with SHT. In this chapter, object with physical properties is named as 'matter element'. Examples are referred to molecule or molecular clusters with certain energy, mass, momentum and angular momentum, particles with specific density and phase, grains with orientation and finite element with density and energy. For the majority of applications, problem can be summarized as constructing new sets of matter element meeting requirements from the existing one. In this section, the applications of general index in shortest path problem, space division, cluster construction, defect atom identification and interface construction are presented.

3.1 Shortest path problem

The shortest path problem (Moore, 1959) is a classical problem in graph theory. It is to find a path between two vertices (or nodes) in a graph such that the sum of the weights of its constituent edges is minimized. The shortest path not only refers to the shortest geographical distance, but also can be extended to other metrics, such as time, cost, and line capacity and so on. The selection and implementation of the shortest path algorithm is the basis of channel route design. It is the research focus of computer science and address information science. Many network-related problems can be incorporated into the category of the shortest path problem. Due to the effective integration of classic graph theory and the continuous development and improvement of computer data structure and algorithm, new shortest path algorithms are emerging with time.

The postman shortest path problem is as follows: Given N cities, one needs to find a shortest loop that the postman can go through all these cities. This is a typical NP problem, because it needs to compare all of the loops to find the optimal path. As a simple application example, the approximate solution is given, it is to find a shorter loop without intersection between its segments to connect all discrete points, and this approximate solution is called a labyrinth. The algorithm of the approximate solution is as follows: add point to the loop one by one in such a way that a previous segment is replaced by two where the newly added point becomes the vertex in between and connecting two previous vertexes, at the same time keep the loop length as shortest as possible.Figure 15 is the schematic for adding a point to loop.

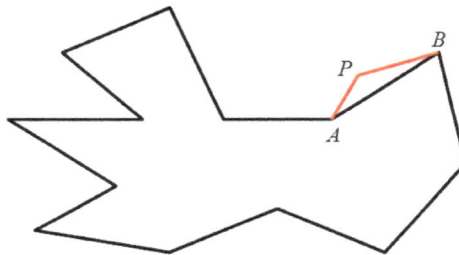

Fig. 15. The change of a loop after adding a point P. The segment AB becomes two segments AP and PB.

The preparatory work for algorithm based on SHT is to design fast searchers. In this section, a design, including three minimum searchers and a conditional searcher, is proposed, where one value-finding function and two range-evaluation functions are defined for each minimum searcher and one conditional function and one identification function are defined for the conditional searcher.

(I)   Minimum searcher MS1: search for the nearest point to a fixed point r_0 in a two-dimensional point tree. The value-finding function is

$$value(p) = \left| \mathbf{r}_p - \mathbf{r}_0 \right| \tag{2}$$

The two range-evaluation functions are as follows:

$$M(b) = \left| \mathbf{c}_b - \mathbf{r}_0 \right| + \sqrt{2}d_b$$
$$m(b) = \max(\left| \mathbf{c}_b - \mathbf{r}_0 \right| - \sqrt{2}d_b, 0) \tag{3}$$

where $\mathbf{r}_p$ is the coordinate of point p, $\mathbf{c}_b$ the coordinate of the center of branch b, $\sqrt{2}d_b$ the circumradius of branch b, $M(b)$ the upper limit and $m(b)$ the lower limit.

(II) Minimum searcher MS2: In a two-dimensional point tree, search for a point with the shortest fold-line connecting the two vertexes of a given segment AB. The value-finding function is

$$value(p) = \left|\mathbf{r}_p - \mathbf{r}_A\right| + \left|\mathbf{r}_p - \mathbf{r}_B\right| \tag{4}$$

The circumcircle of branch b is used to assess the range of b. The range-evaluation functions are as follows:

$$M(b) = \begin{cases} 2\sqrt{2}d_b & A,B \in C \\ 2\sqrt{2}d_b + \left|\mathbf{r}_B - \mathbf{c}_b\right| & A \in C, B \notin C \\ 2\sqrt{2}d_b + \left|\mathbf{r}_A - \mathbf{c}_b\right| & B \in C, A \notin C \\ \left|\mathbf{r}_A - \mathbf{c}_b\right| + \left|\mathbf{r}_B - \mathbf{c}_b\right| & A,B \notin C \end{cases}$$

$$m(b) = \begin{cases} 0 & A \in C \text{ or } B \in C \\ \left|\mathbf{r}_A - \mathbf{c}_b\right| + \left|\mathbf{r}_B - \mathbf{c}_b\right| & A,B \notin C \end{cases} \tag{5}$$

where C is the circumcircle of branch b.

(III) Minimum searcher MS3: Given a point P, in two-dimensional segment tree, search for a segment AB with the minimum value of the length of fold-line APB minus the length of segment AB. The value-finding function is

$$value(p) = \left|\mathbf{r}_p - \mathbf{r}_A\right| + \left|\mathbf{r}_p - \mathbf{r}_B\right| - \left|\mathbf{r}_A - \mathbf{r}_B\right| \tag{6}$$

Ditto, the circumcircle C is used for assessment. The range-evaluation functions are as follows:

$$M(b) = 2(2\sqrt{2}d_b + \left|\mathbf{r}_P - \mathbf{c}_b\right|)$$

$$m(b) = \begin{cases} 0 & P \in C \\ \max(\left|\mathbf{r}_P - \mathbf{c}_b\right| - 3\sqrt{2}d_b, 0)\left|\mathbf{r}_A - \mathbf{c}_b\right| + \left|\mathbf{r}_B - \mathbf{c}_b\right| & P \notin C \end{cases} \tag{7}$$

(IV) Conditional searcher CS1: Given a segment AB, in two-dimensional segment tree, search for a segment ab crossed with segment AB. The conditional function is as follows:

$$condition(ab) = \begin{cases} true & 0 \le x \le 1, 0 \le y \le 1 \\ false & else \end{cases} \tag{8}$$

where x and y are obtain from solving the equation, $x\mathbf{r}_A + (1-x)\mathbf{r}_B = y\mathbf{r}_a + (1-y)\mathbf{r}_b$. The identification function is as follows:

$$maycontain(b) = \begin{cases} true & AB \cap b \ne \phi \\ false & else \end{cases} \tag{9}$$

Up to now, the preparatory work has been finished. The labyrinth construction algorithm is as follows:(I) Initialization: Construct point tree tp from given planar discrete points, cut down a point P_0, search for the nearest point P_1 to point P_0 (P_0 is screened) with MS1 searcher in tree tp, search for point P_2 (P_0 and P_1 are screened) to make the length of fold-line $P_0P_2P_1$ shortest with MS2 searcher in tree tp, generate segment P_0P_1 and construct a segment tree tl from it, generate segment P_2P_0 and P_1P_2, put them on the tree tl. (II)Connection construction: (i) Cut down a point P from tree tp and empty stack s. (ii) If P is null, exit. (iii) Search for a segment L (all segments in stack s are screened) in tree tl with MS3 searcher to make the increased length shortest by adding point P. (iv) Search for a segment L_1 crossed with the segment L with CS1 searcher. (v) If L_1 is null, cut down L from tree tl, generate two segments by connecting the two vertexes of L with point P and put them on the tree tl, go back to step (i), and if the stack s is not empty, then push L into s and go back to step (iii). Figure 16 shows the labyrinth constructed by connecting 40000 random points.

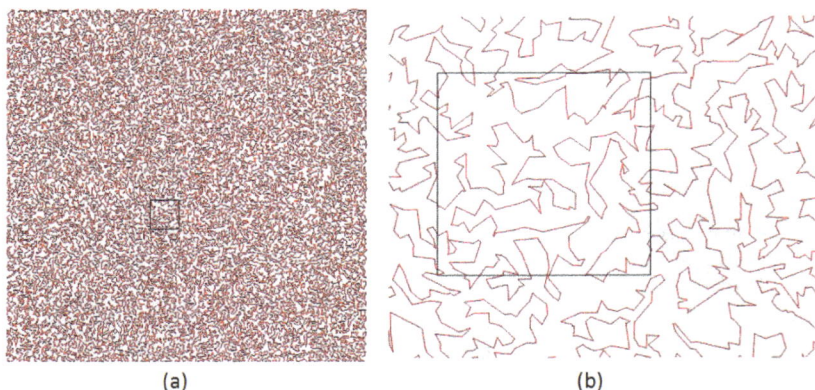

(a)   (b)

Fig. 16. Labyrinth constructed by connecting 40000 random points. Fig. (b) shows the enlarged portion chosen by the small black rectangle in Fig. (a).

3.2 Fast constructing algorithm for Delaunay division

Since the good mathematical characteristics of Delaunay division (de Berg, 2008), it is widely applied in many areas, such as the pre-processing of three-dimensional finite element method, medical visualization, geographical information systems and surface reconstruction. For designing a subdivision algorithm, an important part is to create a mesh with low complexity and good performance. Today, there are lots of algorithms to the construction of Delaunay tetrahedrons in three-dimensional space or Delaunay triangles in two-dimensional space. The complexities of most algorithms are involved with the searching procedures. Here, we propose an algorithm based on the SHT. The algorithm is simple and intuitive. It is convenient to extend to higher-dimensional space.

It's necessary to use two new data or objects, i.e. 'extended-tetrahedron' and 'extended-triangle' in the two-dimensional and three-dimensional algorithms. The 'extended-tetrahedron' is a combination of tetrahedron and its circumsphere, so its data structure contains that of tetrahedron and its circumsphere, and its center and size are those of the circumsphere. The 'extended-triangle' is a combination of triangle and its circumcircle, so its data structure contains that of triangle and its circumcircle, and its center and size are those of the circumcircle.

The algorithm for constructing Delaunay division from discrete points is as follow: when a new point is added to the formed Delaunay division structure, the division is adjusted to meet the condition for Delaunay division. Only tetrahedra whose circumspheres include the added point need to adjust, and they are named to-be-adjusted-polyhedron. The adjustment is to construct new tetrahedron by connecting each outer surface of the to-be-adjusted-polyhedron with the added point. Figure 17 shows the steps for adding a two-dimensional point and re-dividing the space, the red points stands for the newly added point P, the green triangles in figure 7(a) are to-be-adjusted- triangles, the red segments in figure 7(b) are retained boundary segments, the blue segments in figure 7(c) are segments connecting point P with vertexes of boundary. In the three-dimensional case, we need only to replace the triangle with a tetrahedron, replace the line with a triangular face, and replace the extended-triangle with an extended-tetrahedron.

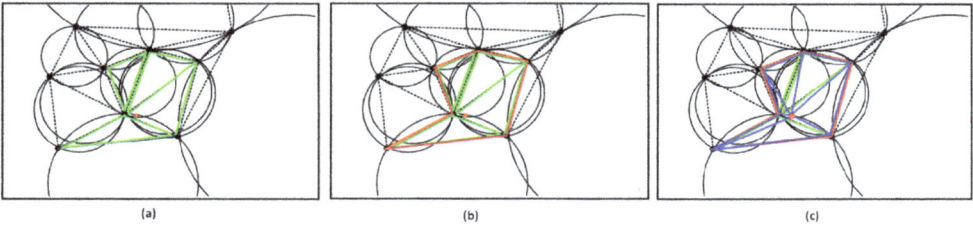

(a)   (b)   (c)

Fig. 17. Three steps to add a new point to a two-dimensional Delaunay division. (a) Finding the triangles whose circumcircle contains the newly added point P; (b) Removing the internal lines of these triangles, retaining the external ones; (c) Connecting each left line with point P to form new triangles.

The preparatory work for algorithm is to design fast searchers. In this section, a design, including two conditional searchers, is proposed.

(I) Conditional searcher CS1: Given a fixed point $\mathbf{r}_0$, search for an 'extended-tetrahedron' CT containing point $\mathbf{r}_0$. The conditional function is as follows:

$$condition(CT) = \begin{cases} true & |\mathbf{c}_{CT} - \mathbf{r}_0| < R_{CT} \\ false & else \end{cases} \tag{10}$$

The circumcircle of branch b is used to identify, the center of b is c_b and the radius is $\sqrt{3}d_b$. The identification function is as follows:

$$maycontain(CT) = \begin{cases} true & |\mathbf{c}_{CT} - \mathbf{r}_0| < \sqrt{3}d_b \\ false & else \end{cases} \tag{11}$$

(II) Conditional searcher CS2: Given three points P_1, P_2 and P_3, search for an extended-tetrahedron CT where P_1, P_2 and P_3 are three of its vertexes. The conditional function is as follow:

$$condition(CT) = \begin{cases} true & three\ vertexes\ are\ P_1, P_2\ and\ P_3 \\ false & else \end{cases} \quad (12)$$

The identification function is as follows:

$$maycontain(CT) = \begin{cases} true & P_1, P_2, P_3 \in b \\ false & else \end{cases} \quad (13)$$

In three-dimensional space, the algorithm for the constructing Delaunay tetrahedron from given discrete points is as follows: (I)Initialization: Generate a sufficiently large tetrahedron to contain all discrete points, record the center and radius of its circumsphere, form an 'extended-tetrahedron' and construct an extended-tetrahedron tree t; (II) Construction of Delaunay tetrahedron: (i) Pick out a discrete point P, search for the extended-tetrahedra whose circumspheres contain P in tree t with searcher CS1, remove these extended-tetrahedra from tree t; put the removed tetrahedra together to form a set named Q; (ii) Add every surface of each tetrahedron in Q to SHT S which is a tree for the external triangular interfaces, search for the surfaces appearing twice with searcher CS2 and remove them because they are interfaces. Triangles in S constitute the external surface of Q; (iii) Pick out each face of S, together with point P, to construct a new tetrahedron; add the newly formed extended-tetrahedra to tree t; (iv) Go back to step (i) until all points are used out.

Up to this step, the group of tetrahedra contained by tree t is just the Delaunay tetrahedra division. Figure 18 shows the Delaunay triangle division of 20000 randomly distributed two-dimensional points. Figure 19 shows the Delaunay tetrahedron division of 20000 randomly distributed points in a three-dimensional sphere.

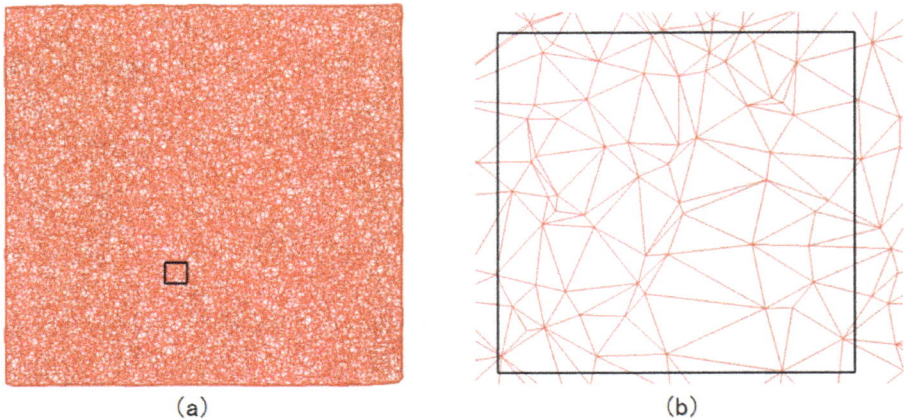

(a)   (b)

Fig. 18. Delaunay division constructed from randomly distributed discrete points in a two-dimensional square area [0,4]×[0,4]. (b) is the enlarged picture of the portion in the small black rectangle in (a).

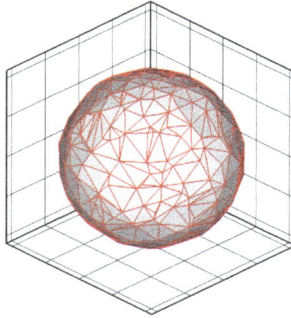

Fig. 19. Delaunay division constructed from randomly distributed discrete points in a three-dimensional spherical region.

The algorithm can be easily extended to n-dimensional space. We need only to replace the tetrahedron with an n-simplex and replace the triangle with an $(n-1)$-simplex. The circumsphere of the n-simplex constructed from $n+1$ point, $\{r_1, r_2, \cdots, r_{n+1}\}$ in n-dimensional space is used in the algorithm. The formula to calculate the center of the circumsphere of the n-simplex is $c_i = A_i/B$, where

$$
A_i = \begin{vmatrix} 1 & -2r_1 & r_1^2 \\ 1 & -2r_2 & r_2^2 \\ \cdots & \cdots & \cdots \\ 1 & -2r_{n+1} & r_{n+1}^2 \\ 0 & e_i & 0 \end{vmatrix}, \quad B = \begin{vmatrix} 1 & -2r_1 & r_1^2 \\ 1 & -2r_2 & r_2^2 \\ \cdots & \cdots & \cdots \\ 1 & -2r_{n+1} & r_{n+1}^2 \\ 0 & 0 & 1 \end{vmatrix}, \tag{14}
$$

and e_i is the unit vector of the i-th direction. The circumsphere radius is $\sqrt{(c - r_i)^2}$.

3.3 Cluster construction and analysis method

Complex configuration and dynamic physical fields are ubiquitous in weapon-physics, astrophysics, plasma-physics, and material-physics. Those structures and their evolutions are characteristic properties of the corresponding physical systems. For example, the interface instability makes significant constraints on the design of an inertial confinement fusion (ICF) device (Ament, 2008), shock waves and jet-flows in high energy physics are common phenomena (de Vriesl et al, 2008), distributions of clouds and nebulae are very concerned issues of astrophysics (Bowle et al., 2009; Hernquist, 1988; Makino, 1990), clusters and filaments occur in the interaction of high-power lasers and plasmas (Hidaka et al., 2008), and structures of dislocation bands determine the material softening in plastic deformation of metals (Nogaret et al., 2008). These structures are also keys to understand the multi-scale physical processes. Laws on small-scales determine the growth, the change and the interactions of stable structures on larger-scales. Description of the evolution of stable structures provides a constitutive relation for larger-scale modeling. Because of the lack of periodicity, symmetry, spatial uniformity or pronounced correlation, the identification and characterization of these structures have been challenging for years.

Existing methods for analyzing complex configurations and dynamic fields include the linear analysis of small perturbations of the background uniform field, characteristic analysis of simple spatial distributions of physical fields, etc. These methods are lacking in a quantitative description of characteristics of the physical domain. For example, the size, the shape, the topology, the circulation and the integral of related physical quantities. Therefore, it is difficult to trace the evolution of the characteristic region or the background. For example, the laws of growth and decline, or the exchange between them.

The difficulties in characteristic analysis are twofold. The first is how to define the characteristic region. The second is how to describe it. The former involves the control equations of the physical system. The latter is related to recovering the geometric structure from discrete points. In recent years, cluster analysis techniques (Kotsiantis & Pintelas, 2004; Fan et al., 2008) in data mining have found extensive applications in identification and the testing of laws of targets. They mainly concern the schemes for data classification. Physicists are concerned more about the nature underlying these structures. Recovering characteristic domains can be attributed to the construction of spatial geometry. The key point is how to connect the related discrete points. The Delaunay grid (Chazelle et al., 2002; Clarkson & Varadarajan 2007) has an excellent spatial neighboring relationship. In this chapter, the Delaunay triangle or tetrahedron is used as the fundamental geometrical element.

For the discrete points in space, there is no strict cluster structure. If the discrete points are considered as objects, such as a molecular ball, lattice or grid, then, the objects can be connected to form clusters. The average size of these assumed objects is the revolution of clusters to be constructed with discrete points. The construction of clusters is very simple. A cluster is formed by connecting all points whose distance in between is less than the revolution length.

After the construction of the Delaunay division for given set of discrete points, remove the lines whose lengths are greater than the revolution length. The remaining spatial structure may have various dimensions. According to connectivity, the structures that are not connected to each other can be decomposed into different clusters. Each cluster may also have structures with various dimensions. For example, a structure consisted of two triangles with a common side, or a structure formed by a triangle and a tetrahedron, etc. In physical problems, the structures with high-dimensional measures play a major role in describing the system. Generally, we need only to analyze clusters with the maximum dimensions.

The cluster construction algorithm consists of three parts. Preparation part: (i) Construct Delaunay tetrahedra from given discrete points. The corresponding SHT is notated as t. (ii) Remove the tetrahedrons whose lengths are greater than the given resolution from t. Single cluster construction part: (iii) Remove tetrahedron T from t if such a T still exists. Create a new cluster named C. Initialize the body tree C->t and face tree C->s as null. Add T to the body tree C->t. Add each of the triangle faces to a triangle tree named i. (iv) Pick out a triangle face S from i. Search tetrahedron Y containing face S from t. If found, add Y to tree C->t and add all faces of Y to tree i. Two faces with opposite directions will annihilate if they meet each other during the adding procedure. If none are found, add S to the tree C->s. (v) Repeat step (iv) until the tree i becomes null. Construct all the clusters: (vi) Add the constructed cluster C to a tree for clusters named c. Repeat the process of constructing single clusters, and add the new cluster to c until t becomes null.

The algorithm for adding a face S to the tree i is as follows. Search and check if a face with the opposite direction of S exists in the tree i. If exists, remove it from the tree i. If does not exist, add S to the tree i.

Up to now, all the constructed clusters are put to the cluster tree c. For each cluster C in the tree c, all tetrahedron elements are placed on the body tree C->t, all the surface triangles are placed on the tree for faces C->s. Figures 20 and 21 show respectively the clusters constructed with random points in two-dimensional and three-dimensional space.

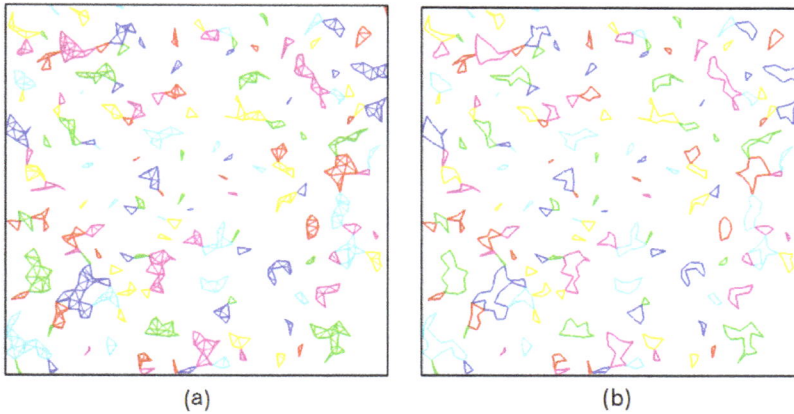

(a)   (b)

Fig. 20. Cluster structure formed from 1000 random discrete points in a two-dimensional square area [0,1]×[0,1]. (a) A cluster; (b) the corresponding cluster boundary.

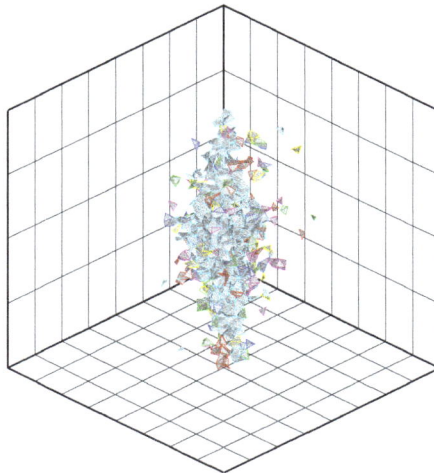

Fig. 21. Cluster structure formed from 5000 three-dimensional random discrete points.

The algorithm is also applicable to n-dimensional discrete points. We need only to replace the tetrahedron with an n-simplex and replace the triangular surface with an (n-1)-simplex.

For space with a dimension higher than three, the number of neighboring points and the connectivity, as well as the number of n-simplex, grow rapidly with the dimension. So, the required memory increases quickly. The Delaunay division can be constructed partition by partition. The main skill in this algorithm is that the space is partitioned according to the main branches of SHT, and points in each partition are added sequentially. After the completion of adding all points in a partition, we need to delete the n-simplex that satisfies two conditions: (i) its external circumsphere is in the completed partition, (ii) at least one side is longer than the given resolution.

3.4 Identification methods of defect atoms

The knowledge about the world is gradually deepened. Before the 19th century, the most prominent theoretical models primarily describe the movement of object. In the 19th century, two important theoretical models are proposed: one is field, another is uniform system. In the 20th century, the outstanding recognition model is about regular structure, represented by energy bond theory and phonon theory on studying crystal structure. Due to the enhanced computing power and increased awareness in the latter half of the 20th century, cognitive models on non-periodic structure and non-equilibrium emerge.

The primary recognition is as follows: During the system evolution, spatial structures of various scales appear, and the evolution of these structures determines the evolution laws of the system, when the system parameter arrives at the critical one, structures of different scales show an overall correlation with each other, and become indistinguishable. The typical structure characteristic is self-similar and renormalization theory can easily handle this singularity.

In the studies on non-periodic structure and non-equilibrium, the phase transition theory and dislocation theory are representatives, and nearly all phenomena can be explained by nonlinear theory. It indicates that the macroscopic properties of non-equilibrium system do not depend on its fast processes, but on the evolution of more stable structure. With different scales of concern, the system structure is not same.

During the technology development with gradually reduced time and space scales, scientific understanding on some processes, which cannot be studies or are unclear before, are gradually obtained, i.e. microscopic mechanism of the fracture process. Meanwhile we've got the multi-scale cognitive model: The overall properties of system are determined by the evolution of large-scale structures, small-scale structures and fast processes determine the large-scale structure and slow processes. In researches, different methods and theories are developed according to structures of different scales. The properties of small-scale structure provide model parameters and constitutive relationships for system with large-scale structures. For the studies on strongly coupled system which is also non-uniform and non-equilibrium, the key issue is to understand the properties and evolution laws of structures.

In particle simulation and 2D or 3D simulations of complex physical systems, how to effectively analyze the spatial and dynamical characteristics the system is the key to understand physical laws. There two problems: how to identify the stable structures existing in the system and how to compute them. For instance, in molecular dynamical simulation studies on the mechanical properties of metal, plastic deformation, phase transitions, and damage processed closely refer to defect structures, such as dislocations, stacking faults, grain boundaries and interfaces. The emergence of those defect structures indicates the change stages of materials, and the evolution of them determines material properties. As defect structures are collection of arranged atoms, they can be appropriately identified with proper analysis methods. When face a huge number of atom coordinates, the key issue of physical analysis is how to recognize various defect structures.

3.4.1 Excess energy method and centro-symmetry parameter method

According to the physical quantities, structure symmetry or local topological connections, defect atoms can be distinguished by a few corresponding methods, for example, the excess energy method, centro-symmetry parameter method (CSP) (Kelchner et al., 1998) and bond-pair analysis (BPA) method (Faken & Jonsson, 1994).

In excess energy method, the atoms with excess energy are selected as defects atoms. Because the lattice equilibrium positions are stable positions for atoms, the potential energies of atoms deviating from their stable positions are higher. This method depends on the physical quantities of particles, so the output data from MD simulation must be completed. However, in many cases, such as, in phase transition represented by symmetric double-well energy function, energy cannot be used to distinguish different structures.

In CSP method, the geometrical symmetry of the collection of nearest atoms of an atom is used to identify defect atoms. All atoms of perfect crystal are in the geometrical center of its nearest atoms, but the defect atoms are not. Therefore, an order parameter is defined as follows:

$$s = \left| \sum_{i \in neighbour} (\mathbf{r}_i - \mathbf{r}_0) \right| \tag{15}$$

Atoms whose order parameter s is greater than a critical value s_c are defect atoms. In the case of strong temperature perturbation, the result of CSP method is not correct, because random thermal motion reduces the lattice symmetry and the order parameter of perfect lattice becomes greater.

The nearest atoms of a specific atom are the atoms within a given sphere region whose center is the specific atom. The radius of the given sphere must be greater than the distance of nearest atoms in perfect crystal and less than the distance of second nearest atoms. In fcc and bcc crystals, as the second nearest distance is $\sqrt{2}$ times of the nearest distance, the sphere center can be given as $(1 + \sqrt{2}) / 2 \approx 1.2$ times as the nearest distance to resist a certain degree of randomness. In the larger deformation of crystal, as the lattice constant alters, the given sphere radius needs to be adjusted. In more complicate cases, as the lattice constant is not known in advance, a better way is first to compute radial distribution function (RDF) and then set the distance corresponding to the first peak of RDF as lattice constant of perfect crystal.

During the computation procedure of radial distribution function and order parameter algorithm, as atoms in given region need to be searched, an index of atoms must be constructed. Since the distribution of atoms is uniform, the background grid index can be used.

3.4.2 Bond-pair analysis method

The CSP and excess energy methods can distinguish defect atoms, but can not easily identify types of defects atoms. The bond-pair analysis (BPA) based on local topological connections can more accurately identify atom type. The idea of BPA is as follows: a bond type is marked in terms of the connections among atoms bonding with the two atoms composing the bond, and an atom type is marked in terms of all bonds of itself.

The 'bond' is defined as the connection between two atoms whose distance is less than a given value R (bonding distance). For convenience, the name of 'bond' proposed here is same as the one used in chemistry, but their meaning is different. The concept of the proposed bond doesn't contain any meaning of quantum chemistry where it indicates overlap between the electric wave functions. The bonding distance is often set as 1.2 times as the nearest distance in perfect lattice. In the case of unknown lattice constant, the lattice constant is usually set as the distance of the first peak of RDF, so to some extent, the topological analysis can tolerate random disturbance.

3.4.2.1 Bond-type identification

For a given bond L, the bond-type of L is determined by the indirect bonding feature of both atoms of L. Pick out all atoms which bond with both the two atoms of L. The bonding feature of picked atoms is used to identify bond-type of L. Assume the two atoms of bond L are A and B. The collection of atoms bonding with both A and B is c. Specifically, a bond-type is marked by a three-digit number. The first digit number is the atom number in c, the second is the number of bond among atoms in c, and the third is the largest coordination number of atoms in c.

Figure 22 shows an example for bond-type identification. The bond-type of bond is 443, where, from left to right, the first number, 4, denotes the number of common neighbor of atoms A and B (i.e. H, I, C, and D), the second number, 4, means the bond number of the common neighbor atoms (i.e. HI, DI, HD, and CD), the third number, 3, indicates the largest coordination number (the coordination number of atom D, i.e. CD, HD, and DI).

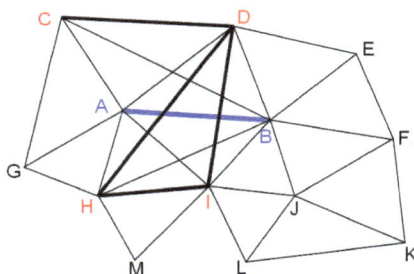

Fig. 22. Scheme for bond-type identification.

In the perfect crystals, the bond-type ID is easy to calculate. For example, all bond types are 421 in fcc crystal. There are two bond-types, 422 and 421, in hcp crystal. There are two bond-types, 441 and 661, in bcc crystal. All bond-types can be gathered as a list which will be referred to bond-type list later. Due to the variety of defect atoms in grain boundaries, there may be a large number of bond-types. For example, {421, 422, 441, 661, 200, 100, 311, 211,

411, 432, 542, 300, 400 ...}. The bond-type list is dynamically increasing except for several special bond-types.

3.4.2.2 Atom-type identification

After identifying the bond-types, each atom gets its own bond-types. A distribution vector of bond-type of an atom can be generated by comparing its bond-types with the system bond-type list, and each component of the vector is number of the corresponding bond-type. For a fcc atom, it contains twelve 421 bonds, the corresponding distribution vector is {12}; for a hcp atom, it contains six 421 bonds and six 422 bonds, the corresponding distribution vector is {6, 6}; for a bcc atom, it contains six 441 bonds and eight 661 bonds, the corresponding distribution vector is {0, 0, 6, 8}; and for a kind of boundary atom, a distribution vector {5, 4, 0, 0, 0, 0, 2, 1} indicates that it contains five 421 bonds, four 422 bonds, two 200 bonds and a 100 bond. The distribution vector accurately describes an atom-type. In calculation, all distribution vectors can be gathered as a system atom-type list, the list is dynamically adjusted.

3.4.2.3 Bond-pair analysis algorithm

The preparatory work for algorithm is to design fast searchers. In this section, a design, including two conditional searchers, is proposed, where design a conditional function and an identification function for each conditional searcher.

(I) Conditional searcher CS1: Given a point p, search for a point whose distance to p is less than r_c. The conditional function is as follows:

$$condition(o) = \begin{cases} true & |\mathbf{r}_o - \mathbf{c}_p| < r_c \\ false & else \end{cases} \tag{16}$$

The circumcircle of branch b is used to identify. The identification function is as follows:

$$maycontain(b) = \begin{cases} true & |\mathbf{r}_P - \mathbf{c}_b| < r_c + d_b \\ false & else \end{cases} \tag{17}$$

(II) Conditional searcher CS2: Given two points P_1 and P_2, search for a points whose distances to P_1 and P_2 are less than r_c . The conditional function is as follows:

$$condition(o) = \begin{cases} true & |\mathbf{r}_{P_1} - \mathbf{r}_o| < r_c \ and \ |\mathbf{r}_{P_1} - \mathbf{r}_o| < r_c \\ false & else \end{cases} \tag{18}$$

The circumcircle of branch b is used to identify. The identification function is as follows:

$$maycontain(b) = \begin{cases} true & |\mathbf{r}_{P_1} - \mathbf{r}_o| < r_c + d_b \ and \ |\mathbf{r}_{P_1} - \mathbf{r}_o| < r_c + d_b \\ false & else \end{cases} \tag{19}$$

The algorithm for bond-pair analysis is as follows: (i) Initialization: set a bond-type array B and an atom-type array A, empty A and B, set a bond-type distribution vector V. Generate a point tree tp from the given discrete points (atoms), calculate the RDF of the system, set bonding distance as r_c according to the distance corresponding to the first maximum of RDF. (ii) For each atom a, empty its bond-type distribution vector V. Search in the tree tp for an atom b bonding to the atom a (the distance between a and b is less than r_c) with searcher CS1. For each a-b bond, search in the tree tp for all atoms bonding to a and b (whose distance to a and b are less than r_c) with searcher CS2, compute the number of those atoms (denoted as l), check the connections between those atoms to get the number of bonds (denoted as m) and the largest coordination number (denoted as n), compute the bond-type of a-b as $100l+10m+n$ and then check whether or not it is a new bond-type by comparing with the bond-type list in B. If yes, add it to array B and then plus one on the corresponding component of V. Complete the loop of all bonds of a, check whether or not the atom-type of a is new by comparing with A. If yes, add it to array A. (iii) Go back to step (ii) until all atoms are used out.

3.4.2.4 Results show

Figure 23 shows the stacking faults and dislocations formed during the low-temperature evolution of gathered point defects (Frank loop) in fcc copper crystal. The defect atoms belonging to dislocations and stacking faults can be accurately identified, where blue atoms are stacking faults, and dislocation atoms are red. Figure 24 shows the growing of the two sphere voids in fcc copper under tension, at the very beginning, dislocations grows from the void surfaces, and different dislocations cross with each other in the late evolution. The atoms belonging to void surface, dislocations and stacking faults (no shown) can be strictly distinguished, they are marked with different colors.

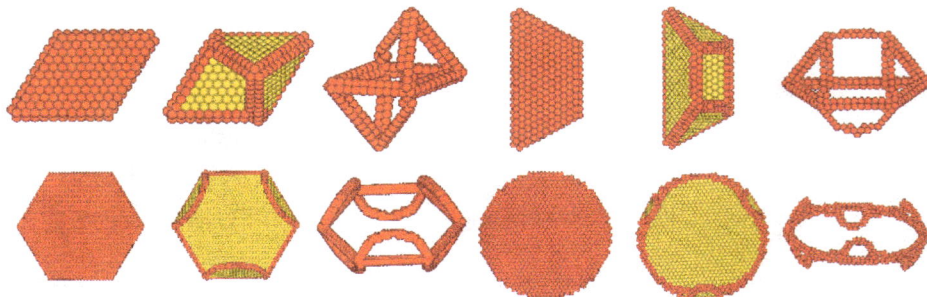

Fig. 23. Stacking faults and dislocations identified by bond-pair analysis.

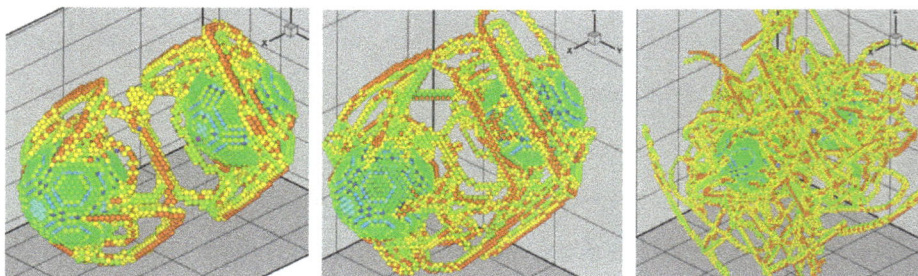

Fig. 24. Voids surfaces and dislocations identified by bond-pair analysis.

3.5 Surface construction algorithm

A key issue in the analysis of complex dynamic system is to form interface of concerned region. For example, in first order phase transition process, particle transportation and structure transition will cause growth and deformation of phase change zone. Interface determination of structural zone is the foundation of analyzing interface movements, cross-interface physical flow and understanding characterization of structure development. Interface construction is similar to surface reconstruction in computational geometry. Surface in computational geometry is formed from sampled points, while physical interface is the results of system evolution.

3.5.1 Packing-sculpting method for constructing object surface from disorder spatial points

It is an important issue in computational geometry to construct object surface from disorder points. The current algorithms can be categorized into four groups (Mencl & Muller, 1997), i.e. space partitioning method (Boissonant, 1984), distance function method (Hoppe et al., 1992), deformation method (Zhao, 2002), and growth method (Bernardini, 1999). Space partitioning is generally based on Delaunay division. The outer surface is generated by removing some Delaunay mesh in the sculpting method. The packing-sculpting method presented below is an intuitive method, the out surface is constructed by directly sculpting the packing convex hull. The basic idea is as follow: First of all, generate the packing convex hull from discrete points, and then sculpt the convex hull to construct the object surface.

3.5.1.1 Packing algorithm

The packing algorithm is to generate a convex hull by packing all given points. In this section, half-plane rotation method is introduced.

For the convenience of description, a new data structure, 'extended-segment', is defined as the combination of one side of triangle and triangle itself, it is an extended side of triangle with the data of the other vertex. The center and length of 'extended-segment' is that of the corresponding side of triangle. In the algorithm, there are various complicate searching for points, lines and surfaces. With the general index based on SHT, fast searching can be designed according to given searching conditions.

The preparatory work for algorithm is to design fast searchers. In this section, a design, including a minimum searcher and a conditional searcher, is proposed.

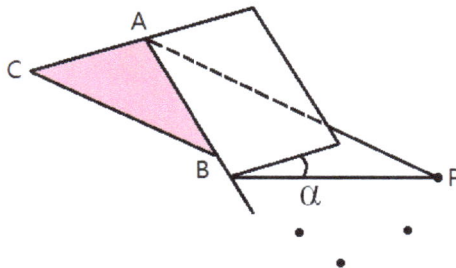

Fig. 25. Scheme for finding the first point during half-plane rotation.

(I)   Minimum searcher MS1: Given a triangle ABC, pick out a segment AB, the extended half-plane of triangle ABC rotates around axis AB, searcher for the first point P meeting the extended half-plane to make smallest the dihedral angle shown in figure 25.

The value-finding function is as follows:

$$value(p) = angle(x, y) \tag{20}$$

Where, $x = \hat{\mathbf{x}} \cdot (\mathbf{r}_P - \mathbf{r}_A)$ and $y = \hat{\mathbf{y}} \cdot (\mathbf{r}_P - \mathbf{r}_A)$ is local coordinate of point P,

$$\hat{\mathbf{x}} = \frac{\mathbf{P}_{xy} \cdot (\mathbf{r}_A - \mathbf{r}_B)}{\left| \mathbf{P}_{xy} \cdot (\mathbf{r}_A - \mathbf{r}_B) \right|}, \hat{\mathbf{z}} = \frac{\mathbf{r}_A - \mathbf{r}_B}{|\mathbf{r}_A - \mathbf{r}_B|}, \hat{\mathbf{y}} = \hat{\mathbf{z}} \times \hat{\mathbf{x}} \tag{21}$$

are respectively the unit vector of direction of three reference axes,

$$\mathbf{P}_{xy} = \mathbf{E} - \hat{\mathbf{z}}\hat{\mathbf{z}} \tag{22}$$

is projection operator and $\mathbf{E}$ is identity operator. The circumsphere S of branch b is used for assessment. The range-evaluation functions are as follows:

$$M(b) = \begin{cases} 2\pi & A \in S \text{ or } B \in S \\ value(\mathbf{c}_b) + arc\sin(\sqrt{3}d_b \ / \left| \mathbf{P}_{xy} \cdot (\mathbf{c}_b - \mathbf{r}_A) \right|) & A, B \notin S \end{cases}$$

$$m(b) = \begin{cases} 2\pi & A \in S \text{ or } B \in S \\ value(\mathbf{c}_b) - arc\sin(\sqrt{3}d_b \ / \left| \mathbf{P}_{xy} \cdot (\mathbf{c}_b - \mathbf{r}_A) \right|) & A, B \notin S \end{cases} \tag{23}$$

where $\mathbf{c}_b$ is the center of branch b.

(II) Conditional searcher CS1: Given a directed segment P_1P_2, search in extended-segment for an extended segment BD being equal to segment P_1P_2. The conditional function is as follows:

$$condition(BD) = \begin{cases} true & B == P_1 \text{ and } D == P_2 \\ false & else \end{cases} \tag{24}$$

The identification function is as follows:

$$maycontain(b) = \begin{cases} true & P_1, P_2 \in b \\ false & else \end{cases} \tag{25}$$

The packing algorithm is as follows: (I) Initialization: Construct a point tree tp from the given discrete points. According to the region size of root tp, pick out two vertexes: $P_1=(-a,-a,a)$ and $P_2=(-a,a,a)$, where a is half of the edge length of root region. Given a point $P_3=(-2a,0,a)$. Search in the tree tp for the first point Q_1 met by rotating triangle $P_1P_2P_3$ around axis P_1P_2 with searcher MS1, generate a new triangle $P_1Q_1P_2$ and cut Q_1 down from tp. Search in the tree tp for the first point Q_2 met by rotating triangle $P_1Q_1P_2$ around axis P_1Q_1 with searcher MS1, generate a new triangle $Q_1Q_2P_1$ and cut Q_2 down from tp. Search in the tree tp

for the first point Q_3 met by rotating triangle $Q_1Q_2 P_1$ around axis Q_1Q_2 with searcher MS1, generate a new triangle $Q_3Q_1Q_2$ and cut Q_3 down from tp. Construct a triangle tree tt from triangle $Q_3Q_1Q_2$, generate extended-segment Q_3Q_1, Q_1Q_2 and Q_2Q_3, construct an extended-segment tree tb from Q3Q1 and put Q1Q2 and Q2Q3 into tb. Figure26 shows the initialization procedure of packing-algorithm.

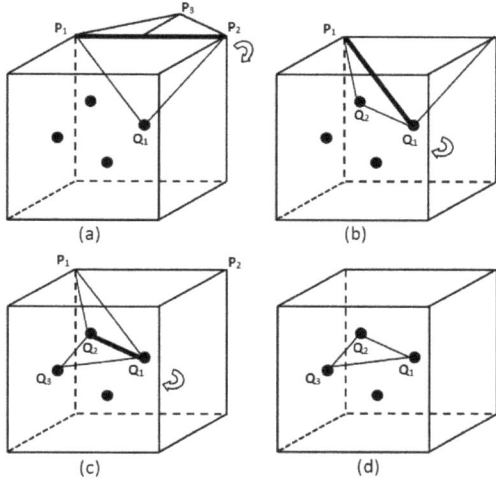

Fig. 26. Scheme for initialization of packing algorithm. (a) the first point Q_1 met by rotating triangle $P_1P_2P_3$ around axis P_1P_2; (b) the first point Q_2 met by rotating triangle $P_1Q_1P_2$ around axis P_1Q_1; (c) the first point Q_3 met by rotating triangle $Q_1Q_2P_1$ around axis Q_1Q_2; (d) initial triangle interface.

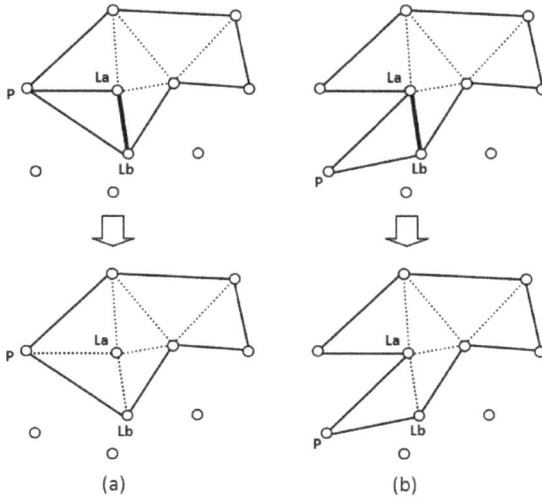

Fig. 27. Two cases. (a) L_aP is in the extended-segment tree; (b) L_aP is not in the extended-segment tree.

(III) Construction of packing convex hull: Cut down an extended-segment L from the tree tb, check whether or not L is empty. If yes, exit. If not, find the corresponding triangle A. Search in tp for the first point P met by rotating A around axis L with searcher MS1, generate a triangle L_bL_aP with point L_a, L_b and P, put the triangle L_bL_aP into the tree tt. Use CS1 to search in tb for an extended-segment G sharing the same segment with L_aP, check whether or not G exist. If yes (see figure 27(a)), cut it down from tb. If not (see figure 27(b)), generate an extended-segment from L_aP and put it into tb. Repeat the same operations to L_bP.

3.5.1.2 Sculpting algorithm

The sculpting algorithm refers to the sculpting method and the sculpting standard. We have to guarantee that no point is carved out during the sculpting procedure and make as smooth as possible the surface after sculpting. The curvature of smooth surface is small, and it indicates that the circumsphere radius of the tetrahedron which is carved is larger. For a triangle ABC under sculpting, the sculpting process means to find a point P in currently packed region to make smallest the height of circumsphere cap $ABCP$. Figure 28 is the scheme for sculpting algorithm.

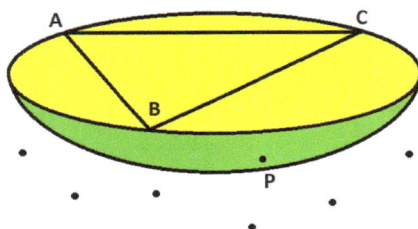

Fig. 28. Scheme for sculpting algorithm.

For a triangular surface, the stopping criterion for sculpting should be that the radius of circumsphere is less than a given value. In the case of uniform sampling of object surface, the criterion meets the requirements. However, in the case of nonuniform sampling (large curvature and dense sample points), the curvature must be related to the distance between local sample points. The ratio of the height of sphere cap and the circumcircle radius of a triangle, i.e. sculpture degree $c=h/r<c_{cri}$, is more suitable for stopping criterion. The criterion can be used in both cases above. Figure 28 shows the sculpting algorithm.

The preparatory work for algorithm is to design fast searchers. In this section, a design, including a minimum searcher, is proposed.

The minimum searcher MS is as follows: Given a triangle ABC, search in point tree for a point P to make smallest the height of circumsphere cap $ABCP$. The construction of value-finding function is as follows: By solving equations,

$$
\begin{aligned}
\mathbf{r}_o &= \alpha\mathbf{r}_A + \beta\mathbf{r}_B + \gamma\mathbf{r}_C \\
r^2 &= (\mathbf{r}_o - \mathbf{r}_A)^2 \\
r^2 &= (\mathbf{r}_o - \mathbf{r}_B)^2 \quad , \mathbf{n} = \frac{(\mathbf{r}_B - \mathbf{r}_A)\times(\mathbf{r}_C - \mathbf{r}_A)}{\left|(\mathbf{r}_B - \mathbf{r}_A)\times(\mathbf{r}_C - \mathbf{r}_A)\right|} \\
r^2 &= (\mathbf{r}_o - \mathbf{r}_C)^2 \\
\alpha + \beta + \gamma &= 1
\end{aligned}
\tag{26}
$$

the circumcircle radius r of triangle ABC, the center position o, and the normal direction $\mathbf{n}$ are obtained. By solving equations,

$$\mathbf{r}_c = \mathbf{r}_o + \lambda\mathbf{n}$$
$$(\mathbf{r}_c - \mathbf{r}_p)^2 = \lambda^2 + r^2 \tag{27}$$

the coordinate of circumsphere center $\mathbf{r}_c$ and the height λ of $\mathbf{r}_c$ to circumcircle of triangle ABC are obtained. The height of circumsphere cap is as follows (i.e. the value-finding function):

$$value(P) = \sqrt{\lambda^2 + r^2} - \lambda \tag{28}$$

The range-evaluation function is calculated according to the tangent cases between circumspheres of branch b and sphere cap $ABCP$. It is as follows:

$$M(b) = \begin{cases} \sqrt{\lambda_M^2 + r^2} - \lambda_M & \text{circumsphere of ABC is out of circumsphere of b} \\ \infty & \text{else} \end{cases}$$
$$m(b) = \begin{cases} \sqrt{\lambda_m^2 + r^2} - \lambda_m & \text{circumsphere of ABC is out of circumsphere of b} \\ \infty & \text{else} \end{cases} \tag{29}$$

where λ_M and λ_m are roots of equation $\left|\mathbf{r}_o + \lambda\mathbf{n} - \mathbf{c}_b\right|^2 = \left(\sqrt{\lambda^2 + r^2} \pm \sqrt{3}d_b\right)^2$, respectively, and

they correspond to the two cases shown in figure 29. Set P as the tangent point of sphere cap and the circumsphere of b. When P becomes point Q_1, the height λ_m of sphere cap $ABCP$ is the smallest. When P becomes point Q_2, the height λ_M of $ABCP$ is the largest..

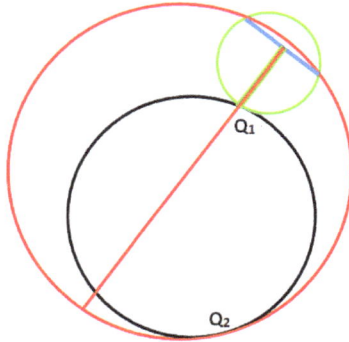

Fig. 29. Two tangent cases between circumspheres of branch b and sphere cap $ABCP$ (cross-section picture), where black circle is for the circumsphere of branch b, blue segment is for the circumcircle of triangle ABC, green and red circle are both for sphere cap $ABCP$, Q_1 and P_2 are for tangent points.

The sculpting algorithm is as follows:(I) Initialization: Take all triangles obtained by packing algorithm as triangle under sculpting, and the corresponding tree as triangle tree tt under sculpting. Set the critical value as 2.0 for stopping sculpting, set surface tree $tt0$ as empty. (II)

Sculpting: cut a triangle T from the tree tt, check whether or not T is empty. If yes, exit. If not, use MS to search for point P. Compute sculpture c, check whether or not c is less than c_{cri}. If yes, stop sculpting and put T into $tt0$. If not, generate three triangles by joining each side of T and point p, and then put them into tt.(III) Go back to step (II).The object surface consists of all surfaces in surface tree $tt0$.

3.5.1.3 Results show

Figure 30 shows the convex hull constructed from discrete points containing two nano-voids and uniform boundary points of a point defect by packing algorithm and object surfaces generated by sculpting. Figure 31 is for the reformed surface from random points sampled from tori. Figure 32 is for the reformed surface from 9000 random points sampled from spheres and tori.

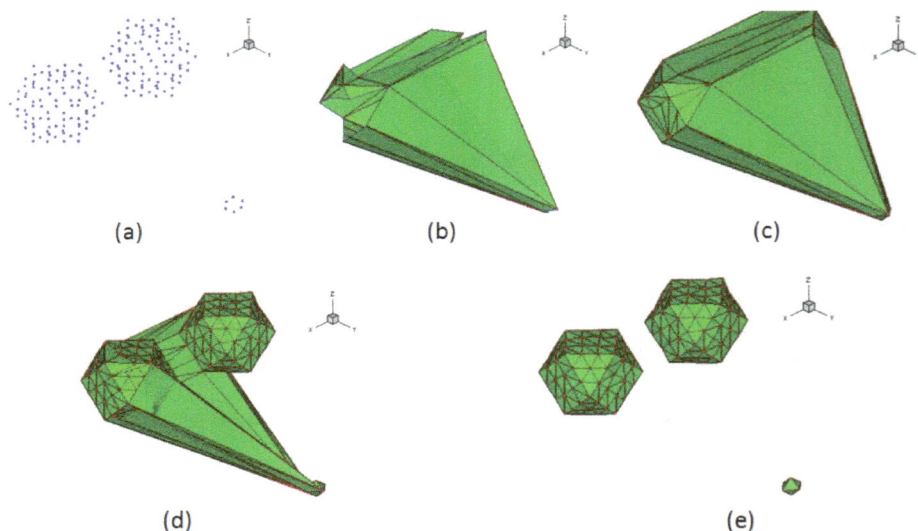

Fig. 30. The procedure of packing-sculpting algorithm. (a) discrete points; (b) the mid of packing procedure; (c) packing convex hull; (d) sculpting procedure; (e) object surface.

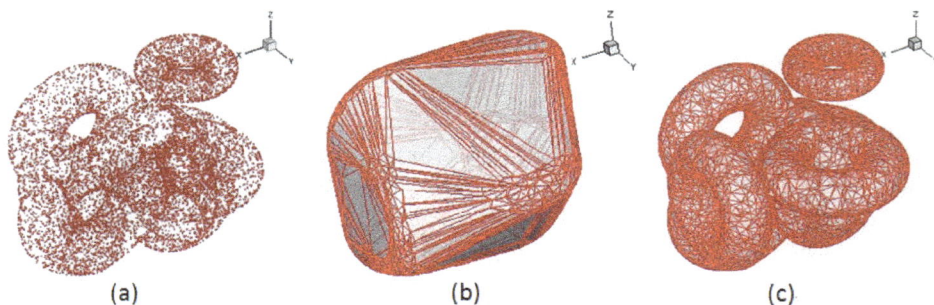

Fig. 31. Reformed surface from random points sampled from tori. (a) discrete points sample; (b) packing convex hull; (c) reconstructed object surfaces.

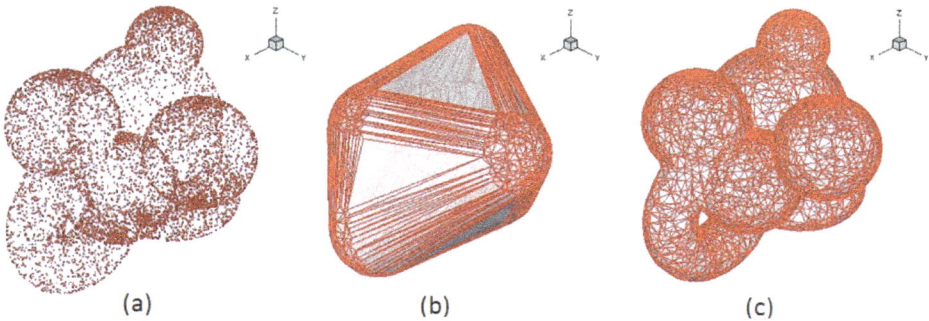

Fig. 32. Reformed surface from 9000 random points sampled from spheres and tori.
(a) discrete points sample; (b) packing convex hull; (c) reconstructed object surfaces.

The value of critical sculpture degree is 2.0 in sculpting algorithm. Application examples show that the algorithm is suitable for the cases of multi-connected or many-body surfaces. In the algorithm, the correctness of topology and geometry are guaranteed and the right surface can be constructed even in regions of greater curvature and at the transition zone of surfaces with different curvature.

3.5.2 Rolling-ball method for finding interfaces of physical regions from disorder spatial point

The difference between searching for interface of physical region from disordered spatial points and constructing surface of object is that the spatial points contain not only the points of physical interface but also points of other structures. The most common approach is to construct physical field on a regular grid, and the contour of physical field is used as the appropriate physical interface. This method is suitable for the case that the distribution of discrete points closed to interface is uniform. In the case of complex distribution of discrete points, it is hard to preserve the smoothness of the constructed interface, the calculated interface is very different from the actual interface. A better way is to use the rolling-ball method without constructing physical fields. The basic idea of rolling-ball method is as follows: roll a ball with fixed size on discrete point group, each rolling goes through three points, and these points constitute a surface element of interface. After the rolling-ball goes through the overall region, the physical interface is constructed. In rolling-ball method, the key parameter is the sphere radius. In the case of sparse sampling, different radiuses define different interfaces. Therefore, the size of the rolling-ball radius is obtained from experience.

3.5.2.1 Rolling-ball algorithm

The preparatory work for algorithm is to design fast searchers. In this section, a design, including three minimum searchers and a conditional searcher, is proposed.

(I) Minimum searcher MS1: Given a directed triangle ABC and pick out a segment AB, search in point tree for the first point met by the rolling-ball above triangle ABC, where the radius of rolling-ball is r and the rotation axis is AB. The construction of value-finding function is as follows: calculate the initial center $\mathbf{r}_o$ of the rolling-ball and directions of local axes $\hat{\mathbf{x}}, \hat{\mathbf{y}}, \hat{\mathbf{z}}$. The calculation formula are as follows:

$$r^2 = (\mathbf{r}_o - \mathbf{r}_A)^2$$
$$r^2 = (\mathbf{r}_o - \mathbf{r}_B)^2 , \hat{\mathbf{x}} = \frac{\mathbf{P}_{xy} \cdot (\mathbf{r}_o - \mathbf{r}_A)}{\left| \mathbf{P}_{xy} \cdot (\mathbf{r}_o - \mathbf{r}_A) \right|} , \hat{\mathbf{y}} = \hat{\mathbf{z}} \times \hat{\mathbf{x}}, \hat{\mathbf{z}} = \frac{\mathbf{r}_B - \mathbf{r}_A}{\left| \mathbf{r}_B - \mathbf{r}_A \right|} \tag{30}$$
$$r^2 = (\mathbf{r}_o - \mathbf{r}_C)^2$$

Calculate the rolling-ball center $\mathbf{r}_n$ after rotating and local coordinate x, y, z. The calculation formula are as follows:

$$r^2 = (\mathbf{r}_n - \mathbf{r}_A)^2$$
$$r^2 = (\mathbf{r}_n - \mathbf{r}_B)^2 , x = \hat{\mathbf{x}} \cdot (\mathbf{r}_n - \mathbf{r}_A), y = \hat{\mathbf{y}} \cdot (\mathbf{r}_n - \mathbf{r}_A), z = \hat{\mathbf{z}} \cdot (\mathbf{r}_n - \mathbf{r}_A) \tag{31}$$
$$r^2 = (\mathbf{r}_n - \mathbf{r}_P)^2$$

Calculate the rotation angle, i.e. the value of value-finding function. Figure 33 shows the scheme for the rotation of triangle ABC.

$$value(P) = angle(x, y) \tag{32}$$

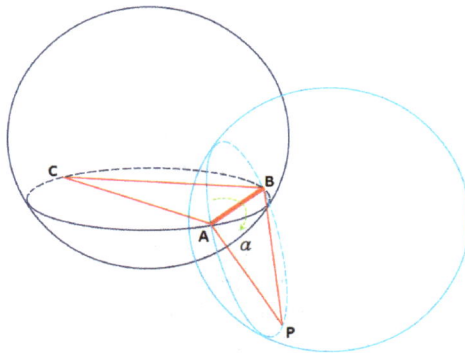

Fig. 33. Scheme for the rotation of triangle ABC.

The procedure for constructing range-evaluation function is as follows: calculate the position of the tangent point T of rolling-ball and the circumsphere of a branch. The calculation formula read

$$r^2 = \left| \mathbf{r}_T - \mathbf{r}_A \right|^2 = \left| \mathbf{r}_T - \mathbf{r}_B \right|^2 , \left| \mathbf{r}_T - \mathbf{c}_b \right| = r + \sqrt{3} d_b \tag{33}$$

The two roots are $\mathbf{r}_{ML}$ and $\mathbf{r}_{mL}$, and the corresponding points are ML and mL. The range-evaluation functions are as follows:

$$M(b) = \begin{cases} value(ML) & \textit{circumsphere of a triangle is out of its circumsphere} \\ \infty & \textit{else} \end{cases}$$
$$m(b) = \begin{cases} value(mL) & \textit{circumsphere of a triangle is out of its circumsphere} \\ \infty & \textit{else} \end{cases} \tag{34}$$

Figure 34 shows the two tangent cases between the circumspheres of branch b and the rolling-ball (cross section picture), where back circle is for the circumsphere of branch b, blue, green and red circle are for rolling-balls, ML and mL are for corresponding tangent points. If the tangent point is ML, the rotation angle of rolling-ball is the smallest. If it is mL, the rotation angle is the largest.

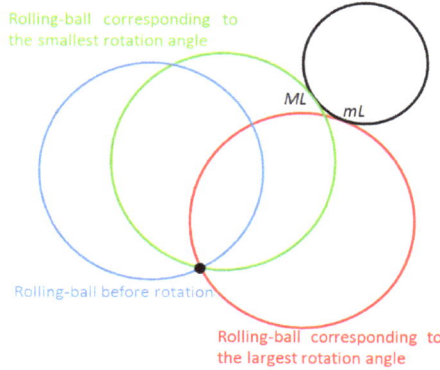

Fig. 34. Scheme for the two tangent cases between the rolling-ball and the circumsphere of branch b.

(II) Minimum searcher MS2: Given a point $\mathbf{r}_0$, search for the nearest point P of the given point in point tree. The value-finding function is

$$value(P) = |\mathbf{r}_P - \mathbf{r}_0| \tag{35}$$

where $\mathbf{r}_P$ is coordinate of point P. The range-evaluation functions are

$$M(b) = |\mathbf{c}_b - \mathbf{r}_0| + \sqrt{3}d_b$$
$$m(b) = \max(|\mathbf{c}_b - \mathbf{r}_0| - \sqrt{3}d_b, 0) \tag{36}$$

(III) Minimum searcher MS3: Given a point and a rotation axis, search in point tree for the first point by the rolling-ball with fixed size. The algorithm is basically the same as MS1. We do not repeat here.

(IV) Conditional searcher CS1: Given two point P_1 and P_2, search in extended-segment tree for segment BD whose vertexes are P_1 and P_2. The conditional function is as follows:

$$condition(BD) = \begin{cases} true & B == P_1 \ and \ D == P_2 \\ false & else \end{cases} \tag{37}$$

The circumsphere S of branch b is used for identification. The identification function is as follows:

$$maycontain(b) = \begin{cases} true & P_1, P_2 \in S \\ false & else \end{cases} \tag{38}$$

The rolling-ball algorithm is as follows: (I) Initialization: Set the radius of rolling-ball and the center as P_0, generate point tree tp from given discrete points, search for the nearest point P_1 to P_0 with searcher MS2. Use searcher MS3 to search in tree tp for a point P_2 which is the first point met by the rolling-ball rotating along x axis. Use searcher MS3 to search in tree tp for a point P_3 which is the first point met by the rolling-ball rotating along the direction of segment P_1P_2. Generate a triangle from P_1, P_2 and P_3, construct a triangle tree tt from triangle P_1P_2 P_3 and put it into extended-segment tree tb. (II) Interface construction: Check whether or not the tree tb is null. If yes, exit. If not, cut down a segment AB of triangle ABC. Search in tree tb for a point P with searcher MS1 to make smallest the rotation angle of circumsphere of triangle ABC, where AB is the rotation axis. Construct triangle BAP and put it into the triangle tree tt. Use CS1 searcher to search in tree tb for an extended-segment L whose vertexes are point B and P. Check whether or not L exist. If yes, cut L down from tb and then delete it. If not, generate an extended-segment PB and put it into tb. Do the same operations to point P and A. (III) Go back to step (II). The collection of triangles contained by tree tt is just the needed physical interface.

3.5.2.2 Results show

In molecular dynamical simulations on voids coalescence in fcc copper, defects atoms include atoms in void walls and dislocations. Figure 35 shows interface of voids constructed from discrete atom positions. Figure 36 shows the corresponding construction process. The radius of rolling-ball is set as 2/3 times of the lattice constant so as to distinguish residual structures on the wall of voids after dislocation slipping. In figure 35(b), the stairs composed by array of atoms can be clearly seen from the constructed surface. When the voids are growing, the trace of dislocation migration and the irregular shape of the void coalescence zone are truly shown up.

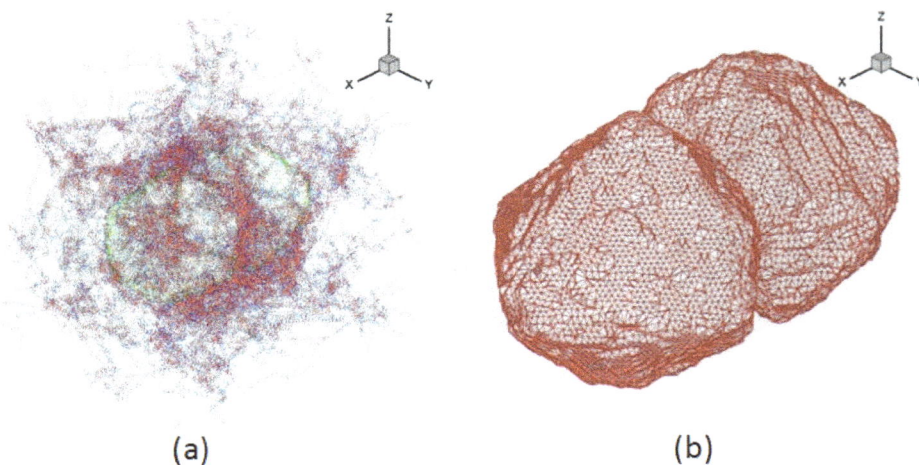

(a)   (b)

Fig. 35. Interface of voids constructed from discrete points. (a) discrete points; (b)constructed interface.

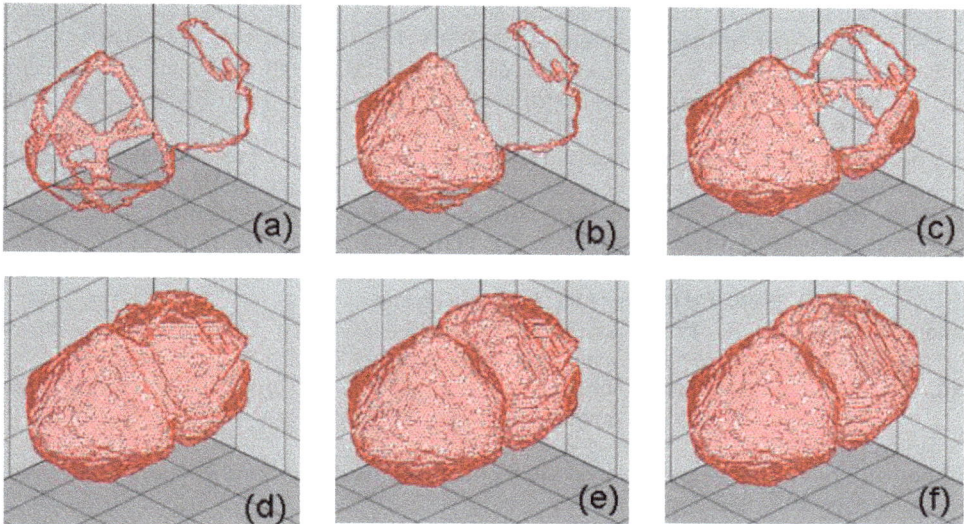

Fig. 36. Process of constructing interface of voids from discrete points.

4. Acknowledgment

Our thanks are due to Profs. Shigang Chen, Yingjun Li, Haifeng Liu, Ping Zhang, Haifeng Song, Guicun Ma, Hui Zheng, and Drs. Feng Chen, Yanbiao Gan, Weiwei Pang, Bo Sun, Yanhong Zhao, Shuaichuang Wang, Gongmu Zhang, Hongzhou Song, Qili Zhang, Mingfeng Tian, for helpful comments and discussions. This work was supported by the National Natural Science Foundation of China (Grant No. 11075021) and the Science Foundations of the Laboratory of Computational Physics and China Academy of Engineering Physics (Grant Nos. 2009A0102005 and 2011A0201002).

5. References

Ackland, G. J. & Jones, A. P. (2006). Physical Review B. *Applications of local crystal structure measures in experiment and simulation*, Vol. 73, No. 5, pp. 054104

Ament, P. (2008). Physical Review Letter, *Effects of ionization gradients on Inertial-Confinement-Fusion capsule hydrodynamics stability*. Vol. 101, No. 11, pp. 11504

Bernardini, F., Mittlelman, J., Rushmeir, H. & Silva, C. (1999). IEEE Trans. Visual. Comput. Graphics, *The ball-pivoting algorithm for surface reconstruction*, Vol. 5, No. 8, pp. 349-359

Boissonant, J. D. (1984). ACM Transactions on Graphics, *Geometric structures for three dimensional shape representation*, Vol. 3, No. 4, pp. 266-286

Bowler, B. P., Waller, W. H., Megeath, S. T., et al. (2009). The Astronomical Journal, *An infrared census of star formation in the horsehead nebula*, Vol. 137, No. 3, pp. 3685−3699

Clarkson, K. L., & Varadarajan, K. (2007). Discrete & Computational Geometry, *Improved approximation algorithms for geometric set cover*. Vol. 37, No. 1, pp.43−58

Chazelle, B., Devillers, O., Hurtado, F., et al. (2002). Algorithmica, *Splitting a Delaunay triangulation in linear time*, Vol. 34, pp.39−46 de Berg, M., Cheong, O., van Kreveld, M. & Overmars M. (2008). *Computational Geometry* (3rd revised ed.), Springer-Verlag, ISBN 978-3-540-77973-5, New York, USA

de Berg, M., Cheong, O., van Kreveld, M. & Overmars M. (2008). *Computational Geometry: Algorithms and Applications*, Springer-Verlag, ISBN 978-3-540-77973-5, New York, USA

de Vriesl, P. C., Hua, M. D., McDonald, D. C., et al. (2008). Nuclear Fusion, *Scaling of rotation and momentum confinement in JET plasmas*,Vol. 48, No. 6, pp. 065006

Dierk, R. (1998). *Computational Materials Science: The Simulation of Materials Microstructures and Properties*, Wiley-VCH , ISBN 978-3527295418, New York, USA

Donald, K. (1998). *The Art of Computer Programming*, Volume 3: Sorting and Searching (2nd Edition), ISBN 978-0201896855, Addison-Wesley

Faken, D. & Jonsson, H. (1994). Computational Materials Science. *Systematic analysis of local atomic structure combined with 3D computer graphics*, Vol.2, No. 2, pp. 279-286

Fan, Y. J., Iyigun, C. & Chaovalitwongse, W. A. (2008). CRM Proc Lecture Notes, *Recent advances in mathematical programming for classification and cluster analysis*, Vol. 45, pp.67 – 93

Hernquist, L. (1988). Computer Physics Communications, *Hierarchical N-body methods*, Vol. 48, pp.107 – 115

Hidaka, Y., Choi, E. M., Mastovsky, I., et al. (2008). Physical Review Letters, *Observation of large arrays of plasma filaments in air breakdown by 1.5-MW 110-GHz gyrotron pulses*, Vol. 100, No. 4, pp.035003

Hoppe, H., DeRose, T., Duchanp, T., Mc-Donald, J. & Stuetzle,W. (1992). ACM Computer Graphics, *Surface reconstruction from unorganized points*, Vol. 26, No. 2,pp. 71-78

Kelchner, C. L., Plimpton, S. J. &Hamilton, J. C. (1998). Physical Review B. *Dislocation nucleation and defect structure during surface indentation*, Vol. 58, No. 17, pp.11085

Kotsiantis, S. B. & Pintelas, P. E. (2004). WSEAS Transactions on Information Science and Applications, *Recent advances in clustering: A brief survey*, Vol. 1, No. 1: 73 – 81

Kumar, K.S., Van Swygenhoven, H. & Suresh, S. (2003). Acta Materialia. *Mechanical behavior of nanocrystalline metals and alloys*, Vol. 51, No. 19, pp. 5743–5774

Makino, J. (1990). Journal of Computational Physics, *Vectorization of a treecode*,Vol. 87, No. 1, pp.148 – 160

Mencl, E. & Muller, H. (1997). Interpolation and approximation of surfaces from three-dimensional scattered data points, *Scientific Visualization Conference 1997*, Dagstuhl, Germany, June 1997

Michael, G., Stephan, K. & Gerhard Z. (2007). *Numerical Simulation in Molecular Dynamics*, Springer-Verlag, ISBN 978-3-540-68094-9,Berlin, Heidelberg

Moore, E.F. (1959). The shortest path through a maze, *Proceedings of an International Symposium on the Theory of Switching* , Cambridge, Massachusetts, April 1957.

Nogaret, T., Rodney, D., Fivel, M., et al. (2008). Journal of Nuclear Materials, *Clear band formation simulated by dislocation dynamics: Role of helical turns and pile-ups*, Vol. 380, No. 1-3, pp. 22 – 29

Zhao, H. K. & Osher, S., Merriman, B. & Kang, M. (2002). Computer Vision and Image Processing, *Implicit, non-parametric shape recontruction from unorganized points using variational level set method*, Vol. 80, No. 3, pp. 295-314

14

Molecular Dynamics Simulations of Proton Transport in Proton Exchange Membranes Based on Acid-Base Complexes

Liuming Yan* and Liqing Xie
Department of Chemistry,
Shanghai University, Shanghai,
China

1. Introduction

Proton exchange membrane fuel cells (PEMFCs) are promising energy conversion devices for the future society for their high energy conversion efficiency, environmental friendliness, and structural compactness. Therefore, PEMFCs have attracted great attention from academic and industry institutions; and billions of dollars of research funding from both private organizations and governmental agencies are invested into this field. However, the large-scale applications of PEMFCs are challenged by the high cost, shortage of infrastructure for the production and distribution of hydrogen, as well as performance inefficiency of the core materials including the proton exchange membranes and electrocatalysts. Though the presently most accepted proton exchange membranes based on poly(perfluorosulfonic acid), such as NAFION®, possess high proton conductivity, high chemical and electrochemical stability, excellent mechanical properties, and long service life; their performances seriously degrade under water deficiency environment or at temperatures above 80°C because of evaporation dehydration. PEMFCs operate at temperatures well above 120°C are essential as the electrocatalytic activity improves greatly at high temperature; and thus the loading of precious platinum on the electrocatalyst could be greatly reduced. Furthermore, the tolerance of electrocatalysts to impurities such as CO in fed gas is also greatly improved at high temperature and the purification cost for fed gas could be significantly reduced. Other benefits of high-temperature PEMFCs include simplified water and heat management systems, and improved overall energy conversion efficiency.

The high temperature performances of proton exchange membranes, specially the high temperature proton conductivity, must be greatly improved in order to elevate the operational temperature of PEMFCs to well above the boiling point of water. One of the methods to improve the high temperature performances of proton exchange membranes is to blend water retention materials into membranes fabricated from poly(perfluorosulfonic acid); however, the operational temperature of PEMFCs could only be increased to about

---

* Corresponding Author

120°C. Since the further increase in operational temperature is difficult using proton exchange membranes based on poly(perfluorosulfonic acid) with the addition of water retention materials, proton exchange membranes being independent on hydration degree need to be developed. Recently, the proton exchange membranes based on acid-base complexes, especially blends of poly(phosphonic acids) and poly(heterocycles), are attracting more and more research attentions owing to the excellent high-temperature proton conductivity under anhydrous states. The proton conduction performances of proton exchange membranes based on acid-base complexes are determined by three essential characteristics: the concentration of transportable protons depending on the acid-base equilibrium, the morphological rearrangement depending on the structural flexibility of the polymeric molecules, and the formation of hydrogen-bond network.

Density functional theory calculations and molecular dynamics simulations of the acid-base complexes are important for the understanding of the proton transport mechanism and could provide essential insight for the rational design of high-temperature proton exchange membranes. In our previous study, we have studied the acid-base equilibrium in complexes of benzimidazole and model acids by density functional theory calculations; and the calculation results are verified by ^{1}H NMR spectra (Zhang & Yan, 2010). We have also studied rotational flexibility of phosphonic groups for some of the phosphonic acids using density functional theory calculations (Yan, Feng et al., 2011), and the hydrogen-bond network in the pristine and phosphoric acid doped polybenzimidazole using molecular dynamics simulations (Zhu, Yan et al., 2011).

In this chapter, density functional theory calculations and molecular dynamics simulations will be applied to the study of the hydrogen bonding characteristics of acid-base complexes, as well as acidic and basic bi- and multifunctional molecules, and the formation of hydrogen-bond network in pristine, hydrated, and heterocycles solvated poly(vinylphosphonic acid). In the first section, we introduce the application background, as well as the major challenges to the proton exchange membranes. In the second section, we describe the accepted proton transport mechanisms including the essential steps for proton transport. In section three, the calculation methods, as well as molecular force field models are developed. In sections four and five, the density functional theory calculation and molecular dynamics simulation results are reported, respectively. In the last section, some of the important conclusions are summarized.

2. The proton transport mechanisms in proton exchange membranes

The proton mobility in liquid water surpasses that of any other cations since protons transport via both vehicle mechanism (molecular diffusion) and hopping mechanism (structural diffusion). However, this simple perspective complicates in proton exchange membranes because of the structural complexity and configurational restriction. For example, the structural diffusion mechanism dominates the proton conducting process in well hydrated sulfonated ionomers where continuous hydrophilic subphase exists. On the other hand, the molecular diffusion mechanism dominates in sulfonated ionomers at low hydration degree where continuous hydrophilic subphase is broken into segregated hydrophilic domains. In sulfonated ionomers at intermediate hydration degree, both vehicle and hopping mechanisms contribute substantially to the proton conducting process (Kreuer, Rabenau et al., 1982).

Although there are many advantages using water as solvent including the fast proton transport dynamics, low cost, and environmental friendliness, the greatest disadvantage is the limitation to the operational temperature of fuel cells as the proton exchange membranes rapidly dehydrate at temperatures above 80°C. Therefore, it is proposed that the operational temperature could be significantly elevated if the solvent water is substituted by nonaqueous protic solvent with high boiling point, such as pyrrole, pyrazole, imidazole, phosphoric acid, and phosphonic acid. The nonaqueous protic solvent molecules usually possess negligible molecular diffusion dynamics and only structural diffusion contributes substantially to proton conductivity. For example, proton transports mainly via the hopping mechanism in concentrated phosphoric acid, where approximately 90% of the total available protons from both phosphoric acid and water (in 85% and 100% H_3PO_4) contribute to proton conducting, as revealed by the ^1H and ^{31}P pulsed gradient spin-echo NMR study (Chung, Bajue et al., 2000). In materials functionalized with phosphonic groups, the structural diffusion via hydrogen bonds of neighboring phosphonic groups facilitates proton transport under dehydrated state at temperatures well above the boiling point of water (Schuster, Rager et al., 2005). Therefore, the elucidation of transport mechanisms is essentially important in the design and development of better proton conducting materials applicable to fuel cells operated at high temperature under anhydrous state (Yan, Feng et al., 2011).

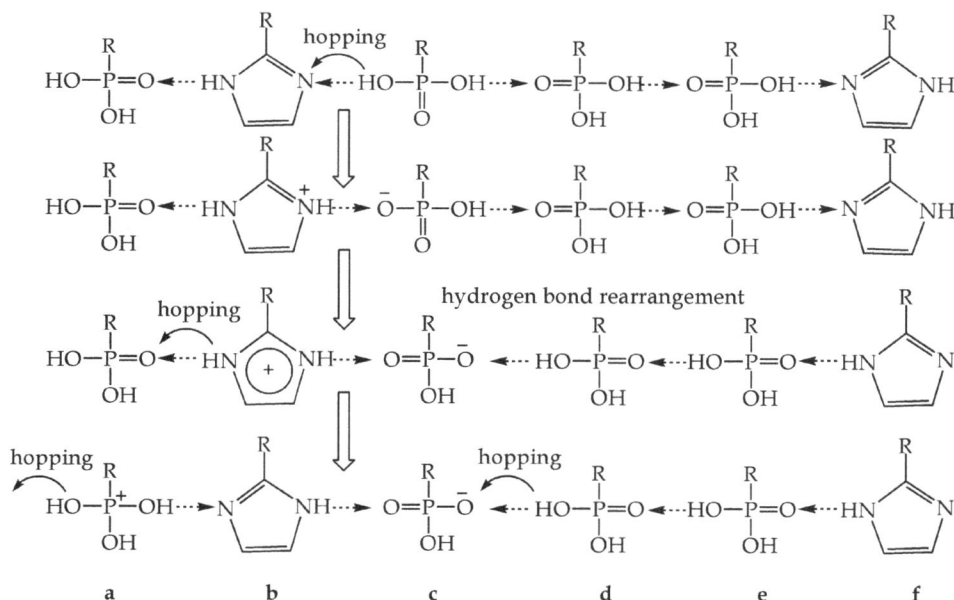

Fig. 1. 1-D schematic diagram for the proton hopping transport in acid-base complex: Firstly, a proton hops from phosphonic group **c** to a neighboring base group **b**. And then, the hydrogen-bond network in **c**, **d**, **e**, and **f** rearranges to accommodate the hopping. If the hydrogen-bond network rearrangement is hindered, the proton will hop back and forth without macroscopic transport. In many systems where pervasive hydrogen-bond networks exist, the hydrogen-bond network rearrangement is hindered, macroscopic proton transport is hampered. Thirdly, proton hops from **b** to **a**, **d** to **c**, … As a result, the protons hop from right to left without the macroscopic transport of any molecules.

In order to facilitate the structural diffusion of protons in nonaqueous proton conducting systems, the following criteria must be met: the existence of dissociable protons and charge defects in terms of excess or missing protons (Kreuer, Fuchs et al., 1998; Joswig and Seifert, 2009), the formation of pervasive hydrogen-bond network, and the rapid rearrangement of the hydrogen-bond network (Fig. 1). The first criterion is accomplished by the complexation of the acidic and basic functional groups, where the acidic functional groups provide the dissociable protons or excess protons, while the basic functional groups accept the dissociated protons and create the charge defects. The second criterion is accomplished by the formation of pervasive hydrogen-bond network between the hydrogen bond donors and acceptors allowing the transport of protons from hydrogen bond donors to hydrogen bond acceptors without macroscopic molecular transport. However, the hopping of protons may only represent the local vibration of protons, and the protons will hop back if no hydrogen-bond network rearrangement occurs to accommodate the hopping of protons (Paddison, Kreuer et al., 2006; Brunklaus, Schauff et al., 2009). For example, in the so-called hop-turn mechanism, the effective proton transport must include the bond rotation or rearrangement of the proton defect groups (the deprotonated phosphonic group, or the hydrogen phosphonate anion) (Münch, Kreuer et al., 2001; Sevil & Bozkurt, 2004; Yamada & Honma, 2005).

3. The calculation methods and molecular force field models

3.1 The density functional theory calculation method

The density functional theory calculations are employed to study the interaction between the acidic and basic functional groups in terms of hydrogen bonding of acid-base complexes. The density functional theory calculations were performed at the B3LYP/6-31G(d)/ 6-311++G(d, p) level of theory (Lee, Yang et al., 1988; Becke 1993) as implemented in the GAUSSIAN 03 suite of programs (Frisch, Trucks et al., 2003). Though the hybrid B3LYP functionals do not always correctly reproduce the polarizabilities and hyper-polarizabilities (Champagne, Perpete et al., 1998; Zhang, Xu et al., 2009), the charge-transfer excitation energies (Chai & Head-Gordon, 2008), the noncovalent bonding interactions (Zhao & Truhlar, 2005), and the hydrogen bonding of aromatic molecules (Elstner, Hobza et al., 2001), the B3LYP functionals have achieved the greatest success in terms of the number of published applications (Yanai, Tew et al., 2004) and are still among the best functionals that provide accurate predictions for geometries and thermochemistry of small covalent systems (Curtiss, Raghavachari et al., 1997; Song, Tokura et al., 2007; Rohrdanz, Martins et al., 2009), especially for the energetics and geometrical properties of the proton transfer and other ion-molecule reactions (Curtiss, Raghavachari et al., 1991, 1993; Merrill & Kass, 1996; Curtiss, Redfern et al., 1998; Friesner, Murphy et al., 1999). Since our major interests are inter- and intramolecular hydrogen bonding interactions and morphologies of acid-base complexes, and acidic and basic bi- and multifunctional molecules, the hybrid B3LYP functionals are still among the best choices of functionals. In addition, the calculations based on hybrid B3LYP functionals and basis set of the split-valence type with diffusive functions have good balance between accuracy and computational cost.

In the density functional theory calculations, the isolated phosphonic acids, heterocyclic compounds, phosphonic acid-heterocycle complexes, and as well as acidic and basic bi- and multifunctional molecules are optimized at the B3LYP/6-31(d) level of theory. And then, vibrational frequency calculations are conducted to verify if local minima have reached. The optimizations are satisfactory since local minima are achieved for all the isolated molecules and acid-base complexes without any imaginary frequencies. Thirdly, hydrogen bonding enthalpies are calculated at the B3LYP/6-311++(d, p) level of theory using the standard procedures taking into account of zero point energies, finite temperature corrections, and the pressure-volume work term (pV) calculated at the B3LYP/6-31(d) level of theory. Finally, various structural and energetic properties are analyzed based on the density functional theory calculations.

3.2 The molecular dynamics simulation method

The molecular dynamics simulations, applied to the study of hydrogen-bond network of the pristine, hydrated, and heterocycles solvated poly(vinylphosphonic acid), are carried out using the DL_POLY program (Smith, Leslie et al., 2003). 3-D periodic boundary conditions are applied to all the simulation cells, and the simulations are carried out in the NPT ensemble with temperature and pressure maintained, respectively, by the Nosé-Hoover thermostat and barostat with the same relaxation time parameters of 0.2 ps (Nosé, 1984; Hoover, 1985). The Verlet integration scheme is used to solve the Newtonian equation of motion for atoms or united-atoms in the simulation cell with an integration step time of 1 fs (Verlet, 1967). The initial simulation cells are constructed by insertion of the molecules randomly. During the initial simulations, the loose initial simulation cells are gradually squeezed and reach to equilibrium densities. After reaching of the equilibrium density, another 1,000,000 steps are carried out for the generation of structural characteristics of the systems.

3.3 The molecular dynamics simulation systems

In order to evaluate the microstructure and formation of hydrogen-bond network in the acid-base complexes, eight molecular systems are simulated. The first molecular dynamics simulation system consists of 24 poly(vinylphosphonic acid) (PVPA) oligomers; and each PVPA oligomer is composed of eight repeat units (Fig. 2). The second molecular dynamics

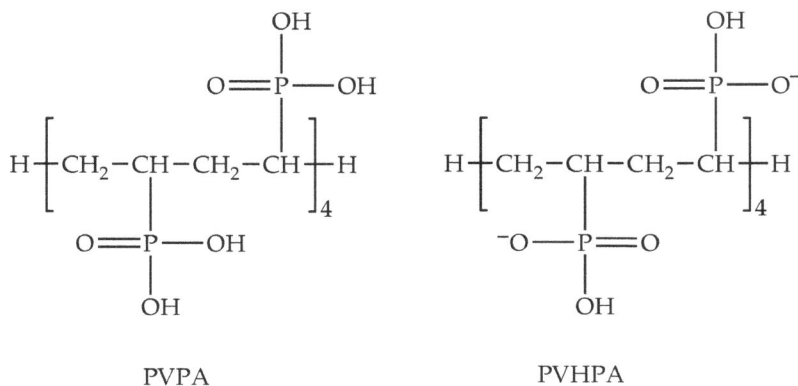

Fig. 2. Molecular structures of poly(vinylphosphonic acid) (PVPA) and poly(vinyl hydrogen phosphonate anion) (PVHPA)

simulation system is hydrated PVPA consisting of 24 poly(vinyl hydrogen phosphonate anion) (PVHPA) oligomers; 192 hydronium cations to compensate the PVHPA (Fig. 2), and 384 water molecules. This system has a hydration degree of three; and all the phosphonic groups are singly ionized.

The other molecular dynamics simulation systems are acid-base complexes consisting of PVPA oligomers and heterocyclic compounds frequently used as nonaqueous solvent for high-temperature proton exchange membranes. In this study, the heterocyclic compounds, including pyrrole, pyrazole, and imidazole, are used as basic components of the acid-base complexes. The solvation degrees, ratios of heterocyclic compound to phosphonic group, are one eighth and three eighths (table 1).

Systems	PVPA	PVHPA	H_2O	H_3O^+	pyrrole	pyrazole	imidazole
I	24						
II		24	384	192			
III	24				24		
IV	24				72		
V	24					24	
VI	24					72	
VII	24						24
VIII	24						72

Table 1. The molecular dynamics simulation systems

3.4 The molecular force field models

In the molecular dynamics simulation, the total potential energy of the molecular system is partitioned into bonded and nonbonded energies. The bonded energy U_{bonded} is further partitioned into bond stretching, bond angle bending, and dihedral torsion energies,

$$U_{bonded} = \sum_l \frac{1}{2} k_l (l - l_0)^2 + \sum_\theta \frac{1}{2} k_\theta (\theta - \theta_0)^2 + \sum_\varphi A_\varphi (1 + \cos(m\varphi - \delta)) \tag{1}$$

In equation 1, the right-hand terms correspond to bond stretching, band angle bending, and dihedral torsion energies, respectively; and the summations run over all bonds, bond angles, and dihedral angles. The l, θ, and φ represent the bond length, bond angle, and dihedral angle; l_0, θ_0, and δ correspond to their equilibrium values; k_l, k_θ, and A_φ corresponds to the potential parameters; and m represents the rotational multiplicity of dihedral angle. The nonbonded energy $U_{nonbonded}$ is partitioned into pairwise Lennard-Jones potential and Coulombic potential,

$$U_{nonbonded} = \sum_{ij} 4\varepsilon_{ij} \left(\left(\frac{\sigma_{ij}}{r_{ij}} \right)^{12} - \left(\frac{\sigma_{ij}}{r_{ij}} \right)^6 \right) + \sum_{ij} \frac{q_i q_j}{r_{ij}} \tag{2}$$

In equation 2, the summations run over all atom and united-atom pairs; r_{ij} are interatomic or inter united-atomic distances; and ε_{ij} and σ_{ij} are Lennard-Jones potential parameters; q_i and q_j are residual atomic charges.

The united-atom models are applied to the PVPA and PVHPA oligomers, where all the hydrogen atoms bonded to carbon atoms are represented implicitly by united-atoms of C1 (with one hydrogen atom) and C2 (with two hydrogen atoms) and the hydrogen atoms bonded to an oxygen atom are represented explicitly. For the hydrated poly(vinylphosphonic acid), all the phosphonic groups are singly ionized resulting hydrogen phosphonate anions. The rigid body models are applied to all the solvent molecules including water, hydronium, pyrrole, pyrazole, and imidazole. The TIP3P model is applied to water molecule (Jorgensen, Chandrasekhar et al., 1983), and the all-atom rigid body force field model is applied to hydronium cation (Urata, Irisawa et al. 2005). The molecular structure of the heterocyclic compounds are optimized by density functional theory at the B3LYP/6-31G(d) level of theory. And all the atomic and united-atomic site names are reported in figure 3.

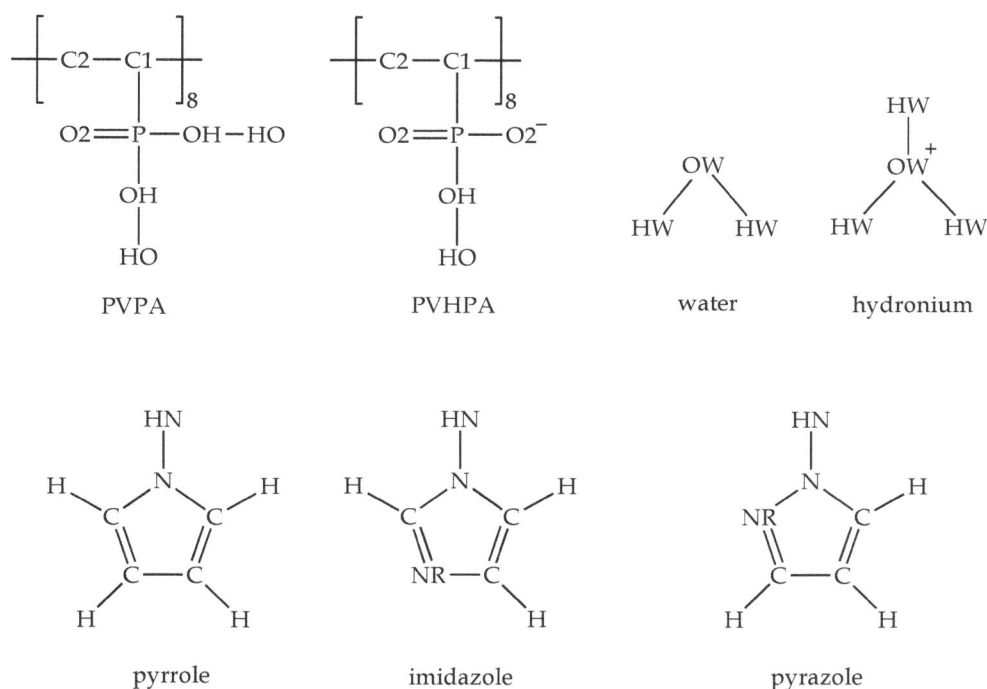

Fig. 3. Atomic and united-atomic site names for the molecular force field models

In table 2, it summarizes the force field parameters for all the simulated molecules. The equilibrium bond lengths l_0, bond angles θ_0, and dihedral angles δ for PVPA and PVHPA are adapted from the optimized structures at the B3LYP/6-31G(d) level of theory (Mayo, Olafson et al., 1990; Cornell, Cieplak et al., 1995; Yan, Zhu et al., 2007; Roy, Ataol et al., 2008). The Lennard-Jones parameters of likely atomic and united-atomic pairs are also summarized in table 2. And the Lennard-Jones parameters for cross-term interactions are evaluated using the Berthelot mixing rules (Allen & Tildesley, 1990).

Bond stretching		
Bonds	l_0 (Å)	k_l (kcal·mol^{-1}·Å^{-2})
C1-C2	1.526	620
C1-P	1.795	700
P-O2	1.480	1050
P-OH	1.610	460
OH-HO	0.960	1106
Bond angle bending		
Bond angles	θ_0 (deg)	k_θ (kcal·mol^{-1}·rad^{-2})
C2-C1-C2	111.0	115.3
C2-C1-P	111.4	115.0
C1-C2-C1	111.0	115.3
C1-P-O2	111.13	145.0
C1-P-OH	108.2	145.0
O2-P-O2	119.9	280.0
O2-P-OH	108.2	90.0
OH-P-OH	105.2	100.3
P-OH-HO	113.8	93.3

Dihedral torsion			
Dihedral angles	A_δ (kcal·mol^{-1})	m	δ (deg)
C1-C2-C1-C2	1.40	3	0
C1-C2-C1-P	1.40	3	0
C2-C1-P-O2	1.25	3	-12
C2-C1-P-OH	1.25	3	-12
C1-P-OH-HO	2.30	1	38
O2-P-OH-HO	0.25	3	0

Lennard Jones parameters		
Atomic or united-atomic pairs	ε_{ii} (kcal·mol^{-1})	σ_{ii} (Å)
C1-C1	0.1450	3.9800
C2-C2	0.2150	4.0800
P-P	0.2150	4.2950
O2-O2 (OH-OH)	0.0957	3.0332
OW-OW	0.1500	3.1500
C-C	0.0950	3.8800
N-N (NR-NR)	0.1450	3.6950
H-H (HN-HN, HO-HO, HW-HW)	0.0000	2.5000

Table 2. Force field parameters for molecular force field models

The residual charges for atoms are evaluated by density functional theory calculations at the B3LYP/6-31G(d) of theory level using the CHELPG method (Breneman & Wiberg, 1990). The residual charges for united-atoms are summation of the carbon and bonded hydrogen atoms. All the residual charges are summarized in figure 4.

PVPA:
$$\begin{bmatrix} 0.00 & 0.02 \\ C2 & C1 \end{bmatrix}_8$$
1.28, 0.47
O=P—O—H
-0.78, -0.73
O -0.730
H 0.47

PVHPA:
$$\begin{bmatrix} -0.036 & -0.265 \\ C2 & C1 \end{bmatrix}_8$$
1.055
O=P—O⁻
-0.702, -0.702
-0.592 O
H 0.242

water:
-0.834
O
H   H
0.417   0.417

hydronium:
0.518
H
-0.554 O⁺
H   H
0.518   0.518

pyrrole:
H 0.300
0.129
H -0.155 N -0.190 0.129
                      H
                -0.155
-0.135   -0.135
0.106 H   H 0.106

imidazole:
H 0.292
0.067   -0.224
H 0.215 N
              H 0.156
          -0.233
N   0.152
-0.491
H 0.066

pyrazole:
H 0.280
-0.034
-0.438 N N -0.045 0.113
                      H
0.190   -0.262
H   H
0.066   0.130

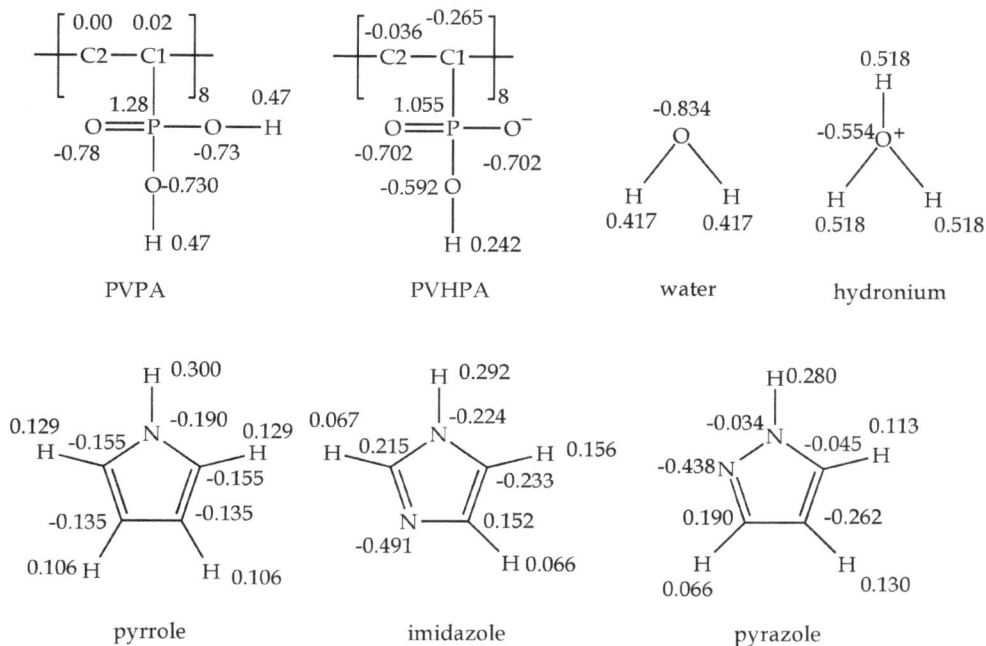

Fig. 4. Residual atomic and united-atomic charges for molecular force field models

Though the all-atom force field models are considered more delicate than the united-atom force field models owing to their complexity and the more force field parameters used to model the same molecular system, the united-atom force field models consist of less force field parameters, thus pertain less parameterization errors. Therefore, it cannot be concluded whether one type of force field models are better than the other type or not. The force field models applied in this study reproduce reliable molecular configurations since the geometrical parameters are based on density functional theory at the B3LYP/6-31+G(d) level of theory. At this level of theory, the calculation errors for most bond lengths and bond angles, respectively, are within 0.02 Å and 0.6° except a few large errors between 0.02~0.04 Å and 0.6~1.0° (El-Azhary & Suter, 1996; Baboul, Curtiss et al., 1999). By combination of the B3LYP/6-31G(d) geometry, CHELPG residual charges, and optimized force field parameters, the force field models used in this study minimize the number of force field parameters and retains the accuracy for the description of molecular structures and hydrogen bonding.

4. Hydrogen bonding in acid-base complexes

4.1 Intermolecular hydrogen bonding

Various polymeric phosphonic acids are widely accepted as proton exchange membranes besides the polymeric sulfonic acids. The phosphonic acids possess intermediate acidity, weaker than sulfonic acids, but stronger than carboxylic acids. The first acidic protons of phosphonic acids dissociate almost completely under hydrated state providing sufficient concentration of dissociable protons for proton conduction. Similar to concentrated phosphoric acid, phosphonic acids self-dissociate under anhydrous state and possess plausible proton conductivity even under anhydrous states and high temperature. In addition, the phosphonic acids are more stable than sulfonic acids as the C-P bonds dissociate at a higher temperature than the C-S bonds do; therefore, the PEMFCs based on polymeric phosphonic acids could operate at higher temperature than that based on sulfonic acids. On the other hand, the carboxylic acids possess much weaker acidity; their dissociation degrees are limited to a few percents; and their dissociation temperatures are lower than that of sulfonic acids. Therefore, phosphonic acids are potential candidates for the proton exchange membranes based acid-base complexes.

Dissimilar to the limited types of organic acidic groups that exist, there exist a great number of types of organic bases possible for the acid-base complexes. In this study, the heterocyclic moieties based on the 5-membered ring containing one or two nitrogen atoms in the ring, including the pyrrole, pyrazole, and imidazole, are used as basic components.

In figure 5, it summarizes the optimized structure of butanephosphonic acid (C4PA) 1, pyrrole 2, pyrazole 3, and imidazole 4, as well as acid-base complexes including C4PA-pyrrole 5, C4PA-pyrazole 6, and C4PA-imidazole 7. For the acid-base complex 5, the pyrrole donates its acidic proton to the phosphonic oxygen forming a hydrogen bond with a bond length of 1.960 Å. And the N-H bond length, 1.008 Å in free pyrrole 2, is stretched to 1.019 Å. The hydrogen bond is relatively weak as the bonding enthalpy is only -5.1 kcal · mol^{-1}. For the acid-base complex 6, where the base component pyrazole consists of two nitrogen atoms, two hydrogen bonds are formed between the phosphonic group and the pyrazole

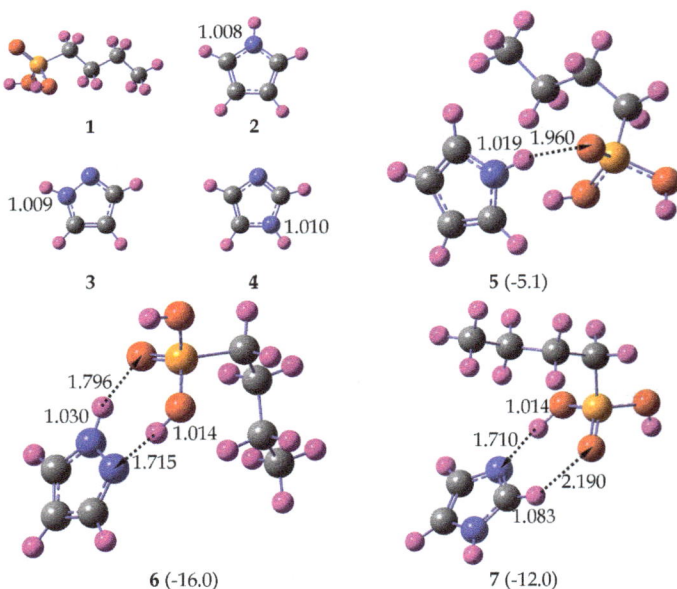

Fig. 5. Intermolecular hydrogen bonding in acid-base complexes at B3LYP/6-31G(d), the number in parenthesis are hydrogen bonding enthalpies in kcal · mol^{-1}

with an overall bonding enthalpy of -16.0 kcal · mol^{-1}. The hydrogen bond lengths, 1.796 and
1.715 Å, respectively, are also shorter than that in **5**. For acid-base complex **7**, the two
nitrogen atoms containing in imidazole are separated by a methylidyne spacer, and can not
form two hydrogen bonds simultaneously with one phosphonic group because of
configuration restriction. However, the hydrogen bond length in **7** is only 1.710 Å,
significantly shorter than that in **5**. The hydrogen bonding enthalpy in **7** is -12.0 kcal · mol^{-1}
and is between that of **5** and **6**. The detailed hydrogen bonding parameters are summarized
in table 3.

In order to evaluate the hydrogen bonding characteristics among poly(vinylphosphonic
acid) and heterocyclic compounds, we also study acid-base complexes based on heptane-1,4-
diphosphonic acid and heterocyclic compounds. From figure 6, it is revealed that the second
phosphonic group separated by two methylene spacers does not have significant impact on
the hydrogen bonding characteristics. The hydrogen bond lengths are similar to the
corresponding system with single phosphonic group, and the hydrogen bonding enthalpies
are almost doubled as the second phosphonic group is incorporated into the same
molecules.

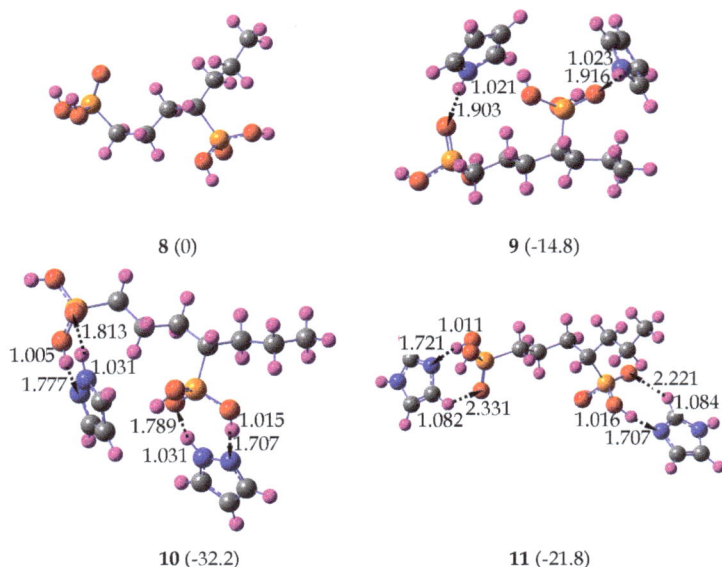

Fig. 6. Intermolecular hydrogen bonding of the acid-base complexes based on heptane-1,4-
diphosphonic acid at B3LYP/6-31G(d), the number in parenthesis are hydrogen bonding
enthalpies in kcal · mol^{-1}

In summary, hydrogen bonds are formed between phosphonic groups and heterocyclic
compounds: pyrrole could form one hydrogen bond; pyrazole could form two hydrogen
bonds simultaneously with the same phosphonic group; and imidazole could form one
hydrogen bond with the same phosphonic group despite the existence of two nitrogen
atoms in the ring. In diphosphonic acids, the second phosphonic group separated by two
methylene spacers does not have significant impact on hydrogen bonding characteristics.

Systems	d_{N-H}	d_{O-H}	$\angle NHO$	υ	ΔH
5	1.019	1.960	144.1	3500	-5.1
6	1.030	1.796	152.6	3327	-16.0
	1.715	1.014	170.3	2945	
7	1.710	1.014	174.4	2934	-12.0
9	1.021	1.903	169.7	3479	-14.8
	1.023	1.916	151.7	3427	
10	1.031	1.813	152.3	3309	-32.2
	1.777	1.005	165.4	3117	
	1.031	1.789	153.2	3313	
	1.707	1.015	170.3	2913	
11	1.721	1.011	173.1	2992	-21.8
	1.707	1.016	169.8	2919	
12	1.017	1.999	153.1	3533	-1.3
13	1.019	1.977	137.8	3509	-0.2
	2.588	0.977	160.4	3671	
14	1.736	1.006	163.6	3087	-3.1
15	1.017	2.046	157.4	3530	-2.3
	1.016	2.133	133.1	3551	
16	1.017	2.051	151.2	3541	-4.7
	1.019	1.927	147.7	3507	
17	1.805	0.997	158.7	3241	-12.0
	1.709	1.012	163.1	2984	

Table 3. Hydrogen bonding parameters for the acid-base complexes (d_{N-H} and d_{O-H} are hydrogen bond lengths in Å, $\angle NHO$ is bond angle in degree, υ is vibrational wave number of the hydrogen bonded proton in cm^{-1}, and ΔH is bonding enthalpy in $kcal \cdot mol^{-1}$).

4.2 Intramolecular hydrogen bonding

There are two processes to incorporate the acidic and basic groups into the same proton exchange membrane: by blending an acidic polymer and a basic polymer together, or by incorporating both acidic and basic groups into the same polymer. Since molecular level of mixing of two polymers, which is essential for the formation of hydrogen-bond network, is difficult, it is difficult to carry out the blending process. On the other hand, it is versatile to incorporate both acidic and basic groups into the same polymer either by copolymerization or by grafting a second functional group into a polymer.

In this section, the intramolecular hydrogen bonding between a phosphonic group and heterocyclyl group in the same molecule is discussed. In figure 7, it summarizes the optimized structure of acidic and basic bifunctional molecules including 3-(2'-pyrrolyl)-butanephosphonic acid 12, 3-(5'-pyrazolyl)-butanephosphonic acid 13, and 3-(4'-imidazolyl)-butanephosphonic acid 14. The hydrogen bond length is 1.999 Å in 12 indicating weak hydrogen bonding. For 13, only one hydrogen bond is formed with a bond length of 1.977 Å despite the possibility forming two hydrogen bonds. The strongest hydrogen bond is formed in 14 with a hydrogen bond length of 1.736 Å. In addition, 12, 13, and 14 are also intentionally optimized to structures without any hydrogen bonding as 12', 13', and 14'. In these structures, the phosphonic group and heterocyclyl group are in opposite position, the intramolecular interaction between the phosphonic group and heterocyclyl group is negligible. By comparison of the two sets of structures, **12** and **12'**, **13** and **13'**, and **14** and **14'**, the intramolecular hydrogen bonding enthalpies are evaluated as the enthalpy difference of the two sets of structures. The calculated hydrogen bonding enthalpies for **12**, **13**, and **14** are only -1.3, -0.2, and -3.1 kcal · mol^{-1}, respectively, indicating marginal intramolecular hydrogen bonding. Therefore, it is concluded that the intramolecular hydrogen bonding between a phosphonic group and a heterocyclyl group separated by two methylene spacers are hindered by configuration restriction. Since these model molecules could represent copolymers of the vinylphosphonic acid and vinyl heterocycles, heterocyclyl grafted poly(vinyl phosphonic acid) or phosphonic group grafted poly(vinyl heterocycles), and polymers grafted with heterocyclyl and phosphonic group, it could be further concluded that intramolecular hydrogen bonding in bi- and multifunctional polymers is hindered by configuration restriction and intramolecular hydrogen-bond network is unlikely to be formed.

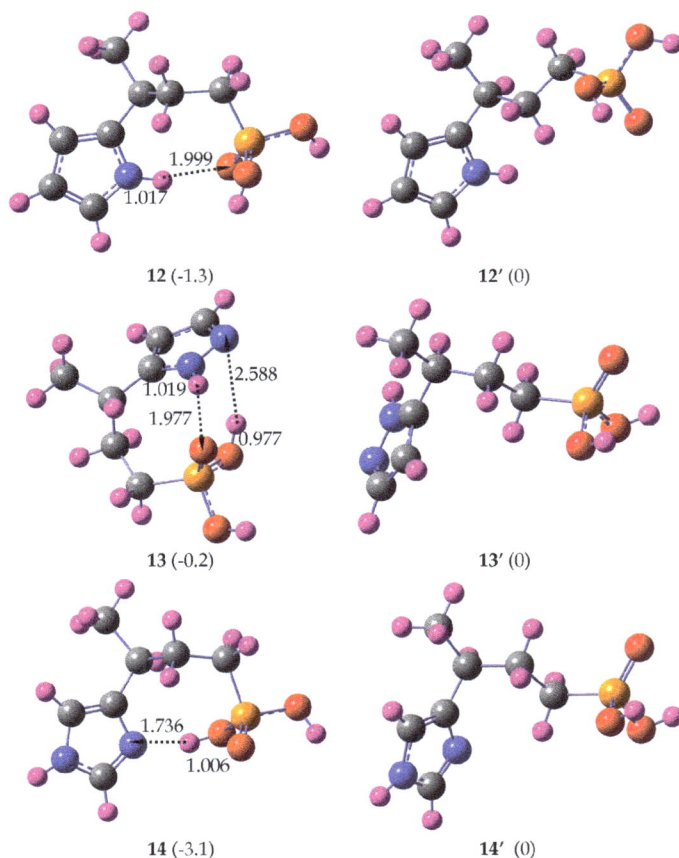

Fig. 7. Intramolecular hydrogen bonding in acid-base complexes based on bifunctional molecules optimized at B3LYP/6-31G(d), the number in parenthesis are hydrogen bonding enthalpies in kcal · mol^{-1}

In figure 8, it summarizes the structures of some multifunctional molecules grafted with two phosphonic groups and two heterocyclyl groups, including the 2,6-di(2'-pyrryl)-heptane-1,4 -diphosphonic acid **15**, 2,6-di (5'-pyrazyl)-heptane-1,4-diphosphonic acid **16**, and 2,6-di(4'- imidazyl)-heptane-1,4-diphosphonic acid **17**. These multifunctional molecules are optimized to form the maximum intramolecular hydrogen bonds between the phosphonic group and heterocyclyl group. The intramolecular hydrogen bonds formed in the multifunctional molecules possess similar bond lengths to that in the bifunctional molecules **12**, **13**, and **14**, indicating that grafting more functional groups to the molecule exerts little impact on the intramolecular hydrogen bonding characteristics. In addition, these multifunctional

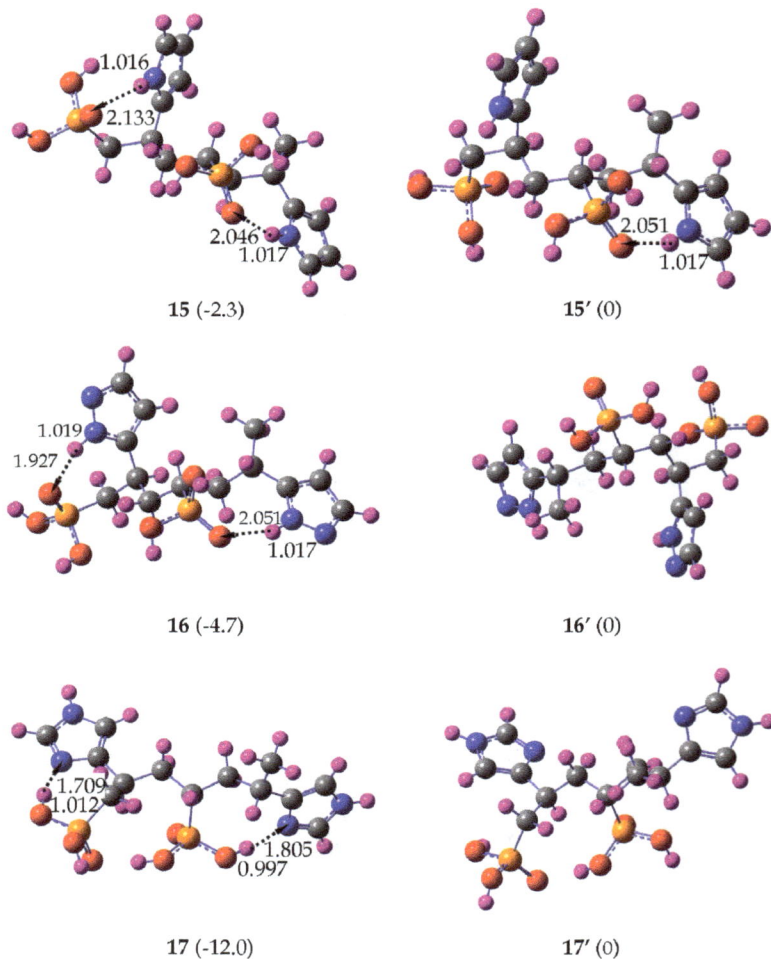

15 (-2.3)   **15'** (0)

16 (-4.7)   **16'** (0)

17 (-12.0)   **17'** (0)

Fig. 8. Intramolecular hydrogen bonding of acid-base complexes based on multifunctional molecules optimized at B3LYP/6-31G(d), the number in parenthesis are hydrogen bonding enthalpies in kcal · mol^{-1}

molecules are intentionally optimized to structures without or only one pair of acidic and basic groups in intramolecular hydrogen bonding configurations shown as **15′**, **16′**, and **17′** in figure 8. The hydrogen bonding enthalpies, enthalpy differences between **15** and **15′**, **16** and **16′**, and **17** and **17′**, are only -2.3, -4.7, and -12.0 kcal·mol⁻¹ for the multifunctional molecules, respectively. Therefore, the intramolecular hydrogen bonding is weak in the multifunctional molecules.

From the density functional theory calculations, it is concluded that the intramolecular hydrogen bonds in the bi- and multifunctional molecules are relatively weak compared to the intermolecular hydrogen bonds in the acid-base complexes with the same acidic and basic functional groups. The advantages to apply bi- and multifunctional molecules for proton exchange membranes is not the formation of intramolecular hydrogen bonds, but is the formation of intermolecular hydrogen bonds as the molecular level of mixing of acidic and basic groups is automatically achieved in these systems.

5. Hydrogen-bond network in acid-base complexes

5.1 The pristine and hydrated PVPA

The PVPA is a highly hydrophilic polymeric material dissolvable in water. Since the phosphonic group is an acid with intermediate strength, the first acidic protons are almost fully dissociated and the second acidic protons are usually not dissociated under hydrated state. For the simplicity of molecular force field model, all the phosphonic groups are supposed to be singly ionized resulting the poly(vinyl hydrogen phosphonate anion) under hydrated state. In order to compare the structural differences between pristine and hydrated states, both pristine and hydrated PVPA are studied using molecular dynamics simulations.

In figure 9, it shows the typical structural snapshots of the pristine and hydrated PVPA revealed by molecular dynamics simulations. For the pristine PVPA, the oligomers are well ordered and densely packed with an equilibrium density at about 1.67 g·cm⁻³. In addition, most of the molecular backbones are stretched in the same direction because of the squeezing from the neighboring PVPA oligomers. The phosphonic groups are interconnected in a hydrogen-bond network by donating the acidic protons HO to the acidic oxygen atoms O2 or OH as shown by dotted lines in figure 9a. Furthermore, the radial distribution functions shown in figure 10a reveal that the hydrogen bonds formed between O2 and HO are much stronger than that formed between OH and HO in terms of peak heights at 7.54 and 1.18 for O2-HO and OH-HO, respectively. The peak positions for the radial distribution functions of O2-HO at 1.425 Å is slightly shorter than that of OH-HO at 1.525 Å. These characteristics are attributed to the strong steric repulsion from OH compared to that from O2. And proton could approach to O2 from all directions but to OH from only half of the directions. In fact, the hydrogen bonds between OH and HO are very weak as a peak height of 1.18 is only slightly higher than the average density of unity.

If water molecules are blended into PVPA, the phosphonic groups are well hydrated forming hydrogen phosphonate anions and separated by coulombic repulsion. Since the squeezing effect from neighboring PVPA oligomers is absent in hydrated states, the molecular backbones become less stretched compared to that in pristine PVPA and trend to distort randomly (Fig. 9b). The water molecules have also great impact on the hydrogen bonding characteristics of PVPA. The strongest hydrogen bonds are formed between O2 and

HO in pristine PVPA; however, the O2-HO hydrogen bonds are greatly weakened and become insignificant in hydrated PVPA. Actually, all the hydrogen bonds formed between two hydrogen phosphonate anions become insignificant compared to that between hydrogen phosphonate anion and water in terms of peak heights of the corresponding radial distribution functions. In hydrated PVPA, the strongest hydrogen bonds are formed between O2 and HW; and the next strongest hydrogen bonds are formed between water molecules (OW and HW). However, the hydrogen bonds between OW and HW are quite weak compared to that between O2 and HW as the peak height of radial distribution function for OW-HW is only 1.21, about one fourth of that for O2-HW. Therefore, it could be concluded that both hydrogen bonding between PVPA oligomers and between water molecules are greatly weakened during hydration.

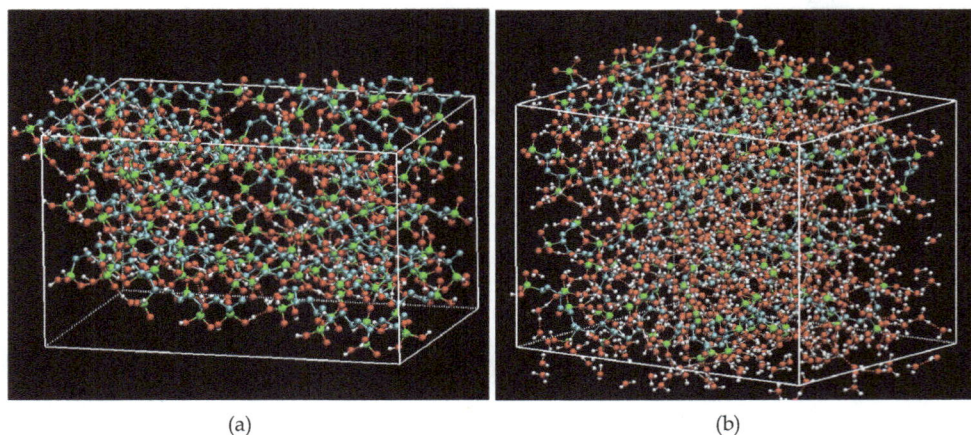

(a)   (b)

Fig. 9. Molecular dynamics simulation snapshots of (a) the pristine and (b) hydrated PVPA. Color code: red for O, white for H, green for P, cyan for C, and the red dotted lines for hydrogen bonds.

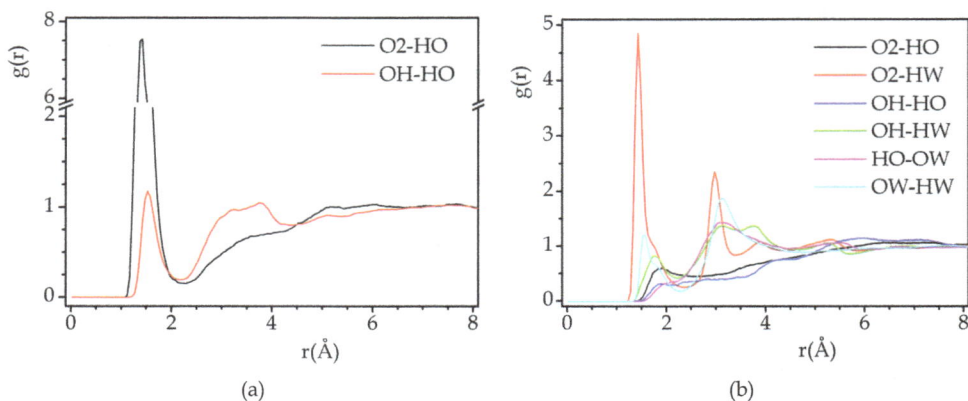

(a)   (b)

Fig. 10. Radial distribution functions showing hydrogen bonding in (a) pristine and (b) hydrated PVPA

The limited hydrogen bonding between water molecules are attributed to the low hydration degree of molecular simulation system as most of the water molecules and hydronium cations are attracted by the hydrogen phosphonate anions, and a few water molecules or hydronium cations are free to form water clusters via intermolecular hydrogen bonding. These characteristics are significantly different from that of hydrated poly(perfluorosulfonic acid) where water molecules, hydronium cations and sulfonate anions aggregate forming hydrophilic clusters and channels; and molecular backbones aggregate forming hydrophobic subphase. The hydrophilic clusters and channels provide uninterrupted proton transport channels facilitating the fast macroscopic transport of protons and the hydrophobic subphase provides the membrane with integrity and mechanical properties. The lack of phase segregation in the hydrated PVPA compared to the hydrated poly(perfluorosulfonic acid) are structural bases explaining the great differences between these two types of proton exchange materials.

The hydrogen bond length also shows significant changes during hydration. For example, the hydrogen bond length for O2 and HW is about 1.425 Å, same as that of O2 and HO in pristine PVPA; on the other hand, the hydrogen bond length for OH and HW is about 1.775 Å compared to 1.525 Å for OH and HO in pristine PVPA. Furthermore, the second coordination layer of O2-HW shows high order as the peak height for the second coordination layer at 2.975 Å is 2.35, almost half of the first peak. Ordered second coordination layer also exists for OW-HW, OW-HO, and OH-HW, but absents for O2-HO and OH-HO. Therefore, it is concluded that two layers of water molecules and hydronium cations surround the hydrogen phosphonate anion and the other hydrogen phosphonate anions are excluded from the first two coordination layers.

5.2 The pyrrole solvated PVPA

In order to increase the operational temperature of proton exchange membrane fuel cells, heterocyclic compounds with high boiling points and low saturated vapor pressures, such as pyrrole, pyrazole, and imidazole, are often employed as protic solvent to substitute water in the solvation of proton exchange membranes. These heterocyclic compounds possess high boiling points and the fuel cells are operational at temperature well above the boiling point of water. On the other hand, water boils at 100°C and begins to evaporate significantly even below 100°C, thus limiting the fuel cell operational temperature to about 80°C.

The microstructures of PVPA solvated by the pyrrole, at solvation degrees of one eighth and three eighths, are shown in figure 11. From the microstructures, it is shown that the PVPA oligomers are well solvated by pyrrole molecules and the pyrrole molecules are evenly dispersed in the PVPA oligomers forming a continuous phase without phase segregation. And hydrogen bonds are formed between different phosphonic groups and between pyrrole and phosphonic group.

For the pyrrole solvated PVPA at solvation degree of one eighth, the strongest hydrogen bonds are formed between acidic oxygen O2 and acidic hydrogen HO of the phosphonic groups in terms of peak height of the corresponding radial distribution functions (fig. 12). For example, the peak height for O2-HO is 9.67, about six times of the second highest peak for OH-HO at 1.54. The second strongest hydrogen bonds are formed between the OH and HO, still between two phosphonic groups. In addition, hydrogen bonds are also formed between a phosphonic group and a pyrrole molecule, where both oxygen O2 and OH accept acidic hydrogen from pyrrole. However, the hydrogen bonds formed between phosphonic group and pyrrole are much weaker than that formed between two phosphonic groups as the peak heights for O2-HN and OH-HN are only 1.10 and 0.84, respectively. The hydrogen bond lengths also show systematic changes. For O2-HO and OH-HO, the hydrogen bond lengths are at 1.375 and 1.525 Å, respectively. These hydrogen bond lengths are significantly closer than the corresponding hydrogen bond lengths in the pristine and hydrated PVPA. The change in hydrogen bond lengths is attributed to the exclusion effect, the interaction between solvent molecules or between PVPA oligomers are more powerful than that between solvent and PVPA, resulting the exclusion of PVPA from the solvent.

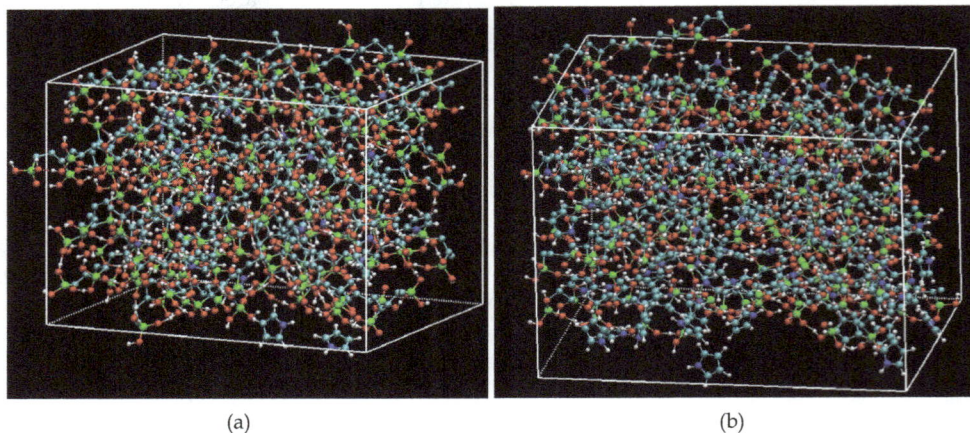

(a)   (b)

Fig. 11. Molecular dynamics simulation snapshots of the pyrrole solvated PVPA at solvation degree of (a) one eighth and (b) three eighths. Color code: red for O, white for H, green for P, cyan for C, blue for N, and the red dotted lines for hydrogen bonds.

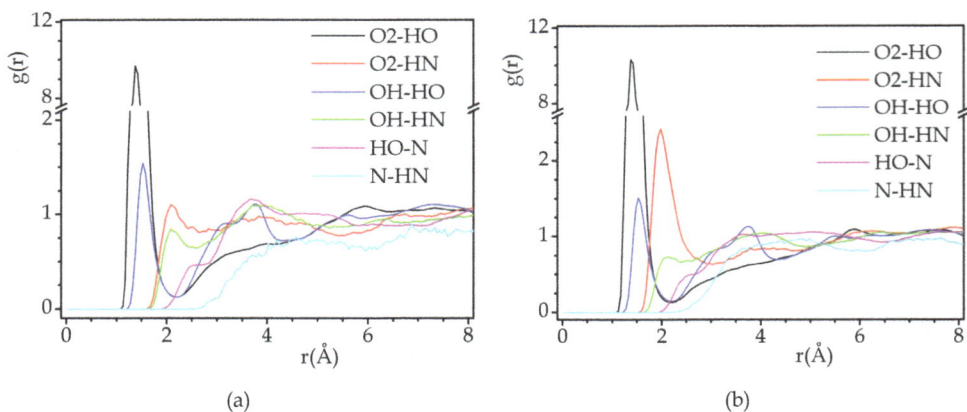

(a)   (b)

Fig. 12. Radial distribution functions showing hydrogen bonding in the pyrrole solvated PVPA at solvation degree of (a) one eighth and (b) three eighths.

For the pyrrole solvated PVPA at solvation degree of three eighths, the overall microstructures do not show significant changes compared to that at solvation degree of one eighth except the relative hydrogen bond strengths and lengths. For example, the peak height for O2-HO increases from 9.67 to 10.30, about 6.5 % increase, as the solvation degree increases from one eighth to three eighths. The exclusion effect reinforces the hydrogen bonds between phosphonic groups in the pyrrole solvated PVPA at high solvation degree compared to that at low solvation degree. In addition, the peak height for O2-HN at 2.43 is now higher than that for OH-HO at 1.51, while at solvation degree of one eighth the peak height for OH-HO is higher than that for O2-HN, despite that the peak position for O2-HN at 1.975 Å is still farther than that for OH-HO at 1.525 Å.

Special attention should be given to the hydrogen bonding characteristics of pyrrole nitrogen. Though amine nitrogen accepts proton in hydrogen bonding, the pyrrole nitrogen does not accept acidic protons from phosphonic groups nor from other pyrrole molecules since the pyrrole nitrogen is in sp^2 hybrid forming a planar structure. The planar structure restrains the approaching of proton in hydrogen bonding.

5.3 The pyrazole solvated PVPA

Although pyrrole is a heterocyclic compound containing one nitrogen heteroatom in the ring, the nitrogen does not accept proton in hydrogen bonding because of its planar structure. In order to relay protons in proton hopping, the compound should not only donate but also accept proton in hydrogen bonding. Therefore, another nitrogen heteroatom which could accept proton in hydrogen bonding has to be included in the heterocyclic compound. Pyrazole and imidazole are two 5-membered heterocyclic compounds containing two nitrogen atoms in the ring; one nitrogen atom accepts proton and the other nitrogen atom donates its proton in hydrogen bonding. In this and the following section, the pyrazole solvated PVPA and imidazole solvated PVPA will be simulated.

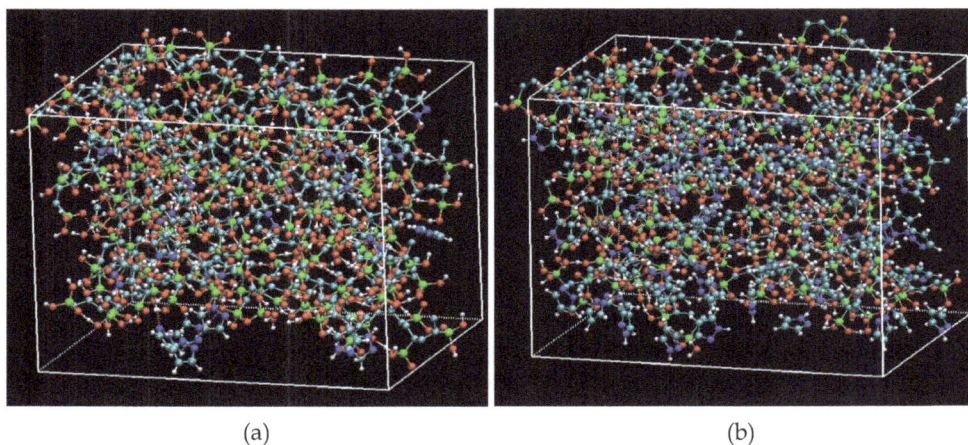

(a)   (b)

Fig. 13. Molecular dynamics simulation snapshots for the pyrazole solvated PVPA at solvation degree of (a) one eighth and (b) three eighths. Color code: red for O, white for H, green for P, cyan for C, blue for N, and the red dotted lines for hydrogen bonds.

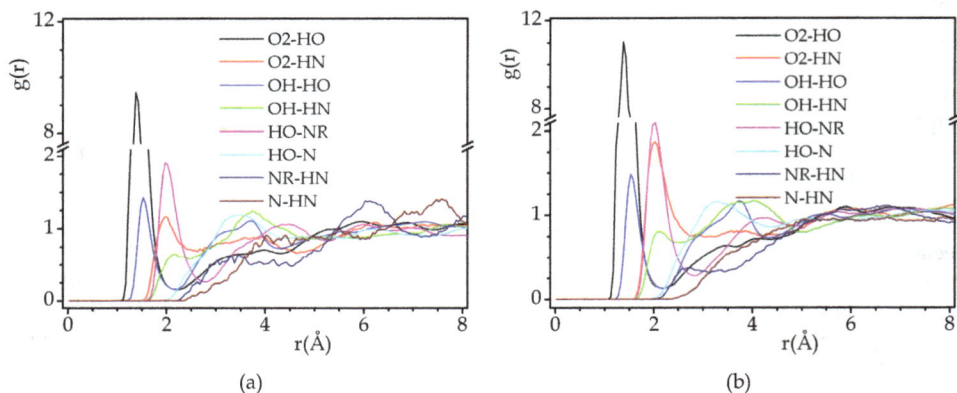

Fig. 14. Radial distribution functions showing hydrogen bonding in the pyrazole solvated PVPA at solvation degree of (a) one eighth and (b) three eighths.

In figure 13, the molecular dynamics simulation snapshots of the pyrazole solvated PVPA are shown at two levels of solvation degrees, and it could be seen that the pyrazole molecules are dispersed evenly in the PVPA for both systems.

From the radial distribution functions (fig. 14), it is concluded that the strongest hydrogen bonds are formed between the acidic oxygen O2 and acidic hydrogen HO of phosphonic groups with a hydrogen bond length of 1.425 Å. The next two types of hydrogen bonds are formed between an acidic proton HO of a phosphonic group and an imine nitrogen NR of a pyrazole molecule, and between an acidic oxygen OH and an acidic hydrogen HO of phosphonic groups. However, the hydrogen bond length between NR and HO is farther than that between OH and HO owing to the steric repulsion between a phosphonic group and a pyrazole ring. The hydrogen bonds are also formed between acidic oxygen O2 and amine hydrogen HN. The hydrogen bond lengths are divided into two groups: the O2-HN and NR-HO, and the O2-HO and OH-HO. For the pyrazole solvated PVPA at solvation degree of three eighths, the exclusion effect which is similar to that of pyrrole solvated PVPA reinforces the hydrogen bonding between phosphonic groups which are excluded by the pyrazole molecules. The increase in hydration degree also enhances the hydrogen bonding between O2 and HN.

Differing from the pyrrole solvated PVPA, there are pervasive hydrogen-bond network formed in pyrazole solvated PVPA: a phosphonic group not only accepts but also donates proton simultaneously; and a pyrazole molecule also forms two hydrogen bonds simultaneously by donating its amine hydrogen HN and accepting a phosphonic acidic proton by the imine nitrogen NR. Therefore, uninterrupted proton transport channels are developed among phosphonic groups and pyrazole molecules in terms of pervasive hydrogen-bond network and proton hopping is facilitated.

5.4 The imidazole solvated PVPA

Imidazole is an isomer of pyrazole containing two nitrogen atoms in the 5-membered heterocyclic ring. Imidazole is more stable than pyrazole owing to the absent of N-N bond in the ring. In addition, the boiling point of imidazole at 256°C is also much higher than that of pyrazole at 187°C. Since the two nitrogen atoms in pyrazole are in neighboring position, two hydrogen bonds are favored between two pyrazole molecules thus restricting the formation of intermolecular hydrogen bonds with other molecules. On the other hand, the two nitrogen atoms in imidazole are separated by a methylene spacer, the formation of two intermolecular hydrogen bonds with other two molecules simultaneously is favored. The structural difference also causes the differences in relative acidity and basicity of the molecules. For pyrazole, the pK_a is 14.0 and the pK_b is 2.5, or the basicity is stronger than its acidity. On the other hand, the imidazole possesses a much balanced pK_a and pK_b at 6.95 and 7.05, respectively, and is almost neutral. Therefore, it is expectable that the hydrogen bonding characteristics in imidazole solvated PVPA significantly differ from that in pyrazole solvated PVPA.

The microstructure of imidazole solvated PVPA is similar to that of pyrazole solvated PVPA. In addition, the hydrogen bonds formed between different phosphonic groups, O2-HO and OH-HO, are similar to that of pyrazole solvated PVPA. Furthermore, the hydrogen bonds formed between the O2 and HN are also similar to that of pyrazole solvated PVPA. The significant difference is the hydrogen bonding characteristics between HO and NR as the imine nitrogen NR in the imidazole rarely accepts acidic proton in hydrogen bonding compared to the powerful proton accepting ability of the pyrazole imine nitrogen NR. This characteristic is important for the proton transport as the hydrogen-bond network formed in imidazole solvated PVPA is less balanced than that formed in the pyrazole solvated PVPA.

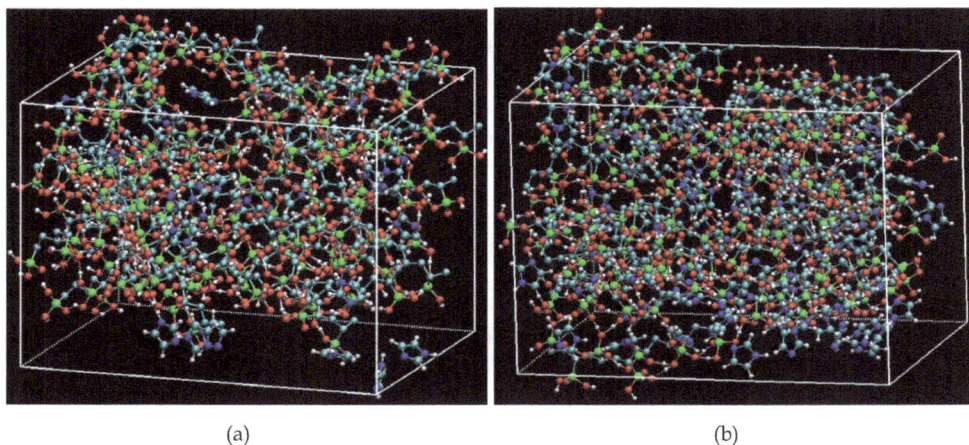

(a)   (b)

Fig. 15. Molecular dynamics simulation snapshots for the imidazole solvated PVPA at solvation degree of (a) one eighth and (b) three eighths. Color code: red for O, white for H, green for P, cyan for C, blue for N, and the red dotted lines for hydrogen bonds.

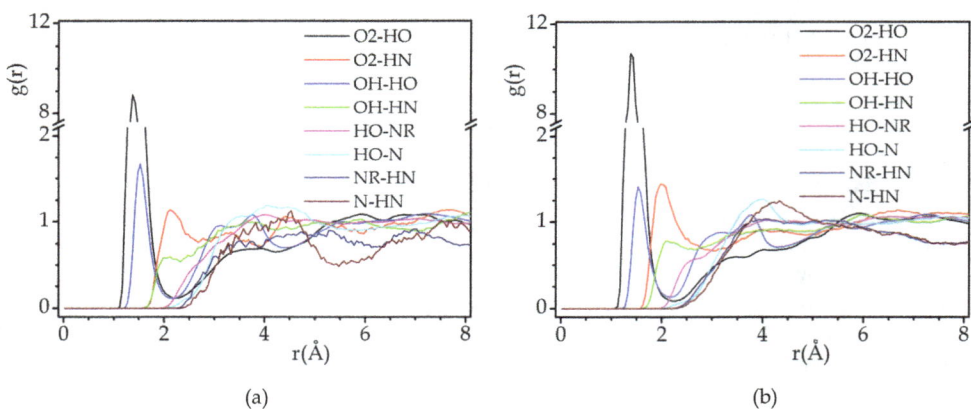

(a)   (b)

Fig. 16. Radial distribution functions showing hydrogen bonding in the imidazole solvated PVPA at solvation degree of (a) one eighth and (b) three eighths

6. Conclusions

Density functional theory calculations and molecular dynamics simulations are applied to the study of proton hopping characteristics in acid-base complexes in terms of hydrogen bonding and hydrogen-bond network. The density functional theory calculations are applied to the study of intra- and intermolecular hydrogen bonding characteristics of acid-base complexes composed of phosphonic acids and heterocyclic compounds, as well as acidic and basic bi- and multifunctional molecules consisting of phosphonic groups and heterocyclyl groups. It is concluded that the intermolecular hydrogen bonds in acid-base complexes are powerful as the entire configuration space could be searched in hydrogen bonding without restriction. On the other hand, the intramolecular hydrogen bonds in the bi- and multifunctional molecules are weak compared to the intermolecular hydrogen bonds because of configurational restriction in hydrogen bonding.

The molecular dynamics simulations based on hybrid force field models, united-atom models for the PVPA and PVHPA and all-atom rigid body models for solvents, are applied to the study of microstructures and hydrogen-bond networks of acid-base complexes including pristine and hydrated PVPA, and pyrrole, pyrazole, or imidazole solvated PVPA. It is revealed that the PVPA oligomers are densely packed with high density in pristine PVPA; while the PVPA oligomers are well hydrated or solvated without phase segregation in the hydrated or solvated PVPA.

It is further concluded that the pervasive hydrogen-bond networks vary significantly for different simulation systems. Firstly, the hydrogen bonds, O2-HO and OH-HO, existing in pristine PVPA, are greatly weakened as the PVPA is hydrated or solvated. In hydrated PVPA, strong hydrogen bonds are formed between phosphonate anion and water molecule even up to well ordered second coordination layer. Thirdly, in heterocycle solvated PVPA, exclusion effect, the exclusion of phosphonic groups from solvent, is recognized. The exclusion of phosphonic groups reinforces the hydrogen-bonding networks. Fourthly, the proton transport channels differ in different heterocycle solvated PVPA systems: the proton transport channels formed in pyrrole solvated PVPA are interrupted, that formed in pyrazole or imidazole solvated PVPA are uninterrupted, and that formed in pyrazole solvated PVPA are more balanced than that formed in imidazole solvated PVPA.

From these researches, it is proposed that special attention should be given to multifunctional polymers or copolymers with both acidic and basic functional groups attached on the same polymeric molecule in future study.

7. Acknowledgements

The authors acknowledge the support from the Chinese National Science Foundation (Nos. 20873081, 21073118), and the Nano project (No. 0952nm01300) of Shanghai Municipal Science & Technology Commission.

8. References

Allen, M. P. & Tildesley, D. J. (1990). *Computer simulation of liquids.* Oxford, Clarendon Press.

Baboul, A. G., Curtiss, L. A. et al. (1999). Gaussian-3 theory using density functional geometries and zero-point energies. *J. Chem. Phys.* 110(16): 7650-7657.

Becke, A. D. (1993). Density-functional thermochemistry. III. The role of exact exchange. J. Chem. Phys. 98(7): 5648-5652.

Breneman, C. M. & Wiberg, K. B. (1990). Determining atom-centered monopoles from molecular electrostatic potentials. The need for high sampling density in formamide conformational analysis. *J. Comp. Chem.* 11(3): 361-373.

Brunklaus, G., Schauff, S. et al. (2009). Proton mobilities in phosphonic acid-based proton exchange membranes probed by ^{1}H and ^{2}H solid-state NMR spectroscopy. *J. Phys. Chem. B* 113(19): 6674-6681.

Chai, J.-D. & Head-Gordon, M. (2008). Systematic optimization of long-range corrected hybrid density functionals. *J. Chem. Phys.* 128(8): 4106-4120.

Champagne, B., Perpete, E. A. et al. (1998). Assessment of conventional density functional schemes for computing the polarizabilities and hyperpolarizabilities of conjugated oligomers: An ab initio investigation of polyacetylene chains. *J. Chem. Phys.* 109(23): 10489-10498.

Chung, S. H., Bajue, S. et al. (2000). Mass transport of phosphoric acid in water: A ^{1}H and ^{31}P pulsed gradient spin-echo nuclear magnetic resonance study. *J. Chem. Phys.* 112(19): 8515-8520.

Cornell, W. D., Cieplak, P. et al. (1995). A second generation force field for the simulation of proteins, nucleic acids, and organic molecules. *J. Am. Chem. Soc.* 117(19): 5179-5197.

Curtiss, L. A., Raghavachari, K. et al. (1991). Gaussian-2 theory for molecular energies of first- and second-row compounds. *J. Chem. Phys.* 94(11): 7221-7230.

Curtiss, L. A., Raghavachari, K. et al. (1993). Gaussian-2 theory using reduced Møller–Plesset orders. *J. Chem. Phys.* 98(2): 1293-1298.

Curtiss, L. A., Raghavachari, K. et al. (1997). Assessment of Gaussian-2 and density functional theories for the computation of enthalpies of formation. *J. Chem. Phys.* 106(3): 1063-1079.

Curtiss, L. A., Redfern, P. C. et al. (1998). Assessment of Gaussian-2 and density functional theories for the computation of ionization potentials and electron affinities. J. Chem. Phys. 109(1): 42-45.

El-Azhary, A. A. & Suter, H. U. (1996). Comparison between optimized geometries and vibrational frequencies calculated by the DFT methods. *J. Phys. Chem.* 100(37): 15056-15063.

Elstner, M., Hobza, P. et al. (2001). Hydrogen bonding and stacking interactions of nucleic acid base pairs: A density-functional-theory based treatment. *J. Chem. Phys.* 114(12): 5149-5155.

Friesner, R. A., Murphy, R. B. et al. (1999). Correlated ab initio electronic structure calculations for large molecules. *J. Phys. Chem. A* 103(13): 1913-1928.

Frisch, M. J., Trucks, G. W. et al. (2003). GAUSSIAN 03, Revision C.2. Wallingford, CT, Gaussian, Inc.

Hoover, W. G. (1985). Canonical dynamics: Equilibrium phase-space distributions. *Phys. Rev. A* 31(3): 1695-1697.

Jorgensen, W. L., Chandrasekhar, J. et al. (1983). Comparison of simple potential functions for simulating liquid water. *J. Chem. Phys.* 79(2): 926-935.

Joswig, J.-O. & Seifert, G. (2009). Aspects of the Proton Transfer in Liquid Phosphonic Acid. *J. Phys. Chem. B* 113(25): 8475-8480.

Kreuer, K. D., Fuchs, A. et al. (1998). Imidazole and pyrazole-based proton conducting polymers and liquids. *Electrochim. Acta* 43(10-11): 1281-1288.

Kreuer, K. D., Rabenau, A. et al. (1982). Vehicle mechanism; a new model for the interpretation of the conductivity of fast proton conductors. *Angew. Chem. Int. Ed. Engl.* 21(3): 208.

Lee, C., Yang, W. et al. (1988). Development of the Colle-Salvetti correlation-energy formula into a functional of the electron density. *Phys. Rev. B* 37(2): 785-789.

Mayo, S. L., Olafson, B. D. et al. (1990). DREIDING: A genericforce field for molecular simulations. *J. Phys. Chem.* 94(26): 8897-8909.

Merrill, G. N. & Kass, S. R. (1996). Calculated gas-phase acidities using density functional theory: Is it reliable? *J. Phys. Chem.* 100(44): 17465-17471.

Münch, W., Kreuer, K. D. et al. (2001). The diffusion mechanism of an excess proton in imidazole molecule chains: first results of an ab initio molecular dynamics study. *Solid State Ionics* 145(1-4): 437-443.

Nosé, S. (1984). A unified formulation of the constant temperature molecular dynamics methods. *J. Chem. Phys.* 81(1): 511-519.

Paddison, S. J., Kreuer, K. D. et al. (2006). About the choice of the protogenic group in polymer electrolyte membranes: Ab initio modelling of sulfonic acid, phosphonic acid, and imidazole functionalized alkanes. *Phys. Chem. Chem. Phys.* 8(39): 4530-4542.

Rohrdanz, M. A., Martins, K. M. et al. (2009). A long-range-corrected density functional that performs well for both ground-state properties and time-dependent density functional theory excitation energies, including charge-transfer excited states. *J. Chem. Phys.* 130(5): 4112-4119.

Roy, S., Ataol, T. M. et al. (2008). Molecular dynamics simulations of heptyl phosphonic acid: A potential polymer component for fuel cell polymer membrane. *J. Phys. Chem. B* 112(25): 7403-7409.

Schuster, M., Rager, T. et al. (2005). About the choice of the protogenic group in PEM separator materials for intermediate temperature, low humidity operation: A critical comparison of sulfonic acid, phosphonic acid and imidazole functionalized model compounds. *Fuel Cells* 5(3): 355-365.

Sevil, F. & Bozkurt, A. (2004). Proton conducting polymer electrolytes on the basis of poly(vinylphosphonic acid) and imidazole. *J. Phys. Chem. Solids* 65(10): 1659-1662.

Smith, W., Leslie, M. et al. (2003). *Computer code DL_POLY_2.14*, CCLRC, Daresbury Laboratory, Daresbury, England.

Song, J. W., Tokura, S. et al. (2007). An improved long-range corrected hybrid exchange-correlation functional including a short-range Gaussian attenuation (LCgau-BOP). *J. Chem. Phys.* 127(15): 4109-4114.

Urata, S., Irisawa, J. et al. (2005). Molecular dynamics simulation of swollen membrane of perfluorinated ionomer. *J. Phys. Chem. B* 109(9): 4269-4278.

Verlet, L. (1967). Computer Experiments on classical fluids. I. Thermodynamical properties of Lennard-Jones molecules. *Phys. Rev.* 159(1): 98-103.

Yamada, M. & Honma, I. (2005). Anhydrous proton conducting polymer electrolytes based on poly(vinylphosphonic acid)-heterocycle composite material. *Polymer* 46(9): 2986-2992.

Yan, L., Feng, Q. et al. (2011). About the choice of protogenic group for polymer electrolyte membrane: Alkyl or aryl phosphonic acid? *Solid State Ionics* 190(1): 8-17.

Yan, L., Zhu, S. et al. (2007). Proton hopping in phosphoric acid solvated NAFION membrane: A molecular simulation study. *J. Phys. Chem. B* 111(23): 6357-6363.

Yanai, T., Tew, D. P. et al. (2004). A new hybrid exchange-correlation functional using the Coulomb-attenuating method (CAM-B3LYP). *Chem. Phys. Lett.* 393(1-3): 51-57.

Zhang, D. & Yan, L. (2010). Probing the acid-base equilibrium in acid-benzimidazole complexes by ^{1}H NMR spectra and density functional theory calculations. *J. Phys. Chem. B* 114(38): 12234-12241.

Zhang, Y., Xu, X. et al. (2009). Doubly hybrid density functional for accurate descriptions of nonbond interactions, thermochemistry, and thermochemical kinetics. *PNAS* 106(13): 4963-4968.

Zhao, Y. & Truhlar, D. G. (2005). Design of density functionals that are broadly accurate for thermochemistry, thermochemical kinetics, and nonbonded interactions. *J. Phys. Chem. A* 109(25): 5656-5667.

Zhu, S., Yan, L. et al. (2011). Molecular dynamics simulation of microscopic structure and
    hydrogen bond network of the pristine and phosphoric acid doped
    polybenzimidazole. *Polymer* 52(3): 881-892.

Permissions

The contributors of this book come from diverse backgrounds, making this book a truly international effort. This book will bring forth new frontiers with its revolutionizing research information and detailed analysis of the nascent developments around the world.

We would like to thank Aurelia Meghea, for lending her expertise to make the book truly unique. She has played a crucial role in the development of this book. Without her invaluable contribution this book wouldn't have been possible. She has made vital efforts to compile up to date information on the varied aspects of this subject to make this book a valuable addition to the collection of many professionals and students.

This book was conceptualized with the vision of imparting up-to-date information and advanced data in this field. To ensure the same, a matchless editorial board was set up. Every individual on the board went through rigorous rounds of assessment to prove their worth. After which they invested a large part of their time researching and compiling the most relevant data for our readers. Conferences and sessions were held from time to time between the editorial board and the contributing authors to present the data in the most comprehensible form. The editorial team has worked tirelessly to provide valuable and valid information to help people across the globe.

Every chapter published in this book has been scrutinized by our experts. Their significance has been extensively debated. The topics covered herein carry significant findings which will fuel the growth of the discipline. They may even be implemented as practical applications or may be referred to as a beginning point for another development. Chapters in this book were first published by InTech; hereby published with permission under the Creative Commons Attribution License or equivalent.

The editorial board has been involved in producing this book since its inception. They have spent rigorous hours researching and exploring the diverse topics which have resulted in the successful publishing of this book. They have passed on their knowledge of decades through this book. To expedite this challenging task, the publisher supported the team at every step. A small team of assistant editors was also appointed to further simplify the editing procedure and attain best results for the readers.

Our editorial team has been hand-picked from every corner of the world. Their multi-ethnicity adds dynamic inputs to the discussions which result in innovative outcomes. These outcomes are then further discussed with the researchers and contributors who give their valuable feedback and opinion regarding the same. The feedback is then collaborated with the researches and they are edited in a comprehensive manner to aid the understanding of the subject.

Apart from the editorial board, the designing team has also invested a significant amount of their time in understanding the subject and creating the most relevant covers. They scrutinized every image to scout for the most suitable representation of the subject and create an appropriate cover for the book.

The publishing team has been involved in this book since its early stages. They were actively engaged in every process, be it collecting the data, connecting with the contributors or procuring relevant information. The team has been an ardent support to the editorial, designing and production team. Their endless efforts to recruit the best for this project, has resulted in the accomplishment of this book. They are a veteran in the field of academics and their pool of knowledge is as vast as their experience in printing. Their expertise and guidance has proved useful at every step. Their uncompromising quality standards have made this book an exceptional effort. Their encouragement from time to time has been an inspiration for everyone.

The publisher and the editorial board hope that this book will prove to be a valuable piece of knowledge for researchers, students, practitioners and scholars across the globe.

List of Contributors

Mirjana Lj. Kijevčanin, Bojan D. Djordjević, Ivona R. Radović, Emila M. Živković, Aleksandar Ž. Tasić and Slobodan P. Šerbanović
Faculty of Technology and Metallurgy, University of Belgrade, Serbia

Abedien Zabardasti
Lorestan University, Iran

Vilma Edite Fonseca Heinzen and Rosendo Augusto Yunes
Department of Chemistry, Federal University of Santa Catarina, University Campus, Trindade, Florianópolis, Santa Catarina, Brazil

Berenice da Silva Junkes
Federal Institute of Education, Science and Technology of Santa Catarina, Florianópolis, SC, Brazil

Carlos Alberto Kuhnen
Department of Physics, Federal University of Santa Catarina, University Campus, Trindade, Florianópolis, Santa Catarina, Brazil

Raj Kumar Gupta and V. Manjuladevi
Department of Physics, Birla Institute of Technology & Science, Pilani, India

Liane Gloria Raluca Stan, Rodica Mirela Nita and Aurelia Meghea
University Politehnica of Bucharest, Romania

Zhi-you Zhou and Shaowei Chen
Department of Chemistry and Biochemistry, University of California, Santa Cruz, CA, USA

Byung I. Kim
Department of Physics, Boise State University, USA

Stephanie J. Benight, Bruce H. Robinson and Larry R. Dalton
University of Washington, USA

Nobuhiro Kamiya
Center for Excellence in Hip Disorders, Texas Scottish Rite Hospital for Children, Dallas, Texas, USA

René van der Ploeg and Tanneke den Blaauwen
University of Amsterdam, Swammerdam Institute for Life Sciences, The Netherland

Youri Timsit
Genomic and Structural Information, CNRS - UPR2589, Institute of Microbiology of the Mediterranean,
Aix-Marseille University, Science Park Luminy, Marseille, France

Akihiko Konagaya
Tokyo Institute of Technology, Japan

Guangcai Zhang, Aiguo Xu and Guo Lu
Institute of Applied Physics and Computational Mathematics, The People's Republic of China

Liuming Yan and Liqing Xie
Department of Chemistry, Shanghai University, Shanghai, China